# Springer Tracts in Modern Physics 81

Ergebnisse der exakten Naturwissenschaften

W0246051

# Springer Tracts in Modern Physics


| | |
|---|---|
| Volume 66 | Quantum Statistics in Optics and Solid-State Physics |
| Volume 67 | Conformal Algebra in Space-Time and Operator Product Expansion |
| Volume 68 | Solid-State Physics |
| Volume 69 | Astrophysics |
| Volume 70 | Quantum Optics |
| Volume 71 | Nuclear Physics |
| Volume 72 | Van der Waals Attraction |
| Volume 73 | Excitons at High Density |
| Volume 74 | Solid-State Physics |
| Volume 75 | Light Scattering by Phonon-Polaritons |
| Volume 76 | Irreversible Properties of Type II Superconductors |
| Volume 77 | Surface Physics |
| Volume 78 | Solid-State Physics |
| Volume 79 | Elementary Particle Physics |
| Volume 80 | Neutron Physics |
| Volume 81 | Point Defects in Metals I: Introduction to the Theory |

G. Leibfried N. Breuer

# Point Defects in Metals I

## Introduction to the Theory

With 138 Figures

Springer-Verlag
Berlin Heidelberg GmbH 1978

Professor Dr. Günther Leibfried †
Dr. Nikolaus Breuer *

Institut für Festkörperforschung der Kernforschungsanlage Jülich
Postfach 1913, D-5170 Jülich

* Present address:
Institut für Theoretische Physik III der Universität Düsseldorf
Universitätsstrasse 1, D-4000 Düsseldorf

---

*Manuscripts for publication should be addressed to:*

Gerhard Höhler
Institut für Theoretische Kernphysik der Universität Karlsruhe
Postfach 6380, D-7500 Karlsruhe 1

*Proofs and all correspondence concerning papers in the process of publication should be addressed to:*

Ernst A. Niekisch
Institut für Grenzflächenforschung und Vakuumphysik der Kernforschungsanlage Jülich
Postfach 1913, D-5170 Jülich

---

ISBN 978-3-662-15448-9 ISBN 978-3-540-37201-1 (eBook)
DOI 10.1007/978-3-540-37201-1

Library of Congress Cataloging in Publication Data. Leibfried, Günther, 1915—. Point defects in metals I: Introduction to the theory. (Springer tracts in modern physics; 81) Bibliography: p. Includes index. 1. Lattice dynamics. 2. Point defects. 3. Metals—Defects. 4. Continuum mechanics. I. Breuer, Nikolaus, 1948—. joint author. II. Title. III. Series. QC1.S797 vol. 81 [QC176.8.L3] 539'.08s [548'.81] 77-24475


Originally published by Springer-Verlag Berlin Heidelberg New York in 1978
Softcover reprint of the hardcover 1st edition 1978


2153/3130 — 543210

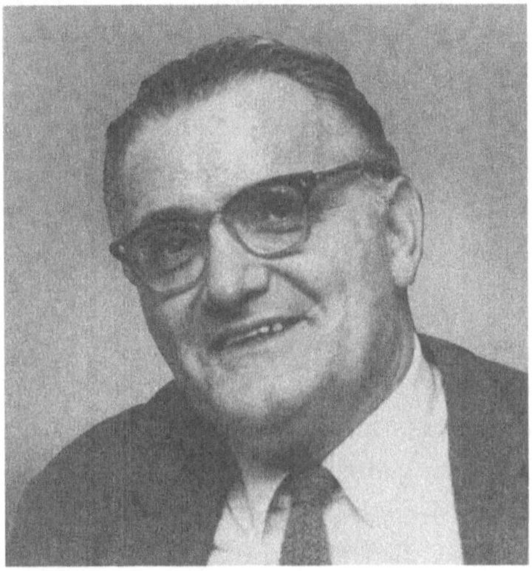

Professor Dr. GÜNTHER LEIBFRIED
(10.6.1915–20.6.1977)

# A Dedication

This book is dedicated to the memory of Günther Leibfried--scientist, educator, and man. For all those who knew and worked with him until his death on June 20, 1977, it is difficult to realize that he did not live to achieve some of his many goals. However, it has been possible to complete this volume, which comprehensively presents the results of his scientific research.

Günther Leibfried was born on June 10, 1915, in Fraulautern/Saar. He studied at the University of Göttingen, where his doctorate was conferred in 1939. He began his research career as an assistant to Richard Becker at the Institute for Theoretical Physics at Göttingen, and became a lecturer there in 1950. After Richard Becker's death, Professor Leibfried was appointed temporary chairman of the Institute, a post which he held from 1955 to 1957, willingly assuming the responsibility for Professor Becker's students.

Professor Leibfried's early scientific work dealt with displacement theory in statistical mechanics and with the quantization of wave fields. In 1954-55 he authored an article on lattice theory for the *Encyclopedia of Physics*. This paper is still required reading for students in the discipline. Encouraged by discussions with Max Born, Professor Leibfried devoted his major efforts to the field of lattice theory and its applications.

Subsequently he accepted the position of Professor for Physics of Reactor Materials at the Technical University of Aachen, a move which coincided with his appointment as Director at the Institute for Reactor Materials at the Nuclear Research Establishment (Kernforschungsanlage "KFA") Jülich. In 1971 he was appointed Director of the newly formed Institute for Solid State Research (Institut für Festkörperforschung) at the KFA Jülich, to the establishment and organization of which he contributed immensely.

At Jülich, Günther Leibfried's studies in the field of radiation-damaged lattices soon brought him international recognition. He performed particularly successful research with the Oak Ridge National Laboratory in Tennessee, where he regularly spent one month each year. Many of his German students who went with him remained at Oak Ridge for extended periods of time, and many Oak Ridge researchers visited in Jülich.

Günther Leibfried demanded a great deal from his co-workers, because he was himself an inexhaustible model for all. He was pleasant and easy to deal with, although quick to anger when presented with incomplete or improperly formulated theories. He aggressively promoted the further education of his students and colleagues, many of whom are now university professors and chairmen of research institutions.

In addition to his purely scientific side, Günther Leibfried had a human touch that was extremely important to his students. He never failed to respond to their personal problems and often shared recreational time with them. This close relationship between him and his students is best exemplified by their annual get-togethers, usually on his birthday or at Christmas, in which almost all of his former students participated. He frequently performed musically with his colleagues, and was always eager for a celebration of any kind.

His death came too soon for all of us. The editors and publishers of "Springer Tracts in Modern Physics", as well as his students and colleagues, will not forget Günther Leibfried.

GERHARD HÖHLER
ERNST A. NIEKISCH
HELMUT K.V. LOTSCH

January 1978

# Preface

During the last decade, our understanding of point defects in metals has improved greatly. Although schools and conferences have been held on this subject, no comprehensive and simple review has appeared; therefore, it seemed worthwhile to have a summary of the basic theory as well as of the experimental results. It turned out to be more appropriate to split such a survey into two parts, the present volume and a forthcoming one of this series entitled *Point Defects in Metals II, Atomic Structures and Vibrational Behaviour* by W. SCHILLING, P.H. DEDERICHS and H. TRINKAUS. Whereas the latter part will represent an actual review on both experimental and theoretical work including (in contrast to the present part) comprehensive references to literature, we here concentrate on the basic theoretical concepts and methods; in a sense this *Introduction to the Theory* may be viewed as a kind of textbook with particular emphasis on simplicity.

The physics of point defects in metals is rather different from that of semiconductors and insulators. In a metal the defect structure is simple, in contrast to the variety of charged and excited states of an atomic defect in a semiconductor or insulator. On the other hand, in a transparent non-metal one can use with great success the interaction with an electromagnetic field to investigate structure and dynamics of a defect; this is not possible in metals because of the small penetration. Instead, one can use the interaction with an elastic strain field and indeed the influence of defects on elastic behaviour and on lattice waves, i.e., the influence of defects on the mechanical properties of a crystal (which is also of great technological importance) has been used extensively in experimental investigations of metals. Even more powerful methods are X-ray and neutron scattering by defect crystals; the scattering of X-rays has been used widely to determine the static structure of defects, while neutron scattering exhibits the changes of lattice modes. Furthermore, the Mößbauer effect can give information about the vibrational (and migrational) behaviour of defects.

Although the electronic (microscopic) theory of defects in metals is not well established so far, one can develop realistic phenomenological models on an atomistic scale, because the atomic interaction is well screened by the metal electrons and is, therefore, of short range. Furthermore, one can concentrate on simple lattices, the crystal structure of many relevant metals. By employing these atomistic models it is possible to discuss, for example, the mechanical properties of metals with point defects in a simple manner. We have tried to achieve simplicity by supporting the text with many illustrations, by demonstrating some useful mathematics in an appendix and by employing particularly simple examples; furthermore we have restricted ourselves essentially to cubic Bravais crystals and treat only the most symmetrical and simple defects. We believe that anyone of graduate student level will be able to handle the theory if he is willing to spend a reasonable amount of time.

We gratefully acknowledge many clarifying discussions with W. SCHILLING, F.W. YOUNG and P.H. DEDERICHS; we are particularly indebted to H.R. SCHOBER, who calculated Green's functions and spectra which we needed to illustrate some important points. We enjoyed steady criticism from Mrs. B. SPLETTSTÖSSER, who has been reading the manuscript with a critical eye. Mrs. M. SPATZEK has been most helpful in preparing the manuscript with many secret and not so secret sighs. Last but not least, we are very grateful to Mrs. G. HAHN, who prepared the camera-ready manuscript most carefully bearing the last corrections with great patience (and no fewer sighs). One of the authors (G.L.) would like to dedicate this volume to the Solid State Division of Oak Ridge National Laboratory (Tennessee, USA) and its staff; discussions during the time spent there over the years have laid the ground for a great part of the work presented here.

After the original manuscript was completed, the draft of the above preface was written by both authors. However, because of his sudden death in June 1977, Prof. G. LEIBFRIED never saw the final, camera-ready version of the manuscript. To honour G. LEIBFRIED, I could sorrowfully repeat almost exactly the words he himself used in honour of R. BECKER in the preface to R. BECKER's *Theory of Heat*. Being well aware of G. LEIBFRIED's impatience with glorifying but idle words, I think that I feel like many others of his students and co-workers: we have lost an extraordinary personality, a passionate scientist, an exciting and inspiring teacher, and a reliable and benign advisor.

Jülich, September 1977 N. BREUER

# Contents

# 1. Introduction and Survey

This volume is an introduction to the theory of defects in metals. The emphasis will be on the change of mechanical properties by defects in small concentration. The physics is based on the properties of a single defect, which determine in turn the property changes produced by many defects (in small concentration). We, therefore, must start with the physics of a single defect.

Close to the defect, say within a few atomic spacings, the lattice structure is essential and one must use microscopic lattice theory. Far from the defect, one still can use lattice theory; however, lattice structure becomes less and less important with increasing distance and the appropriate description becomes then the macroscopic theory of an elastic continuum. Consequently, defect physics requires knowledge both of lattice *and* continuum theory. For this reason, Chapters 2 and 3 contain a thorough introduction to lattice theory, Chapter 4 lays the foundation of elastic continuum theory and Chapter 5 establishes the link between the two. It is characteristic of defect physics that one needs both descriptions; the physicist dealing with defects uses terms from both theories, e.g., change of a (microscopic) defect doubleforce tensor by a (macroscopic) strain.

To simplify matters we have restricted[1] ourselves to the "harmonic approximation", a linear theory which is easy to handle mathematically. This requires some knowledge of linear algebra which we have tried to demonstrate in a simple and condensed fashion in the appendix. We have made extensive use of projectors (a specialty of linear theory) because their use allows a very concentrated and transparent representation. Projectors are explained in the appendix first in three dimensions where the concept is trivial. The theory makes also extensive use of the concept of eigenvectors and eigenvalues for which simple examples can also be found in the appendix. The use of the much-dreaded Green's function (linear response) is, of course, unavoidable, it is the backbone of the whole of defect physics. In this field this concept is easily understood; the Green's function is simply the displacement pattern produced by a single force acting on one atom of the crystal. The harmonic approximation

[1] Except for a brief treatment of anharmonicity in Section 4.9, needed to estimate anharmonic effects in the change of elastic data by defects.

is equivalent to a set of independent harmonic oscillators. We have therefore started with a treatment of *one* harmonic oscillator, and have introduced the Green's function right at this point. The extrapolation to many oscillators is then not difficult. The mathematics here involves Fourier transformation and complex integration; again we have simple examples in the appendix.

In summary, Chapter 2 to 5 establish the ground on which defect physics is based. They contain a complete review on the dynamics of lattice and continuum. We have tried to keep the theory simple by treating in detail only cubic lattices and the elastic theory only for cubic crystals. The extension to lower symmetries is straightforward in principle. We should add that group theory is used extensively. However, the symmetry arguments used on many occasions need not be supported by formal group theory.

We would like to make some general remarks on the introductory Chapters 2 to 5. Chapter 2, which treats the most general finite, harmonic ensemble, is unfortunately the most difficult to read. Because it is so general, it is rather abstract, formal and concentrated, and it does not contain any useful examples and applications. However, the representation of the harmonic approximation in terms of eigenvectors of the interatomic coupling, which is introduced there, is essential for all later discussions. The discussion of the one-dimensional oscillator, in particular its thermal averages, is also quite dry, but needed later to full extent. Except for neutron scattering theory, this is the only place where quantum mechanics is used; the single steps in calculating the averages are so given that the initiated reader should be able to follow; the uninitiated must more or less trust the relatively illustrative results, which are necessary only for the theory of neutron scattering. Chapter 3 contains two points which, ordinarily, are not treated at all. One is the discussion of force constant models in which contributions by direct two body interaction and by indirect (electronic) shell and bond charge effects are treated in detail; this is absolutely necessary to obtain self-consistent defect models. Secondly, the asymptotic expansion of Green's functions (displacements far from the producing force) is treated in length; though this expansion is not needed explicitly, we have included it for completeness. Also, the usual reasons given for its validity are not well founded and we thought it worthwhile to review the matter. In Chapter 4 on elasticity we have used a new notation (besides the old) for strain, stress and elastic data. This notation is particularly adapted to cubic symmetry (including isotropy); it is very transparent and simplifies the treatment of cubic crystals a great deal. Chapter 5 contains a brief discussion of the problem of how to define stress microscopically; although that problem is still unsolved, we have included the discussion because of its general importance.

After this extensive introduction to the theory of perfect crystals, single lattice defects are treated in Chapter 6. Here we treat only the simplest defects of highest symmetry to keep calculations short and physics transparent. The main content is the response of a defect to external strains (either to incoming waves or to static strains), in other words scattering theory. Here, the isotopic defect (change of mass of a single atom) is treated in great detail. From the single defect response one can often obtain, somewhat crudely, the combined response of many defects in small concentration by linear superposition. A more thorough theory of crystals with small defect concentration is given in Chapter 9; there the change of elastic data and the change of lattice dispersion curves are also discussed in detail.

One powerful method of investigating defects is scattering of X-rays and neutrons. For this reason Chapter 7 contains an abbreviated theory of scattering by crystals, the results of which are needed to discuss experiments in the review part and in the final Chapter 9.

Chapter 8 on statistics is short. It contains some remarks on the use of averages. The distribution of defects in a crystal is subject to statistics. As a rule one calculates the average of a physical quantity and compares this with experiment. However, many people fail to realize that there are fluctuations about averages, and that comparisons with "unique" experimental results are reasonable only if these fluctuations are not too large. Employing X-ray scattering by an alloy as a simple example, we show that one must be careful when betting on averages. We thought it necessary to point out the statistical problems of defect physics more thoroughly because, mostly, this important aspect is neglected and not treated at all.

Each Chapter and most of the sections have short introductions of their own. The reader should see where he is going and why. We have kept the references to the literature to a minimum; there is no claim to completeness.

# 2. Harmonic Approximation and Linear Response (Green's Function) of an Arbitrary System

/2.1-7/

In this chapter we treat an assembly of atoms which undergo *small displacements* from their equilibrium positions. While this assumption is restrictive it is nevertheless valid in a broad area of metal physics. For these small amplitudes the theory is very simple, because it is linear in displacements and applied forces; it is called the *harmonic approximation*. The equilibrium positions are given by the minimum of an *"adiabatic" potential*.

In treating an assembly of atoms we have in mind a *finite* crystal in contrast to a periodic and therefore *infinite* crystal lattice. The rigorous treatment of a finite crystal is not possible because of surface effects. The infinite crystal, in contrast, has such a high symmetry that one can calculate rigorously many of its properties (e.g., its elastic data) which in fact agree with the bulk properties of a finite (macroscopic) crystal. Therefore later we will mostly consider an infinite lattice, but in order to understand these relationships and to get an estimate of surface effects one must start with the finite crystal.

The harmonic approximation can be represented by a set of one-dimensional oscillators. We therefore first treat the one-dimensional oscillator, its linear response and its thermal averages very extensively. These results can immediately and easily be applied to the most general problem.

The *static* situation is treated in great detail; it is analogous to and ought to be compared with the corresponding static case in continuum theory. The response to a force on a single atom (Green's function) is discussed more carefully because this concept has to be used with great caution in a finite crystal. The variational methods developed here to obtain approximate Green's functions can as well be applied in infinite lattices and in continuum theory.

## 2.1 The Adiabatic Potential and Its Invariances[1]

In the adiabatic approximation one assumes that the electrons follow the nuclear motion adiabatically, i.e., the electrons are always in the ground state corresponding to the actual nuclear positions. This assumption allows one to eliminate the electronic degrees of freedom and to define a potential

$$\Phi(\underline{r}^1 \ldots \underline{r}^m \ldots \underline{r}^N) \; ; \quad \underline{r}^m = (x_i^m) = (x_1^m, x_2^m, x_3^m) = (x^m, y^m, z^m) \tag{2.1}$$

which depends only on the position of the N nuclei, $\underline{r}^m$, $m = 1 \ldots N$. This potential must be invariant against common translations and rotations (see App. A)

$$\Phi(\ldots \underline{r}^m \ldots) = \Phi(\ldots \tilde{\underline{r}}^m \ldots) \tag{2.1a}$$

where $\underline{r}^m$ and $\tilde{\underline{r}}^m$ are connected by a common translation $\underline{T} = (T_i)$ and a common rotation $D = (D_{ik})$ (including inversion $D = -1 = (-\delta_{ik})$)

$$\tilde{\underline{r}}^m = D\underline{r}^m + T \; ; \quad \tilde{x}_i^m = D_{ik} x_k^m + T_i \; . \tag{2.1b}$$

For *small* ("infinitesimal") changes

$$\tilde{\underline{r}}^m = \underline{r}^m + \varphi \hat{\underline{D}} \times \underline{r}^m + \underline{T} \; , \tag{2.2}$$

where the angle of rotation $\varphi$ about the axis $\hat{\underline{D}}$ and the translation $\underline{T}$ are "small",

or in components

$$\tilde{x}_i^m - x_i^m = \varphi (\hat{\underline{D}} \times \underline{r}^m)_i + T_i = \omega_{ik} x_k^m + T_i \; , \tag{2.2a}$$

where the antisymmetric matrix $\omega_{ik} = (-\omega_{ki})$ and $T_i$ are "small",

the invariance (2.1a) leads to relations between the *first derivatives* of the potential.

Expanding (2.1a) up to first order, one obtains for arbitrary $\omega$ (or $\hat{\underline{D}}$ and $\varphi$) and arbitrary $\underline{T}$

$$0 = \Phi(\ldots \tilde{\underline{r}}^m \ldots) - \Phi(\ldots \underline{r}^m \ldots) = \sum_m (\omega_{ik} x_k^m + T_i) \Phi_i^m \tag{2.2b}$$

---

[1] In the following we will adopt the summation convention. If an index appears twice the sum over this index has to be taken, e.g.,

$$\Phi_i^m x_k^m = \sum_{m=1}^{N} \Phi_i^m x_k^m \; , \quad \omega_{ik} \Phi_i^m x_k^m = \sum_{m,i,k} \omega_{ik} \Phi_i^m x_k^m \; .$$

Exceptions will be denoted by a crossed sum: $\not\Sigma$ .

$$= \sum_m [\varphi(\hat{\underline{D}}, \underline{r}^m \times \underline{\Phi}^m) + (\underline{T}, \underline{\Phi}^m)] = 0 \; , \tag{2.2b}$$

where $-\underline{\Phi}^m = -(\Phi_i^m) = -(\partial\Phi/\partial x_i^m)$ is the internal force on atom m due to the internal potential. Eq. (2.2b) implies then that the internal forces do not possess a total force

$$\sum_m - \underline{\Phi}^m = 0 \; , \quad \sum_m - \Phi_i^m = 0 \tag{2.2c}$$

nor a total torque

$$\sum_m \underline{r}^m \times \underline{\Phi}^m = 0 \quad \text{or} \quad \underline{r}^m \times \underline{\Phi}^m = 0 \; ,$$

$$\sum_m \omega_{ik} x_k^m (-\Phi_i^m) = 0 \quad , \quad \sum_m x_k^m \Phi_i^m = \sum_m x_i^m \Phi_k^m \; . \tag{2.2d}$$

Forces with the properties (2.2c,d) cannot produce a translation or a rotation of the system; they conserve momentum and angular momentum. Only "external" forces violating (2.2c,d) can change these momenta. Force patterns obeying (2.2c,d) we call invariant (Fig. 2.1)

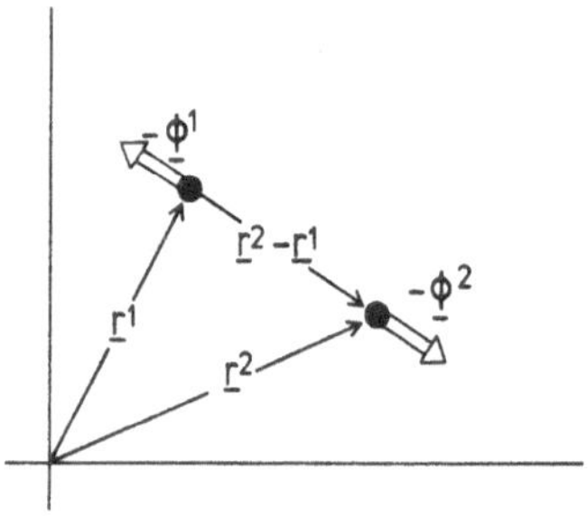


Fig. 2.1. Invariant internal force patterns for two atoms. Because of translational invariance the potential between two atoms depends only on the vectorial distance $\underline{r}^2 - \underline{r}^1$; because of rotational invariance only the absolute distance $|\underline{r}^2 - \underline{r}^1| = r_{21}$ remains:

$$-\underline{\Phi}^1 = -\frac{\partial}{\partial \underline{r}^1}\Phi(r_{21}) = \frac{\underline{r}^2 - \underline{r}^1}{r_{21}} \frac{\partial}{\partial r_{21}}\Phi = -(-\underline{\Phi}^2)$$

In principle, electronic structure determines the adiabatic potential. However, we will mainly employ phenomenological and central force (two body) interactions fitted to experimental data because, in particular for defect states, the basic theory is not sufficiently developed as yet.

## 2.2 The Harmonic Approximation

### 2.2.1 Expansion in Powers of the Displacements

Except for special cases one can use classical theory where equilibrium is given by the "minimum" of the potential energy where $\underline{\Phi}^m = 0$. Because of (2.1a) the equilibrium positions $\underline{R}^m$ are determined only except for a common translation and rotation. The atomic positions are given by the displacements $\underline{s}^m$ from equilibrium (Fig. 2.2). If the displacements are small, a good approximation for $\Phi(\ldots\underline{r}^m = \underline{R}^m + \underline{s}^m\ldots)$ are the leading terms in an expansion in powers of the displacements

$$\Phi(\ldots\underline{R}^m + \underline{s}^m\ldots\underline{R}^n + \underline{s}^n\ldots) = \qquad (2.3)$$

| | |
|---|---|
| | 0th order |
| $\Phi(\ldots\underline{R}^m\ldots\underline{R}^n\ldots) +$ | potential energy in equilibrium; does not enter dynamics |
| | 1st order |
| $+\ \Phi_i^m(\underline{R}^1\ldots\underline{R}^N)s_i^m +$ | equilibrium condition: $\Phi_i^m(\underline{R}^1\ldots) = 0$, no internal forces, "minimum" of potential energy |
| | 2nd order |
| $+\ \frac{1}{2}\underbrace{\Phi_{ik}^{mn}(\underline{R}^1\ldots)}_{\frac{\partial^2}{\partial x_i^m \partial x_k^n}\Phi(\underline{R}^1\ldots)} s_i^m s_k^n +$ | deformation (harmonic) energy U, vanishing for common small translations and rotations; first term entering dynamics; symmetry $\Phi_{ik}^{mn} = \Phi_{ki}^{nm}$ |
| | 3rd and higher order |
| + higher order terms | anharmonic terms. |

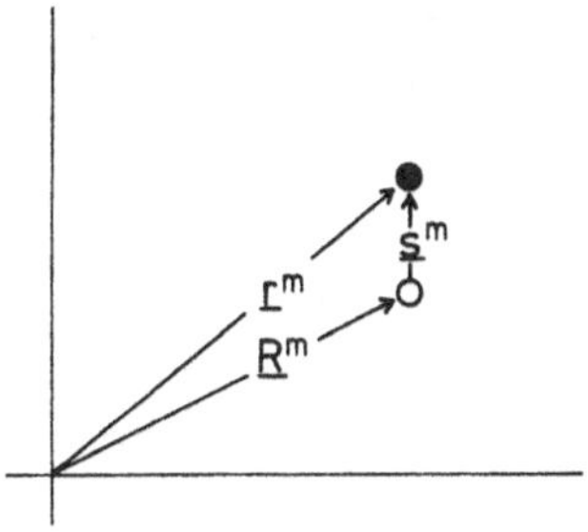


Fig. 2.2. Equilibrium position $\underline{R}^m$ (open circle) and displacement $\underline{s}^m$: actual position at $\underline{R}^m + \underline{s}^m$ (solid circle)

In the harmonic approximation only the 2nd order term is taken into account. It determines the internal forces $-\underline{\Phi}^n(\underline{R}^1 + \underline{s}^1 \ldots)$ which for small deviations from equilibrium are proportional to the displacements. If for convenience one introduces 3N-dimensional vectors $\underline{s} = (s_i^m)$, $\underline{R} = (\underline{R}^m)$, etc., and a 3N×3N symmetrical matrix[2] $\Phi = (\Phi_{ik}^{mn})$, one has in the harmonic approximation[3]

$$U = \frac{1}{2}\, s_i^m\, \Phi_{ik}^{mn}\, s_k^n = \frac{1}{2}\, (\underline{s},\Phi\underline{s}) = \frac{1}{2}\, \langle\underline{s}|\Phi|\underline{s}\rangle \quad , \tag{2.4a}$$

$$-\Phi_i^m(\underline{R} + \underline{s}) = -\frac{\partial U}{\partial s_i^m} = -\Phi_{ik}^{mn}\, s_k^n \;, \quad -\underline{\Phi}(\underline{R} + \underline{s}) = -\Phi\underline{s} \;. \tag{2.4b}$$

## 2.2.2 Symmetries and Meaning of the Force Constants $\Phi_{ik}^{mn}$

From (2.3) one has

$$\Phi_{ik}^{mn} = \Phi_{ki}^{nm} \;. \tag{2.5a}$$

The invariances (2.1a) imply vanishing internal forces (2.4b) for translational and rotational displacements and lead therefore also to relations between *second* derivatives:

$$\text{translation}\quad s_k^n = T_k \;, \quad \sum_{n(\text{or } m)} \Phi_{ik}^{mn} = 0 \;, \quad -\Phi_{ik}^{mm} = \sum_{n(\neq m)} \Phi_{ik}^{mn} \tag{2.5b}$$

$$\text{rotation}\quad s_k^n = \omega_{kl} X_l^n \;, \quad \Phi_{ik}^{mn} \omega_{kl} X_l^n = 0 \;, \quad \Phi_{ik}^{mn} X_l^n = \Phi_{il}^{mn} X_k^n \;. \tag{2.5c}$$

These are the only symmetries of a finite crystal and they already guarantee the correct symmetry of its elastic data. For this reason they are discussed here in detail.

The quantities $\Phi_{ik}^{mn}$ are called *coupling parameters* or *force* or *spring constants*. Their meaning is clear from (2.4b):

$-\Phi_{ik}^{mn}$ = the internal force on atom m in direction i if only atom n is displaced by unit length in direction k and all other atoms remain in their equilibrium positions (Fig. 2.3). (2.6)

---

[2] Here $\Phi$ denotes the force constant matrix rather than the adiabatic potential.

[3] A vector is denoted by $\underline{a}$, $|\underline{a}\rangle$ or $|a\rangle$ and the scalar product by $(\underline{a},\underline{b})$ or $\langle\underline{a}|\underline{b}\rangle$ or $\underline{a}\,\underline{b}$, whatever is more convenient (see also App. B).

Because of (2.5b) $\Phi^{mn}_{ik}$ also is the contribution of the spring m-n to the self-restoring force $-\Phi^{mm}_{ik}$.

As an example consider a simple spring f between m and n (Fig. 2.4) where[4]

$$\Phi^{mn}_{ik} = -f\,\hat{X}^{mn}_i \hat{X}^{mn}_k \,, $$ 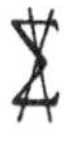 $$, \quad \underline{R}^{mn} = \underline{R}^m - R^n \,, \quad \hat{\underline{R}}^{mn} = \frac{\underline{R}^{mn}}{|\underline{R}^{mn}|} \,. \qquad (2.6a)$$

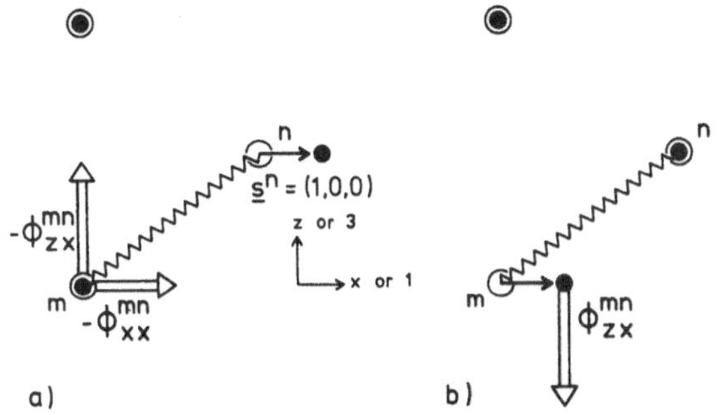


Fig. 2.3a and b. Meaning of the force constants (the springs are always drawn in their equilibrium positions).
a) Internal forces on m exerted by "m-n spring" if only atom n is displaced
b) Contribution of spring m-n to self-restoring force on m if m alone is displaced

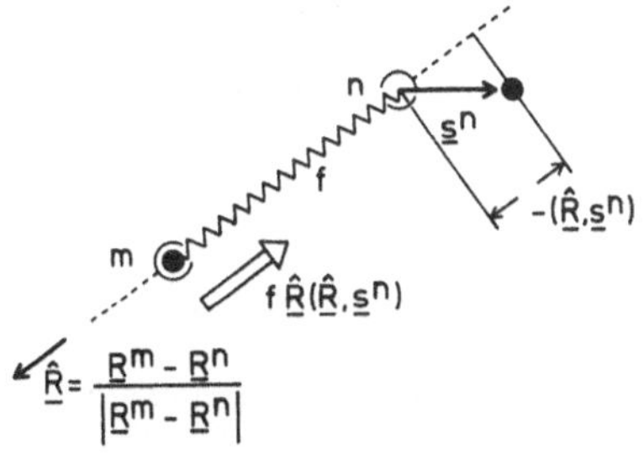


Fig. 2.4. Longitudinal spring f between atoms m and n: the force is in spring direction $\hat{\underline{R}}$ and proportional to the elongation in spring direction

## 2.2.3 Representation of the Coupling by Its Eigenvectors and Eigenvalues, Stability

Because $\Phi$ is symmetrical, it possesses a complete orthonormal system of eigenvectors $\overset{\nu}{\underline{s}} = |\nu\rangle$, $\nu = 1 \ldots 3N$, with eigenvalues $\overset{\nu}{\Phi}$ (App. B)

$$\Phi \overset{\nu}{\underline{s}} = \overset{\nu}{\Phi}\overset{\nu}{\underline{s}} \,, \quad \langle\mu|\nu\rangle = \delta^{\mu\nu} \,, \quad \overset{\nu}{\Phi} \text{ real} \qquad (2.7)$$

and one can express $\underline{s}$, $\Phi$ and U in the following way:

---

[4] Unit vectors are denoted by a caret: $\hat{\underline{R}} = \underline{R}/|\underline{R}|$ .

$$\underline{s} = |\nu\rangle\langle\nu|\underline{s}\rangle = |\nu\rangle s_\nu , \; s_\nu = \text{components in the eigensystem of } \Phi; \; |\nu\rangle \text{ and } s_\nu \text{ can be chosen as real} \tag{2.7a}$$

$$\Phi = \sum_{\nu=1}^{3N} |\nu\rangle \overset{\nu}{\Phi} \langle\nu| = \sum_\nu \overset{\nu}{\Phi}\overset{\nu}{P} , \quad \overset{\nu}{P} = |\nu\rangle\langle\nu| \text{ is the projector onto } |\nu\rangle \tag{2.7b}$$

$$U = \frac{1}{2}\langle\underline{s}|\Phi|\underline{s}\rangle = \frac{1}{2}(\underline{s},\Phi\underline{s}) = \frac{1}{2}\sum_\nu s_\nu \overset{\nu}{\Phi} s_\nu . \tag{2.7c}$$

The *stability* of the system can now be discussed. Stability would require a *minimum* of the potential energy in equilibrium: $U > 0$ for all $\underline{s} \neq 0$, or $\overset{\nu}{\Phi} > 0$ for all $\nu$. Actually these requirements cannot be met completely because of translational-rotational invariance of the potential. One has three independent translations (say $\nu = 1,2,3$) and three independent rotations (say $\nu = 4,5,6$) which are eigenvectors of $\Phi$ with eigenvalues zero[5]:

$$\overset{\nu}{\Phi} = 0 \quad \text{for } \nu = 1\ldots6. \tag{2.8a}$$

Consequently one can only ask that $U > 0$ for genuine deformations

$$\overset{\nu}{\Phi} > 0 \quad \text{for } \nu = 7\ldots3N . \tag{2.8b}$$

2.2.4 Dynamics

If one includes external forces $\underline{F}$ in addition to internal and inertial forces the equation of motion becomes[6]

$$M^m\partial_t^2 s_i^m(t) = M^m\ddot{s}_i^m(t) = -\Phi_i^m(\underline{R}+\underline{s}) + F_i^m = -\Phi_{ik}^{mn}s_k^n + F_i^m , \quad M\ddot{\underline{s}} + \Phi\underline{s} = \underline{F}(t) , \tag{2.9}$$

a linear, inhomogeneous differential equation for the displacement $\underline{s}$, $M^m$ being the mass of atom m and $M = (M^m\delta^{mn}\delta_{ik})$, being a positive diagonal matrix.

The basic physical quantities are

$$(\underline{s},\Phi\underline{s})/2 = U , \quad \text{the internal potential energy} \tag{2.9a}$$

---

5 The translational and rotational modes can be represented by three translations along and three infinitesimal rotations about the axes of any Cartesian coordinate system. If the origin of this system is given by $\sum_m X_i^m = 0$ and if its axes are the principal axes of $X_i^m X_k^m$, the six states are mutually orthogonal.

6 Mostly we abbreviate differentiations as follows:
$\partial/\partial t = \partial_t$ , $\partial/\partial x_i = \partial_{x_i} = \partial_i$ , etc.

$-(\underline{s},\underline{F}) = U_{ex}$ , the potential energy of external forces (2.9b)

$(\dot{\underline{s}},M\dot{\underline{s}})/2 = K$ , the kinetic energy. (2.9c)

## 2.3 Dynamics of the One-Dimensional Harmonic Oscillator /2.8/

Every harmonic system such as (2.9) can be represented by a set of 3N independent one-dimensional oscillators. For that reason it is convenient to discuss the behaviour of *one* oscillator first (Fig. 2.5). The notation can practically be taken over to represent the most general case.

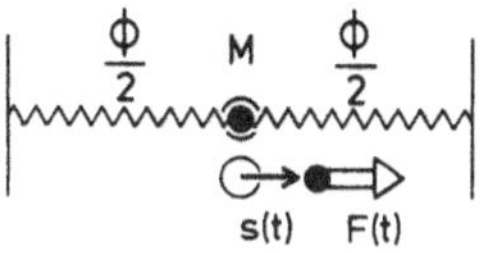


Fig. 2.5. One-dimensional oscillator, mass M, springs $\Phi/2$, displacement s(t) and external force F(t). The restoring force is $-2(\Phi/2)s = -\Phi s$

### 2.3.1 Equation of Motion and Solutions

The equation of motion is

$$M\ddot{s}(t) = -\Phi s(t) + F(t) \, . \tag{2.10}$$

Its solution is not unique without additional requirements, e.g., initial conditions, say at t = 0. The homogeneous equation

$$M\ddot{s} + \Phi s = 0 \quad \text{or} \quad \ddot{s} + \frac{\Phi}{M} = 0 = \ddot{s} + \Omega^2 s \, , \quad \Omega^2 = \frac{\Phi}{M} > 0 \, , \tag{2.11}$$

where[7] $\Omega = \sqrt{\Phi/M}$ is the eigenfrequency,

has as solutions $e^{\pm i\Omega t}$ or $\cos \Omega t$ and $\sin \Omega t$, and the solution of (2.11) with initial conditions at t = 0 is

$$s(t) = s(0) \cos \Omega t + \dot{s}(0)\frac{\sin \Omega t}{\Omega} . \tag{2.11a}$$

---

[7] Even though it does not matter here, let us always take $\Omega$ as the *positive* root of $\Phi/M$.

The inhomogeneous equation can be solved by superposition of solutions for specialized forces, for instance $\delta(t)$, a suddenly applied force at $t = 0$, which transfers momentum 1. If $S(t)$ is a solution of

$$M\ddot{S} + \Phi S = \delta(t) \ , \ M\dot{S}(t) \Big|_{-0}^{+0} = M[\dot{S}(+0) - \dot{S}(-0)] = 1 \quad (2.12)$$

[8]

then

$$s(t) = \int_{-\infty}^{+\infty} dt' \ S(t - t')F(t')$$

solves the inhomogeneous equation (2.10).

The solutions S to a $\delta$-force are also called Green's functions or linear response functions. Usually one specifies these Green's functions in the following way: the retarded (causal) Green's function

$$S(t) = g(t) = g^{+}(t) = \frac{\sin \Omega t}{M\Omega}\Theta(t) = \frac{\sin \Omega t}{M\Omega}\begin{cases} 0 \text{ for } & t < 0 \\ 1 \text{ for } & t \geq 0 \end{cases} \ , \quad (2.13a)$$

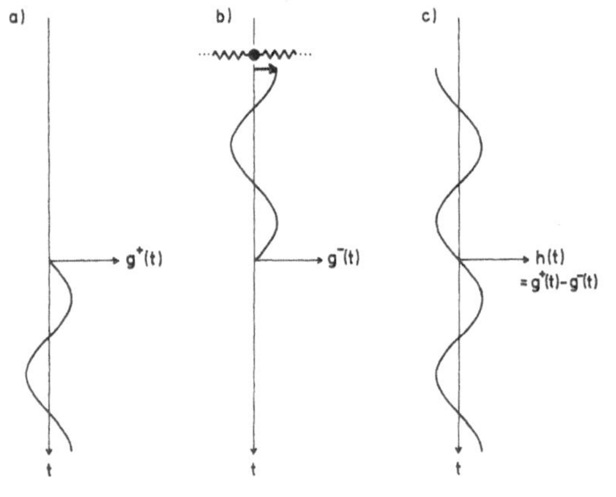


Fig. 2.6a-c.

a) Retarded $g(t) = g^{+}(t)$, (2.13a): to the oscillator at rest momentum 1 is transferred at $t = 0$.

b) Advanced $g^{-}(t)$, (2.13b): the oscillator is stopped by momentum transfer 1 at $t = 0$.

c) $h(t) = g^{+} - g^{-} = \frac{\sin \Omega t}{M\Omega}$, (2.13c), is a solution of the *homogeneous* equation and therefore *not* a Green's function

where the response occurs *after* the force has been applied (Fig. 2.6a), and the advanced Green's function

[8] The symbol +0 (-0) denotes the limit from positive (negative) values to zero.

$$S(t) = g^-(t) = -\frac{\sin \Omega t}{M\Omega}\Theta(-t) = g^+(-t) = -\frac{\sin \Omega t}{M\Omega}\begin{cases} 1 & \text{for} \quad t \le 0 \\ 0 & \text{for} \quad t > 0 \end{cases}, \tag{2.13b}$$

where the oscillator is stopped by a $\delta$-force at $t = 0$ (Fig. 2.6b). Any combination $\alpha g^+ + \beta g^-$ with $\alpha + \beta = 1$ would also be a solution of (2.12), whereas in the combination $g^+ - g^- = h(t)$ the forces cancel and

$$h(t) = g^+ - g^- = \frac{\sin \Omega t}{M\Omega} \quad \text{for} \ -\infty < t < +\infty \tag{2.13c}$$

is a solution of the homogeneous equation (Fig. 2.6c). Therefore one can rewrite (2.11a) as

$$s(t) = Ms(0)\dot{h}(t) + M\dot{s}(0)h(t) \quad \left(= Ms(0)\dot{g}(t) + M\dot{s}(0)g(t) \quad \text{for } t > 0\right). \tag{2.11b}$$

### 2.3.2 Green's Function

For physical reasons *the* Green's function is always taken as the retarded one; the general solution for $t > 0$ and initial conditions at $t = 0$ can be written as

$$s(t) = Ms(0)\dot{g}(t) + M\dot{s}(0)g(t) + \int_0^\infty g(t - t')F(t')dt' \ , \quad t \ge 0. \tag{2.14}$$

The use of the *retarded* g can be conveniently represented by introducing a small genuine friction $2M\eta\dot{s}$ ($\eta$ small) and an even smaller additional restoring force proportional to $\eta^2$. The equation of motion, instead of (2.10), is then

$$M\ddot{s}(t) + 2M\eta\dot{s}(t) + (M\eta^2 + \Phi)s = F(t) \ , \quad \eta > 0. \tag{2.15}$$

The introduction of $\eta > 0$ favors one direction in time and rules out all advanced aspects. Indeed, the solution of

$$M\ddot{g}(\eta,t) + 2M\eta\dot{g} + (M\eta^2 + \Phi)g = \delta(t) \tag{2.15a}$$

is now unique[9]

$$g(\eta,t) = \frac{\sin \Omega t}{M\Omega} e^{-\eta t}\Theta(t) = g(t)e^{-\eta t} \ , \tag{2.15b}$$

and passes into the retarded g for $\eta \to +0$. Only the introduction of the small, genuine friction ($\eta > 0$) is important here; the factor 2 and the $M\eta^2$-term are for convenience (to simplify the solution).

---

[9] The solution $g^-(t)e^{-\eta t}$ ($t < 0$), diverging for $t \to -\infty$, has to be ruled out.

The introduction of $\eta$ is also most convenient in discussing the Fourier transform[10] of g (App. C)

$$g(t) = \int_{-\infty}^{+\infty} \frac{d\omega}{2\pi} G(\omega) e^{-i\omega t} \; , \quad G(\omega) = \int_{-\infty}^{+\infty} dt \; g(t) e^{i\omega t} \; . \tag{2.16}$$

From the Fourier transform of (2.15a) one has (App. C)

$$G(\omega) = \frac{1}{M[\Omega^2 - (\omega + i\eta)^2]} = \frac{1}{M} \frac{\Omega^2 + \eta^2 - \omega^2}{(\Omega^2 + \eta^2 - \omega^2)^2 + 4\eta^2\omega^2} + i \frac{1}{M} \frac{2\eta\omega}{(\Omega^2 + \eta^2 - \omega^2)^2 + 4\eta^2\omega^2}$$

$$= \mathrm{Re}\{G\} + i \, \mathrm{Im}\{G\} \; . \tag{2.17}$$

In the limit $\eta \to +0$

$$\mathrm{Re}\{G(\omega)\} = \frac{1}{M} \frac{P}{\Omega^2 - \omega^2} \; , \quad \mathrm{Im}\{G(\omega)\} = \frac{\pi \, \mathrm{sgn} \, \omega}{M} \delta(\Omega^2 - \omega^2) \; , \tag{2.17a}$$

where P denotes Cauchy's principal value. For a more detailed discussion see Appendix C.

If one employs $\eta < 0$, the limit $\eta \to -0$ results in

$$G^-(\omega) = G^*(\omega) \; , \tag{2.18a}$$

and the Fourier transform of $h = \frac{\sin \Omega t}{M\Omega}$ becomes

$$H(\omega) = G^+(\omega) - G^-(\omega) = G(\omega) - G^*(\omega) = 2i \; \mathrm{Im}\{G(\omega)\} = \frac{2\pi i}{M} \mathrm{sgn}\,\omega \; \delta(\Omega^2 - \omega^2) . \tag{2.18b}$$

Of course, $H(\omega)$ can be obtained by direct Fourier transformation of $h(t)$. However, most theoretical work starts with the retarded Green's function G.

### 2.3.3 Thermal Averages, Displacement-Displacement Correlation Function

In this section we present briefly some thermal averages, which we need in Chapter 7 to discuss neutron and X-ray scattering. Because quantum effects are important, we first summarize the quantum theory of the harmonic oscillator.

From the classical Hamiltonian

$$\mathscr{H} = K + U = \frac{p^2}{2M} + \frac{M\Omega^2}{2} q^2 \; , \text{ coordinate } q = s, \text{ momentum } p = M\dot{s} \; , \tag{2.19}$$

one obtains the Hamilton operator upon replacing p by $(\hbar/i)\partial_q$

$$H = -\frac{\hbar^2}{2M} \partial_q^2 + \frac{M\Omega^2}{2} q^2 \; . \tag{2.19a}$$

[10] $g(t)$ is the displacement for $F(t) = \delta(t)$, $G(\omega)e^{-i\omega t}$ is the displacement for $F(t) = e^{-i\omega t}$.

Its eigenfunctions $\psi_n$ and eigenvalues $\varepsilon_n$ are given by

$$H\psi_n = \varepsilon_n\psi_n(q) \; , \quad \varepsilon_n = \hbar\Omega(n+1/2) \; , \quad n = 0,1,2... \; . \tag{2.19b}$$

In the ground state

$$\psi_o(q) = \left(\frac{M\Omega}{\pi\hbar}\right)^{1/4} \exp(-\frac{M\Omega q^2}{2\hbar}), \quad \varepsilon_o = \frac{\hbar\Omega}{2} \tag{2.20}$$

we can easily calculate some averages (expectation values), $\langle ... \rangle_o$.

$$\langle K\rangle_o = \langle U\rangle_o = \int dq \; U(q)\psi_o^2(q) = \frac{\varepsilon_o}{2}$$

$$\langle p\rangle_o = \langle q\rangle_o = 0 \quad , \quad \langle qp\rangle_o = -\frac{\hbar}{2i}$$

$$\langle \exp \alpha q\rangle_o = \exp \frac{\langle(\alpha q)^2\rangle_o}{2} \tag{2.20a}$$

$$\langle \exp A\rangle_o = \exp \frac{\langle A^2\rangle_o}{2}$$ valid for all operators A linear in p and q with $\langle A\rangle_o = 0$ such as p , q , $\alpha q + \beta p$.

These relations remain valid for thermal averages $\langle ... \rangle_{th}$ if $\varepsilon_o$ is replaced by the average thermal energy

$$\varepsilon_{th}(\Omega) = \langle H\rangle_{th} = \frac{\sum_n \varepsilon_n \exp(-\frac{\varepsilon_n}{kT})}{\sum_n \exp(-\frac{\varepsilon_n}{kT})} = \hbar\Omega\left[n_{th}(\Omega) + 1/2\right]$$

$$n_{th}(\Omega) = \left[\exp(\frac{\hbar\Omega}{kT}) - 1\right]^{-1} , \tag{2.21}$$

for instance,

$$\langle U\rangle_{th} = \frac{M\Omega^2}{2} \langle q^2\rangle_{th} = \frac{\varepsilon_{th}}{2} \; , \quad \langle qp\rangle_{th} = -\frac{\hbar}{2i} , \quad \text{etc.} \tag{2.21a}$$

With the aid of (2.21a) we can now calculate the *displacement - displacement correlation function* (Sec. 7.6),

$$l(t) = \langle s(0)\; s(t)\rangle_{th} \tag{2.22}$$

where $s(0) = q$ and $s(t) = \exp(i\frac{Ht}{\hbar})\; q \; \exp(-i\frac{Ht}{\hbar})$ is the corresponding Heisenberg operator which - as in classical theory - can be expressed by

$$s(t) = q \cos \Omega t + p\frac{\sin \Omega t}{M\Omega} = Mq\dot{h}(t) + p\,h(t) \; . \tag{2.22a}$$

Consequently,

$$l(t) = \langle Mq^2\rangle_{th}\; \dot{h}(t) + \langle qp\rangle_{th}\; h(t)$$

$$= \frac{\varepsilon_{th}(\Omega)}{\Omega^2}\; \dot{h}(t) + \frac{i\hbar}{2}\; h(t) \; , \tag{2.23a}$$

$$1(0) = \langle s(0)^2 \rangle_{th} = \langle q^2 \rangle_{th} = \frac{\varepsilon_{th}(\Omega)}{M\Omega^2} \quad . \tag{2.23b}$$

Later we will need also the Fourier transform of $1(t)$

$$L(\omega) = -iH(\omega)\left[\frac{\omega\varepsilon_{th}(\Omega)}{\Omega^2} - \frac{\hbar}{2}\right] = \frac{2\pi}{M}\,\delta(\Omega^2 - \omega^2)\ \mathrm{sgn}\ \omega\left[\frac{\omega\varepsilon_{th}(\Omega)}{\Omega^2} - \frac{\hbar}{2}\right] \tag{2.24}$$

Because of the $\delta$-function, $\Omega$ can be replaced by $|\omega|$

$$L(\omega) = -iH(\omega)\left[\frac{\varepsilon_{th}(|\omega|)}{\omega} - \frac{\hbar}{2}\right] = -iH(\omega)\hbar\cdot\begin{cases} n_{th}(\omega) & \text{for } \omega > 0 \\ -[n_{th}(\bar{\omega}) + 1] & \text{for } \omega = -\bar{\omega} < 0 \end{cases} \tag{2.24a}$$

or

$$L(\omega) = \frac{2\pi\hbar}{M}\,\delta(\Omega^2 - \omega^2)\cdot\begin{cases} n_{th}(\omega) & \text{for } \omega > 0 \\ -[n_{th}(\bar{\omega}) + 1] & \text{for } \omega < 0 \end{cases} \quad . \tag{2.24b}$$

## 2.4 Dynamics of an Assembly of Atoms

The difference between equations

(2.10) $M\ddot{s} + \Phi s = F$ and (2.9) $M\ddot{\underline{s}} + \Phi\underline{s} = \underline{F}$

is the number of degrees of freedom, 1 in (2.10) and 3N in (2.9) for N atoms. In (2.9) $\underline{s}$ and $\underline{F}$ are 3N-dimensional vectors, M and $\Phi$ are 3N×3N matrices, whereas in (2.10) s, F, M and $\Phi$ are numbers. The diatomic molecule is the simplest nontrivial example for (2.9), N = 2. This example is discussed in Appendix D.

### 2.4.1 Dynamical Matrix

The 3N-dimensional analogous to the one-dimensional $\Omega^2$ is not simply "$\Phi/M$", but rather, because $\Phi$ and M do not commute, the symmetrized version

$$\frac{1}{\sqrt{M}}\,\Phi\,\frac{1}{\sqrt{M}} = D = \Omega^2 = D' \quad , \tag{2.25}$$

called the dynamical matrix. Operating with $M^{-1/2}$ on (2.9) one has

$$\sqrt{M}\ddot{\underline{s}} + \underbrace{\frac{1}{\sqrt{M}}\,\Phi\,\frac{1}{\sqrt{M}}}_{D\,=\,\Omega^2}\sqrt{M}\,\underline{s} = \frac{1}{\sqrt{M}}\underline{F} \tag{2.26}$$

or

$$\ddot{\tilde{\underline{s}}} + D\tilde{\underline{s}} = \tilde{\underline{F}} \quad \text{where} \quad \tilde{\underline{s}} = \sqrt{M}\,\underline{s} \ , \quad \tilde{\underline{F}} = \frac{1}{\sqrt{M}}\underline{F} \ . \tag{2.26a}$$

In the eigenbasis of D (compare Sec. 2.2.3), $D|\nu\rangle = \overset{\nu}{D}|\nu\rangle = \overset{\nu}{\Omega}{}^2|\nu\rangle$, $\tilde{s}_\nu = \langle\nu|\tilde{\underline{s}}\rangle =$ *normal coordinates*, $\tilde{F}_\nu = \langle\nu|\tilde{F}\rangle$, eq. (2.26a) separates into 3N independent equations for oscillators with eigenfrequencies[11] $\overset{\nu}{\Omega}$ ( $\geqslant 0$) and mass 1:

$$\ddot{\tilde{s}}_\nu + \overset{\nu}{\Omega}{}^2\tilde{s}_\nu = \tilde{F}_\nu(t) \ . \tag{2.26b}$$

If M is a pure number, the eigenbasis of $\Phi$ and $\Omega^2$ are the same and $\overset{\nu}{\Omega} = \sqrt{\overset{\nu}{\Phi}/M}$. Later we will mostly discuss this simple case.

### 2.4.2 Green's Functions

The methods of Section 2.3 can now be applied and one finds

$$g(t) = \frac{1}{\sqrt{M}}\frac{\sin \Omega t}{\Omega}\frac{1}{\sqrt{M}}\,\Theta(t) \text{ corresponding to (2.13a) ,} \tag{2.27}$$

$$G(\omega) = \frac{1}{\sqrt{M}}\frac{1}{\Omega^2 - (\omega + i\eta)^2}\frac{1}{\sqrt{M}} = \frac{1}{\Phi - M(\omega + i\eta)^2} \text{ corresponding to (2.17)} \tag{2.27a}$$

$$h(t) = \frac{1}{\sqrt{M}}\frac{\sin \Omega t}{\Omega}\frac{1}{\sqrt{M}} \text{ corresponding to (2.13c), and} \tag{2.27b}$$

$$\underline{s}(t) = M\underline{s}(0)\dot{h}(t) + M\dot{\underline{s}}(0)h(t) \quad \text{corresponding to (2.11b).} \tag{2.27c}$$

The physical meaning of Green's functions is:

$g^{mn}_{ik}(t)$ = the displacement of atom m in direction i, if at t = 0 unit momentum in direction k is transferred to atom n[12] (for $t \leq 0$ all atoms are at rest), and (2.28)

$G^{mn}_{ik}(\omega)e^{-i\omega t}$ = the displacement of atom m in direction i due to a force $e^{-i\omega t}$ on n in direction k. (2.29)

For applications, later we will need mainly $G(\omega)$ and also G(0), the static Green's function; G(0) describes the response to static (time independent) forces and is closely related to the elastic macroscopic behaviour.

---

[11] From $\Phi|\nu = 1\ldots6\rangle = 0$ (comp. footnote to (2.8a)) one has, according to (2.26), $D\sqrt{M}|\nu = 1\ldots6\rangle = 0$ corresponding to six states with vanishing $\overset{\nu}{D}$ or $\overset{\nu}{\Omega}$. To ensure orthogonality one has now to use the following coordinate system: the origin is the center of mass, $M^m X^m_i = 0$, and the axes are that of the moment of inertia, $\sum_m M^m X^m_i X^m_k$.

[12] Note that in contrast to the definition of the force constants here *all* atoms are free to move.

Unfortunately it is impossible to evaluate all the eigenvalues $\overset{\nu}{\Omega}$ and the eigenvectors $|\nu\rangle$ needed in

$$g = \frac{1}{\sqrt{M}} \frac{\sin \Omega t}{\Omega} \frac{1}{\sqrt{M}} = \sum_{\nu} \frac{1}{\sqrt{M}} |\nu\rangle \frac{\sin \overset{\nu}{\Omega} t}{\overset{\nu}{\Omega}} \langle\nu| \frac{1}{\sqrt{M}}$$

for a finite system. However, the high symmetry of an infinite crystal leads to so much information about its eigenvectors that for simple metals one is left, at worst, with cubic equations to obtain $|\nu\rangle$ and $\overset{\nu}{\Omega}$.

### 2.4.3 Kramers-Kronig Relations

As in (2.17), one has (equal masses):

$$G(\omega) = \underbrace{\frac{1}{M} \frac{P}{\Omega^2 - \omega^2}}_{G_1(\omega^2) = \mathrm{Re}\{G\}} + \underbrace{i \,\mathrm{sgn}\,\omega \frac{\pi}{M} \delta(\Omega^2 - \omega^2)}_{G_2(\omega^2) = \mathrm{sgn}\,\omega\; \mathrm{Im}\{G\}} \;. \tag{2.30}$$

The real part is symmetric, the imaginary part antisymmetric in $\omega$. Because $G_2$ contains only a $\delta$-function, one can express $G_1(\omega^2)$ by $G_2$

$$\begin{aligned} \mathrm{Re}\left\{G(\omega^2)\right\} = \frac{1}{M} \frac{P}{\Omega^2 - \omega^2} &= \int_0^\infty \frac{1}{M} \frac{P}{\omega'^2 - \omega^2} \frac{M}{\pi} G_2(\omega'^2) d\omega'^2 \\ &= \int_0^\infty \frac{d\omega'^2}{\pi} \frac{P}{\omega'^2 - \omega^2} G_2(\omega'^2) \\ &= \int_{-\infty}^{+\infty} \frac{d\omega'}{\pi} \frac{P}{\omega'^2 - \omega^2} \omega' \,\mathrm{Im}\left\{G(\omega')\right\} \;. \end{aligned} \tag{2.31}$$

These equations (2.31) are called Kramers-Kronig relations[13]. Because the eigenfrequencies $\overset{\nu}{\Omega}$ have a lower and upper bound, $G_2$ vanishes outside this range. Therefore, the interval of integration in (2.31) is finite.

### 2.4.4 Thermal Averages, Displacement-Displacement Correlation Functions

In complete analogy to (2.22) we obtain for the correlation function

$$l(t) = \dot{h}(t) \sqrt{M} \frac{\varepsilon_{th}(\Omega)}{\Omega^2} \frac{1}{\sqrt{M}} + \frac{i\hbar}{2} h(t) \tag{2.32}$$

---

[13] For finite $\eta$ one obtains from $G_2(\eta)$ via the Kramers-Kronig-relations $G_1(2\eta)$. As a rule G must be calculated numerically; because of (2.31), this problem is reduced to the evaluation of $G_2$.

$$1(0) = \frac{1}{\sqrt{M}} \frac{\varepsilon_{th}(\Omega)}{\Omega^2} \frac{1}{\sqrt{M}} \ . \tag{2.32a}$$

Because of (2.27a)

$$H(\omega) = 2\pi i \ \mathrm{sgn}\ \omega \frac{1}{\sqrt{M}} \delta(\Omega^2 - \omega^2) \frac{1}{\sqrt{M}} \ ; \qquad \omega \geq 0 \ , \tag{2.33}$$
$$\delta(\Omega^2 - \omega^2) = \frac{1}{\pi} \sqrt{M}\ \mathrm{Im} \left\{ G(\omega)\sqrt{M} \right\};$$

for any function $f(\Omega)$, $\Omega \geq 0$, say $\varepsilon_{th}(\Omega)$ or $\varepsilon_{th}(\Omega)/\Omega^2$,

$$f(\Omega) = \int_0^\infty f(\omega)\ \delta(\Omega^2 - \omega^2) d\omega^2 = \int_0^\infty f(\omega) \frac{\sqrt{M}}{\pi} \mathrm{Im} \left\{ G(\omega) \right\} \sqrt{M}\, d\omega^2 \ . \tag{2.33a}$$

Therefore one can again replace $\Omega$ by $|\omega|$ in $L(\omega)$:

$$L(\omega) = -iH(\omega)\hbar \begin{cases} n_{th}(\omega) & \text{for } \omega > 0 \\ \left[ n_{th}(\bar{\omega}) + 1 \right] & \text{for } \omega = -\bar{\omega} < 0 \end{cases} \ . \tag{2.32b}$$

Of course all quantities above such as $1, h, M, \Omega$ are matrices, e.g.,

$$1^{mn}_{ik}(t) = \langle s^m_i(0)\ s^n_k(t) \rangle_{th} = \sum_{p,l} \dot{h}^{mp}_{il} \sqrt{M^p} \left[ \frac{\varepsilon_{th}(\Omega)}{\Omega^2} \right]^{pn}_{lk} \frac{1}{\sqrt{M^n}} + \frac{i\hbar}{2} h^{mn}_{ik}(t) \tag{2.32c}$$

$$1^{mn}_{ik}(0) = \langle s^m_i(0)\ s^n_k(0) \rangle_{th} = \frac{1}{\sqrt{M^m}} \left[ \frac{\varepsilon_{th}(\Omega)}{\Omega^2} \right]^{mn}_{ik} \frac{1}{\sqrt{M^n}} \ , \quad \not\Sigma \ . \tag{2.32d}$$

The quantity $L(\omega)$ will be needed for neutron scattering; for X-ray scattering $1(0)$ is required (Chapter 7). The intensity of the Mößbauer line is determined by the "Debye-Waller-Factor" $\exp -\langle (\underline{p}_\gamma, \underline{s}^o)^2 \rangle_{th}$, where $\underline{p}_\gamma$ is the momentum of the emitted $\gamma$-quantum and $\underline{s}^o$ the displacement of the Mößbauer atom (at site 0). The decisive quantity is

$$1^{oo}_{11}(0) = \frac{1}{\overset{o}{M}} \left[ \frac{\varepsilon_{th}(\Omega)}{\Omega^2} \right]^{oo}_{11} = \int d\omega^2 \frac{\varepsilon'_{th}(\Omega)}{\pi\omega^2}\ \mathrm{Im} \left\{ G^{oo}_{11} \right\} . \tag{2.32e}$$

Another important thermal average is the thermal energy $E(t)$, the thermal average of kinetic and potential energy. Obviously

$$E(t) = \sum_{\nu=1}^{3N} \varepsilon_{th}(\overset{\nu}{\Omega}) = \mathrm{tr} \left\{ \varepsilon_{th}(\Omega) \right\} = \left[ \varepsilon_{th}(\Omega) \right]^{mm}_{ii} \tag{2.34}$$
$$= \int d\omega^2\ \varepsilon_{th}(\omega) \sum_{m,i} \frac{M^m}{\pi} \mathrm{Im} \left\{ G^{mm}_{ii} \right\} .$$

Clearly $\sum_{m,i} (M^m/\pi)\ \mathrm{Im} \left\{ G^{mm}_{ii} \right\} = \mathrm{tr} \left\{ \delta(\Omega^2 - \omega^2) \right\} = \sum_{\nu} \delta(\overset{\nu}{\Omega}{}^2 - \omega^2) = \tilde{Z}(\omega^2)$ is the distribution of eigenfrequencies, the "spectrum" (comp. Sec. 3.5.3). It is

$$\int_0^\infty d\omega^2\, \tilde{Z}(\omega^2) = 3N = \text{number of degrees of freedom} ;$$

$$\int_{\omega_1^2}^{\omega_2^2} d\omega^2\, \tilde{Z}(\omega^2) = \text{number of eigenfrequencies in the interval } \omega_1^2 \leq \Omega^2 \leq \omega_2^2 ;$$

$$\int_{\omega_1^2}^{\omega_2^2} d\omega^2 \frac{\tilde{Z}(\omega^2)}{3N} = \text{fraction of eigenfrequencies in the interval } \omega_1^2 \leq \Omega^2 \leq \omega_2^2 ;$$

$\tilde{Z}/3N = Z(\omega^2)$ is a normalized distribution: $\int_0^\infty Z(\omega^2)d\omega^2 = 1.$

## 2.5 Statics

### 2.5.1 Discussion of Static Green's Function

The response for static (time independent) forces where

$$\Phi\underline{s} = \underline{F} \tag{2.35}$$

ought to be given by

$$\underline{s} = G(\omega = 0)\underline{F} \quad \text{where} \quad G(0) = \frac{1}{\Phi} = \Phi^{-1}. \tag{2.35a}$$

In this section we denote static Green's function by G instead of G(0). Consequently, one would have

$$s_i^m = G_{ik}^{mn} F_k^n \quad , \quad G_{ik}^{mn} = \text{displacement of atom m in i-direction if unit force in k-direction is applied to atom n.} \tag{2.35b}$$

In terms of eigenvectors and eigenvalues of $\Phi$, one would have

$$G = \frac{1}{\Phi} = \sum_\nu |\nu\rangle \frac{1}{\overset{\nu}{\Phi}} \langle\nu| \quad , \quad \underline{s} = G\underline{F} = \sum_\nu |\nu\rangle \frac{1}{\overset{\nu}{\Phi}} \langle\nu|\underline{F}\rangle \ . \tag{2.35c}$$

As we have seen in Section 2.2, the six eigenvalues $\overset{\nu}{\Phi}$ belonging to translational-rotational motion $|\nu = 1...6\rangle$ are zero. Consequently G, according to (2.35c), does not exist; it is obvious that the application of a force to a single atom must result in translational motion of the whole system and cannot lead to static equilibrium. Therefore, when discussing the static response one must take precaution against translational-rotational contributions.

### 2.5.2 Green's Function for Invariant Force Patterns

A *static* response exists only if an invariant force pattern (2.2c,d) is applied,

$$\langle\nu|\underline{F}\rangle = 0 \quad \text{for } \nu = 1\ldots 6 , \tag{2.36}$$

because then the divergent terms in (2.35c) drop out

$$\underline{s} = \sum_{\nu\geq 7} |\nu\rangle \frac{1}{\overset{\nu}{\Phi}} \langle\nu|\underline{F}\rangle . \tag{2.36a}$$

Eq. (2.36) is equivalent to the statement that $\underline{F}$ must not have components in the subspace $\overset{\nu}{\Phi} = 0$ (or $\Phi = 0$) determined by the states $|\nu = 1\ldots 6\rangle$ or by the projector

$$P = \sum_{\nu=1}^{6} |\nu\rangle\langle\nu| \tag{2.37}$$

Invariant force patterns obey

$$P\underline{F} = 0 \quad \text{or} \quad Q\underline{F} = \underline{F} \quad \text{where } Q = 1 - P . \tag{2.38}$$

Hence from (2.36a) one recognizes that

$$G = \sum_{\nu\geq 7} |\nu\rangle \frac{1}{\overset{\nu}{\Phi}} \langle\nu| = \frac{1}{Q\Phi Q} \; ; \quad Q\Phi Q = \sum_{\nu\geq 7} |\nu\rangle \overset{\nu}{\Phi} \langle\nu| = \Phi \tag{2.38a}$$

is the proper Green's function for invariant $\underline{F}$'s. Note that G operates only in the subspace Q, where $\Phi \neq 0$:

$$GQ = QG = G \quad , \quad GP = PG = 0 . \tag{2.38b}$$

Moreover, G as defined by (2.38a) can be applied even to a noninvariant force pattern $\underline{F}$ (for instance to a force on a single atom), and the displacements are those of an invariant force pattern $Q\underline{F}$ instead of $\underline{F}$.

Any invariant force pattern can be resolved into single forces: $\sum_m \underline{F}^m$ (comp. Fig. 2.7a). With G from (2.38a) the response can also be decomposed into responses to "single forces" $Q\underline{F}^m$ , $\underline{s} = G\underline{F} = \sum_m GQ\underline{F}^m$. The transition from $\underline{F}^m$ to $Q\underline{F}^m$ means a "renormalization" of the single force which removes its translational and rotational legs. The renormalized single forces of Fig. 2.7a are shown in Fig. 2.7b,c. Definition (2.35b) can now be maintained for renormalized forces.

### 2.5.3 Response for Given Surface Displacements

A more general static problem is to calculate the response if the displacements of a set of atoms are given. In particular we are interested in the bulk response for given surface displacements. These boundary conditions guarantee static equilibrium,

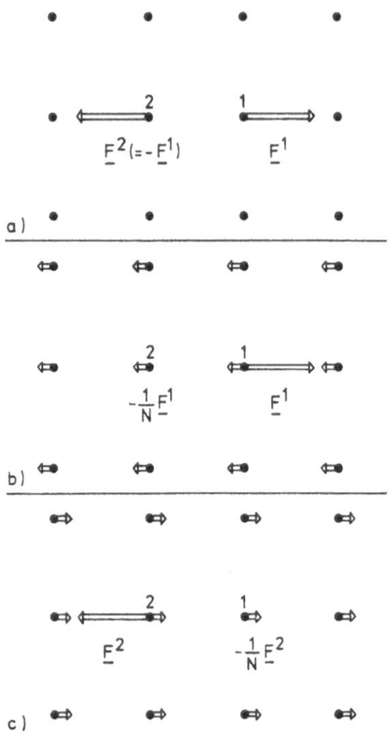


Fig. 2.7a-c. Invariant force patterns and renormalized single forces. a) Simple invariant force pattern consisting of two opposite single forces $\underline{F}^1$ and $\underline{F}^2 = -\underline{F}^1$. b) The renormalized single force $Q\underline{F}^1$ is an invariant force pattern; in addition to the original single force one has forces $-\frac{1}{N}\underline{F}^1$ on all atoms (note that N is large and that the drawing is not to scale). c) Renormalized single force $Q\underline{F}^2$. By adding the forces of b) and c), the renormalization cancels and one obtains the pattern a)

even though the displacements can contain (static) translational-rotational components (Fig. 2.8).

We introduce projectors

$$\text{onto the surface: } p \quad ; \quad \text{into the bulk: } q = 1 - p \quad ; \tag{2.39}$$

$$p^{mn}_{ik} = \delta_{ik} \sum_{\mu}{}' \delta^{m\mu}\delta^{n\mu} \quad , \tag{2.39a}$$

where the sum extends only over the surface atoms (Fig. 2.8a). The problem is now to determine $q\underline{s}$, the displacement in the bulk, from

$$\Phi\underline{s} = \underline{F} \quad \text{with given } p\underline{s}\ ,\ q\underline{F} \quad , \tag{2.40}$$

where $p\underline{s}$ are the surface displacements and $q\underline{F}$ are the forces in the bulk. Operating on (2.40), $\Phi(q + p)\underline{s} = (q + p)\underline{F}$, with q one obtains (because $qp = 0$) an equation in the subspace q (the bulk)

$$q\Phi q\underline{s} + q\Phi p\underline{s} = q\underline{F} \quad , \tag{2.41}$$

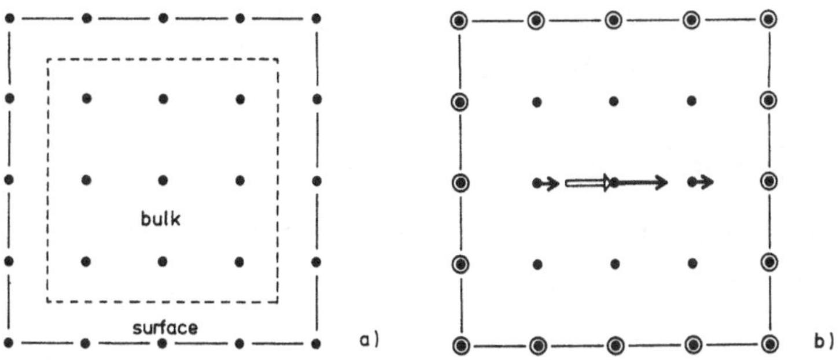


Fig. 2.8a and b. a) Definition of bulk and surface (schematically). b) Fixed surface atoms indicated by ◉. Obviously, the response contains a translational contribution

which is solved by

$$q\underline{s} = \frac{1}{q\Phi q}(q\underline{F} - q\Phi p\underline{s}) = G(q\underline{F} - q\Phi p\underline{s}) \tag{2.41a}$$

with

$$G = \frac{1}{q\Phi q} \quad , \quad qG = Gq = G \ . \tag{2.41b}$$

One realizes from (2.41a) that G according to (2.41b) is Green's function for vanishing surface displacements ($p\underline{s} = 0$).

For *infinite* crystals vanishing displacements in infinity are required. Therefore the definition of G by (2.41a) is most convenient for a transition from finite to infinite crystals.

In order to illustrate the meaning of (2.39 - 41), we rewrite them more explicitly in terms of 3N dimensional vectors and 3N×3N matrices. By suitably labelling the components of the vector $\underline{s}$ we obtain

$$\left(\underline{s}\right) = \left(\frac{p\underline{s}}{q\underline{s}}\right) = \left(\begin{matrix}\text{"surface"} \\ \text{"bulk"}\end{matrix} > \text{components}\right) \tag{2.42}$$

where the upper components $p\underline{s}$ refer to the atoms in the surface and the lower ones to those in the bulk. The projectors p , q are in this notation

$$p = \left[\begin{array}{c|c} 1 & 0 \\ \hline 0 & 0 \end{array}\right] \quad , \quad q = \left[\begin{array}{c|c} 0 & 0 \\ \hline 0 & 1 \end{array}\right] \tag{2.42a}$$

and $\Phi$ can be split up into blocks

$$\Phi = \left[\begin{array}{c|c} p\Phi p & p\Phi q \\ \hline q\Phi p & q\Phi q \end{array}\right] \tag{2.42b}$$

where, e.g., $-q\Phi p$ gives the internal forces on atoms in the bulk due to displacements in the surface. Hence (2.40) becomes

$$\left[\begin{array}{c|c} p\Phi p & p\Phi q \\ \hline q\Phi p & q\Phi q \end{array}\right]\begin{pmatrix} p\underline{s} \\ q\underline{s} \end{pmatrix} = \begin{pmatrix} p\Phi p\underline{s} + p\Phi q\underline{s} \\ q\Phi p\underline{s} + q\Phi q\underline{s} \end{pmatrix} = \begin{pmatrix} p\underline{F} \\ q\underline{F} \end{pmatrix} \quad , \tag{2.42c}$$

the q-components of which are equivalent to (2.41). For Green's function according to (2.41b) one has in this notation

$$G = \left[\begin{array}{c|c} 0 & 0 \\ \hline 0 & 1/q\Phi q \end{array}\right] \quad . \tag{2.42d}$$

As an application of (2.40 - 42), needed later, we can express the elastic energy $U = (\underline{s},\Phi\underline{s})/2$ for given surface displacements $p\underline{s}$ and vanishing forces in the bulk, $q\underline{F} = 0$, in terms of $p\underline{s}$. From (2.41a) we obtain

$$q\underline{s} = -G\Phi p\underline{s} \quad , \quad \text{hence } \underline{s} = (p + q)\underline{s} = (1 - G\Phi)p\underline{s}$$

and therefore

$$2U = (\underline{s},\Phi\underline{s}) = \Big(\underline{s},(\Phi - \Phi G\Phi)p\underline{s}\Big) \quad . \tag{2.43}$$

One recognizes immediately from the explicit matrix notation (2.42b,d) that

$$\Phi - \Phi G\Phi = \left[\begin{array}{c|c} p\Phi p - p\Phi q(1/q\Phi q)q\Phi p & 0 \\ \hline 0 & 0 \end{array}\right] = p(\Phi - \Phi G\Phi)p$$

operates only on p and one can on the right-hand side of (2.43) replace $\underline{s}$ by $p\underline{s}$:

$$2U = \Big(p\underline{s},(\Phi - \Phi G\Phi)p\underline{s}\Big) \quad . \tag{2.43a}$$

Of course the above formalism is not restricted to a bulk-surface partition. For instance, one can distinguish a single atom, say $\mu$:

$$\overset{\mu}{p} \quad , \quad \overset{\mu}{q} = 1 - \overset{\mu}{p} \quad , \quad \overset{\mu}{p}{}^{mn} = \delta^{m\mu}\delta^{n\mu} \quad , \quad (\overset{\mu}{p}\Phi\overset{\mu}{p})^{mn}_{ik} = \Phi^{\mu\mu}_{ik}\,\overset{\mu}{p}{}^{mn} \quad .$$

For rigid environment $\overset{\mu}{q}\underline{s} = 0$, the atom $\mu$ behaves like a three-dimensional "Einstein"-oscillator with the 3×3 force constant matrix $\Phi^{\mu\mu}_{ik}$.

### 2.5.4 Variational Methods

Variational methods can be employed to obtain approximate solutions of (2.35), $\Phi\underline{s} = \underline{F}$. In order to avoid restrictions on $\underline{F}$, we assume an infinite crystal, i.e., the displacements must vanish in infinity. The variational method makes use of the fact that the total energy L including the potential energy of the external forces, $L = U - U_{ext}$ , is a minimum in static equilibrium.

For arbitrary displacements $\underline{s} + \underline{\eta}$ where $\underline{\eta}$ is a deviation from the exact solution $\underline{s}$, the total energy is

$$L(\underline{s} + \underline{\eta}) = \left(\underline{s} + \underline{\eta}, \Phi(\underline{s} + \underline{\eta})\right)/2 - (\underline{s} + \underline{\eta}, \underline{F}) \tag{2.44}$$

$$L(\underline{s} + \underline{\eta}) = L(\underline{s}) + (\underline{\eta}, \Phi\underline{s} - \underline{F}) + \underbrace{(\underline{\eta}, \Phi\underline{\eta})/2}_{\geq 0} \quad . \tag{2.44a}$$

Obviously $L(\underline{s} + \underline{\eta})$ is minimal for $\underline{\eta} = 0$ if $\underline{s}$ satisfies (2.35), and

$$L(\underline{s}) = -(\underline{s}, \Phi\underline{s})/2 = -(\underline{F}, G\underline{F})/2 \tag{2.45}$$

is the lowest possible value of the total energy for any displacement:

$$L(\underline{s} + \underline{\eta}) \geq L(s) \quad . \tag{2.45a}$$

Consequently, if one considers a variational, trial displacement, $\underline{s}_{var} = \underline{s}_v = \underline{s} + \underline{\eta}_v$, that depends on variational parameters, $L(\underline{s}_v)$ must always be larger than $L(\underline{s})$: $L(\underline{s}_v) \geq L(\underline{s})$, $U(\underline{s}_v) \leq U(\underline{s})$. This fact can be used to determine optimal variational parameters by requiring $L(\underline{s}_v)$ to be as low as possible. This will give an optimal approach to $L(\underline{s})$ and implies an approximation $\underline{s}_v$ to $\underline{s}$, which becomes better as $L(\underline{s}_v)$ lowers.

This method can be used to calculate approximate G's by considering a unit force in direction $\hat{\underline{a}}$ applied to a single atom $\mu$:

$$\underline{F}^m = \hat{\underline{a}}\delta^{m\mu} \ , \quad \underline{s}^m = G^{m\mu}\hat{\underline{a}} \ , \quad 2L(\underline{s}) = -(\hat{\underline{a}}, G^{\mu\mu}\hat{\underline{a}}) \quad . \tag{2.46}$$

Most simply, one can restrict $\underline{s}_v$ to atom $\mu$ alone (Einstein approximation, local response):

$$\underline{s}_v^m = \underline{b}\delta^{m\mu} \ , \quad 2L(\underline{s}_v) = (\underline{b}, \Phi^{\mu\mu}\underline{b}) - 2(\underline{b}, \hat{\underline{a}}) \quad . \tag{2.47}$$

The minimum of $L(\underline{s}_v)$ is given by

$$\Phi^{\mu\mu}\underline{b} = \hat{\underline{a}} \ , \quad \underline{b} = \frac{1}{\Phi^{\mu\mu}}\hat{\underline{a}} \ , \tag{2.47a}$$

and the approximate G is[14]

$$G_v^{m\mu} = \delta^{m\mu}\frac{1}{\Phi^{\mu\mu}} \ . \tag{2.47b}$$

Eq. (2.45a) becomes

$$(\hat{\underline{a}}, \frac{1}{\Phi^{\mu\mu}}\hat{\underline{a}}) \leq (\hat{\underline{a}}, G^{\mu\mu}\hat{\underline{a}}) \ , \tag{2.48}$$

[14] Note, that $1/\Phi^{\mu\mu}$ is the reciprocal of the 3×3 matrix $\Phi^{\mu\mu}$ and not $(1/\Phi)^{\mu\mu}$, which is the exact $G^{\mu\mu}$.

which means that the approximate displacement of $\mu$ parallel to the force is always smaller than the exact one, or roughly speaking that the approximate response is smaller than the exact one. If *more* atoms are included in $\underline{s}_v$, then $L(\underline{s}_v)$ will decrease, and $G_v$ (now nonlocal) will be a better approximation.

# 3. Lattice Theory

In the preceding chapter the framework of harmonic theory was established, and the most general case was discussed. Emphasis was put on the implication of finite systems, but only formal solutions could be obtained. To work out the "details", namely the eigenvectors of $\Phi$ and D needed for the linear response G, proved to be prohibitively difficult due to the tremendous number of degrees of freedom. In this chapter we treat periodic structures (infinite crystals). Though one now faces infinite degrees of freedom in these structures, infinite crystals possess an infinite number of symmetries (Sec. 3.2), which allows easy calculation of these details. Only simple Bravais lattices are treated. Here the eigenvectors of $\Phi$ and D are identical; their calculation requires at most the solution of a cubic equation, and from these results one can represent G by relatively simple integrals which can be calculated numerically (Sec. 3.5).

The central quantity in defect physics is the Green's function of the host lattice; it can be replaced by that of the ideal infinite lattice if the defect is in the bulk. For this reason G is discussed in great detail (Sec. 3.5) to give a feeling for its overall behaviour and also for the numerical values. To provide a thorough understanding of the lattice Green's function is the main aim of this chapter. The results will be used later to demonstrate the properties of simple defects.

It has already been pointed out that the calculation of force constants from first principles is difficult and not very reliable. Often, models are used and fitted to experimental data. These models cannot be employed uncritically for defects because rotational invariance is not guaranteed. For this reason we have included a separate section (3.4) on force constant models (2-body forces, bond-charge and shell models), which can be used also in defect physics.

## 3.1 Periodic Structures

### 3.1.1 The One-Dimensional Lattice

The one-dimensional lattice (linear chain) can be most simply visualised as a periodic arrangement of like atoms (Fig. 3.1a) with rest positions

$$X^m = am \; ; \quad m = 0,\pm1,\pm2,\ldots \text{ integer} \; ; \quad a = \text{lattice distance.} \tag{3.1}$$

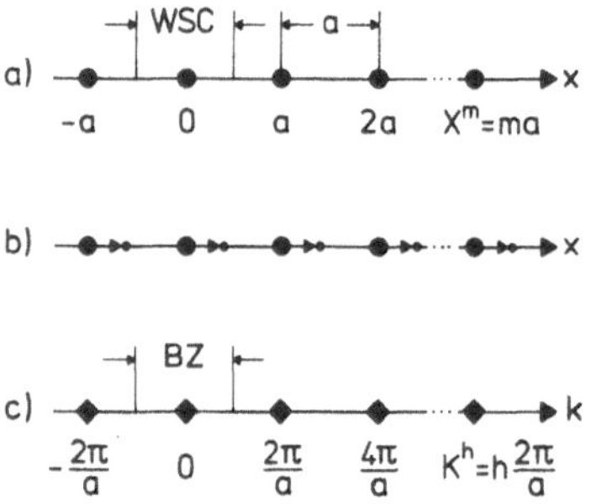


Fig. 3.1a-c. Periodic one-dimensional structures. a) Bravais lattice with atoms at $X^m = ma$; a is the lattice distance or the period. The Wigner-Seitz-Cell (WSC), $-a/2 \leqslant x \leqslant a/2$, shown here is symmetrical about atom 0. b) Non-Bravais lattice with period a consisting of diatomic molecules ( ● · ) or oriented dipole moments ( → ). c) Reciprocal lattice: $K^h = h2\pi/a$. The interval in the reciprocal lattice which corresponds to the Wigner-Seitz-Cell is the (1st) Brillouin-Zone (BZ)

The physics has the period a. We restrict ourselves to unstructured atomic constituents, without directional properties such as dipole moments. We will treat only lattices of this kind which are called Bravais lattices[1].

Any physical quantity in equilibrium, $p(x)$, must be periodic and obey

$$p(x - X^m) = p(x - ma) = p(x - a) = p(x) \; , \tag{3.2}$$

for example the linear mass density

$$\rho_0(x) = \rho_0(x - a) = M \sum_m (x - X^m) \; .$$

A periodic $p(x)$ is already determined by its values in one basic interval (cell) which can be any length a. A convenient choice is the Wigner-Seitz-Cell which surrounds one atom symmetrically as shown in Fig. 3.1a.

Periodic functions can be expanded into a Fourier series (App. E):

$$p(x) = \sum_h \tilde{p}^h e^{-iK^h x} \; , \quad \tilde{p}^h = \int_{-a/2}^{+a/2} \frac{dx}{a} p(x) e^{iK^h x} \; , \tag{3.3}$$

---

1 Examples for non-Bravais lattices are: Periodic arrangements of equally oriented dipoles or diatomic lattices; they have directional properties (Fig. 3.1b).

in particular

$$\sum_m \delta(x - X^m) = \sum_h \frac{e^{-iK^h x}}{a} , \qquad (3.3a)$$

$$K^h = 2\pi bh = \frac{2\pi}{a} h , \quad ba = 1 , \quad h = 0,\pm1,\pm2\ldots , \text{ integer (reciprocal lattice points).} \qquad (3.3b)$$

The Fourier series (3.3) is a special case of a Fourier integral, $p(x) = \int_{-\infty}^{+\infty} dk\, \tilde{p}(k) \exp(-ikx)\, dx$. One distinguishes between real space, x, and "reciprocal" space, k. For periodic p(x) the Fourier transform vanishes unless k is a point of the reciprocal lattice (3.3b) with period $2\pi b$; the interval in the reciprocal lattice corresponding to the Wigner-Seitz-Cell is called the (1st) Brillouin-Zone (Fig. 3.1c).

A periodic function in reciprocal space can be expanded analogously:

$$\tilde{p}(k - K^h) = \tilde{p}(k) = \sum_m P^m e^{-iX^m k} , \qquad (3.4a)$$

$$P^m = \int_{-\pi b}^{+\pi b} \frac{dk}{2\pi b} \tilde{p}(k) e^{iX^m k} , \text{ and} \qquad (3.4b)$$

$$\sum_h \delta(k - K^h) = \sum_m \frac{1}{2\pi b} e^{-iX^m k} = \delta_p(k) \qquad (3.4c)$$

(the subscript p indicates that this $\delta$-function is periodic in the reciprocal lattice).

The integrals in (3.4b) or (3.3) need not extend over the Wigner-Seitz-Cell or the Brillouin-Zone; they can be taken over any respective basic interval. Note that the relation between lattice and reciprocal lattice is symmetrical: if $K^h$ is the reciprocal lattice to $X^m$, then $X^m$ is the reciprocal lattice to $K^h$.

Another relationship worth mentioning is indicated in (3.4c). One needs the Fourier coefficients $P^m$ up to infinite order in m $\left[P^m = 1/(2\pi b)\right]$ in order to represent the sharp $\delta$-peaks in the original function. One can make a more general statement, seen from the example:

$$\tilde{p}(k) = \frac{1}{\sqrt{2\pi\kappa^2}} \sum_h \exp\left(-\frac{(k - K^h)^2}{2\kappa^2}\right) , \quad P^m = \frac{1}{2\pi b} \exp\left(-\frac{(\kappa X^m)^2}{2}\right) = \frac{1}{2\pi b} \exp\left(-\frac{(\kappa a)^2}{2} m^2\right) ,$$

a generalisation of (3.4c), the $\delta$-functions being replaced by Gaussians of width $\kappa$; the range of $P^m$ is approximately $1/(\kappa a)$. Consequently, the finer the details in $\tilde{p}(k)$ given by a width $\kappa$, the more Fourier coefficients $P^m$ are needed.

### 3.1.2 Three-Dimensional Bravais Lattices

These lattices are given by 3 basis periods represented by three basis (nonplanar) vectors $\underline{a}^{(j)}$ j = 1,2,3. The atomic equilibrium positions are given by superposition of integer multiples of the basis vectors,

$$\underline{R}^{\underline{m}} = \underline{a}^{(j)} m_j = A\underline{m} = (A_{ij} m_j) \ , \quad X_i^{\underline{m}} = a_i^{(j)} m_j = A_{ij} m_j \tag{3.5}$$

where the "integer vector" $\underline{m} = (m_j) = (m_1, m_2, m_3)$ has integer components. The 3 basis vectors $\underline{a}^{(j)}$ define the basis A of the Bravais lattice, which as a rule is not orthonormal (for representation of vectors in such a basis compare App. F).

In particular we discuss the 3 cubic Bravais lattices[2] (Fig. 3.2a-c):

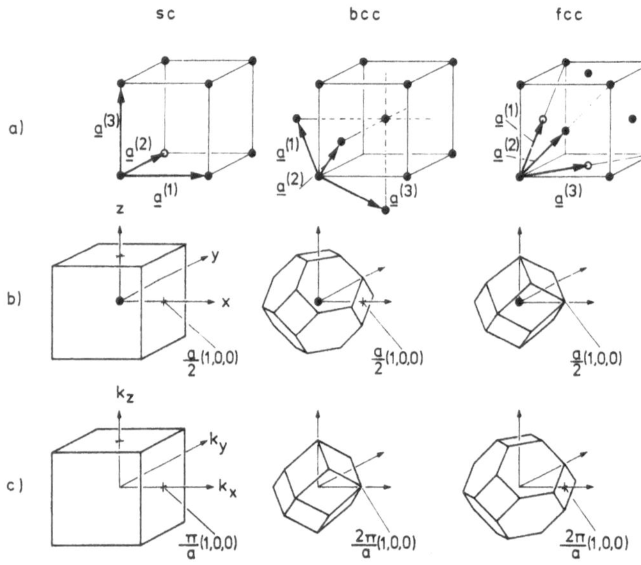


Fig. 3.2a-c. Geometry of the three cubic Bravais lattices. a) The 3 basis vectors $\underline{a}^{(j)}$ $j = 1,2,3$. b) The Wigner-Seitz-Cell with one atom in the center, volume $V_c$. c) The Brillouin-Zone (the Wigner-Seitz-Cell of the reciprocal lattice) with the origin of the reciprocal lattice in the center, volume $V_B = (2\pi)^3/V_c$

| simple cubic | body centered cubic | face centered cubic |
|---|---|---|
| sc | bcc | fcc |

$$A = a\begin{bmatrix}1&0&0\\0&1&0\\0&0&1\end{bmatrix}, \qquad \frac{a}{2}\begin{bmatrix}\bar{1}&1&1\\1&\bar{1}&1\\1&1&\bar{1}\end{bmatrix}, \qquad \frac{a}{2}\begin{bmatrix}0&1&1\\1&0&1\\1&1&0\end{bmatrix} \tag{3.6}$$

2 Negative values are denoted by a bar, e.g., $\bar{1} = -1$, $(1,\bar{1},0) = (1,-1,0)$, etc.

where a is the cubic lattice distance (Fig. 3.2a). The three basis vectors span a parallelepipedon, the elementary cell, which contains *one* atom for Bravais lattices and has the volume

$$V_c = \text{Det } A = \left(\underline{a}^{(3)}, \underline{a}^{(2)} \times \underline{a}^{(1)}\right) = a^3, \frac{a^3}{2}, \frac{a^3}{4} \ . \tag{3.6a}$$

The elementary cell contains all nonequivalent points. A periodic function

$$p(\underline{r}) = p(\underline{r} - \underline{R}^{\underline{m}}) \quad \text{for all } \underline{m} \ , \tag{3.7}$$

is completely determined by its values in one elementary cell. Translated cells are equivalent just as in one dimension (e.g., the sc cell in Fig. 3.2b, translated such that the atom is in the middle of the cell, is equivalent to the elementary cube of Fig. 3.2a). The volume equivalent to the elementary cell with highest symmetry is the Wigner-Seitz-Cell shown in Fig. 3.2b. It is constructed as follows: the connection is drawn between atom 0 and all its neighbours; all bisecting planes are drawn but only the innermost faces are kept; they form the Wigner-Seitz-Cell with one atom in the center. This cell also contains all nonequivalent points, has therefore the volume $V_c$, and it exhibits furthermore the full symmetry of the lattice which the original cells formed by the $\underline{a}^{(j)}$ do not.

Periodic functions (3.7) can be expanded into a Fourier series (App. E),

$$p(\underline{r}) = \sum_{\underline{h}} \tilde{p}^{\underline{h}} e^{-i\underline{K}^{\underline{h}}\underline{r}} \ , \quad \tilde{p}^{\underline{h}} = \int_{V_c} \frac{d\underline{r}}{V_c} \, p(\underline{r}) e^{i\underline{K}^{\underline{h}}\underline{r}} \ , \tag{3.8}$$

in particular

$$\sum_{\underline{m}} \delta(\underline{r} - \underline{R}^{\underline{m}}) = \sum_{\underline{h}} \frac{1}{V_c} e^{-i\underline{K}^{\underline{h}}\underline{r}} \ , \quad \text{where} \tag{3.8a}$$

$$\underline{K}^{\underline{h}} = B\underline{h} = 2\pi (A')^{-1}\underline{h} = 2\pi \underline{b}_{(j)} h_j \ , \quad \underline{h} \text{ integer (reciprocal lattice points)}, \tag{3.8b}$$

$$\left(\underline{b}_{(j)}, \underline{a}^{(i)}\right) = \delta_{ji} \ , \quad \underline{b}_{(1)} = \frac{\underline{a}^{(2)} \times \underline{a}^{(3)}}{V_c} \quad \text{and cyclic}^3 \ ,$$

$$(\underline{K}^{\underline{h}}, \underline{R}^{\underline{m}}) = 2\pi(\underline{h}, \underline{m}) = \text{integer multiple of } 2\pi.$$

The points $\underline{K}^{\underline{h}}$ form the reciprocal lattice[4] :

---

[3] The relation between basis A and reciprocal basis B is discussed in Appendix F.

[4] The atomic positions can be envisioned as being located on discrete planes with normal $\hat{\underline{K}}^{\underline{h}}$: $(\hat{\underline{K}}^{\underline{h}}, \underline{R}^{\underline{m}}) = (\underline{h}, \underline{m}) 2\pi / |\underline{K}^{\underline{h}}|$. If the three integers $h_{1,2,3}$ have no common divisor, the distance between those planes is $2\pi / |\underline{K}^{\underline{h}}|$.

A:sc A:bcc A:fcc

$$B: \quad \frac{2\pi}{a}\begin{bmatrix}1 & 0 & 0\\ 0 & 1 & 0\\ 0 & 0 & 1\end{bmatrix}, \quad \frac{2\pi}{a}\begin{bmatrix}0 & 1 & 1\\ 1 & 0 & 1\\ 1 & 1 & 0\end{bmatrix}, \quad \frac{2\pi}{a}\begin{bmatrix}\bar{1} & 1 & 1\\ 1 & \bar{1} & 1\\ 1 & 1 & \bar{1}\end{bmatrix}, \quad V_B = \frac{(2\pi)^3}{V_c} = \det B. \tag{3.9}$$

One sees that the fcc and bcc structure are reciprocal. The Wigner-Seitz-Cells of the reciprocal lattice are called (first) Brillouin-Zones. They are shown in Fig. 3.2c. For a periodic function in reciprocal space:

$$\tilde{p}(\underline{k} - \underline{K}^{\underline{h}}) = \tilde{p}(\underline{k}) = \sum_{\underline{m}} p^{\underline{m}} e^{-i\underline{R}^{\underline{m}}\underline{k}} ,$$

$$p^{m} = \int_{V_B} \frac{d\underline{k}}{V_B} \tilde{p}(\underline{k}) e^{i\underline{R}^{\underline{m}}\underline{k}} ; \qquad \sum_{\underline{h}} \delta(\underline{k} - \underline{K}^{\underline{h}}) = \sum_{\underline{m}} \frac{1}{V_B} e^{-i\underline{R}^{\underline{m}}\underline{k}} = \delta_p(\underline{k}) \tag{3.10}$$

[the subscript p means "periodic", (3.4c)].

The integrals in (3.8) or (3.10) extend over any elementary cell $V_c$ or $V_B$. For the parallelelepipedons defined by the $\underline{a}^{(j)}$ of Fig. 3.2a one has in (3.8):

$\underline{r} = \underline{a}^{(j)}\xi_j = A\underline{\xi}$ , $0 \leqslant \xi_j \leqslant 1$ , $\int_{V_c} d\underline{r}\; p(\underline{r}) = \iiint_0^1 V_c d\underline{\xi}\; p(A\underline{\xi})$. Of course, one also can integrate over an elementary cube but then has to account for the enlarged volume: $\int_{cube} d\underline{r}\; p = a^3/V_c \int_{V_c} d\underline{r}\; p$.

## 3.2 Lattice Symmetries

We have seen in Chapter 2 that the potential is invariant against arbitrary translations, $\underline{T}$ , and rotations, D. Symmetry operations, S, consist of a special translation and/or rotation $S(\underline{T},D)$, which leave not only the potential but also the lattice itself invariant:

$$S\underline{R}^{\underline{m}} = \underline{T} + D\underline{R}^{\underline{m}} = \underline{R}^{\underline{m}_s} \quad \text{where } \underline{m}_s \text{ is also integer,} \tag{3.11}$$

i.e., with $\underline{R}^{\underline{m}}$ also $\underline{R}^{\underline{m}_s}$ covers all lattice sites.

These symmetries are very important; they allow one to reduce the number of independent force constants and to obtain the eigenvectors of the coupling matrix. Therefore, we have to discuss them in detail.

### 3.2.1 The Linear Bravais Lattice (Fig. 3.1a)

Its symmetries are simply (Fig. 3.3):

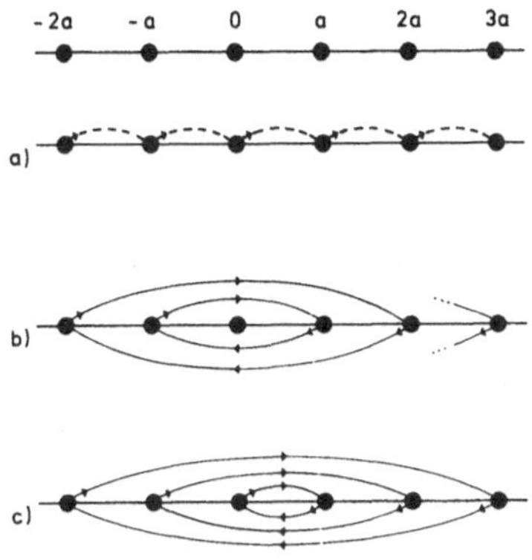


Fig. 3.3a-c. Symmetry operations of the linear Bravais lattice (indicated by arrows):
a) translation by a, T(1),
b) inversion about 0, I(0),
c) inversion about a/2, I(1/2)

translation by the lattice distance ha:

$$SX^m = X^{h+m} , \quad m_s = h + m , \quad T = ha = T(h) , \quad D = 1 ; \tag{3.12a}$$

inversion about the lattice point $X^h$:

$$SX^m = X^{2h-m} , \quad m_s = 2h - m , \quad T = 0 , \quad D = I(h) , \tag{3.12b}$$

$$\text{if } h = 0: \; SX^m = X^{-m} = -X^m , \quad m_s = -m ; \quad T = 0 , \quad D = I(0) = -1 ;$$

inversion about a midpoint $X^h + a/2$:

$$SX^m = X^{2h+1-m} , \quad m_s = 2h + 1 - m , \quad T = 0 , \quad D = I(h + 1/2) . \tag{3.12c}$$

It suffices to consider the translational symmetry T(1) and the "point" symmetries I(0), I(1/2). The general symmetries (3.12) follow by repeated application, e.g., $T(h) = [T(1)]^h$. It is also obvious that only the periodic (infinite) structure can have all the symmetry elements (3.12), whereas for a finite assembly at best *one* inversion is left. As a rule we will not need (3.12c).

The translational-rotational invariance (2.1),

$$\Phi(\ldots X^m + s^m \ldots) = \Phi(\ldots T + DX^m + Ds^m \ldots) = \Phi(\ldots X^{m_s} + Ds^m \ldots) ,$$

means clearly that the expansion of the potential[5] about $X^m$ with displacement $s^m$ is identical to the expansion about $X^{m_s}$ with displacements $Ds^m$,

$$2U = s^m \Phi^{mn} s^n = (Ds^m)\Phi^{m_s n_s}(Ds^n) = s^m(D\Phi^{m_s n_s}D)s^n \ . \tag{3.13}$$

This holds for arbitrary displacements, leading to

$$\Phi^{mn} = \Phi^{m_s n_s} \quad \text{because } Is^m = -s^m \ , \quad Ds^m = \pm s^m \ , \quad D^2 = 1, \text{ for the chain.} \tag{3.14}$$

Translational symmetry (3.12a),

$$\Phi^{mn} = \Phi^{m+h,n+h} = \Phi^{m-n,0} = \Phi^{(m-n)} \ , \tag{3.14a}$$

shows that the coupling depends only on the difference m-n, and inversion symmetry (3.12b) with I(0),

$$\Phi^{mn} = \Phi^{-m,-n} = \Phi^{\pm(m-n)} \ , \tag{3.14b}$$

means that $\Phi^{mn}$ depends only on $|m-n|$. These results are trivial as can be seen from Fig. 3.4a-c.

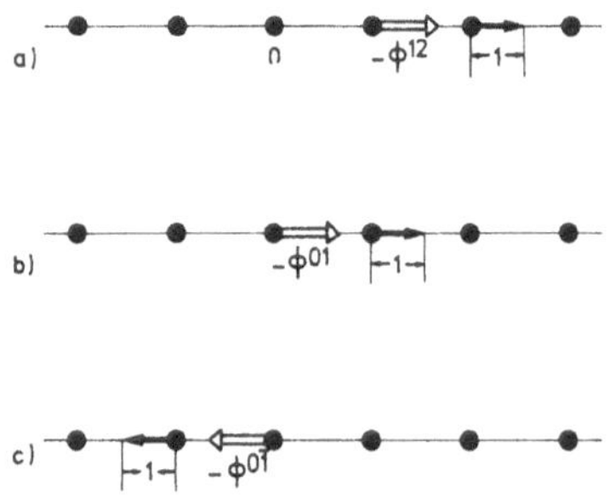


Fig. 3.4a-c. Equivalent displacement-force patterns. The above patterns are connected by symmetry operations: a and b by the translation T(1), c and b by the inversion I(0)

The symmetries (3.14) reduce the number of independent force constants drastically. If, for instance, one assumes purely 1st neighbour interaction, $\Phi^{(0)}$, $\Phi^{(1)} = \Phi^{(\bar{1})} \neq 0$, one obtains from translational invariance (2.5) that $\Phi^{(0)} = -2\Phi^{(1)}$ and a

---

5 In the expansion of the potential in powers of the displacements the 1st order term vanishes for symmetry reasons: $\Phi^{m_s} = \Phi^m$ leads to $\Phi^m = \Phi^o$ from (3.12a) and $I\Phi^o = -\Phi^o = \Phi^o = 0$ from (3.12b). Hence, for a lattice arrangement the internal forces vanish whatever the lattice distance a; to determine a one has to use other procedures (comp. Sec. 3.4.1).

single parameter is left, $\Phi^{(1)}$, which can be represented by a longitudinal spring $-\Phi^{(1)}$ between nearest neighbours.

Furthermore, from (3.14a) one obtains immediately the eigenvectors and eigenvalues of the infinite matrix $\Phi$, (comp. App. E), because

$$\Phi^{mn} e^{ikX^n} = \sum_n \Phi^{(m-n)} e^{ik(X^{n-m}+X^m)} = \underbrace{\sum_h \Phi^{(h)} e^{-ikX^h}}_{\Phi^k = \tilde{\Phi}(k)} e^{ikX^m} . \tag{3.15}$$

Therefore $\tilde{\Phi}(k)$ is an eigenvalue of $\Phi$ (the index k replaces the index $\nu$ of (2.7), (for $\langle m|k\rangle$ comp. App. B.7),

$$\Phi|k\rangle = \tilde{\Phi}(k)|k\rangle , \quad \langle m|k\rangle = \frac{e^{ikX^m}}{\sqrt{2\pi b}} ,$$
$$\langle k'|k\rangle = \frac{1}{2\pi b}\sum_m e^{i(k'-k)X^m} = \delta_p(k' - k). \tag{3.15a}$$

The eigenvalues $\tilde{\Phi}(k)$ and the "normalized" eigenvectors[6] $|k\rangle$ are periodic in the reciprocal lattice. Therefore one restricts k to the Brillouin-Zone (or to an equivalent period of the reciprocal lattice),

$$-\pi b \leqslant k \leqslant \pi b . \tag{3.15b}$$

Because of inversion symmetry (3.14b), $\Phi^{(h)} = \Phi^{-(h)}$, it is

$$\tilde{\Phi}(k) = \tilde{\Phi}(-k) = \Phi^*(k) . \tag{3.15c}$$

Because of (3.15c) the eigenvectors $|\pm k\rangle$ are degenerate and combinations are also eigenvectors; besides $e^{\pm ikX}$ one can choose $\cos kX^m$ and $\sin kX^m$ or any combination as eigenvectors to $\Phi$. These solutions are called lattice waves; the wavelength is $2\pi/k$.

Because $G(\omega) = \left[\Phi - M(\omega + i\eta)^2\right]^{-1}$, it is clear that $\Phi$ and G have the same symmetries,

$$G^{mn} = G^{\pm(m-n)} . \tag{3.14c}$$

Evidently $\tilde{\Phi}(k)$ and $\Phi^{(h)}$ are related by the Fourier transformation (3.4),

$$\Phi^{(h)} = \int_{-\pi b}^{+\pi b} \frac{dk}{2\pi b}\,\tilde{\Phi}(k)e^{ikX^h} , \quad G^{(h)}(\omega) = \int_{-\pi b}^{+\pi b} \frac{dk}{2\pi b}\,\underbrace{\frac{1}{\tilde{\Phi}(k) - M(\omega + i\eta)^2}}_{\tilde{G}(k,\omega)}\,e^{ikX^h} , \tag{3.14d}$$

6 The "normal coordinates" (Sec. 2.4.1) are $\langle k|\sqrt{M}\, s\rangle$.

or

$$G = \int_{-\pi b}^{+\pi b} dk \frac{|k\rangle\langle k|}{\tilde{\Phi}(k) - M(\omega + i\eta)^2} ,$$

$$G^{\underline{m}\underline{n}} = \int_{-\pi b}^{+\pi b} dk \frac{\langle m|k\rangle\langle k|n\rangle}{\tilde{\Phi}(k) - M(\omega + i\eta)^2} = \int_{-\pi b}^{+\pi b} \frac{dk}{2\pi b} \frac{e^{ik(m-n)}}{\tilde{\Phi}(k) - M(\omega + i\eta)^2} . \tag{3.14e}$$

### 3.2.2 Three-Dimensional Bravais Lattices

The three-dimensional case can be discussed in much the same way. Instead of (3.13) we have now

$$2U = (\underline{s}^{\underline{m}}, \Phi^{\underline{m}\underline{n}}\underline{s}^{\underline{n}}) = (D\underline{s}^{\underline{m}}, \Phi^{\underline{m}_s\underline{n}_s} D\underline{s}^{\underline{n}}) = (\underline{s}^{\underline{m}}, D^{-1}\Phi^{\underline{m}_s\underline{n}_s} D\underline{s}^{\underline{n}}) \tag{3.16}$$

which leads to

$$\Phi^{\underline{m}\underline{n}} = D^{-1}\Phi^{\underline{m}_s\underline{n}_s} D , \quad \Phi^{\underline{m}\underline{n}}_{ik} = D_{i'i}\Phi^{\underline{m}_s\underline{n}_s}_{i'k'} D_{k'k} . \tag{3.17}$$

Translational symmetry, $\underline{m}_s = \underline{m} + \underline{h}$, $D = 1$, means

$$\Phi^{\underline{m}\underline{n}} = \Phi^{\underline{m}+\underline{h},\underline{n}+\underline{h}} = \Phi^{(\underline{m}-\underline{n})} . \tag{3.17a}$$

Inversion[7] about a lattice point, $D = -1$, $\underline{m}_s = -\underline{m}$, is also a symmetry operation for every Bravais lattice,

$$\Phi^{\underline{m}\underline{n}} = \Phi^{\pm(\underline{m}-\underline{n})} = \Phi^{\underline{n}\underline{m}} , \quad \Phi^{\underline{m}\underline{n}}_{ik} = \Phi^{\underline{m}\underline{n}}_{ki} . \tag{3.17b}$$

Besides inversion the point symmetries contain rotations about lattice points. They depend on the type of lattice and will be treated later. As in the linear crystal, G and $\Phi$ have the same symmetries,

$$G^{(\underline{h})} = G^{\pm(\underline{h})} . \tag{3.17c}$$

Also, the eigenvalue equation can be compiled from the linear chain, with two differences: the lattice waves have a vectorial amplitude, the *polarization* $\underline{A}$, and k must be replaced by the *wave vector* $\underline{k}$, i.e., instead of (3.15) the eigenvalue equation is now

---

7 Here we only consider inversion about a lattice point. As in the linear case there are other inversion centers in the elementary parallel epiped, namely half edges, e.g., $\underline{a}^{(1)}/2$; face centers, e.g., $\left[\underline{a}^{(1)} \pm \underline{a}^{(2)}\right]/2$; body center $\left[\underline{a}^{(1)} + \underline{a}^{(2)} + \underline{a}^{(3)}\right]/2$.

$$\Phi^{\underline{mn}}\underline{A}e^{i\underline{kR}^{\underline{m}}} = \underbrace{\sum_{\underline{h}} \Phi^{(\underline{h})} e^{-i\underline{kR}^{\underline{h}}}}_{\tilde{\Phi}(\underline{k}) \,=\, \tilde{\Phi}(-\underline{k}) \,=\, \tilde{\Phi}^*(\underline{k})} \underline{A}e^{i\underline{kR}^{\underline{m}}} \tag{3.18}$$

Therefore the eigenvectors of $\Phi$ are determined by the eigenvectors $\underline{e}(\underline{k},\sigma)$, $\sigma = 1,2,3$, of the 3×3 symmetrical matrix $\tilde{\Phi}(\underline{k})$ $\left[M\Omega^2(\underline{k}) = \tilde{\Phi}(\underline{k})\right]$,

$$\tilde{\Phi}(\underline{k})\underline{e}(\underline{k}\sigma) = \tilde{\Phi}(\underline{k}\sigma)\underline{e}(\underline{k}\sigma) \; , \quad \Big(\underline{e}(\underline{k}\sigma),\underline{e}(\underline{k}\sigma')\Big) = \delta_{\sigma,\sigma'} \; . \tag{3.18a}$$

Finally ($\nu$ is replaced by $\underline{k}\sigma$, for $<\frac{\underline{m}}{i}|\underline{k}\sigma>$ comp. App. B.7)[8]

$$\Phi \; |\underline{k}\sigma> = \tilde{\Phi}(\underline{k}\sigma)|\underline{k}\sigma> \; , \quad <\tfrac{\underline{m}}{i}|\underline{k}\sigma> = e_i(\underline{k}\sigma)\,\frac{e^{i\underline{kR}^{\underline{m}}}}{V_B^{1/2}} \; , \quad <\underline{k}'\sigma'|\underline{k}\sigma> = \delta_p(\underline{k}' - \underline{k})\delta_{\sigma,\sigma'} \, . \tag{3.18b}$$

All quantities are periodic in the reciprocal lattice, i.e., $\underline{k}$ can be restricted to the Brillouin-Zone

$$\underline{k} \text{ in } V_B \; . \tag{3.18c}$$

The equivalent to (3.14d) is evidently

$$\Phi^{(\underline{h})} = \int_{V_B} \frac{d\underline{k}}{V_B} \tilde{\Phi}(\underline{k}) e^{i\underline{kR}^{\underline{h}}} \; , \quad G^{(\underline{h})}(\omega) = \int_{V_B} \frac{d\underline{k}}{V_B} \underbrace{\frac{1}{\tilde{\Phi}(\underline{k}) - M(\omega + i\eta)^2}}_{\tilde{G}(\underline{k},\omega)} \; . \tag{3.19a}$$

Here $\tilde{G}(\underline{k},\omega)$ is still a 3×3 matrix which can be written in terms of the polarizations, $\underline{e}$,

$$\tilde{G}_{ij}(\underline{k},\omega) = \sum_{\sigma} \frac{e_i(\underline{k}\sigma)e_j(\underline{k}\sigma)}{\tilde{\Phi}(\underline{k}\sigma) - M(\omega + i\eta)^2} = \sum_{\sigma} e_i(\underline{k}\sigma)\tilde{G}(\underline{k}\sigma;\omega)e_j(\underline{k}\sigma) \; . \tag{3.19b}$$

In summary, one obtains an integral representation of $\Phi$ (and G) in terms of its eigenvectors and eigenvalues:

$$\Phi = \sum_{\sigma} \int_{V_B} d\underline{k} \; |\underline{k}\sigma> \tilde{\Phi}(\underline{k}\sigma)<\underline{k}\sigma| \quad , \quad G(\omega) = \sum_{\sigma} \int_{V_B} d\underline{k} \; |\underline{k}\sigma> \tilde{G}(\underline{k}\sigma;\omega) <\underline{k}\sigma| \quad . \tag{3.20}$$

8 The "normal coordinates" (Sec. 2.4.1) are $<\underline{k}\sigma|\sqrt{M}\;\underline{s}>$.

Note that due to translational symmetry one is only left with the three-dimensional[9] problem (3.18a) and an integration over $V_B$ in order to obtain $G(\omega)$.

### 3.2.3 Point Symmetries

Point symmetries consist of pure rotations ($\underline{T} = 0$) about lattice points (including inversion). Number and character depend on the special lattice. Many examples demonstrating the use and the usefulness of point symmetries will be given later. Here we will treat only two examples of direct interest.

First we discuss (3.17),

$$\Phi^{(\underline{h})} = D^{-1}\Phi^{(\underline{h}_s)}D , \quad D\underline{R}^{\underline{h}} = \underline{R}^{\underline{h}_s} , \tag{3.21}$$

or in another notation,

$$\Phi(\underline{R}^{\underline{h}}) = D^{-1}\Phi(D\underline{R}^{\underline{h}})D . \tag{3.21a}$$

If for fixed $\underline{R}^{\underline{h}}$ one considers only rotations which leave $\underline{R}^{\underline{h}}$ invariant, then

$$\Phi(\underline{R}^{\underline{h}}) = D^{-1}\Phi(\underline{R}^{\underline{h}})D , \quad \Phi^{(\underline{h})} = D^{-1}\Phi^{(\underline{h})}D \tag{3.21b}$$

restricts the form of $\Phi^{(\underline{h})}$ to make it compatible with the particular point symmetry. After having found a compatible $\Phi^{(\underline{h})}$, one finds the coupling matrix for equivalent $\underline{h}_s$ from (3.21).

The other example is $\tilde{\Phi}(\underline{k})$ for which from (3.18) and (3.21),

$$\tilde{\Phi}(\underline{k}) = D\tilde{\Phi}(D^{-1}\underline{k})D^{-1} , \tag{3.22}$$

which restricts $\tilde{\Phi}(\underline{k})$ in the same way as $\Phi(\underline{R}^{\underline{h}})$.

We will discuss only structures of highest symmetry. Fig. 3.5a again shows the linear lattice, with the obvious point symmetries unity and inversion, corresponding to $\tilde{x} = (Dx) = \pm x$. The symmetries of the plane square lattice, Fig. 3.5b, consist of all rotations which leave the elementary square (the Wigner-Seitz-Cell) invariant. These are the 8 operations $(\tilde{x},\tilde{z}) = (D\underline{r}) = (\pm x,\pm z) = (\pm z,\pm x)$.

The highest symmetry in three-dimensional lattices is cubic. It contains all 48 rotations about the center of a cube which transform that cube into itself,

$$D\underline{r} = (D_{ik}x_k) = \tilde{\underline{r}} = (\tilde{x}_i) = (\pm x_1,\pm x_2,\pm x_3) \text{ and permutations of } 1,2,3. \tag{3.23}$$

---

[9] The dimension of the problem is three times the number of atoms in one basic cell: 1,2,3 for 1,2,3 dimensional Bravais lattices; 6 for a three-dimensional lattice with 2 atoms per cell such as hexagonal close packed metals (Zn) or alkali halides (NaCl).

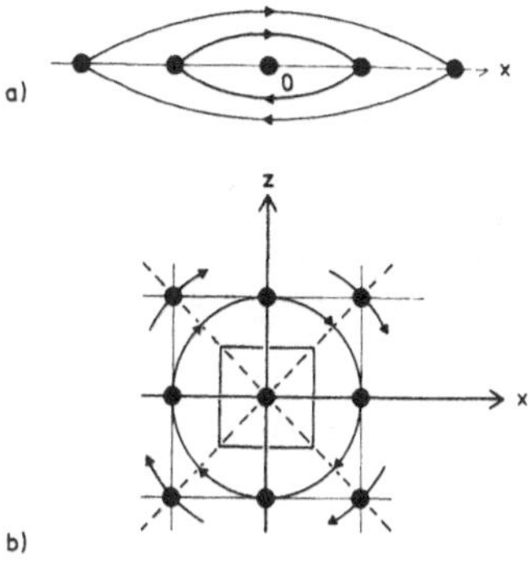


Fig. 3.5 a and b. Point symmetries of Bravais lattices
a) One-dimensional lattice: unity and inversion, $\tilde{x} = \pm x$; inversion indicated by arrows.
b) Two-dimensional square lattice: all 8 operations which leave the square invariant,

$$(\tilde{x},\tilde{z}) = (\tilde{x}_1,\tilde{x}_3) = (\pm x_1,\pm x_3),(\pm x_3,\pm x_1).$$

The symmetries contain the $\pi/2$-rotation D about the $x_2$-axis (perpendicular to the drawing plane), $(\tilde{x}_1,\tilde{x}_3) = (-x_3,x_1)$, indicated by arrows, and its powers: $D^0$ = unity, $D^2$ = inversion, and furthermore reflections R at $x_1 \pm x_3 = 0$, $x_1 = 0$; the 8 symmetry operations can be represented by $D^n$, n = 1,2,3,4, and $D^nR$, where R is the reflection at $x_1 = 0$

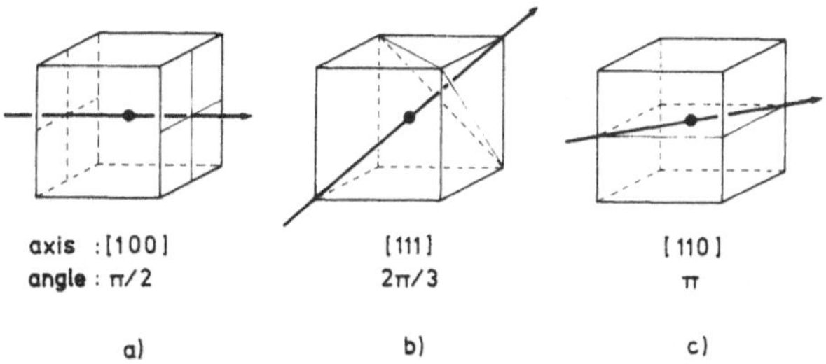


Fig. 3.6a-c. Cubic symmetry operations. a) fourfold axis ∥ cube edge, [100], b) threefold axis ∥ body diagonal, [111], c) twofold axis ∥ face diagonal, [110]

The various rotations are shown in Fig. 3.6a-c; there are 3 (fourfold) axes of type ⟨100⟩ (Fig. 3.6a), and therefore $3 \cdot 3 = 9$ rotations (excluding unity), 4 (threefold) axes of type ⟨111⟩ corresponding to $4 \cdot 2 = 8$ rotations and 6 (twofold) axes of type ⟨110⟩ resulting in 6 rotations. Consequently one has 9 + 8 + 6 + 1 = 24 proper rotations (including unity) as symmetry elements, and 48 if one includes inversion.

The cube in question can be visualized as the Wigner-Seitz-Cell of the simple cubic lattice in Fig. 3.2b. However, the symmetries also hold about body- and face-centered positions of the sc lattice. Consequently, the symmetries (3.23) are valid for bcc and fcc lattices as well.

### 3.2.4 Reduction of the Force Constant Matrix and Its Fourier Transform for Cubic Symmetry

The relations (3.21b), $M = \tilde{M}$, $\tilde{M} = D^{-1}MD$, or $M_{ik} = \tilde{M}_{ik} = D_{si}M_{st}D_{tk}$, valid for a 3×3 matrix M and for a certain set of rotations, restrict the elements of M. For cubic symmetry (3.23) equations (3.21b) are simple to discuss, for example

$$\tilde{\underline{r}} = (-x,y,z) \, , \quad D = \begin{bmatrix} \bar{1} & 0 & 0 \\ 0 & 1 & 0 \\ 0 & 0 & 1 \end{bmatrix} : \quad \begin{array}{ll} \tilde{M}_{ik} = -M_{ik} \, , & \text{if } i,k \text{ contain a single } x \, , \\ \tilde{M}_{ik} = M_{ik} & \text{otherwise ;} \end{array}$$

therefore $\tilde{M}_{yx} = -M_{yx} = M_{yx} = 0$, all elements containing one subscript x vanish. Analogously,

$$\tilde{\underline{r}} = (y,x,z) \, , \quad D = \begin{bmatrix} 0 & 1 & 0 \\ 1 & 0 & 0 \\ 0 & 0 & 1 \end{bmatrix} \quad \text{yields symmetry in } x,y\text{:} \quad M_{xx} = M_{yy} \, , \; M_{xz} = M_{yz} \, .$$

Let us first treat $\Phi^{(0)} = \Phi(0)$ which, obviously, is invariant against all operations (3.23): the off-diagonal elements must vanish, the diagonal elements are equal and $\Phi^{(0)}$ is scalar[10]

$$\Phi^{(0)}_{ik} = f\delta_{ik} \, , \quad 1 \text{ independent parameter } f. \tag{3.24}$$

If $\underline{R}^h \parallel [100]$ the rotations $\tilde{\underline{r}} = (x,\pm y,\pm z)$, $(x,\pm z,\pm y)$ don't change $\underline{R}^h$. The coupling must have the form

$$\Phi(\underline{R}^h \parallel [100]) = \Phi^{[100]} = - \begin{bmatrix} f_l & 0 & 0 \\ 0 & f_t & 0 \\ 0 & 0 & f_t \end{bmatrix} , \quad 2 \text{ independent parameters.} \tag{3.25}$$

It can be represented by a longitudinal (spiral) spring $f_l$ and an isotropic transversal (leaf) spring $f_t$, (Fig. 3.7a-c),

$$-\Phi^{[100]} = f_l P_l^{[100]} + f_t(1 - P_l^{[100]}) = f_l P_l^{[100]} + f_t P_t^{[100]} \tag{3.25a}$$

with the longitudinal projector in direction $\underline{R}^h$,

$$P_l^{[100]} = |\hat{\underline{R}}^h\rangle \langle \hat{\underline{R}}^h| = \begin{bmatrix} 1 & 0 & 0 \\ 0 & 0 & 0 \\ 0 & 0 & 0 \end{bmatrix} . \tag{3.25b}$$

By the rotation $\tilde{\underline{r}} = (-y,x,z)$, where [100]⟶[010], one obtains from (3.21a)

---

[10] This holds for all tensors of second rank with full cubic symmetry.

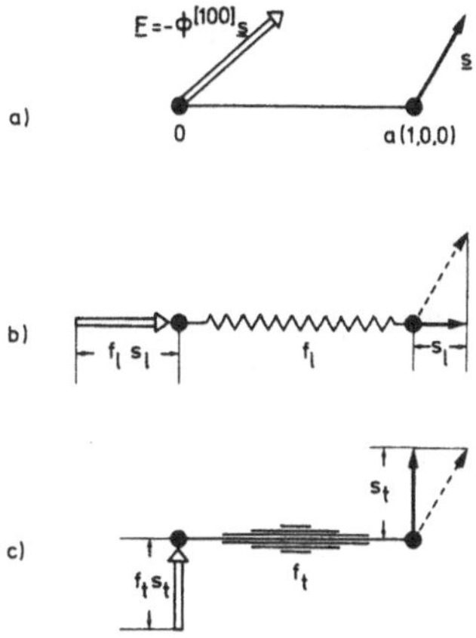


Fig. 3.7a-c. Coupling in [100]-direction
a) Force $\underline{F}$ on atom 0 due to an arbitrary displacement $\underline{s}$ of an atom at $\underline{R}^{\underline{h}} \parallel [100]$, here $\underline{R}^{\underline{h}} = (1,0,0)$.
b) The longitudinal component of $\underline{s}$, $s_l$, leads to a force $f_l s_l$ in [100]-direction, which can be represented by a longitudinal (spiral) spring $f_l$.
c) The transversal component of $\underline{s}$, $s_t$, leads to a force $f_t s_t$ perpendicular to the [100]-direction, which can be represented by an isotropic transversal (leaf) spring $f_t$

$$-\Phi^{[010]} = \begin{bmatrix} f_t & 0 & 0 \\ 0 & f_l & 0 \\ 0 & 0 & f_t \end{bmatrix} = f_l P_l^{[010]} + f_t P_t^{[010]} \ , \qquad P_l^{[010]} = \begin{bmatrix} 0 & 0 & 0 \\ 0 & 1 & 0 \\ 0 & 0 & 0 \end{bmatrix} \ ;$$

this shows how to extend (3.25a) to other $\langle 100 \rangle$ couplings.

If $\underline{R}^{\underline{h}} \parallel [111]$ all permutations of (x,y,z) leave $\underline{R}^{\underline{h}}$ unchanged and $\Phi^{(h)}$ has a representation equivalent to that of (3.25a)

$$-\Phi^{[111]} = f_l P_l^{[111]} + f_t(1 - P_l^{[111]}) \ , \qquad P_l^{[111]} = \frac{1}{3}\begin{bmatrix} 1 & 1 & 1 \\ 1 & 1 & 1 \\ 1 & 1 & 1 \end{bmatrix} \ , \tag{3.26}$$

2 independent parameters,

although the form of $\Phi^{(\underline{h})}$ is changed (Fig. 3.8a-d).

If $\underline{R}^{\underline{h}} \parallel [110]$ all operations $\underline{\tilde{r}} = (x,y,\pm z)$, $(y,x,\pm z)$ are admitted. The coupling is represented by

$$-\Phi^{[110]} = f_l P_l^{[110]} + f_{t'} P_{t'}^{[110]} + f_t P_t^{[110]} \ , \quad \text{3 independent parameters,} \tag{3.27}$$

i.e., a longitudinal spring and two different transversal springs; $P_t$ projects onto a cubic edge

$$P_l^{[110]} = \frac{1}{2}\begin{bmatrix} 1 & 1 & 0 \\ 1 & 1 & 0 \\ 0 & 0 & 0 \end{bmatrix} \ , \qquad P_t^{[110]} = \begin{bmatrix} 0 & 0 & 0 \\ 0 & 0 & 0 \\ 0 & 0 & 1 \end{bmatrix} = P_l^{[001]} \ ,$$

$$P_{t'}^{[110]} = \frac{1}{2}\begin{bmatrix} 1 & \bar{1} & 0 \\ \bar{1} & 1 & 0 \\ 0 & 0 & 0 \end{bmatrix} = P_l^{[1\bar{1}0]} \ . \tag{3.27a}$$

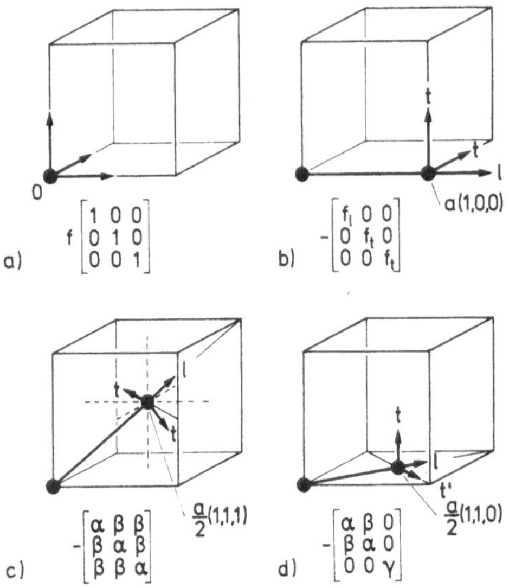


Fig. 3.8a-d. Coupling matrices for cubic symmetry
a) $\underline{R}^{\underline{h}} = 0$, self-coupling $\Phi^{(0)}$ is isotropic (scalar),
b) $\underline{R}^{\underline{h}} \parallel [100]$, (3.25), $\Phi^{[100]}$ contains 2 parameters, one longitudinal spring $f_l$ and one isotropic transversal spring $f_t$,
c) $\underline{R}^{\underline{h}} \parallel [111]$, (3.26), $\Phi^{[111]}$ as in b), $f_l = \alpha + 2\beta$, $f_t = \alpha - \beta$,
d) $\underline{R}^{\underline{h}} \parallel [110]$, (3.27), $\Phi^{[110]}$ contains 3 parameters, one longitudinal spring $f_l$ and two different transversal springs $f_{t',t}$: $f_l = \alpha + \beta$, $f_{t'} = \alpha - \beta$, $f_t = \gamma$

The coupling is still longitudinal and transversal in character, but the transversal behaviour is no longer isotropic (Fig. 3.8a-d).

The directions in (3.25), (3.26) and (3.27) are those of highest symmetry. For those directions the point symmetries already fix the eigenvectors and the degeneracy of the eigenvalues for the coupling[11]. The same considerations apply to $\tilde{\Phi}(\underline{k})$. Consequently, the eigenvectors of $\tilde{\Phi}(\underline{k})$, the polarizations, are already given by symmetry if $\underline{k}$ is parallel to a high symmetry direction.

Because of translational invariance (2.5b),

$$\sum_{\underline{h}} \Phi^{(h)} = 0 \quad \text{or} \quad \Phi^{(0)} = -\sum_{\underline{h}(\neq 0)} \Phi^{(\underline{h})} . \tag{3.28}$$

In cubic crystals where $\Phi^{(0)}$ is scalar,

$$\Phi^{(0)}_{ik} = \delta_{ik} \frac{\mathrm{tr}\{\Phi^{(0)}\}}{3} = -\frac{\delta_{ik}}{3} \sum_{\underline{h}(\neq 0)} \mathrm{tr}\{\Phi^{(h)}\} = f^{(0)} \delta_{ik} . \tag{3.28a}$$

A very crude model, where only $\Phi^{(0)}$ enters, is the Einstein model. Here it is assumed that each atom moves independently, coupled to the rigid environment $\Phi^{(\underline{h})} = \Phi^{(0)} \delta^{\underline{h}0}$.

---

11 For $\underline{R}^{\underline{h}}$ of type [1,2,0] one is left with operations

$$\tilde{\underline{r}} = (x,y,\pm z): \quad -\Phi = \begin{bmatrix} \alpha & \beta & 0 \\ \beta & \gamma & 0 \\ 0 & 0 & \delta \end{bmatrix} , \quad \text{4 independent parameters.}$$

Here one transversal eigenvector (0,0,1) is fixed; the two remaining ones depend on the parameters and need not be purely longitudinal or transversal.

Each atom is represented by the same harmonic oscillator, which is isotropic for cubic symmetry, characterized by 1 parameter, the eigenfrequency $\Omega_E$ or $f^{(O)} = M\Omega_E^2$.

A first neighbour model in cubic Bravais crystals contains then only the parameters of 1st neighbour coupling (2 for sc, ⟨100⟩, and bcc, ⟨111⟩, $f_l$, $f_t$; 3 for fcc, ⟨110⟩: $f_l$, $f_{t'}$, $f_t$) while $\Phi^{(O)}$ is given by (3.28a):

$$f^{(O)} = \begin{matrix} 2(f_l + 2f_t) \\ \frac{8}{3}(f_l + 2f_t) \\ 4(f_l + f_{t'} + f_t) \end{matrix} \quad , \quad \begin{matrix} -\mathrm{tr}\{\Phi^{(\underline{h})}\} = f_l + 2f_t, & \mathrm{sc} \\ -\mathrm{tr}\{\Phi^{(\underline{h})}\} = f_l + 2f_t, & \mathrm{bcc}\ . \\ -\mathrm{tr}\{\Phi^{(\underline{h})}\} = f_l + f_{t'} + f_t, & \mathrm{fcc} \end{matrix} \tag{3.28b}$$

(for any first neighbour)

Fig. 3.9 illustrates the 1st neighbour model for the fcc lattice.

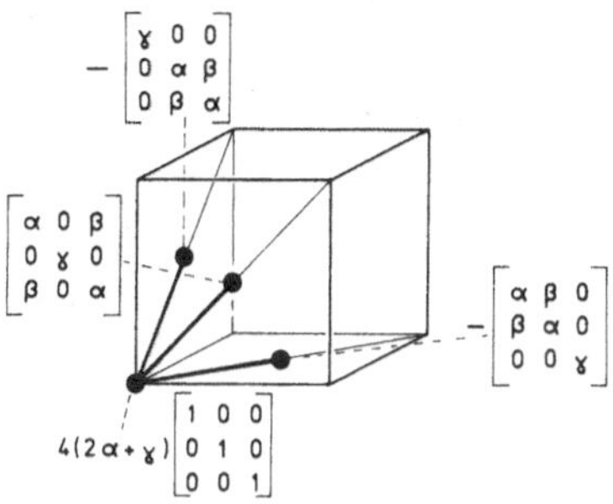


Fig. 3.9. Nearest neighbour model for a fcc lattice with 3 parameters α, β, γ or $f_l$, $f_{t'}$, $f_t$

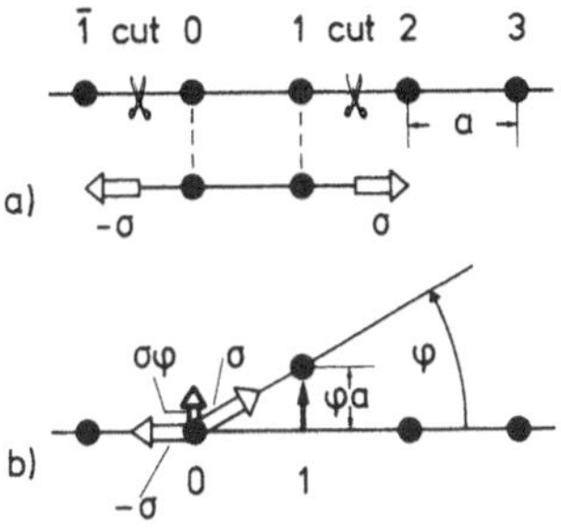


Fig. 3.10a and b. Longitudinal spring $f_l$ under tension σ. a) Cutting a spring requires forces ±σ in order to maintain equilibrium, b) for a transversal displacement, φa, of atom 1 the net force on atom 0 is σφ; this can be represented by a transversal spring, $f_t = \sigma\varphi/\varphi a = \sigma/a$

One ought to realize that transversal springs between two atoms do not really represent an interaction between those two atoms alone but rather many body interactions. This is most easily seen for a longitudinal spring $f_l$ under tension σ which produces a transversal spring $f_t = \sigma/a$ (Fig. 3.10). The tension is carried by the environment. One, therefore, has to treat transversal springs with some precaution. As an example, let us discuss a simple model for the vacancy in a nearest neighbour coupling model. In this model one just cuts the springs from the "vacancy" to its nearest neighbours: cutting the transversal springs violates rotational invariance, as shown

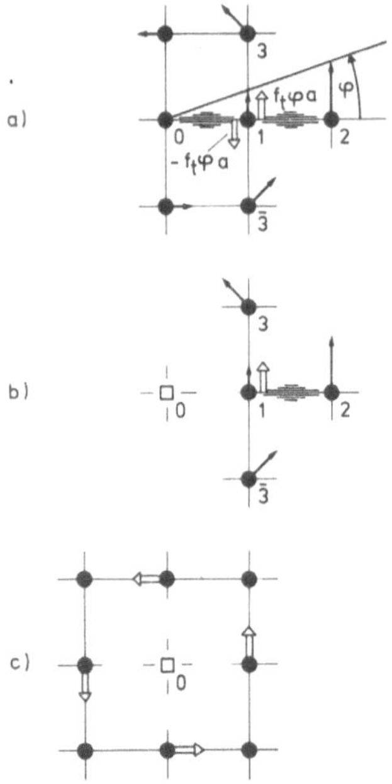


Fig. 3.11a-c. Violation of rotational invariance for a vacancy model (sc structure with lattice distance a) a) Displacement pattern for rotation by the angle $\varphi$ about atom 0. The forces on 1 by the spring between 0 and 1, $-f_t\varphi a$, and the spring between 1 and 2, $f_t\varphi a$, cancel. The same holds for the forces from 3 and $\bar{3}$. b) The springs to the vacancy ( □ ) are cut. The force on 1 by 2 is no longer canceled. c) Force pattern around a vacancy due to rotation; the resulting torque is $4f_t\varphi a^2$

in Fig. 3.11 for a sc structure. This demonstrates that transversal springs have to be treated with more caution than longitudinal springs, which can be cut with impunity (comp. Sec. 3.4.3).

## 3.3 Lattice Waves

In Section 3.2 we have discussed the eigenvectors and eigenvalues of $\Phi$:

$$|\underline{k}\sigma\rangle \ , \quad \text{and } \tilde{\Phi}(\underline{k}\sigma) \text{ as eigenvalues of } \tilde{\Phi}(\underline{k}) = \sum_{\underline{h}} \Phi^{(\underline{h})} e^{-i\underline{k}\underline{R}^{\underline{h}}} \ , \ (3.18). \tag{3.29}$$

According to (2.11, 26b)[12] ,

$$|\underline{k}\sigma\rangle \, e^{-i\Omega(\underline{k}\sigma)t} \ , \quad \Omega(\underline{k}\sigma) = \sqrt{\tilde{\Phi}(\underline{k}\sigma)/M} \ ,$$

is an (eigen-) solution of the equation of motion without forces: $M\ddot{\underline{s}}^{\underline{m}} + \Phi^{\underline{mn}}\underline{s}^{\underline{n}} = 0$
It represents a propagating lattice wave,

$$s_i^{\underline{m}} = \frac{e_i(\underline{k}\sigma)}{\sqrt{V_B}} \exp\left\{ i[\underline{k}\underline{R}^{\underline{m}} - \Omega(\underline{k}\sigma)t] \right\} \ , \tag{3.29a}$$

with wave vector $\underline{k}$ and frequency $\Omega(\underline{k}\sigma)$ which depends on wave vector and polarization.

---

[12] exp $i\Omega t$, cos $\Omega t$, sin $\Omega t$ and combinations thereof are solutions as well.

One can rationalize these solutions by stating that $\tilde{G}\left(\underline{k}\sigma\ ;\ \omega = \Omega(\underline{k}\sigma)\right)$ diverges. This indicates that no force is needed to maintain the eigensolutions (3.29a).

The curves $\Omega(\underline{k}\sigma)$ versus $\underline{k}$ are called *dispersion curves*. They can be measured, e.g., in neutron scattering (Chap. 7); they give directly $\tilde{\Phi}(\underline{k})$, from which the spatial coupling $\Phi^{(\underline{h})}$ can be obtained; it reveals details about the kind of interactions in the crystal under investigation. Furthermore, changes in the dispersion curves by defects yield microscopic details about defect properties. Therefore a treatment of dispersion curves is of central importance.

### 3.3.1 The Linear Lattice

The linear chain is most simple. The polarization is longitudinal (displacement in chain direction), and it is

$$\Omega(k) = \sqrt{\tilde{\Phi}(k)/M}\ , \tag{3.30}$$

$$\tilde{\Phi}(k) = \sum_{h=-\infty}^{\infty} \Phi^{(h)} e^{-ikah} = \sum_{h} \Phi^{(h)}(e^{-ikah} - 1) \tag{3.31}$$

$$= \sum_{h} \Phi^{(h)}(\cos kah - 1) = -2\sum_{h} \Phi^{(h)} \sin^2\frac{kah}{2}\ ,$$

where we have used translational invariance, $\sum_h \Phi^{(h)} = 0$, and inversion symmetry, $\Phi^{(h)} = \Phi^{(-h)}$.

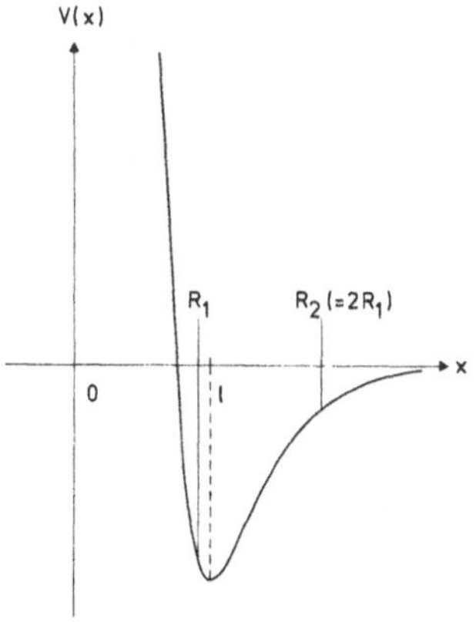


Fig. 3.12. Dispersion curve of linear lattice with 1st neighbour spring f

If one assumes a spring between 1st neighbours only, $f = -\Phi^{(1)}$ (see Sec. 3.2.1),

$$\tilde{\Phi}(k) = 4f \sin^2\frac{ka}{2}\ ,\qquad \Omega(k) = \sqrt{\frac{4f}{M}}\sin\frac{ka}{2}\quad \text{for } 0 \leqslant k \leqslant \frac{\pi}{a}\ ,$$

as shown in Fig. 3.12, which is typical also for 3-dimensional dispersion curves.

For small k, as a rule, $\tilde{\Phi}$ is proportional to $k^2$ and $\Omega$ proportional to k,

$$\Omega \cong k\left[-\sum_h \Phi^{(h)} \frac{a^2h^2}{2}\right]^{1/2} = kc \quad \text{for small k (long waves)}, \tag{3.33}$$

and the lattice wave,

$$\exp ik(X^m - ct) \;, \tag{3.33a}$$

for small k (small changes from atom to atom, $ka \ll 1$) becomes a wave of continuum theory,

$$\exp ik(x - ct) \text{ with sound velocity}^{13}\ c. \tag{3.33b}$$

*Stability* (dynamical) requires

$$\tilde{\Phi}(k) > 0 \;, \text{ except for } k = 0 \text{ (translation), or } \Omega(k) \text{ real and} > 0. \tag{3.34}$$

If this linear chain is a member of a three-dimensional crystal with small interaction between chains, one must also consider polarizations perpendicular to the chain, perpendicular displacements $s_\perp$. In the above model, with a 1st neighbour spiral spring f, there is no transversal restoring force, $\Omega_\perp(k) = 0$; this chain, considered in 3 dimensions, is not stable.

The 1st neighbour model above can be obtained by considering two-body interactions, where the total potential consists of translational-rotational-invariant interactions $V(|\underline{r}^m - \underline{r}^n|)$ between two atoms alone; in 3 dimensions

$$\Phi = \frac{1}{2} \sum_{\substack{m,n \\ m \neq n}} V(|\underline{r}^m - \underline{r}^n|) \;, \tag{3.35}$$

for the linear chain, $|\underline{r}^m - \underline{r}^n| = |x^m - x^n|$ ,

$$\Phi = \frac{1}{2} \sum_{m \neq n} V(|x^m - x^n|) = \sum_{m>n} V(x^m - x^n) \;, \quad x^m > x^n \text{ if } m > n. \tag{3.36}$$

The potential $V(|x|)$ between two atoms at distance x has typically, /3.1,2/, the form of Fig. 3.13, repulsive for short distances, attractive for long distances and minimal at l. Here the question arises how to define the equilibrium, because for a lattice arrangement the internal forces vanish whatever the lattice distance a (inversion symmetry).

---

[13] In general one must distinguish between phase velocity $c_{ph} = \Omega(k)/k$ (here $2c/ka \sin \frac{ka}{2}$ ) and group velocity $v = \partial\Omega(k)/\partial k$ (here $c \cos \frac{ka}{2}$ ). For small k both agree. At the BZ boundary, $ka = \pi$, one has $c_{ph} = 2c/\pi$, $v = 0$.

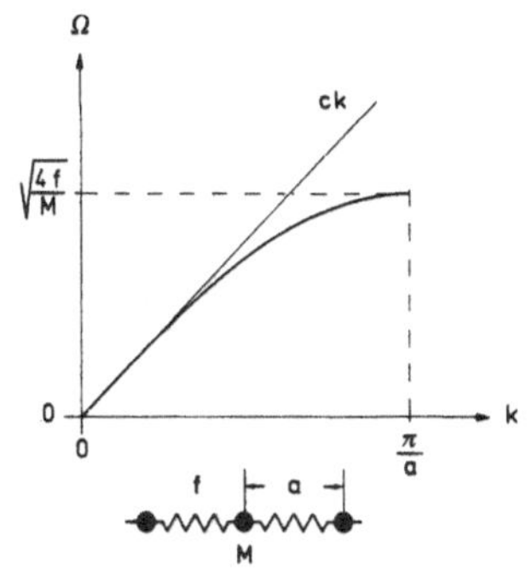


Fig. 3.13. Potential $V(x)$ between two atoms. For 1st neighbour interaction the equilibrium is determined by $V'(R_1 = a) = 0$, $a = 1$, (3.38a). For 1st and 2nd neighbour interaction the equilibrium is determined by: $V'(R_1) + 2V'(R_2) = 0$, $R_1 = a$, $R_2 = 2a$, (3.39a)

Let us first discuss purely 1st neighbour interaction where the potential falls off so fast that only interaction between 1st neighbours need be taken into account:

$$\sum_m V(x^m - x^{m-1}) \ . \tag{3.37}$$

One term in the sum is the energy of the bond between m and m-1. For a lattice the total energy diverges but the energy, $E_c$, per atom (or per bond or per elementary cell) exists,

$$E_c = V(a) \ , \tag{3.38}$$

and its minimum,

$$V'(a) = 0 \ , \quad a = 1 \tag{3.38a}$$

determines the lattice distance. The expansion[14] of (3.37),

$$U = \sum_m \left[V(a + s^m - s^{m-1}) - V(a)\right] \cong \sum_m \frac{1}{2} V''(1)(s^m - s^{m-1})^2 \ , \tag{3.38b}$$

leads to the 1st neighbour model with $f = V''(1) = V''(a)$.

If one includes transversal displacements, $s_\perp$, besides the longitudinal displacements, $s$,

$$V\left(\sqrt{(a + s)^2 + s_\perp^2}\right) = V(a) + V'(a)s + \frac{V'(a)}{2a} s_\perp^2 + \frac{V''(a)}{2} s^2 + \dots \ , \tag{3.38c}$$

the transversal contribution vanishes because $V'(a) = V'(1) = 0$, i.e., there are no

---

14 The total potential (3.37) is obviously infinite. However, the expansion (3.38b) from equilibrium is finite as long as the displacements are properly restricted. Further, $\tilde{\Phi}(k)/2$ is the energy per atom for a plane wave $s^m = \exp(ikam)$, the total energy again being infinite.

transversal springs: transversal deformations require no energy (transversal instability).

If, in a next step, one includes 2nd neighbour interaction, one obtains from

$$\Phi = \sum_m V(|\underline{r}^m - \underline{r}^{m-1}|) + V(|\underline{r}^m - \underline{r}^{m-2}|) \;, \quad E_c = V(a) + V(2a) \;, \tag{3.39}$$

the equilibrium condition

$$V'(a) + 2V'(2a) = 0 \;; \quad a < 1 \;; \quad V'(a) < 0 \;, \quad V'(2a) = -\frac{V'(a)}{2} > 0 \;, \quad \text{(Fig. 3.13)} \;, \tag{3.39a}$$

and the expansion (3.38b) becomes

$$U \cong \sum_m \Big\{ \underbrace{\frac{V'(a)}{a}}_{f_t^{(1)}} (s_\perp^m - s_\perp^{m-1})^2 + \underbrace{V''(a)}_{f_l^{(1)}}(s^m - s^{m-1})^2 + \underbrace{\frac{V'(2a)}{2a}}_{f_t^{(2)}} (s_\perp^m - s_\perp^{m-2}) + \underbrace{V''(2a)}_{f_l^{(2)}}(s^m - s^{m-2})^2 \Big\} \;. \tag{3.39b}$$

Evidently one has longitudinal springs, $f_l^{(1,2)}$, and transversal springs, $f_t^{(1,2)}$, between 1st and 2nd neighbours. The longitudinal springs are given by the second derivatives of V, whereas the transversal springs are related to the first derivatives. The first derivatives can be viewed as tensions; the resulting transversal springs are, according to Fig. 3.10, given by tension over distance: $f_t^{(n)} = V'(na)/na$.
Eq. (3.39b) shows that longitudinal and transversal motions are not coupled. Therefore, now one has two branches in the dispersion curve, a longitudinal branch

$$M\Omega_l^2 = 4f_l^{(1)} \sin^2 \frac{ka}{2} + 4f_l^{(2)} \sin^2 ka \;, \tag{3.40a}$$

which is not changed appreciably by the 2nd neighbour contribution $\big(V''(a) \gg |V''(2a)|$, Fig. 3.13$\big)$, and a transversal (twofold degenerate) branch $\big($from (3.39a): $f_t^{(1)} + 4f_t^{(2)} = 0\big)$,

$$M\Omega_t^2 = 4f_t^{(1)} \sin^2 \frac{ka}{2} + 4f_t^{(2)} \sin^2 ka = f_t^{(1)} (4\sin^2 \frac{ka}{2} - \sin^2 ka) \;,$$

$$M\Omega_t^2 = f_t^{(1)} \begin{cases} k^4a^4/4 & \text{for small } k \\ 4 & \text{for } k = \pi/a \end{cases} \;. \tag{3.40b}$$

The second neighbour contribution vanishes at the BZ boundary for both branches; for small k the expansion of $\Omega_t^2$ is proportional[15] to $k^4$ in contrast to $\Omega_l^2 \cong k^2$. Moreover, because $f_t^{(1)}$ is negative according to Fig. 3.13, the transversal motion is unstable throughout, $\Omega_t^2(k) < 0$ for all k. This instability is caused by the negative value of $f_t^{(1)}$, $V'(a) < 0$, which one can view as a compressed spiral spring: a transversal displacement is enhanced ($f_t < 0$), because the spiral spring tension is negative (comp. Fig. 3.10). This can also be directly seen from the potential, because a transversal displacement for $a < 1$ *reduces* the deviation from equilibrium, lowering the potential energy.

To obtain a stable description of a linear arrangement, 2 body forces are not sufficient. One must include bending forces, typical for covalent bonds, e.g., the single (isotropic) bonds between C-atoms in polyethylene, $C_nH_{2n+2}$. Most simply, the bending can be represented by a 3 body potential[16],

$$B(\underline{r}^{m-2},\underline{r}^{m-2},\underline{r}^{m}) = B(\varphi) = B(-\varphi) \cong B(0) + \frac{B''(0)}{2}\varphi^2 = B(0) + \frac{B''}{2}\varphi^2 ,$$

$$\frac{B''}{2}\varphi^2 \cong \frac{B''}{2a^2}(s_\perp^m + s_\perp^{m-2} - 2s_\perp^{m-1})^2 , \qquad (3.41)$$

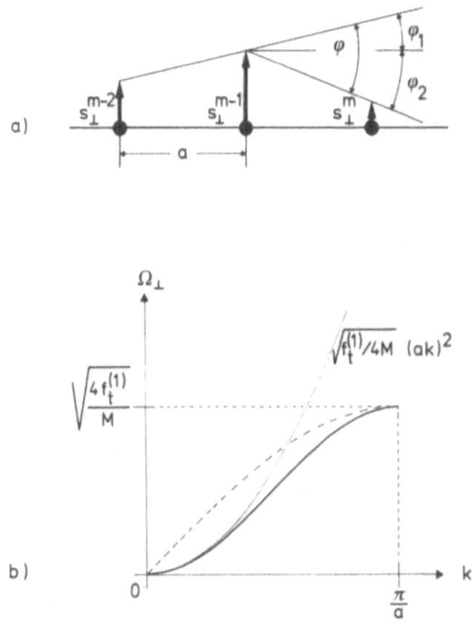


Fig. 3.14a and b. Bending of chemical bonds a) The three body potential depends on the bending angle, $\varphi = \varphi_1 + \varphi_2$, which requires three atomic coordinates for its definition; if the displacements are small, one has

$\varphi_1 \cong (s_\perp^{m-1} - s_\perp^{m-2})/a$ , $\varphi_2 \cong (s_\perp^{m} - s_\perp^{m-1})/a$

b) The transversal dispersion curve $\Omega_\perp(k) = \Omega_t(k)$ according to (3.40b) (full line); $\Omega_\perp$ extrapolated from small k (weak line); the longitudinal dispersion curve of Fig. 3.12 (dashed line)

---

15 This behaviour is rather extraordinary. The transversal sound velocities vanish. It is easy to see that the proportionality to $k^4$ still holds if one includes even higher order neighbours because of the equilibrium condition $\partial E_c/\partial a = 0$.

16 This potential cannot be split into 2 body potentials. It is the simplest example of a "many body" potential.

where $\varphi$ is the angle between two adjacent bonds, Fig. 3.14. The bending energy, $U_B$, becomes

$$U_B = \frac{B''}{2a^2}\sum_m (s_\perp^m + s_\perp^{m-2} - 2s_\perp^{m-1})^2 = \frac{B''}{2a^2}\sum_m \left[4(s_\perp^m - s_\perp^{m-1})^2 - (s_\perp^m - s_\perp^{m-2})^2\right] \quad (3.41a)$$

which corresponds to transversal springs as in (3.39b),

$$f_t^{(1)} = \frac{4B''(0)}{a^2} \ , \quad f_t^{(1)} + 4f_t^{(2)} = 0 \ , \quad (3.41b)$$

and to a transversal dispersion curve $\Omega_\perp(k)$ as in (3.40b), this time with positive $f_t^{(1)}$ (Fig. 3.14b). This can serve as an exotic example of a dispersion curve where the sound velocity is zero. Actually, transversal modes of this type are observed in crystals with chain-like structure /3.3/.

### 3.3.2 Dispersion Curves in Three-Dimensional Lattices

The three-dimensional case can be treated along the same lines. From given $\Phi^{(\underline{h})}$ one determines according to (3.18) the 3×3 matrix $\tilde{\Phi}(\underline{k}) = \sum_{\underline{h}} \Phi^{(\underline{h})}\exp(-i\underline{k}\underline{R}^{\underline{h}}) = -2\sum_{\underline{h}} \Phi^{(\underline{h})}\sin^2(\underline{k}\underline{R}^{\underline{h}}/2)$ and its eigenvalues

$$\tilde{\Phi}(\underline{k}\sigma) = M\Omega^2(\underline{k}\sigma) = \left(\underline{e}(\underline{k}\sigma),\tilde{\Phi}(\underline{k}\sigma)\underline{e}(\underline{k}\sigma)\right) \ , \quad (3.42)$$

which represent twice the potential energy per atom for a wave $\underline{s}^{\underline{m}} = \underline{e}(\underline{k}\sigma)\exp(i\underline{k}\underline{R}^{\underline{m}})$. The symmetries of $\tilde{\Phi}(\underline{k})$ and $\Phi(\underline{R}^{\underline{h}})$ are identical, according to (3.22). For *cubic* crystals we can take over the results of Section 3.2.4 where the eigenvectors of $\Phi(\underline{R}^{\underline{h}})$ were derived for $\underline{R}^{\underline{h}}$ parallel to the main symmetry directions. The eigenvectors of $\Phi(\underline{R}^{\underline{h}})$ illustrated in Fig. 3.8 can be viewed as polarizations for the corresponding three $\underline{k}$-directions. The eigenvectors are all either parallel to $\underline{k}$ (longitudinal) or perpendicular to $\underline{k}$ (transversal). However, in other directions the polarizations need not be longitudinal or transversal.

For $\underline{k} = k_x(1,0,0)$ one has longitudinal polarization, $\underline{e}_l = \hat{\underline{k}} = (1,0,0)$, and transversal polarization, $\underline{e}_t \perp \underline{k}$, twofold degenerate:

$$\tilde{\Phi} = \begin{bmatrix} \tilde{\Phi}_{11} & 0 & 0 \\ 0 & \tilde{\Phi}_{22} & 0 \\ 0 & 0 & \tilde{\Phi}_{22} \end{bmatrix} \ , \quad \begin{aligned} M\Omega_l^2 &= \tilde{\Phi}_{11} = (\underline{e}_l,\tilde{\Phi}\underline{e}_l) \\ M\Omega_t^2 &= \tilde{\Phi}_{22} = \tilde{\Phi}_{33} \end{aligned} \ . \quad (3.43)$$

Dynamical stability requires

$$\tilde{\Phi}(\underline{k}) > 0 \quad \text{except for translations,} \quad \underline{k} = 0, \quad \text{where } \tilde{\Phi} = 0. \quad (3.44)$$

The eigenvectors $|\underline{k}\sigma\rangle$ form a complete basis for displacements with $(\underline{s},\underline{s}) < \infty$; the eigenvectors themselves cannot be normalized, but their displacements remain at least bounded at infinity. This does not hold for rotations. Consequently, rotations are not even admitted as proper displacement fields.

To avoid the difficulties connected with the normalization of eigenvectors, one often imposes periodic boundary conditions to restrict the degrees of freedom. If one requires periodicity of the displacements with the 3 periods $N_1\underline{a}^{(1)}$, $N_2\underline{a}^{(2)}$, $N_3\underline{a}^{(3)}$, the number of atoms in the periodicity volume is $N_1N_2N_3 = N$. The permitted $\underline{k}$ values are obviously $\underline{k}^{\underline{\nu}} = \sum_j 2\pi\underline{b}_{(j)}\nu_j/N_j$, $\underline{\nu}$ integer, because, e.g., $\underline{k}^{\underline{\nu}}N_1\underline{a}^{(1)} = 2\pi\nu_1$. The discrete $\underline{k}^{\underline{\nu}}$ values form a sublattice, which for $N_1 = N_2 = N_3$ is a diminutive reciprocal lattice and has N points in the elementary reciprocal cell. The $\underline{k}^{\underline{\nu}}$ points are distributed uniformly with density $N_1N_2N_3/V_B$. The normalized eigenvectors are then

$$<{}^{\underline{m}}_{i}|\underline{k}^{\underline{\nu}}\sigma> = e_i(\underline{k}^{\underline{\nu}}\sigma)\exp(i\underline{k}^{\underline{\nu}}\underline{R}^{\underline{m}})/\sqrt{N} \quad ,$$

if the $\underline{R}^{\underline{m}}$ are restricted to the periodic volume. Solutions obeying physical boundary conditions, such as fixed or free surface, are not even known (except for one--dimensional problems), in contrast to the simple periodic solutions. Because it can be shown that the bulk properties do not depend on the surface conditions, the periodic condition is widely used; it can also be employed for numerical calculations, e.g., for Green's functions, where the integral over the BZ becomes a sum. Note that in the periodic scheme rotations again are not admitted because they are incompatible with periodicity.

### 3.3.3 Born-von Karman Models in Cubic Lattices

Because a theory of the potential and of the force parameters is rather complicated and it is difficult to obtain results from first principles, one often uses models with fitting parameters to represent the dispersion curves. The number of parameters should be small. In metals, as a rule, the electrons screen the interaction between ions such that one expects interactions only between near neighbours. Consequently one can use near neighbour force constants as fitting parameters. This is called a Born- von Karman model. One can, of course, fit any dispersion curve with the many parameters available from far neighbours. However, if one really has to use distant[17] neighbour springs for a fit, the Born- von Karman model becomes complicated and difficult to assess. In that case one should use other physical properties to describe the long range behaviour, such as peculiarities in the electronic

[17] As pointed out in Section 3.1.1, one can judge from the appearance of the dispersion curves whether or not one needs distant neighbour springs. If $\Omega(\underline{k})$ is smooth, only near neighbours contribute, whereas structure indicates contributions from more distant neighbours.

screening, Coulomb interactions or electronic polarization. Often the long range part can be represented by other models with only few parameters; this will be discussed briefly in Section 3.4.

Here we will restrict ourselves to short range interactions, and for demonstration we choose the fcc lattice with 1st neighbour interaction only (3.27), (Fig. 3.9), for which

Table 3.1. Dispersion curves in main symmetry directions for fcc crystals with first neighbour interaction, $(\alpha,\beta,\gamma)$

| $\underline{k}$ | $\underline{e}$ | $M\Omega^2(\underline{k}\sigma)$ | $\rho_o c_\sigma^2$ [18] |
|---|---|---|---|
| $k_x(1,0,0)$ | l: $(1,0,0)$ | $\tilde{\Phi}_{11}(\underline{k}) = 16\alpha \sin^2 \frac{k_x a}{4}$ | $\frac{4\alpha}{a}$ |
| | t: $(0,1,0)$ $(0,0,1)$ | $\tilde{\Phi}_{22}(\underline{k}) = 8(\alpha+\gamma) \sin^2 \frac{k_x a}{4}$ | $\frac{2(\alpha+\gamma)}{a}$ |
| $k_x(1,1,0)$ | l: $\frac{(1,1,0)}{\sqrt{2}}$ | $\tilde{\Phi}_{11} + \tilde{\Phi}_{12} = 8(\alpha+\gamma) \sin^2 \frac{k_x a}{4} + 4(\alpha+\beta) \sin^2 \frac{k_x a}{2}$ | $\frac{(3\alpha+2\beta+\gamma)}{a}$ |
| | t': $\frac{(1,\bar{1},0)}{\sqrt{2}}$ | $\tilde{\Phi}_{11} - \tilde{\Phi}_{12} = 8(\alpha+\gamma) \sin^2 \frac{k_x a}{4} + 4(\alpha-\beta) \sin^2 \frac{k_x a}{2}$ | $\frac{(3\alpha-2\beta+\gamma)}{a}$ |
| | t: $(0,0,1)$ | $\tilde{\Phi}_{33} = 16\alpha \sin^2 \frac{k_x a}{4} + 4\gamma \sin^2 \frac{k_x a}{2}$ | $\frac{2(\alpha+\gamma)}{a}$ |
| $k_x(1,1,1)$ | l: $\frac{(1,1,1)}{\sqrt{3}}$ | $\tilde{\Phi}_{11} + 2\tilde{\Phi}_{12} = (8\alpha+8\beta+4\gamma) \sin^2 \frac{k_x a}{2}$ | $\frac{(8\alpha+8\beta+4\gamma)}{3a}$ |
| | t: $\frac{(1,\bar{1},0)}{\sqrt{2}}$ $\frac{(1,1,\bar{2})}{\sqrt{6}}$ | $\tilde{\Phi}_{11} - \tilde{\Phi}_{12} = (8\alpha-4\beta+4\gamma) \sin^2 \frac{k_x a}{2}$ | $\frac{(8\alpha-4\beta+4\gamma)}{3a}$ |

[18] These quantities are linear combinations of the elastic data (comp. Fig. 4.15); $\rho_o = M/V_c = 4M/a^3$ is the macroscopic mass density.

$$\tilde{\Phi}_{11}(\underline{k}) = -2\sum_{\underline{h}} \Phi_{11}^{(\underline{h})} \sin^2 \frac{(\underline{k},\underline{R}^{\underline{h}})}{2} = 4\gamma\left(\sin^2 \frac{k_2 + k_3}{4} a + \sin^2 \frac{k_2 - k_3}{4} a\right)$$

$$+ 4\alpha\left(\sin^2 \frac{k_1 + k_2}{4} a + \sin^2 \frac{k_1 - k_2}{4} a + \sin^2 \frac{k_1 + k_3}{4} a + \sin^2 \frac{k_1 - k_3}{4} a\right)$$

$$4\alpha = 2(f_l + f_{t'}) \ , \quad 4\gamma = 4f_t \ , \qquad (3.45)$$

$$\tilde{\Phi}_{12}(\underline{k}) = 4\beta\left(\sin^2 \frac{k_1 + k_2}{4} a - \sin^2 \frac{k_1 - k_2}{4} a\right) \ , \quad 4\beta = 2(f_l - f_{t'}) \ , \quad \text{and cyclic.}$$

Table 3.1 gives a summary of the dispersion curves in the main symmetry directions, where the polarizations are longitudinal and transversal. For long waves $\tilde{\Phi}(\underline{k})$ becomes quadratic in $\underline{k}$ and the sound velocities c of macroscopic theory are $c_\sigma^2 = \Omega^2(\underline{k}\sigma)/k^2$ for small k. Again, one must distinguish between phase- and group velocity, $\underline{c} = \hat{\underline{k}}\Omega/k$ and $\underline{v} = \partial\Omega/\partial\underline{k}$. In the main symmetry directions $\underline{c}$ and $\underline{v}$ are parallel, and for small k one has $\underline{v} = \underline{c}$.

We will illustrate these dispersion curves in Fig. 3.15 by three examples.

1) In Fig. 3.15b the experimental dispersion curves of copper /3.4/ and a Born-von Karman fit is shown. The 3 springs are fitted to the slopes of $\Omega$ at $\underline{k} = 0$, i.e., to the sound velocities. It will be shown later that with this fit the macroscopic elastic behaviour (with 3 independent elastic data) is described correctly. One sees that the fit is rather good even at large $\underline{k}$ near the boundary of the Brillouin Zone which confirms the expected short range coupling, at least for Cu. In some other metals, e.g., Pb, this fit is not so satisfactory (see Tab. 3.2). Elastic isotropy means that $c_l^2$ and $c_t^2$ do not depend on direction: $\alpha - 2\beta - \gamma = 0$ according to Tab. 3.1, or $f_l + 2f_t - 3f_{t'} = 0$.

2) Fig. 3.15c shows the dispersion curves with a purely longitudinal coupling: only $f_l \neq 0$; $f_t = f_{t'} = 0$. Just as in the linear chain this corresponds to two body 1st neighbour coupling. Still the fit is not too bad. Macroscopically the crystal is *not* isotropic.

3) In addition Fig. 3.15d shows an oversimplified model; here it is assumed that all 3 springs are equal, $f_l = f_t = f_{t'} = f$ ($\alpha = \gamma$, $\beta = 0$), whereupon the coupling becomes scalar, $\tilde{\Phi}_{ik}(\underline{k}) = \delta_{ik}\tilde{\Phi}_{11}(\underline{k})$. The eigenvalue problem is completely degenerate. The elastic behaviour is isotropic, and in addition $c_l^2 = c_t^2 = a^2f/M$. This model is indeed very simple, and, even though it turns out to be unstable elastically, one can employ it to demonstrate the mathematical structure of the theory, Green's functions in particular.

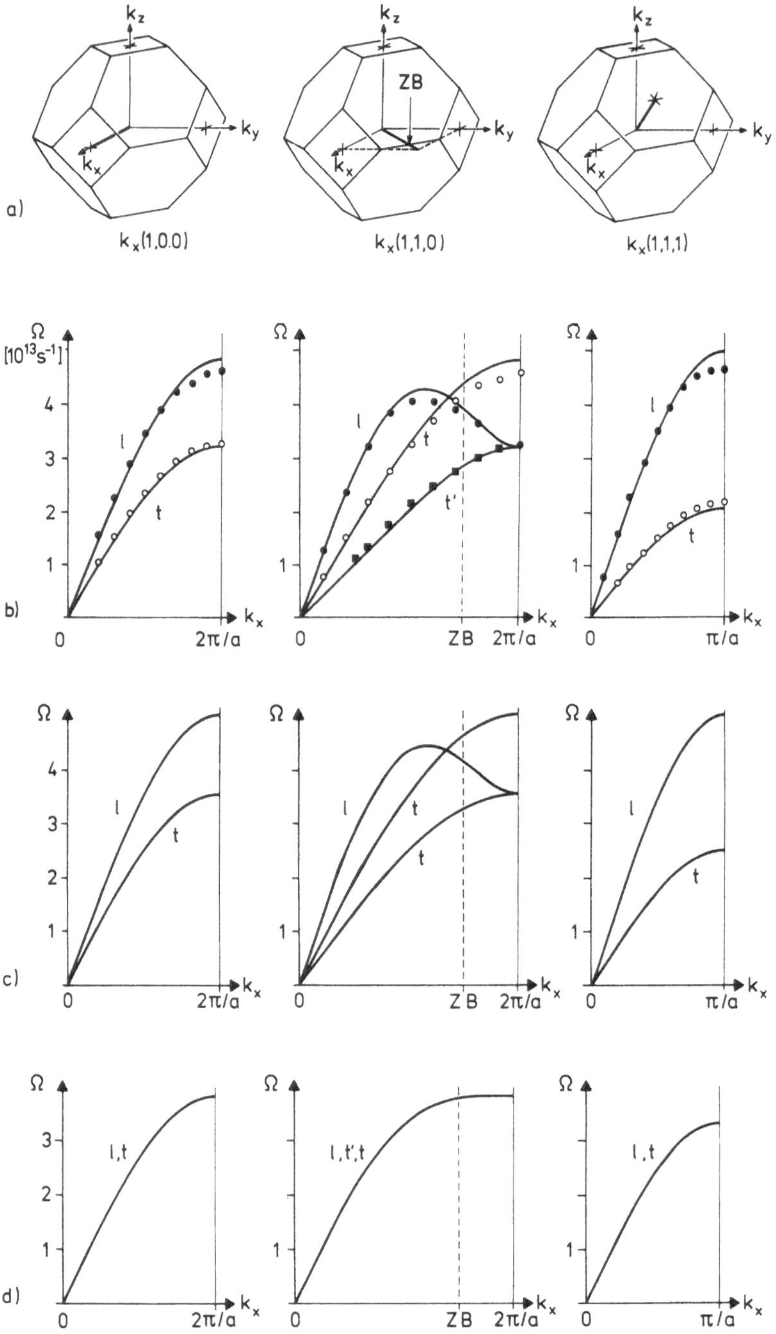
a)
$k_x(1,0,0)$
$k_x(1,1,0)$
$k_x(1,1,1)$
ZB
b)
Ω
[$10^{13}s^{-1}$]
l
t
t'
c)
d)
l,t
l,t',t
0
ZB
2π/a
π/a
$k_x$
$k_y$
$k_z$

Fig. 3.15a-d

Table 3.2. First neighbour springs for fcc crystals, fitted to the initial slopes of $\Omega(\underline{k}\sigma)$

$$\Phi^{[110]} = -\begin{bmatrix} \alpha & \beta & 0 \\ \beta & \alpha & 0 \\ 0 & 0 & \gamma \end{bmatrix}, \quad f_l = \alpha + \beta , \quad f_{t'} = \alpha - \beta , \quad f_t = \gamma$$

| Substance | $f_l$ [$10^4$ dyne/cm] | $f_{t'}/f_l$ | $f_t/f_l$ | Fit |
|---|---|---|---|---|
| Cu /3.4/ | 3.31 | -0.078 | -0.049 | good |
| Al /3.5/ | 1.98 | 0.092 | -0.258 | good |
| Ag /3.6/ | 2.66 | -0.055 | -0.124 | satisfactory |
| Au /3.7/ | 4.01 | -0.034 | -0.268 | satisfactory |
| Pb /3.8/ | 1.32 | -0.072 | -0.185 | poor |

## 3.4 Models for Force Constants

We have seen above how dispersion curves can be fitted to near neighbour Born- von Karman models. The model is rather abstract, in the sense that one does not see the cause of the interaction. In particular this refers to the transversal springs; their origin is not clear, and one cannot remove them without violating rotational invariance. Consequently, the use of such models for defect calculations is rather re-

◁ Fig. 3.15a-d. Dispersion curves of an fcc lattice with 1st neighbour coupling in the main symmetry directions. The $k_x$-scales for the three directions are such that the slopes at $k_x = 0$, the sound velocities, can be directly compared.
a) $\underline{k}$-vectors in the main symmetry directions, the zone boundary (ZB) is indicated by ( + ).
b) The experimental points ( • , o , ■) for Cu /3.4/, and a Born- von Karman fit with three different 1st neighbour springs, values form Table 3.2.
c) Purely longitudinal coupling with $f_l$ as in b.
d) Scalar coupling: The longitudinal and the two transversal springs are equal, f. The value of $f = (f_l + f_{t'} + f_t)/3$ with $f_{l,t',t}$ from Table 3.2 is chosen such that the Einstein-frequency is identical to that of b

stricted[19]. One can, however, explore other couplings which are based on more physical models, and which then must obey rotational invariance. We will not enter into a discussion of sophisticated models which represent the screening of ionic charges by electrons. Rather, we will review briefly the simplest models presently available.

### 3.4.1 Two Body Forces

Two body forces are based on two body potentials, $V(r) = V_{rep} + V_{att}$ (mentioned already in Section 3.3.1, Fig. 3.13). The potentials are not very well known. Often the repulsive part $V_{rep}$ is represented by a "Born-Mayer" potential $\propto \exp(-r/r_c)$ or by a power potential $\propto r^{-12}$, and the attractive part, $V_{att}$, by a van der Waals potential $\propto -r^{-6}$. Fig. 3.13 is typical for all these potentials which are essentially of short range, say a 12-6 Lennard-Jones potential $\propto [(r/r_o)^{-12} - (r/r_o)^{-6}]$. The substances best described by these potentials are the noble gases. The parameters can be fitted to, e.g., lattice energy, lattice distance, elastic data and dispersion curves.

By including only neighbours up to 1st or 2nd order, one obtains the simplest models suitable also for defect calculations. We will demonstrate this for cubic lattices. The energy per atom or per cell is given by

$$E_c = \frac{1}{2} \sum_{\underline{m}(\neq 0)} V(R^{\underline{m}}) = \frac{1}{2} \sum_{\nu \geq 1} Z_\nu V(R_\nu) \ . \tag{3.46}$$

The sums refer to atom 0; in the 2nd sum the atoms are arranged in shells about the origin, where $Z_\nu$ is the number of atoms in shell $\nu$ and $R_\nu$ is the distance; for an fcc lattice:

$$Z_1 = 12 \ , \quad Z_2 = 6 \ ; \quad R_1 = a/\sqrt{2} \ , \quad R_2 = a \ .$$

The equilibrium condition is vanishing $\partial_a E_c$ or $\partial_{R_1} E_c$,

$$\partial_a E_c = \sum_\nu Z_\nu V'(R_\nu) \frac{R_\nu}{a} = 0 \ . \tag{3.47}$$

We know that the springs connecting atoms of distance $R_\nu$ are

$$f_l^{(\nu)} = V''(R_\nu) \ , \quad \text{longitudinal} \ , \tag{3.48a}$$

[19] Violations of rotational invariance can be remedied via the following (not absolutely convincing) procedure, demonstrated for a "vacancy" in an fcc lattice with 1st neighbour interaction, $f_l$, $f_{t'}$, $f_t$. After having cut out the 12 1st neighbour springs, rotational invariance is recovered by correcting the 24 first neighbour transversal springs surrounding the vacancy: $f_{t'} \to f_{t'} - f_t/2$, $f_t \to f_t - f_{t'}/2$.

$$f_t^{(\nu)} = V'(R_\nu)/R_\nu \ , \quad \text{transversal isotropic ;} \tag{3.48b}$$

according to (3.47), the transversal springs obey

$$\sum_\nu Z_\nu f_t^{(\nu)} R_\nu^2 = 0 \ . \tag{3.48c}$$

A *1st neighbour model*[20] contains then

$$f_1^{(1)} = V''(R_1) \ , \quad f_t^{(1)} = 0 \ , \quad 1 \text{ parameter ,} \tag{3.49}$$

and a *2nd neighbour model*

$$f_1^{(1)} = V''(R_1) \ , \quad f_1^{(2)} = V''(R_2) \ , \quad -\frac{f_t^{(1)}}{f_t^{(2)}} = \frac{Z_2 R_2^2}{Z_1 R_1^2} = \begin{cases} 4 & \text{for sc} \\ 1 & \text{for bcc} \\ 1 & \text{for fcc} \end{cases} \ , \quad 3 \text{ parameters.} \tag{3.50}$$

Let us discuss again the simple vacancy model shown in Fig. 3.11. In the 1st neighbour model the distance $R_1$ is such that the force on one atom by any neighbour vanishes, $V'(R_1) = 0$. Consequently, one can take out one atom without disturbing equilibrium. This corresponds exactly to cutting the 12 springs from the vacancy to its former 1st neighbours. The situation changes when one considers 1st and 2nd neighbour interaction. In equilibrium the atom in the origin exerts radial forces $-V'(R_1)$ and $-V'(R_2)$ on its 1st and 2nd neighbours. These forces are cancelled by the contributions of other atoms. If one forms the vacancy (in the origin) these forces are removed, and net forces $V'(R_1)$, $V'(R_2)$ act on the atoms in the unrelaxed positions (Fig. 3.16a). If the 2nd neighbour contributions $f_t^{(1)}$, $f_t^{(2)}$, $f_1^{(2)}$, $V'(R_1)$, $V'(R_2)$ are "small", the forces will produce only small displacements[21] which can in this "order" be calculated with the response of the 1st neighbour model ($f_1^{(1)}$ alone, springs to vacancy removed). If one neglects the small displacements, when considering a rotation, one obtains the forces illustrated in Fig. 3.16b. The neighbours of the vacancy are not force-free under a rotation, $\underline{s}^{\underline{h}} = \varphi\hat{\underline{D}}\times\underline{R}^{\underline{h}}$; the forces are $\underline{F}^{\underline{h}} = f_{\underline{t}}^{\underline{h}}\underline{s}^{\underline{h}}$, and the rotational invariance is violated; but at least the total torque vanishes, $\sum_{\underline{h}} \underline{R}^{\underline{h}}\times\underline{F}^{\underline{h}} = 24/3\cdot\hat{\underline{D}}\sum_\nu Z_\nu f_t^{(\nu)} R_\nu^2 = 0$. Therefore, invariance can be saved by taking into account small deviations from the original structure. If the 2nd and higher order neighbours are important, the displacements are large; the new equilibrium and the new springs have to be determined by a special and analytically difficult calculation.

---

[20] The 1st neighbour model gives unstable sc and bcc lattices. The dispersion curves for fcc lattices are shown in Fig. 3.15c.

[21] This problem will be treated in Section 6.3.2. The displacements are short range and confined to the neighbourhood of the vacancy.

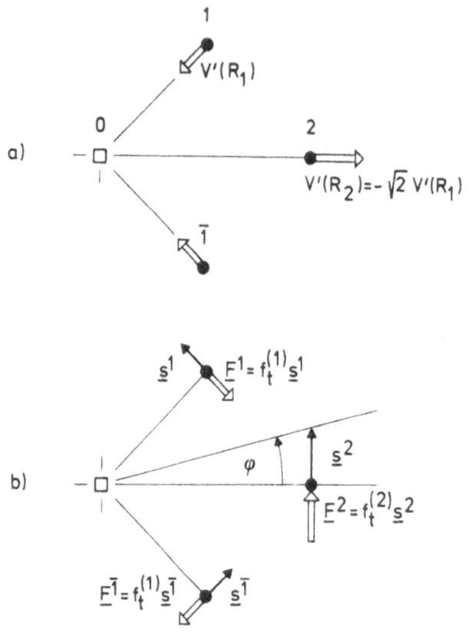


Fig. 3.16a and b. Vacancy model in fcc lattice for central forces including 1st and 2nd neighbours. a) Forces around a vacancy. b) Forces produced by a rotation: the single forces do not vanish, but the total torque, $\sum_{\underline{h}} \underline{R}^{\underline{h}} \times \underline{F}^{\underline{h}}$, becomes zero

In discussing two body potentials we always have had in mind potentials such as in Fig. 3.13, where $V'(R_1)$ and $f_t^{(1)}$ are negative. Negative springs always indicate a tendency towards *instability*, because a displacement is enhanced by the spring forces. In the perfect crystal these forces are counteracted and overcompensated by the action of other springs for which the displacement is partly longitudinal. One also sees from Fig. 3.13 that the transversal spring becomes very large and negative for distances much smaller than the equilibrium distance l (large negative $f_t$). A special example is a "self-interstitial" where *two* atoms of the same kind are present in *one* basic volume. Even though the relaxation (the displacements relaxing the forces caused by the additive atom) is large near the interstitial, the distances will be appreciably smaller than l. Therefore, one has large negative $f_t$'s which can cause small restoring forces for certain motions. The region around the interstitial can become quite soft (almost unstable), and can therefore greatly influence the static and dynamical response of the medium. This is discussed in great detail in the review volume.

Two body potentials alone cannot explain the dispersion curves. A 1st neighbour approximation gives only *one* longitudinal spring instead of *three* possible force constants. To achieve more flexibility, one must consider more general (many body) interactions.

### 3.4.2 Many Body Potentials

A many body potential depends at least on three atomic positions. An example is the three body potential $V_3(\varphi) = B(\varphi)$, (3.41), where the bending angle $\varphi$ is determined by the position of three atoms. The basic interaction in solids is of Coulomb type and therefore of two body character. The many body character is introduced by eliminating the electronic degrees of freedom via the adiabatic approximation and expressing the potential energy by the positions of the nuclei alone. An electron interacts with many nucleons and leaves a many body interaction upon its elimination. We will treat in Section 3.4.3 simple examples of this kind. The many body character also becomes obvious, if one considers the screening of the Coulomb interaction between two ions by electrons. The screening depends, of course, on the environment, i.e. on the positions of the surrounding nuclei, and, therefore, carries many body character. Polarization of core electrons is just a special case of screening. Here we will only discuss briefly some phenomenological many body potentials. A simple three body bending potential $V_3(\varphi)$ has been considered in Section 3.3.1. The most general three body potential would be a function of the three invariants of the triangle given by the three atomic positions. It is, of course, difficult to use such completely general potentials without any more physical guidance.

A *simple* model would be to introduce "volume dependent" electronic contributions; in the free electron picture the Fermi wave vector is proportional to $n^{1/3}$ (n = density of electrons), the energy per electron is $Cn^{2/3}$, and the electronic energy density is $Cn^{5/3} = Cv^{-5/3}$, where v is the volume per electron. In a monovalent metal v equals $V_c$ in equilibrium. In a deformed lattice one can define an atomic volume, $V^{\underline{m}}$, for each lattice site $\underline{m}$ by the (deformed) volume of the Wigner-Seitz-Cell, which, if the deformations are still small, is determined by the position of $\underline{m}$ and its 12 nearest neighbours; in the undeformed lattice: $V^{\underline{m}} = V_c = v$. If one assumes that in the deformed state one has still one electron in $V^{\underline{m}}$ and that its contribution is as before $C\cdot(V^{\underline{m}})^{-5/3}$, the electronic contribution becomes $\sum_{\underline{m}} C\cdot(V^{\underline{m}})^{-5/3}$. This can be considered as a sum of 1 + 12 = 13 body potentials.

In nonmetals, where the charges are not screened, one must, in addition, consider long range Coulomb forces. We will not further enlarge on these models, but rather concentrate on two very simple and successful models, which can also be used for defect calculations.

### 3.4.3 The Bond Charge Model /3.9/

The bond charge model is based on the fact that the electronic charge, which piles up in the center of chemical bonds, can be viewed as a dynamical quantity which influences the interatomic coupling. We will employ the bond charge in the following way: we neglect Coulomb interaction (screening) but rather introduce longitudinal short range springs between bond charges and atoms; the bond charges are eliminated adiabatically and one obtains effective "many body springs" between the atoms, i.e., as if obtained from a many body atomic potential. The effective springs can be long range. Because only longitudinal (two body) springs are used originally, the cutting of springs when introducing defects does not violate rotational invariance.

a) *In one dimension* the simplest situation is shown in Fig. 3.17a. This figure shows "bond charges" with displacements $\underline{u}$ coupled to their neighbour atoms by w and to their neighbour bond charges by v. If at first one includes the small ("electronic") mass m of the bond charge, one has

$$M\ddot{\underline{s}} = -\Phi^{ss}\underline{s} - \Phi^{su}\underline{u} + \underline{F}^{s} ,$$
$$m\ddot{\underline{u}} = -\Phi^{us}\underline{s} - \Phi^{uu}\underline{u} + \underline{F}^{u} . \qquad (3.51)$$

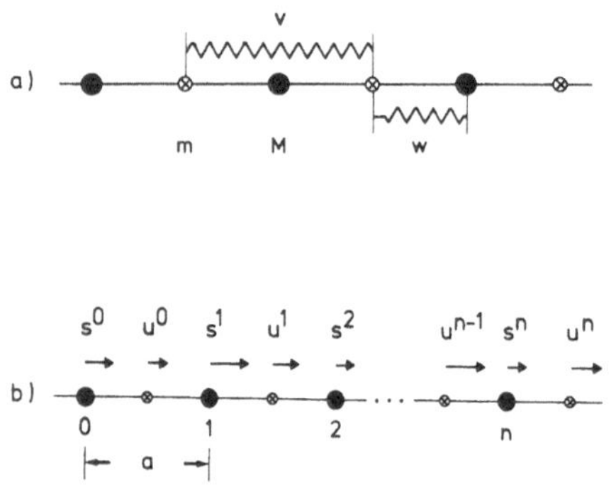


Fig. 3.17a and b. Simple bond charge model. a) Interactions: spring w between bond charge (mass m) and neighbouring atoms (mass M), and spring v between 1st neighbour bond charges; springs between atoms are not contained in this model but can easily be included. b) Displacements

Here $\Phi^{ss}$ represents the coupling within the sublattice of atoms while the bond charge sublattice is fixed, $\underline{u} = 0$; note that $\Phi^{ss}$ contains the coupling of the atoms to the *rigid* bond charges, i.e., a common translation of the atoms alone is an eigenstate of $\Phi^{ss}$ with *non*-vanishing eigenvalue: $\sum_n (\Phi^{ss})^{mn} \neq 0$. The same holds for $\Phi^{uu}$, the coupling of bond charges to fixed atoms. Therefore, when calculating the static Green functions of the sublattices $(1/\Phi^{ss}, 1/\Phi^{uu})$, one does not encounter the difficulties with translations discussed in Section 2.5. The coupling between the two sublattices is denoted by $\Phi^{us}$ and $\Phi^{su}$ (note $\Phi = \Phi'$). For vanishing m (adiabatic approximation) and vanishing $\underline{F}^{u}$ one has

$$\Phi^{uu}\underline{u} = -\Phi^{us}\underline{s} , \qquad (3.52a)$$

and $\underline{u}$ can be expressed by $\underline{s}$ ,

$$\underline{u} = \frac{1}{\Phi^{uu}}(-\Phi^{us}\underline{s}) = -L\Phi^{us}\underline{s} , \tag{3.52b}$$

via the static Green function L of the bond charge lattice. The final equation for $\underline{s}$ alone becomes

$$M\ddot{\underline{s}} = -(\Phi^{ss} - \Phi^{su}L\Phi^{us})\underline{s} + \underline{F}^s = -\Phi^{ss}_{eff}\underline{s} + \underline{F}^s . \tag{3.53}$$

For the model of Fig. 3.17 the forcefree equations are

$$\begin{aligned} M\ddot{s}^n &= -w(2s^n - u^n - u^{n-1}) , \\ m\ddot{u}^n &= -w(2u^n - s^n - s^{n+1}) - v(2u^n - u^{n+1} - u^{n-1}) . \end{aligned} \tag{3.54}$$

Eqs. (3.54) for $m \neq 0$ are those of a diatomic chain. With $s^n$, $u^n \propto \tilde{s}$, $\tilde{u}\cdot\exp(ikan)$ one has

$$M\Omega^2(k)\tilde{s} = w\,[\,2\tilde{s} - \tilde{u}(1 + e^{-ika})\,] , \tag{3.55a}$$

$$m\Omega^2(k)\tilde{u} = w\,[\,2\tilde{u} - \tilde{s}(1 + e^{ika})\,] + 2v\tilde{u}(1 - \cos ka) . \tag{3.55b}$$

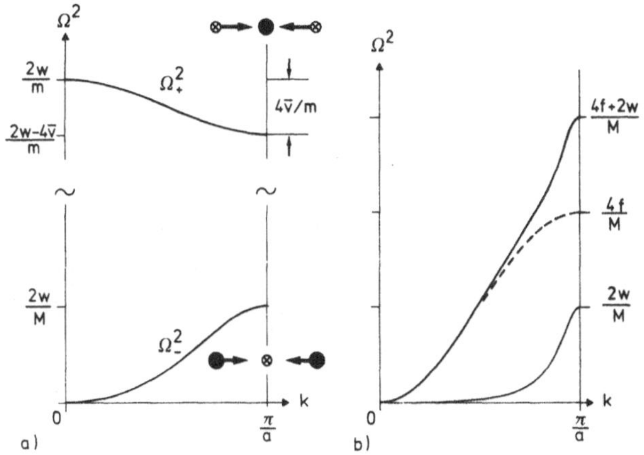


Fig. 3.18a and b. Dispersion curves in a lattice with bond charges.
a) Dispersion curves ($m \ll M$) for $\bar{v} = -v = w/4$. In the acoustical branch: $\tilde{u}/\tilde{s} > 0$, $\tilde{u} = 0$ for $ka = \pi$. In the optical branch: $\tilde{s}/\tilde{u} < 0$, $\tilde{s} = 0$ for $ka = \pi$. The indicated frequencies at $ka = \pi$ are correct for all m and M.
b) Peak produced by bond charges on a normal dispersion curve; the bond charge contribution (lower line), $\Omega^2_{max} = 2w/M$, and the normal dispersion curve (dashed line), $\Omega^2_{max} = 4f/M$, are shown separately for $w = f$

These equations have two solutions, $\Omega_\pm^2$ (Fig. 3.18a), which for $m \ll M$ are given by a *low frequency* "acoustical" solution ($m = 0$ in (3.55b), the light bond charges follow adiabatically)

$$M\Omega_-^2 \cong w(1 - \cos ka) \frac{w + 2v}{w + v(1 - \cos ka)} , \tag{3.56a}$$

and by a *high frequency* "optical" solution[22] $(\tilde{s} \cong 0$ in (3.55b), the heavy atoms do not follow the fast motion of bond charges$)$

$$m\Omega_+^2 \cong 2w + 2v(1 - \cos ka) . \tag{3.56b}$$

As a rule, only the acoustical branch, $\Omega_- = \Omega$, is accessible to neutron scattering and will be discussed further. Stability of the bond charge lattice requires $M\Omega_+^2 > 0$:

$$2w + 2v(1 - \cos ka) > 0 , \quad \text{leading to } w > 0 , \quad w + 2v > 0 ,$$
$$\text{i.e., } 2\bar{v} < w \quad \text{for } \bar{v} = -v. \tag{3.57}$$

The *acoustical* dispersion curve (3.56a) is for $v \ll w$ that of a chain with 1st neighbour spring $w/2$, $M\Omega^2 \cong w(1 - \cos ka)$. If, however, the denominator in (3.56a) becomes important one can produce strange dispersion curves corresponding to far reaching effective springs between atoms. For $v \gg w$ the dispersion curve is flat: $M\Omega^2 \cong 2w$ for $(ka)^2 \gg w/v$. If $v < 0$ and $w - 2v = \eta w \gtrsim 0$ $(\eta \ll 1)$, the bond charge lattice is nearly unstable[23] for $ka = \pi$. The dispersion curve

$$M\Omega^2 = 2w \sin^2 \frac{ka}{2} \frac{w + 2v}{w - 2\bar{v} \sin^2 \frac{ka}{2}} = 2w \sin^2 \frac{ka}{2} \frac{\eta}{\eta + (1 - \eta)\cos^2 \frac{ka}{2}} \tag{3.58}$$

exhibits a sharp peak at $ka = \pi$ of width $2\sqrt{\eta}$ (in $ka$, Fig. 3.18b). If one adds a direct atomic contribution $4f \sin^2 ka/2$ to $M\Omega^2$, the bond charge contribution produces a peak at $ka = \pi$ on an otherwise normal dispersion curve, Fig. 3.18b. Peaks at other k-values can be produced by employing springs between more distant neighbours. The situation becomes better elucidated if one determines directly the induced effective springs, $f_{eff}^{(h)} = f_{eff}^{(-h)}$,

$$M\Omega^2(k) = \sum_h f_{eff}^{(h)}(1 - \cos kah) = \sum_{h \geq 1} 4f_{eff}^{(h)} \sin^2 \frac{ka}{2} h . \tag{3.59}$$

The effective springs are most easily expressed by the static Green function

---

[22] For $ka \ll 1$ the acoustical branch represents long (acoustic) waves, whereas the optical modes ($\tilde{u}$ and $\tilde{s}$ have opposite sign) describe macroscopic dipole oscillations if the two constituents are charged.

[23] Instability of the $ka = \pi$-mode means that two bond charges would pile up at every second atom, corresponding to a "charge density wave" of period 2a (Fig. 3.18a).

$L^{mn} = L^{\pm(m-n)}$ of the bond charge lattice,

$$2wL^{(m)} + v\left[2L^{(m)} - L^{(m+1)} - L^{(m-1)}\right] = \delta^{mo} , \qquad (3.60)$$

or its Fourier transform, $L(\tilde{k})$,

$$2w + 2v(1 - \cos ka)\, \tilde{L}(k) = 1, \qquad (3.60a)$$

$$L^{(h)} = \frac{a}{2\pi} \int_{-\pi/a}^{\pi/a} dk\, \tilde{L}(k) e^{ikah} = \int_{-\pi}^{\pi} \frac{d(ka)}{4\pi} \frac{e^{ikah}}{w + v(1 - \cos ka)}$$
$$= \frac{1}{4\pi\bar{v}} \int_{-\pi}^{\pi} d\xi \frac{e^{i\xi h}}{\cos\xi - \cos\kappa} , \text{ where } \cos\kappa = \frac{w + v}{v} = -\frac{w - \bar{v}}{\bar{v}} . \qquad (3.60b)$$

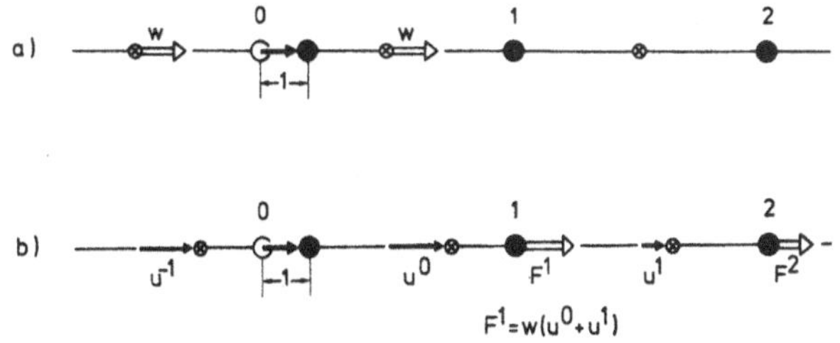


Fig. 3.19a and b. Forces transferred by the bond charge lattice.
a) Displacement 1 of atom 0 leads to forces w on the fixed neighbouring bond charges.
b) Subsequent relaxation of the bond charge lattice with fixed atoms leads to displacements $u^h = w\left[L^{(h)} + L^{(h+1)}\right]$ and to forces $w(u^{h-1} + u^h)$ on atom h, e.g., $F^1 = w(u^o + u^1)$. Note that in the harmonic approximation the relaxation of the bond charge lattice for fixed atoms is represented by $\Phi^{uu}$, independent of the actual positions of atoms

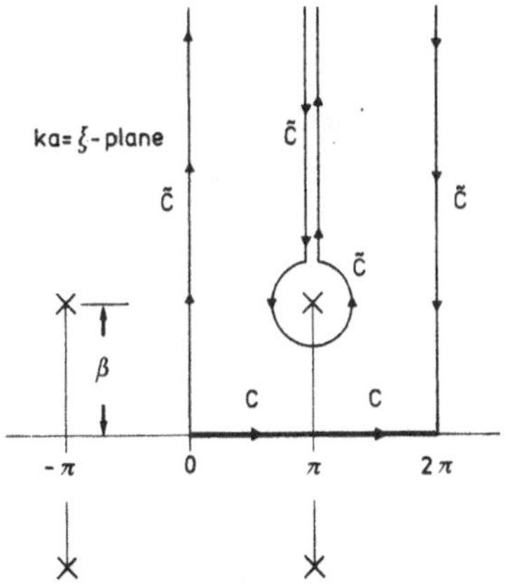


Fig. 3.20. Evaluation of the integral (3.60b). The integrand and the poles (indicated by ×) have the period $2\pi$, hence $\int_{-\pi}^{\pi}\ldots = \int_{0}^{2\pi}\ldots = \int_{C}\ldots$ . The path of integration, C, is the interval $(0,2\pi)$ on the real axis. For $h > 0$ one can shift the path of integration to $+i\infty$, $\tilde{C}$; the paths parallel to the imaginary axis ($\mathrm{Re}\,\xi = 0,2\pi$) cancel because of periodicity. Only the contribution of the pole remains, near the pole $\cos\xi - \cos\kappa \cong i(\xi - \kappa)\sinh\beta$

The effective spring $f_{eff}^{(h)}$ is obtained by the following procedure (Fig. 3.19): Displacement 1 of atom 0 produces the force w on the two neighbouring atoms, if all other positions are fixed. Now let the bond charge lattice relax and keep all the atoms at rest: the displacements are $u^h = w\left[L^{(h)} + L^{(h+1)}\right]$, the force on atom h is $w(u^h + u^{h-1}) = w^2\left[2L^{(h)} + L^{(h+1)} + L^{(h-1)}\right]$, corresponding to an effective spring between atom 0 and h,

$$f_{eff}^{(h)} = w^2\left[2L^{(h)} + L^{(h+1)} + L^{(h-1)}\right] . \tag{3.61}$$

For v = 0 one has the trivial case of purely *local response*, where $L^{(h)} = (1/2w)\delta_{ho}$, corresponding to an effective 1st neighbour spring $f_{eff}^{(1)} = w/2$. If $v \neq 0$, the springs are in principle long range, and $L^{(h)}$ has to be evaluated according to (3.60b). For simplicity we treat only negative $v = -\bar{v} < 0$. In the complex ka = $\xi$-plane the integrand has poles at $\kappa$ (Fig. 3.20),

$$\cos\kappa = -\frac{w-\bar{v}}{\bar{v}} < -1 , \quad \kappa = \pi \pm i\beta + 2\nu\pi , \cosh\beta = \frac{w-\bar{v}}{\bar{v}} = \frac{1+\eta}{1-\eta}, \ \beta > 0, \tag{3.62}$$

and we obtain

$$L^{(h)} = L^{(-h)} = \frac{1}{2\bar{v}\sinh\beta}(-1)^h e^{-\beta h} , \quad h \geqslant 0. \tag{3.60c}$$

One verifies easily that in the limit $v \to 0$ the response is local: $\beta \to \infty$, $e^{-\beta h} \to \delta_{ho}$ and $2\bar{v}\sinh\beta \to 2w$. In the "resonance" case, $\bar{v} \gtrsim w/2$, $\eta \ll 1$, $\beta^2 \cong 4\eta$, the pole is close to the real axis and is responsible for the sharp peak at ka = $\pi$.

The effective springs are for $h \geqslant 1$

$$f_{eff}^{(h)} = \frac{w^2}{\bar{v}}(-1)^h e^{-\beta h}\frac{1-\cosh\beta}{\sinh\beta} \cong -w\beta(-1)^h e^{-\beta h} , \quad \beta \cong 2\sqrt{\eta} \text{ for } \eta \ll 1 ; \tag{3.61a}$$

they alternate in sign and have very long range, (about $1/\beta$) if $\beta$ is small; note that this range and the width of the peak are reciprocal (comp. Sec. 3.1.1). After having discussed possible long range effects we will now concentrate on local coupling which is more important in metals.

b) *A simple two-dimensional example* is shown in Fig. 3.21, a bond charge in the center of an equilateral triangle connected to the corner atoms by a spring w. The effective (three body) spring between two atoms, say 0 and 1, mediated by the bond charge is calculated as above: Atom 1 is displaced by $\underline{s}^1$, the force on the fixed bond charge is then $\underline{F}^B = w|\underline{b}\rangle\langle\underline{b}|\underline{s}^1\rangle$, the displacement $\underline{u}$ of the bond charge due to $\underline{F}^B$ is $\underline{u} = \underline{F}^B/(3w/2)$ (its response is isotropic corresponding to a spring 3w/2); the force on atom 0 caused by $\underline{u}$ is $\underline{F}^o = w|\underline{a}\rangle\langle\underline{a}|\underline{u}\rangle = \frac{2w}{3}|\underline{a}\rangle\langle\underline{a}|\underline{b}\rangle\langle\underline{b}|\underline{s}^1\rangle$, $\langle\underline{a}|\underline{b}\rangle = -1/2$.

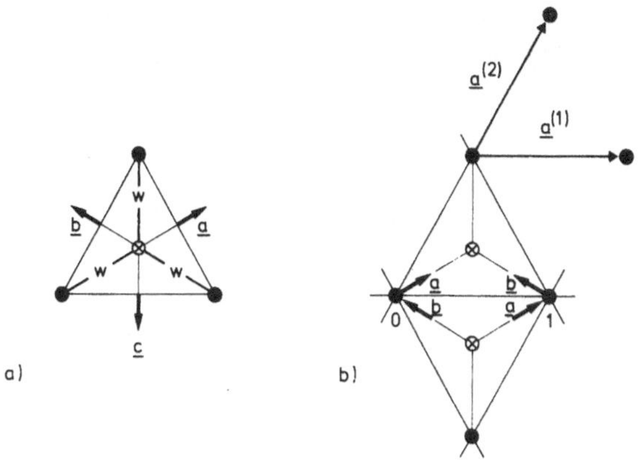


Fig. 3.21a and b. Bond charge in equilateral triangle.
a) The bond charge is connected to the three corners by longitudinal springs w in directions a, b, c (a + b + c = 0).
b) Hexagonal two-dimensional lattice with basis vectors $a^{(1,2)}$. The spring between 0 and 1 is now caused by *two* bond charges in the indicated triangles

Consequently, the coupling

$$f^{01} = -\frac{w}{3}\,|\underline{a}><\underline{b}|\ , \qquad f^{01}_{ik} = -\frac{w}{3}\,a_i b_k \tag{3.63}$$

is not symmetrical[24]. In a two-dimensional close packed (hexagonal) lattice (Fig. 3.21b), two bond charges contribute to the spring between 0 and 1, for the second bond charge $\underline{a}$ and $\underline{b}$ are interchanged. The coupling becomes symmetrical[25]

$$f = -\frac{w}{3}\,(|\underline{a}><\underline{b}| + |\underline{b}><\underline{a}|) = \frac{w}{6}\,(3P_l - P_t)\ , \tag{3.63a}$$

and separates into a longitudinal and a transversal part, $f_{l,t}$, in the ratio of 3 to -1. One recognizes that, although all the springs are longitudinal and positive, negative $f_t$ can be obtained quite naturally. Further one can cut all the (longitudinal) springs without violating rotational invariance. As an example, Fig. 3.22 shows a "vacancy" in that lattice where all the six bond charges around the vacancy have been removed. The six springs $f^{12}$, $f^{23}$, etc., around the vacancy are changed, because one bond charge is missing, e.g., the spring between 2 and 3 passes from

24 The effective coupling parameters or spring constants as defined by (2.3) and (2.6) are $-f^{01}_{ik}$; of course, $f^{01}_{ik}$ obeys the general symmetry (2.5a), i.e. $f^{01}_{ik} = f^{10}_{ki}$, but in (3.63) $f^{01}_{ik} \neq f^{01}_{ki}$.

25 The eigenvectors of f are $\underline{a} \pm \underline{b}$ with eigenvalues $(\underline{a},\underline{b}) \pm 1 = 1/2, -3/2$ (+: transversal, -: longitudinal).

$-(w/3)(|\underline{a}\rangle\langle\underline{b}| + |\underline{b}\rangle\langle\underline{a}|)$ to $-(w/3)\ |\underline{a}\rangle\langle\underline{b}|$ ; the six springs from the vacancy to its former neighbours are cut and all other springs do not change. This should also reflect the general situation quite well.

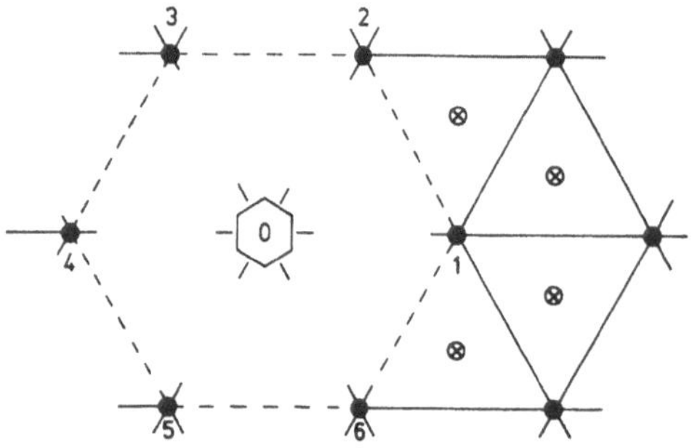


Fig. 3.22. Vacancy in two-dimensional hexagonal lattice. To form a vacancy at 0, the six surrounding bond charges are removed. This corresponds to cutting the six interior springs and to changing the six springs (---) on the hexagon surrounding the vacancy; all other springs (——) do not change

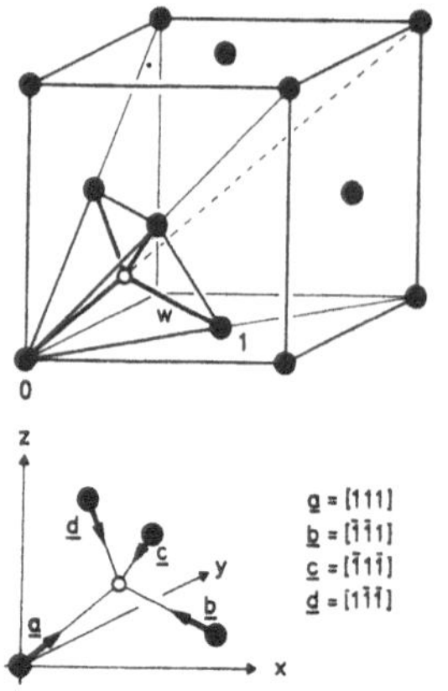


Fig. 3.23. First neighbour tetrahedron in fcc lattice with bond-charge (o) in center. The atoms, 0,1 have in common one other tetrahedron, obtained by reflection at $z = 0$, $\underline{a} \to -\underline{b}$, $\underline{b} \to -\underline{a}$

c) *Three-dimensional bond charge lattices with purely local response* are illustrated using a fcc lattice. Here four 1st neighbours form a tetrahedron (Fig. 3.23), and the bond charge is supposed to be in the center of the tetrahedron, connected to the four corners by w. The response of the bond charge is isotropic, corresponding to a spring 4w/3. As above we obtain ($\langle\underline{a}|\underline{b}\rangle = -1/3$) $f^{01} = (3w/4)\ |\underline{a}\rangle\langle\underline{a}|\underline{b}\rangle\langle\underline{b}| = -(w/4)\cdot$ $\cdot|\underline{a}\rangle\langle\underline{b}|$ . Including the second tetrahedron, shared by 0 and 1 (Fig. 3.23), we obtain

$$f^{01} = -\frac{w}{4}\left(|\underline{a}\rangle\langle\underline{b}| + |\underline{b}\rangle\langle\underline{a}|\right) = \frac{w}{6}\left(2P_l - P_t\right),\quad P_l = P^{[110]},\quad P_t = P^{[001]}, \qquad (3.64)$$

which again contains one negative transversal spring. As an example one can take Au, where according to Table 3.2, $f_l \cong 4$, $f_{t'} \cong 0$, $f_t \cong -1.1$, corresponding to $w = 6.6$ and a direct atomic longitudinal coupling of 1.8 (in units of $10^4$ dyn/cm).

One can also consider triangular forces (as in Fig. 3.21) mediated by bond charges centered in the surfaces of the tetrahedra. Fig. 3.23 contains two triangles which contribute to the spring between 0 and 1; two more triangles are due to the second (reflected) tetrahedron. The four triangles form two pairs as in Fig. 3.21b, with transversal directions (perpendicular to [110] and in the planes of the triangles) $[1\bar{1}2]$ and $[1\bar{1}\bar{2}]$. From (3.63a) one obtains now

$$f^{01} = \frac{w}{6}\,(2\times 3P_1 - P^{[1\bar{1}2]} - P^{[1\bar{1}\bar{2}]}) = w(P_1 - \frac{1}{9}\,P_{t'} - \frac{2}{9}\,P_t) \;; \qquad (3.64a)$$

here the tranversal part containing the two non-orthogonal directions $[1\bar{1}2]$ and $[1\bar{1}\bar{2}]$, can be expressed by the orthogonal projectors $P_{t'}$ and $P_t$ (eigenrepresentation of the transversal part of $f^{01}$). Again the transversal springs are negative.

### 3.4.4 Simple Shell Models

Shell models are supposed to represent the polarizability of the electronic shells surrounding the nuclei. Again, we neglect Coulomb interaction. We will see that, as in Section 3.4.3, by shell interaction long range springs are induced and that for negative coupling the shell lattice can become soft, which causes a dip in the dispersion curve rather than a peak as produced by bond charges.

a) *In one dimension* the simplest model is shown in Fig. 3.24. One shell is bound to its nucleus by a spring 2w and by a spring v to its 1st neighbour shells. The equations of motion,

$$\begin{aligned} M\ddot{s}^n &= -f(2s^n - s^{n+1} - s^{n-1}) - 2w(s^n - u^n) \;, \\ m\ddot{u}^n &= -2w(u^n - s^n) - v(2u^n - u^{n+1} - u^{n-1}) \;, \end{aligned} \qquad (3.65)$$

can be treated as before. The notation is such that the static Green function of the shell lattice is identical to that of the bond charge lattice. A direct nuclear interaction (spring f) has been included. From the Fourier transform of (3.65)

$$\begin{aligned} M\Omega^2\tilde{s} &= 2f\tilde{s}(1 - \cos ka) + 2w(\tilde{s} - \tilde{u}) \;, \\ m\Omega^2\tilde{u} &= 2w(\tilde{u} - \tilde{s}) + 2v\tilde{u}(1 - \cos ka) = 0 \quad (m = 0,\ \text{adiabatic approximation}) \end{aligned} \qquad (3.66)$$

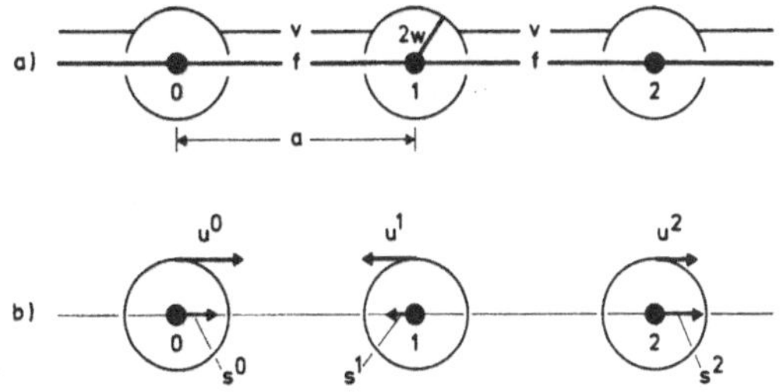


Fig. 3.24a and b. Linear shell lattice. a) Interactions: spring 2w between shell and its atom, v between 1st neighbour shells, f between 1st neighbour atoms. b) Displacements

one obtains the dispersion curve

$$M\Omega^2 = 2(1 - \cos ka)\left[f + \frac{wv}{w + v(1 - \cos ka)}\right] . \qquad (3.66a)$$

The stability condition for the shells is again $w + 2v > 0$, and near instability ($\bar{v} = w(1 - \eta)/2$, $\eta \ll 1$)

$$M\Omega^2 = 4\sin^2\frac{ka}{2}\left[f - \frac{w(1 - \eta)/2}{\eta + (1 - \eta)\cos^2\frac{ka}{2}}\right] = D(k) . \qquad (3.67)$$

For small $\eta$ the dispersion curve exhibits a sharp dip (*negative* shell contribution), $D(\pi/a) = 4f - 2w/\eta$. A dip of 50% is obtained with $w = \eta f$ which shows that an appreciable dip can be produced with w and v both small (Fig. 3.25). The effective springs are long range

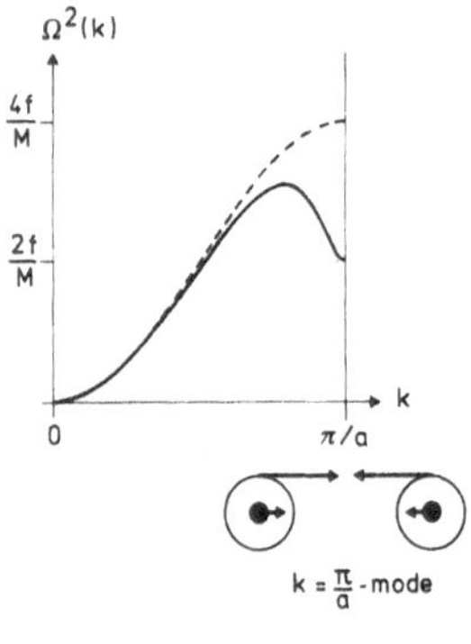


Fig. 3.25. Dispersion curve of linear shell model ($\eta = 1/25$, $v/w = -12/25$, $w/f = 1/25$). Dashed line: without shell contribution ($v = 0$); full line: with shell contribution. For the $ka = \pi$-mode the amplitudes are indicated (not to scale), $\tilde{u}/\tilde{s} \cong 1/\eta$; in the optical branch $\tilde{u}/\tilde{s} \cong M\eta/m$, in contrast to the bond charge model, where $\tilde{u} = 0$ or $\tilde{s} = 0$, respectively

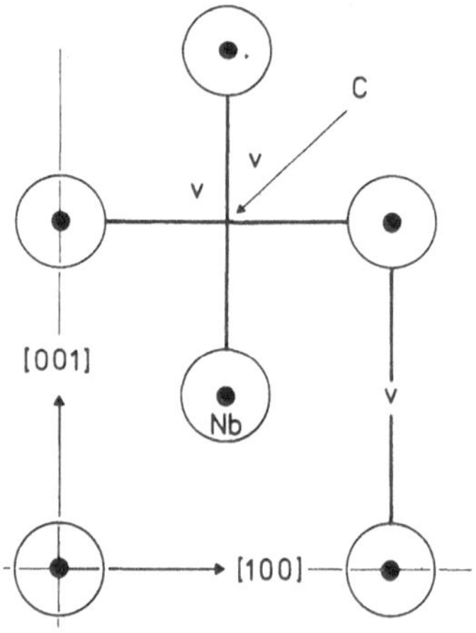


Fig. 3.26. Shell coupling in the (010)-plane of a fcc lattice (Niobium Carbide). Only the shell coupling of the Nb lattice is shown. The position of C-atoms is indicated

$$h > 0: \quad f_{eff}^{(h)} - f\delta_{h,1} = 2wL^{(h)}2w = \frac{(2w)^2(-1)^h e^{-\beta h}}{2v \sinh \beta}$$

$$\left[\cong \frac{2w}{\sqrt{\eta}}(-1)^h e^{-2\sqrt{\eta}h} \quad \text{for } \eta \ll 1\right] . \tag{3.68}$$

b) *Three-dimensional shell lattices* can be treated analogously. As an example we consider the above coupling (f,2w,v) in ⟨100⟩-directions of cubic lattices (Fig. 3.26). Because all the springs of this coupling are longitudinal and have ⟨100⟩ symmetry, their contribution to $\tilde{\Phi}(\underline{k})$ is diagonal:

$$\begin{bmatrix} D(k_x) & 0 & 0 \\ 0 & D(k_y) & 0 \\ 0 & 0 & D(k_z) \end{bmatrix} .$$

For $\underline{k} = k_x(1,0,0)$ there is no contribution to $M\Omega_t^2 = \tilde{\Phi}_{22}$ since $D(0) = 0$; to $M\Omega_l^2 = \tilde{\Phi}_{11}$ one obtains the full dip contribution $D(k_x)$. This shows that the occurence of dips depends on polarization. The t-modes in ⟨100⟩ and the t-mode in ⟨110⟩ have no dip; all other branches in the main symmetry directions exhibit the same contribution. The dips are located on the planes $k_x$, $k_y$, $k_z = \pm\pi/a$ of the reciprocal lattice. For ⟨111⟩ the dip is at the zone boundary and in the other directions inside the Brillouin-Zone (comp. Fig. 3.15a). Dips of this kind are observed in the acoustical branches of some transition metal-carbides or -nitrides, e.g., in NbC which has NaCl (fcc) structure; here the dips[26] can be described by employing only Nb-shells.

## 3.5 Green's Functions, $G^{(\underline{h})}(\omega)$

Green's functions have been discussed before for the linear oscillator (Sec. 2.3.2), for an arbitrary assembly (Sec. 2.4.2, 3) and for Bravais lattices (Sec. 3.2.2). In this section we will use the results of Chapter 2, but apply them only to Bravais lattices, where all masses are equal and where the eigenvectors of $\Phi$ and D coincide.

### 3.5.1 The Physical Meaning of Green's Functions

In Table 3.3 Green's functions are explained as responses to special force patterns. For defect physics we will need exclusively $G(\omega)$, in particular $G(0)$ for static problems. The representation of G by $\tilde{G}$ is mostly used for computational purposes;

[26] The dips are very sensitive to the parameters w,v, in particular to the smallness of $\eta$ or of $\beta$, the distance of the pole from the real axis (Fig. 3.20). Consequently, small changes, for instance by defects, can produce drastic changes in the dip. In fact one finds that already 5 at% missing C-atoms reduce the dip appreciably /3.10/. One can explain this by assuming that with each C-atom the three corresponding v-springs in Fig. 3.26 are removed.

Table 3.3. Green's functions in Bravais crystals

| Force $\underline{F}^{\underline{m}}(t)$ | Response $\underline{s}^{\underline{m}}(t)$ |
|---|---|
| $\underline{K}\delta^{\underline{m}o}\delta(t)$ | $g^{(\underline{m})}(t)\underline{K}$ , $g^{(\underline{m})}(t) = \int_{-\infty}^{\infty} \frac{d\omega}{2\pi} e^{-i\omega t} G^{(\underline{m})}(\omega) = \Theta(t)\, \frac{\sin \Omega t}{M\Omega}$ $= \frac{\Theta(t)}{M} \int_{V_B} \frac{d\underline{k}}{V_B} \sin\sqrt{\frac{\tilde{\Phi}(\underline{k})}{M}}\, t \cdot e^{i\underline{k}\underline{R}^{\underline{m}}}$ |
| $\underline{K}\delta^{\underline{m}o}e^{-i\omega t}$ | $G^{(\underline{m})}(\omega)\underline{K}e^{-i\omega t}$ , $G^{(\underline{m})}(\omega) = \int_{V_B} \frac{d\underline{k}}{V_B} e^{i\underline{k}\underline{R}^{\underline{m}}} \tilde{G}(\underline{k},\omega)$ $= \left[\frac{1}{\Phi - M(\omega + i\eta)^2}\right]^{(\underline{m})}$ |
| $\underline{K}e^{i(\underline{k}\underline{R}^{\underline{m}}-\omega t)}$ | $\tilde{G}(\underline{k},\omega)\underline{K}e^{i(\underline{k}\underline{R}^{\underline{m}}-\omega t)}$ , $\tilde{G}(\underline{k},\omega) = \frac{1}{\tilde{\Phi}(\underline{k}) - M(\omega + i\eta)^2}$ $= \sum_{\sigma=1}^{3} \vert\underline{e}(\underline{k}\sigma)\rangle \tilde{G}(\underline{k}\sigma,\omega) \langle\underline{e}(\underline{k}\sigma)\vert$ |
| $\underline{e}(\underline{k}\sigma)e^{i(\underline{k}\underline{R}^{\underline{m}}-\omega t)}$ | $\tilde{G}(\underline{k}\sigma,\omega)\underline{e}(\underline{k}\sigma)e^{i(\underline{k}\underline{R}^{\underline{m}}-\omega t)}$ , $\tilde{G}(\underline{k}\sigma,\omega) = \frac{1}{M\Omega^2(\underline{k}\sigma) - M(\omega + i\eta)^2}$ |
| $\underline{K}\delta^{\underline{m}o}$ | $G^{(\underline{m})}(\omega = 0)\underline{K}$ , $G^{(\underline{m})}(0) = \left[\frac{1}{\Phi}\right]^{(\underline{m})} = \int_{V_B} \frac{d\underline{k}}{V_B} e^{i\underline{k}\underline{R}^{\underline{m}}} \frac{1}{\tilde{\Phi}(\underline{k})}$ |

from the 3×3 matrix $\tilde{\Phi}(\underline{k})$ with eigenvectors $\underline{e}(\underline{k}\sigma)$ and eigenvalues $M\Omega^2(\underline{k}\sigma)$ one obtains

$$\tilde{G}(\underline{k},\omega) = \left[\tilde{\Phi}(\underline{k}) - M(\omega + i\eta)^2\right]^{-1} = \sum_{\sigma} \vert\underline{e}(\underline{k}\sigma)\rangle \frac{1}{M\Omega^2(\underline{k}\sigma) - M(\omega + i\eta)^2} \langle\underline{e}(\underline{k}\sigma)\vert$$

and calculates $G(\omega)$ by numerical integration over the Brillouin-Zone. It has already been pointed out that[27,28]

$$G(\omega) = G_1(\omega^2) + i \operatorname{sgn}\omega\, G_2(\omega^2) , \tag{3.69a}$$

$$G_2(\omega^2) = \frac{\pi}{M}\,\delta(\Omega^2 - \omega^2), \quad G_2^{(\underline{h})}(\omega^2) = \int_{V_B} \frac{d\underline{k}}{MV_B}\, e^{i\underline{k}\underline{R}^{\underline{h}}}\, \underbrace{\delta\left(\Omega^2(\underline{k}) - \omega^2\right)}_{\sum_\sigma |\underline{e}(\underline{k},\sigma)\rangle\, \delta\left(\Omega^2(\underline{k}\sigma) - \omega^2\right) \langle\underline{e}(\underline{k}\sigma)|} , \tag{3.69b}$$

$$G_1(\omega^2) = \frac{1}{M}\,\frac{P}{\Omega^2 - \omega^2} = \int \frac{d\omega'^2}{\pi}\,\frac{P}{\omega'^2 - \omega^2}\, G_2(\omega'^2) , \tag{3.69c}$$

where the Kramers-Kronig relations (3.69c) are most convenient for numerical purposes. The Fourier transforms $\tilde{G}(\underline{k}\sigma,\omega)$ are needed for the neutron scattering cross section which is proportional to $G_2$. The divergence of $\tilde{G}(\underline{k}\sigma,\omega = \Omega(\underline{k}\sigma))$ indicates that the lattice waves $|\underline{k}\sigma\rangle$ are force-free solutions.

### 3.5.2 A Simple One-Dimensional Example

The 1st neighbour coupling model of Section 3.5.1, where

$$M\Omega^2(k) = 2f(1 - \cos ka) , \quad M\Omega^2_{max} = 4f , \quad \Omega_{max} = \text{maximum frequency},$$

yields

$$G^{(h)}(\omega) = \frac{1}{4\pi f}\int_{-\pi}^{+\pi} dz\, \frac{e^{ihz}}{1 - 2\hat{\omega}^2 - \cos z - i\eta\,\frac{\omega}{f}} , \quad \eta \to +0 , \quad ka = z,\ \hat{\omega} = \omega/\Omega_{max}. \tag{3.70}$$

($\frac{\omega}{f}$ can be replaced by $\operatorname{sgn}\omega$, if $\eta \to +0$).

The integrand has period $2\pi$; it has poles $z_p$ close to the real axis for $0 \le \omega \le \Omega_{max}$ and at $z_p = \pi \pm i\kappa$ for $\omega \le \Omega_{max}$. The poles are shown in Fig. 3.27. The integration is performed as in Fig. 3.20; for $h \ge 0$ the path from $-\pi$ to $+\pi$ can be shifted to $+i\infty$ and only the residuum of one pole in the upper half plane remains:

---

27 The matrix $\Phi/M$ is denoted by $\Omega^2$, $\tilde{\Phi}(\underline{k})/M$ by $\Omega^2(\underline{k})$ and $\tilde{\Phi}(\underline{k}\sigma)/M$ by $\Omega^2(\underline{k}\sigma)$. Here we have omitted the tilde in $\Omega^2(\underline{k})$ and $\Omega^2(\underline{k}\sigma)$. From the context in which $\Omega^2$ is used the meaning of $\Omega^2$ will always be clear.

28 Note that $G_2^{(\underline{h})}$ is real; $\exp(i\underline{k}\underline{R}^{\underline{h}})$ can be replaced by $\cos \underline{k}\underline{R}^{\underline{h}}$ because $\Omega^2(\underline{k})$ and the integration volume are invariant under $\underline{k}$-inversion.

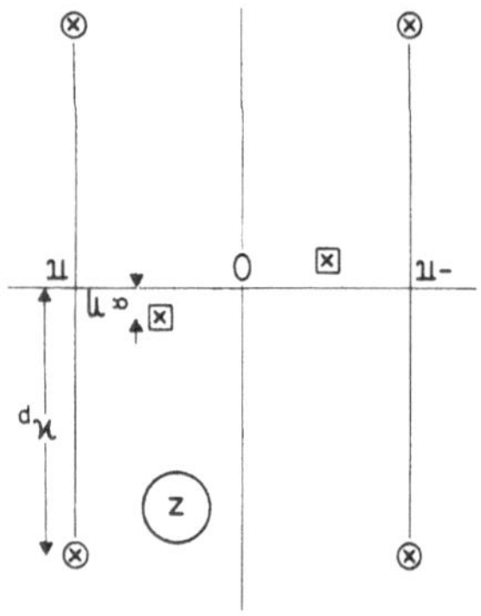


Fig. 3.27. Evaluation of $G^{(h)}(\omega)$ for the linear chain. For $0 \le \hat{\omega} \le 1$ the poles ( ⊠ ) are given by $\cos z_p = 1 - 2\hat{\omega}^2 - i\eta \,\mathrm{sgn}\, \hat{\omega}$; they are just above and below the real axis. For $\hat{\omega} \ge 1$ the poles ( ⊗ ) are given by $z_p = \pi + i\kappa_p$, $\cosh \kappa_p = 2\hat{\omega}^2 - 1$

$$G^{(h)}(\omega) = \frac{i}{2f}\frac{e^{iz_p h}}{\sin z_p} = \frac{ie^{iz_p h}}{4f\sqrt{\hat{\omega}^2(1-\hat{\omega}^2)}} \quad \text{for } 0 \le \hat{\omega} \le 1 \;; \qquad \cos z_p = 1 - 2\hat{\omega}^2 \,, \quad z_p, h \ge 0; \tag{3.71a}$$

$$G^{(h)}(\omega) = -\frac{(-1)^h e^{-\kappa_p h}}{4f\sqrt{\hat{\omega}^2(\hat{\omega}^2 - 1)}} \quad \text{for } 1 \le \hat{\omega} \;; \quad \cosh \kappa_p = 2\hat{\omega}^2 - 1 \,, \quad \kappa_p, h \ge 0; \tag{3.71b}$$

in particular,

$$G^{(o)}(\omega) = \begin{cases} \dfrac{i}{4f}\dfrac{1}{\sqrt{\hat{\omega}^2(1-\hat{\omega}^2)}} = iG_2^{(o)}(\omega^2) & \text{for } \; 0 \le \hat{\omega} \le 1 \\[2ex] -\dfrac{1}{4f}\dfrac{1}{\sqrt{\hat{\omega}^2(\hat{\omega}^2 - 1)}} = G_1^{(o)}(\omega^2) & \text{for } \; 1 \le \hat{\omega} \end{cases} \tag{3.72}$$

which is plotted in Fig. 3.28. One recognizes that in

$$g^{(h)}(t) = \int \frac{d\omega \; e^{-i\omega t}}{2\pi} G^{(h)}(\omega) \tag{3.73}$$

the integrand for $0 \le \omega \le \Omega_{max}$ is proportional to $\exp i(k_p ah - \omega t)$ ; this represents waves which move away from the center at $h = 0$, called outgoing waves. The integration over $\omega$ can be carried out with the result,

$$\dot{g}_1^{(h)}(t) = \frac{\Theta(t)}{M} J_{2h}(\Omega_{max} t) \,, \qquad g_1^{(h)}(t) = \frac{\Theta(t)}{M\Omega_{max}} \int_0^{\Omega_{max} t} J_{2h}(\alpha)\, d\alpha \,, \tag{3.73a}$$

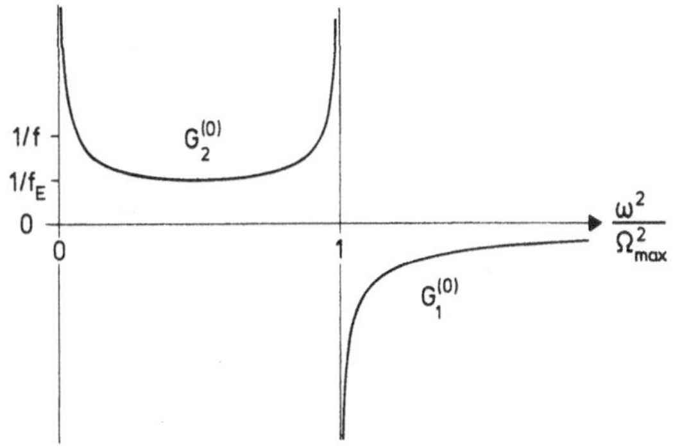


Fig. 3.28. $G^{(o)}(\omega) = G_1^{(o)}(\omega^2) + i \operatorname{sgn}\omega\, G_2^{(o)}(\omega^2)$ for the linear chain. The spectrum of the chain is $Z(\omega^2) = \frac{M}{\pi} G_2^{(o)}(\omega^2) = \frac{1}{\pi\Omega^2_{max}} \frac{1}{\sqrt{\xi(1-\xi)}}$ , $\xi = \hat{\omega}^2 = \omega^2/\Omega^2_{max}$, comp. following subsection; $f_E$ = Einstein spring (here $f_E = 2f$)

where

$$J_{2h}(\alpha) \cong \begin{cases} \sqrt{\frac{2}{\pi\alpha}} \cos[\alpha - (h + 1/4)\pi] & \text{for } \alpha \gg 2h \\ \frac{1}{(2h)!} \left(\frac{\alpha}{2}\right)^{2h} & \text{for } \alpha \ll 2h \end{cases} \tag{3.73b}$$

are Bessel functions of order 2h. The response even for large h is *immediate* (though proportional to $t^{2h}$ in the beginning) because the inertia of the springs has been neglected. However, the response moves out with *finite* velocity for all practical purposes. This is seen from (3.73b) for large h: the response at first increases with time, $(\Omega_{max} t/2)^{2h}/(2h)!$, and finally decreases, $\sqrt{2/(\pi\Omega_{max} t)}$, and must have a sharp maximum at about $\Omega_{max} t \cong 2h$, or $h = \Omega_{max} t/2 = ct/a$, where $c = a\Omega_{max}/2 = a\sqrt{f/M}$ is the sound velocity, which here decides the velocity of the response.

### 3.5.3 The Motion of One Lattice Atom Under a Force

The motion of one lattice atom, say atom 0, under a force $\underline{F}^o(t)$, acting only on atom 0, is determined by $G^{(o)}(\omega)$; the Fourier transformed equation of motion yields

$$\underline{s}^o(\omega) = G^{(o)}(\omega)\underline{F}^o(\omega) \ , \quad \underline{F}^o(t) = \int \frac{d\omega}{2\pi} \underline{F}^o(\omega) e^{-i\omega t} \ , \tag{3.74a}$$

$$\frac{1}{G^{(o)}(\omega)} \underline{s}^o(\omega) = \underline{F}^o(\omega) \ . \tag{3.74b}$$

For given $\underline{s}^o$ the motion of the other lattice atoms ($\underline{m} \neq 0$) is obtained from

$$\underline{s}^{m}(\omega) = G^{(m)}(\omega)\underline{F}^{o}(\omega) = G^{(m)}(\omega)\frac{1}{G^{(o)}(\omega)}\underline{s}^{o}(\omega). \tag{3.74c}$$

Inverting the Fourier transformation in (3.74a,b) we have

$$\underline{s}^{(o)}(t) = \int dt' \int \frac{d\omega G^{(o)}(\omega)}{2\pi} e^{-i\omega(t-t')}\underline{F}(t') = G^{(o)}(i\partial_t)\underline{F}^{o}(t) \ , \tag{3.75a}$$

$$\int dt' \int \frac{d\omega}{2\pi}\frac{1}{G^{(o)}(\omega)} e^{-i\omega(t-t')}\underline{s}(t') = \underline{F}^{o}(t) = \frac{1}{G^{(o)}(i\partial_t)}\underline{s}^{o}(t). \tag{3.75b}$$

Here $G^{(o)}$ is a 3×3 matrix in general; in cubic crystals it is scalar. In (3.75a,b) $\omega$ has been replaced formally by $i\partial_t$, because $i\partial_t e^{-i\omega t} = \omega e^{-i\omega t}$: application of $G^{(o)}(i\partial_t)$ on $\underline{F}^{o}(t)$ is identical to application of $\int_{-\infty}^{\infty} dt'\ g^{(o)}(t - t')$ on $F^{o}(t')$, i.e., $G^{(o)}(i\partial_t)\underline{F}^{o}(t) = \int_{-\infty}^{\infty} dt'\ g^{(o)}(t - t')\underline{F}^{o}(t')$. Eq. (3.75b) is then a "differential equation" for the amplitude of atom 0 under a force $\underline{F}^{o}$ with the whole lattice hanging on.

Equation (3.74b) is the equation of motion of atom 0 under forces restricted to 0. The equation of motion in time would be, $M\ddot{\underline{s}}^{o}(t) = \underline{F}^{o}(t) + \underline{F}^{e}(t)$, where the force $\underline{F}^{e}$, exerted by the environment of atom 0, must be included (for simplicity we do not denote the time dependent quantities and their Fourier transforms differently). Under the circumstances considered here, $\underline{F}^{e}(t)$ depends linearly on $\underline{s}^{o}(t')$. After Fourier transformation one obtains: $-M\omega^2\underline{s}^{o} = \underline{F}^{o} + \underline{F}^{e}$. By comparison with (3.74b) it is: $\underline{F}^{e} = -[M\omega^2 + 1/G^{(o)}(\omega)]\underline{s}^{o}$. Because $\underline{F}^{e}(t)$ can only depend on the displacement $\underline{s}^{o}$ and not on the particular mass of atom 0, $\underline{F}^{e} = \underline{F}^{e}(\underline{s}^{o})$, the equation of motion of an "isotopic defect"[29] at 0, meaning change of mass from M to $M_d = M + m$, is given by $M_d\ddot{\underline{s}}^{o} = \underline{F} + \underline{F}^{e}(\underline{s}^{o})$ or

$$-M_d\omega^2\underline{s}^{o} = \underline{F}^{o} - \left[M\omega^2 + \frac{1}{G^{(o)}(\omega)}\right]\underline{s}^{o} \ ; \quad \left[-m\omega^2 + \frac{1}{G^{(o)}(\omega)}\right]\underline{s}^{o} = \underline{F}^{o}. \tag{3.76}$$

Eq. (3.76) can be immediately obtained from a more general point of view. The masses are given by M + mP where P projects out atom 0. The general response is[30]

$$\underline{s} = \frac{1}{\Phi - M(\omega + i\eta)^2 - mP(\omega + i\eta)^2}\underline{F} = G\frac{1}{1 - Pm\omega^2 G}\underline{F} \ . \tag{3.76a}$$

If $\underline{F}$ acts only on 0, $P\underline{F} = \underline{F} = \underline{F}^{o}$, and if we ask for $\underline{s}^{o} = P\underline{s}$, we have

---

29 Comp. Section 6.3.1

30 We use here: $\frac{1}{A + B} = \frac{1}{(1 + BA^{-1})A} = \frac{1}{A}\cdot\frac{1}{1 + BA^{-1}}$. Further the factor $(\omega + i\eta)^2$ of m can be replaced by $\omega^2$ in the equation of motion (3.76).

$$\underline{s}^{\circ} = PG \frac{1}{1 - m\omega^2 PG} P\underline{F}^{\circ} = G^{(o)} \frac{1}{1 - m\omega^2 G^{(o)}} \underline{F}^{\circ} , \tag{3.76b}$$

because G is always enclosed by two P's and PGP = $G^{(o)}$. This is identical with (3.76). Consequently $G^{(o)}(\omega)$ contains complete information on the dynamics of an isotopic defect surrounded by a perfect lattice.

We will discuss in detail only cubic lattices, where $G^{(o)}$ is scalar:

$$G^{(o)}_{ik}(\omega) = \delta_{ik}\, \mathrm{tr}\left\{G^{(o)}\right\}/3 , \quad \mathrm{tr}\left\{G^{(o)}\right\}/3 = G^{(o)}_{xx} = G^{(o)}_{yy} = G^{(o)}_{zz} . \tag{3.77}$$

The imaginary part of $G^{(o)}$, $\mathrm{Im}\left\{G^{(o)}\right\} = \mathrm{sgn}\,\omega\, G^{(o)}_2(\omega^2)$, has a simple meaning,

$$\frac{1}{3}\,\mathrm{tr}\left\{G^{(o)}_2(\omega^2)\right\} = \frac{\pi}{M}\sum_{\sigma}\frac{1}{3}\int\frac{d\underline{k}}{V_B}\,\delta\left(\Omega^2(\underline{k}\sigma) - \omega^2\right) = \frac{\pi}{M} Z(\omega^2) , \quad \int_0^{\infty} d\omega^2 Z(\omega^2) = 1, \tag{3.78}$$

according to (3.69b). Evidently $\int_{\omega_1^2}^{\omega_2^2} d\omega^2 Z(\omega^2)$ is the fraction of $\Omega^2$'s in the interval $\omega_1^2 \leqslant \Omega^2(\underline{k}\sigma) \leqslant \omega_2^2$; Z is called the spectral distribution (normalized), or simply the *spectrum*. Often one uses the distribution z of $\Omega$ rather than of $\Omega^2$,

$$\omega > 0: \quad z(\omega)d\omega = Z(\omega^2)d\omega^2 , \quad z(\omega) = Z(\omega^2)2\omega , \quad \int_0^{\infty} d\omega z(\omega) = 1 , \tag{3.78a}$$

where $\int_{\omega_1}^{\omega_2} d\omega z(\omega)$ is the fraction of frequencies between $\omega_1$ and $\omega_2$.

For non cubic crystals, where $G^{(o)}$ is not scalar, quantities such as

$$G^{(o)}_{2,xx} = \frac{\pi}{M}\sum_{\sigma}\int\frac{d\underline{k}}{V_B} e_x^2(\underline{k}\sigma)\,\delta\left(\Omega^2(\underline{k}\sigma) - \omega^2\right) = \frac{\pi}{M} Z_{xx}(\omega^2) , \quad \int d\omega^2 Z_{xx} = 1 \tag{3.78b}$$

are considered, where $Z_{xx}$ is called a projected spectrum weighted by $e_x^2$. In defect crystals $G^{\underline{m}\underline{m}}$ does not equal $G^{(o)}$. As an example let us consider again one isotopic defect at 0. Then one can discuss "local" spectra such as

$$Z^{\underline{m}}(\omega^2) = M^{\underline{m}}/3\pi\, \mathrm{tr}\left\{\overset{D}{G}{}^{\underline{m}\underline{m}}_2(\omega^2)\right\} , \quad \not\Sigma ,$$

where $\overset{D}{G}$ is the defect G, including the mass. The isotopic defect destroys translational symmetry. Therefore $Z^{\underline{m}}$ depends on position; far from the defect $Z^{\underline{m}}$ approaches the ideal spectrum. Spectra like this are needed in, for instance, the Mössbauer effect, which is determined by the spectrum of the Mössbauer atom (comp. Sec. 2.4.4).

The displacement of atom 0 due to a real force,

$$\underline{F}^{(o)} = \underline{\kappa} \cos \omega t = \frac{\underline{\kappa}}{2}(e^{-i\omega t} + e^{i\omega t}), \quad \text{is} \tag{3.79}$$

$$\begin{aligned} \underline{s}^{(o)} &= \left[G^{(o)}(\omega)e^{-i\omega t} + G^{(o)}(-\omega)e^{i\omega t}\right]\frac{\underline{\kappa}}{2} \\ &= \Big[G^{(o)}_1(\omega^2)\cos\omega t + \underbrace{\mathrm{sgn}\,\omega\, G^{(o)}_2}_{\mathrm{Im}\{G^{(o)}\}}\sin\omega t\Big]\,\underline{\kappa} , \end{aligned} \tag{3.79a}$$

$$\underline{\dot{s}}^{(o)} = \omega \left[ -G_1^{(o)} \sin \omega t + \text{sgn}\, \omega\, G_2^{(o)} \cos \omega t \right] \underline{\kappa} \; . \tag{3.79b}$$

In cubic crystals the power, averaged over one period $2\pi/\omega$, is

$$\overline{(\underline{\dot{s}}^{o}, \underline{\kappa} \cos \omega t)} = \frac{\omega \,\text{sgn}\, \omega}{2} \left( \underline{\kappa}, G_2^{(o)} \underline{\kappa} \right) = \underbrace{|\omega|\, G_{2,xx}^{(o)}(\omega^2)}_{\frac{\pi}{M} |\omega|\, Z(\omega^2) = \frac{\pi}{2M} z(\omega)} \frac{(\underline{\kappa}, \underline{\kappa})}{2} \; . \tag{3.79c}$$

The energy input by the external force in one period, $\pi^2 Z(\omega^2)\kappa^2/M$, is directly proportional to the spectrum. This is a very illustrative result, because this input is transformed into outgoing lattice waves which carry the energy to infinity. The out-of-phase part of $\underline{s}^{(o)}$, which corresponds to damping and is proportional to $G_2$, is caused by the emission of energy by lattice waves. Consequently, it is not surprising that the damping is proportional to the density of available frequencies. The damping vanishes if $\omega$ exceeds the maximum lattice frequency $\Omega_{max}$.

In contrast to the infinite lattice where the interpretation of the atomic response is simple and straightforward, the same problem in finite crystals meets conceptual difficulties. For a finite crystal, say with vanishing boundary displacements, the procedure is as above and results in $\underline{s}^{\underline{m}} = G^{\underline{m}\underline{m}} \underline{F}^{\underline{m}}$, $\not{\Sigma}$, for atom $\underline{m}$. In the bulk one expects independence of $\underline{m}$ and arrives at (3.74a). The spectrum is no longer smooth, as in the infinite lattice, but consists of a sequence of $\delta$-functions at the discrete eigenfrequencies $\overset{\nu}{\Omega}$ of the finite system.

Consequently, if $\omega$ lies between the eigenfrequencies the response is finite and without damping. If $\omega = \overset{\nu}{\Omega}$ the response diverges in analogy to the linear oscillator[31]. For macroscopic crystals the distance $\Delta\Omega$ between two subsequent peaks is of the order $\Delta\Omega \cong c/L$, where c is a typical sound velocity and L is a typical macroscopic length. The maximum frequency $\Omega_{max}$ is of order c/a (a = lattice constant), and $\Delta\Omega/\Omega_{max} = a/L \ll 1$. In fact, forces with sharp frequencies are as ficticious as is the infinite crystal. Rather, one has forces with a width $\Delta\omega$ in $\underline{F}(\omega)$, and one must realize that $\Delta\omega$ is usually very large as compared with $\Delta\Omega$, i.e., many frequencies contribute to the damping. If $z_d$ is the discrete spectrum, the damping is not given by $z_d(\omega)$, but rather by the integrated quantity $1/\Delta\omega \int_{\omega}^{\omega+\Delta\omega} d\omega' z_d(\omega')$. This is a smooth function of $\omega$, well represented by $z(\omega)$.

In the infinite lattice the energy is radiated to and lost in infinity. In a finite crystal (again with vanishing boundary displacements) the outgoing waves are reflected in a complicated way from the surface, and there is no actual energy loss.

31 The divergence there is as follows. If $\omega = \Omega$ and $F = \Theta(t) \sin \Omega t$, one obtains $s(t) = \frac{1}{2M\Omega}\left(\frac{\sin \Omega t}{\Omega} - t \cos \Omega t\right)$ which diverges for $t \to \infty$.

The reflected displacements will be out of phase with the acting force such that no net energy change of atom 0 results; instead, the rest of the crystal will heat up. The situation is not stationary in principle, but it will take such a long time, until deviations from the simple picture of the atom surrounded by an infinite lattice become important, that it is of no practical consequence.

In the following figures we will give some typical examples for cubic crystals. Fig. 3.29 shows numerical results for the three models used for the dispersion curves of Fig. 3.15. The spectra $Z(\lambda)$ must all start proportional to $\sqrt{\gamma} = \sqrt{\omega^2}$. This can easily be seen from (3.78) because for small $\omega^2$, k must also be small, and then, $\Omega^2(\underline{k}) \propto k^2$, $Z(\omega^2) \propto \int k^2 dk\ \delta(k^2 - \omega^2) \propto \sqrt{\omega^2}$. In three dimensions, as a rule, one has singularities of the type $\sqrt{\lambda - \lambda_c}, \sqrt{\lambda_c - \lambda}$ near critical values $\lambda_c$ (comp. App. G), in particular $\sqrt{\lambda}$ for $\lambda_c = 0$ and $\sqrt{\Omega^2_{max} - \lambda}$ for $\lambda_c = \Omega^2_{max}$. For $\lambda$ near 0 the $\sqrt{\ldots}$-behaviour is seen in all the spectra; for $\lambda \lesssim \Omega^2_{max}$ the more physical spectra of Fig. 3.29a,b show the $\sqrt{\ldots}$-behaviour. The spectrum of Fig. 3.29c diverges logarithmically at $\Omega^2_{max}$; this, however, is an artifact and does not commonly occur. Another uncommon singularity in $G_2^{(o)}$ is the step in Fig. 3.29b at $\omega^2 = \Omega^2_{max}/2$, which in turn yields a logarithmic singularity in $G_1^{(o)}$. The reasons for these singulari-

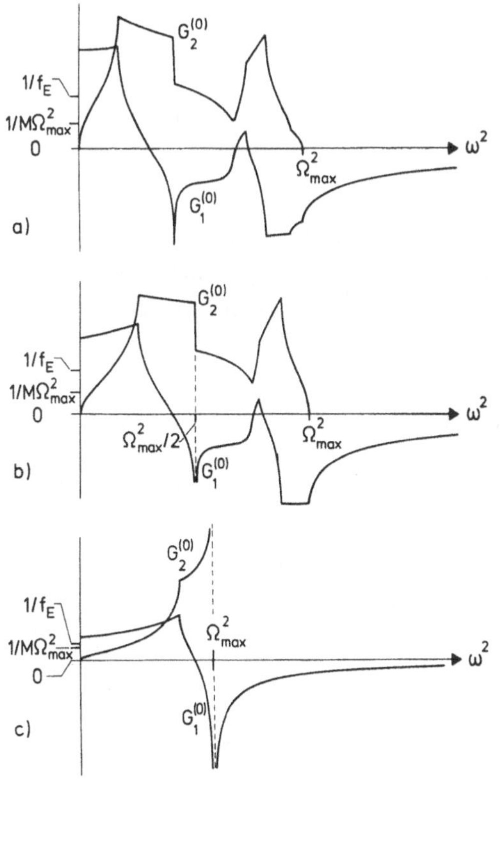


Fig. 3.29a-c. $G^{(o)}(\omega) = G_1^{(o)}(\omega^2) + i \operatorname{sgn} \omega\, G_2^{(o)}(\omega^2)$ for the three models of Fig. 3.15; $Z(\omega^2) = MG_2^{(o)}/\pi$, $f_E$ = Einstein spring.
a) First neighbour Born-von Karman model for Cu (Fig. 3.15b): $f_l$, $f_{t'}/f_l = -0.078$, $f_t/f_l = -0.049$; $\Omega^2_{max} = (8f_l + 4f_t)/M = 2.41 \cdot 10^{27} s^{-2}$.
b) First neighbour longitudinal spring $f = f_l$ (Fig. 3.15c), no transversal springs, $\Omega^2_{max} = 8f_l/M = 2.51 \cdot 10^{27} s^{-2}$.
c) First neighbour isotropic coupling, $f = (f_l + f_{t'} + f_t)/3$ (Fig. 3.15d), all springs equal; $\Omega^2_{max} = 16(f_l + f_{t'} + f_t)/3M = 1.46 \cdot 10^{27} s^{-2}$

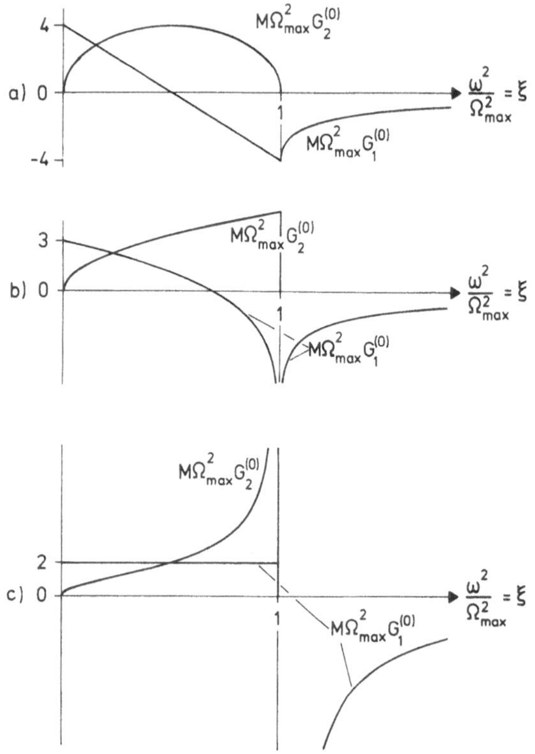


Fig. 3.30a-c. $G^{(o)}(\omega)$ for the model spectra of Table 3.4.
a) Regular spectrum,
b) Debye spectrum,
c) Singular spectrum

ties are discussed in App. G. The real parts, $G_1^{(o)}$, are also shown. They are calculated from $G_2^{(o)}$ via the Kramers-Kronig-relations (3.69c).

Fig. 3.30 shows $G^{(o)}$ for three model spectra for which the integral (3.69c) can easily be evaluated (Table 3.4); in particular one can see the singularities in $G_1^{(o)}$ induced by the singularities of $G_2^{(o)}$. The "regular spectrum" (Fig. 3.30a) has the correct $\sqrt{\ldots}$-behaviour at both ends and $G^{(o)}$ resembles Fig. 3.29a,b; in particular it has finite $G_1^{(o)}(\lambda_{max})$. The "Debye spectrum" is often used as an approximate spectrum; it merely extrapolates the starting $\sqrt{\ldots}$-behaviour of a given spectrum up to a cutoff determined by normalization (App. G). The result for $G_1^{(o)}$ is reasonable, except near $\lambda_{max}$, where it diverges logarithmically, which is caused by the step in the Debye-spectrum. The "singular spectrum" behaves correctly for small $\lambda$ but has a $1/\sqrt{\ldots}$-divergence at $\lambda_{max}$. Here[32] $G_1^{(o)}$ is curiously constant for $\lambda \leq \lambda_{max}$ and exhibits a $1/\sqrt{\ldots}$-divergence for $\lambda \geq \lambda_{max}$.

[32] Spectra with divergencies have to be treated with caution with respect to approximations. If one would approximate the singular spectrum by Debye spectrum, the approximate results would be very misleading. This would be even worse for the simple one-dimensional spectrum (Fig. 3.28) with two divergencies.

Table 3.4. Model spectra, $G_2^{(o)}$, and the corresponding $G_1^{(o)}$ obtained via the Kramers-Kronig-relations (cubic crystals, $\lambda = \omega^2$)

$$MG^{(o)} = \pi Z(\lambda) \ , \quad MG_1^{(o)}(\lambda) = \int_0^\infty d\lambda' \frac{P}{\lambda' - \lambda} Z(\lambda') = \int_0^\infty d\xi' \frac{P}{\xi' - \xi} \tilde{Z}(\xi')$$

$$\xi = \lambda/\lambda_{max} \ , \quad \lambda_{max} Z(\lambda) = \tilde{Z}(\xi = \lambda/\lambda_{max}) \ .$$

The integrals can be evaluated directly as real integrals by using the definition of Cauchy's principal value or by complex integration (App. C.4). Averages: $<\lambda/\lambda_{max}> = <\xi> = \int d\xi \ \xi\tilde{Z}(\xi)$ etc.

| $\lambda_{max}Z(\lambda) = \tilde{Z}(\xi)$ | $<\xi>$ | $<\frac{1}{\xi}>$ | $M\lambda_{max}G_1^{(o)}$ | Expansions |
|---|---|---|---|---|
| regular spectrum $\frac{8}{\pi}\sqrt{\xi(1-\xi)}$ | $\frac{1}{2}$ | 4 | $\xi \leqslant 1$: $4 - 8\xi$<br>$\xi \geqslant 1$: $4 - 8\xi + 8\sqrt{\xi(\xi - 1)}$ | $4 - 8\xi \quad \xi \ll 1$<br>$-\frac{1}{\xi} - \frac{3}{5\xi^2} \quad \xi \gg 1$ |
| Debye spectrum $\frac{3\sqrt{\xi}}{2}\Theta(1 - \xi)$ | $\frac{3}{5}$ | 3 | $\xi \leqslant 1$: $3 - \frac{3\sqrt{\xi}}{2}\ln\frac{1+\sqrt{\xi}}{1-\sqrt{\xi}}$<br>$\xi \geqslant 1$: $3 - \frac{3\sqrt{\xi}}{2}\ln\frac{\sqrt{\xi}+1}{\sqrt{\xi}-1}$ | $3 - 3\xi \quad \xi \ll 1$<br>$-\frac{1}{\xi} - \frac{3}{4\xi^2} \quad \xi \gg 1$ |
| singular spectrum $\frac{2}{\pi}\sqrt{\frac{\xi}{1 - \xi}}$ | $\frac{3}{4}$ | 2 | $\xi \leqslant 1$: $2$<br>$\xi \geqslant 1$: $2 - 2\sqrt{\frac{\xi}{\xi - 1}}$ | $2 \quad \xi \ll 1$<br>$-\frac{1}{\xi} - \frac{1}{2\xi^2} \quad \xi \gg 1$ |

Actually the value $G_1^{(o)}(\lambda_{max})$ determines the minimum mass change m required for a "localized" vibration of an isotopic defect. Eq. (3.76) tells us how to obtain the "eigenfrequency" of an undamped ($\lambda \geqslant \lambda_{max}$, $G_2^{(o)} = 0$), stationary, force-free oscillation of the isotopic defect[33]. The condition is clearly

$$-m\lambda + \frac{1}{G^{(o)}(\lambda)} = 0 \ , \quad \lambda \geqslant \lambda_{max} \ , \quad \text{or } m\lambda G_1^{(o)}(\lambda) = 1 \ , \tag{3.80}$$

33 After having determined the eigenfrequency, one obtains the amplitudes of the surrounding lattice from (3.74c). Because $\lambda > \lambda_{max}$, the amplitudes decrease exponentially with distance from the isotope (discussed in the next section). The displacements are confined to the vicinity of the defect; hence the name "localized".

which can only be solved for $m = -\bar{m} < 0$, i.e., if the mass of the isotope $(M - \bar{m})$ is smaller than the ideal mass (M), because $G^{(o)}(\lambda \geq \lambda_{max}) = G_1^{(o)}(\lambda) < 0$:

$$-\bar{m}\lambda G^{(o)}(\lambda) = \frac{\bar{m}}{M}\lambda \int_0^{\lambda_{max}} \frac{Z(\lambda')d\lambda'}{\lambda - \lambda'} = 1 . \tag{3.80a}$$

A localized vibration exists, if m is larger than a critical $m_c$, which is determined by $G_1^{(o)}(\lambda_{max})$ (Fig. 3.31). For the spectra of Fig. 3.29a,b the critical mass is finite; for the spectrum of Fig. 3.29c or for a Debye spectrum $\bar{m}_c$ vanishes, and an

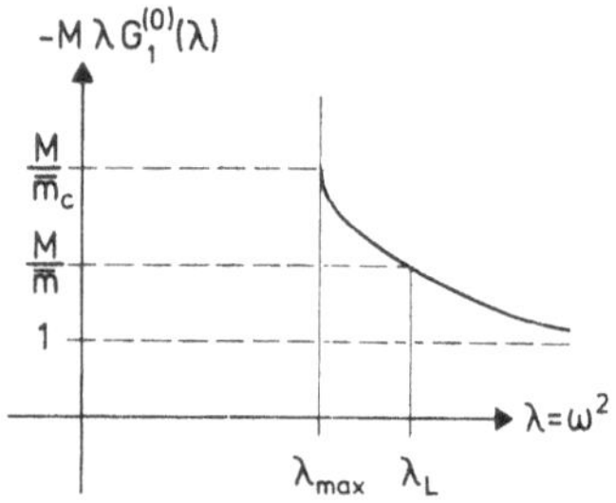


Fig. 3.31. How to determine the frequency, $\omega_L$, of a localized vibration. (3.80a): $\frac{M}{\bar{m}} = \lambda \int_0^{\lambda max} \frac{Z(\lambda)d\lambda}{\lambda - \lambda'} = - M\lambda G_1^{(o)}(\lambda)$, the right-hand side decreases monotonically. One obtains a localized $\lambda_L = \omega_L^2$ whenever $\bar{m}$ is larger than a critical $\bar{m}_c$, $\bar{m}_c/M = -\left\{M\lambda_{max}G_1^{(o)}(\lambda_{max})\right\}^{-1}$. For the regular spectrum of Table 3.4, $\bar{m}_c/M = 1/4$. For $\bar{m} \gtrsim \bar{m}_c$ the localized frequency is close to $\Omega_{max}$; for $\bar{m} \lesssim M$ $(M + m \ll M$, small mass of the isotopic defect) the localized frequency is very large, $\omega_L \gg \Omega_{max}$, $\lambda_L \gg \lambda_{max}$

only slightly smaller isotopic mass than M would produce a localized frequency. Clearly a Debye approximation to a reasonable spectrum, e.g., Fig. 3.29a,b, is completely unreliable with respect to an approximate $\bar{m}_c$.

### 3.5.4 Expansions for High and Low Frequencies

a) *The expansion of* $G^{(o)}(\omega)$ *for high frequencies* ($\omega \gg \Omega_{max}$, $\mathrm{Im}\left\{G^{(o)}\right\} = 0$) is simple (in cubic lattices):

$$\begin{aligned} G^{(o)}(\omega) = G_1^{(o)}(\omega^2 = \lambda) &= \frac{1}{M}\int \frac{d\lambda' Z(\lambda')}{\lambda' - \lambda} = -\frac{1}{M\lambda}\int d\lambda' Z(\lambda') \sum_{\mu=0}^{\infty} \frac{(\lambda')^\mu}{\lambda^\mu} \\ &= -\frac{1}{M\lambda}\sum_{\mu=0}^{\infty} \frac{\lambda^{(\mu)}}{\lambda^\mu} , \end{aligned} \tag{3.81}$$

where

$$\lambda^{(\mu)} = \int d\lambda Z(\lambda)\lambda^\mu = \langle\lambda^\mu\rangle = \langle\omega^{2\mu}\rangle \tag{3.81a}$$

are the "moments" of the distribution Z. The averages, $<\ >$, taken over the spectrum Z, can be rewritten as averages over the Brillouin-Zone and over polarization, for instance

$$\langle\lambda\rangle = \underbrace{\int \lambda\, Z(\lambda) d\lambda}_{\frac{1}{3V_B}\sum_\sigma \int_{V_B} d\underline{k}\, \delta\left(\Omega^2(\underline{k}\sigma) - \lambda\right)} = \frac{1}{3V_B} \int_{V_B} d\underline{k} \underbrace{\sum_\sigma \Omega^2(\underline{k}\sigma)}_{\frac{\mathrm{tr}\,\Phi(\underline{k})}{M}} = \mathrm{tr}\left\{\int_{V_B} d\underline{k} \sum_{\underline{h}} \Phi^{(\underline{h})} e^{-i\underline{k}\underline{R}^{\underline{h}}}\right\}/3MV_B = \mathrm{tr}\left\{\Phi^{(o)}\right\}/3M = \Omega_E^2\ , \tag{3.82}$$

where $\Omega_E$, the Einstein frequency, corresponds to the coupling with fixed neighbours (Sec. 3.2.4).

For very high $\omega$, or very fast motion, we need only consider the first two terms,

$$G^{(o)} \cong -\frac{1}{M\omega^2}\left(1 + \frac{\Omega_E^2}{\omega^2}\right)\ , \quad \frac{1}{G^{(o)}} \cong -M\omega^2 + M\Omega_E^2 = M\delta_t^2 + M\Omega_E^2 \quad \text{(comp. (3.75a,b))},$$

which is equivalent to the equation of motion

$$M(\ddot{\underline{s}}^o + \Omega_E^2 \underline{s}^o) = \underline{F}^o(t)\ . \tag{3.83}$$

For high frequencies the atom behaves like an Einstein-oscillator, which means that the amplitudes, $G^{(\underline{h}\neq o)}$, of the surrounding atoms must be negligbly small[34].

b) *The expansion of* $G^{(\underline{h})}(\omega)$ *for large* $\omega$ shows this directly. We have

$$G = \frac{1}{\Phi - M\omega^2} = -\frac{1}{M\omega^2}\sum_{\mu=o}^{\infty}\left(\frac{\Phi}{M\omega^2}\right)^\mu\ , \tag{3.84}$$

$$G^{(\underline{h})}(\omega) = -\frac{1}{M\omega^2}\left[\delta^{\underline{h}o} + \frac{\Phi^{(\underline{h})}}{M\omega^2} + \frac{[\Phi^2]^{(\underline{h})}}{M^2\omega^2} + \dots\right]\ , \tag{3.84a}$$

which shows that the amplitudes for $\underline{h} \neq 0$ start at least with $\omega^{-4}$, whereas $G^{(o)} \propto \omega^{-2}$. One recognizes from (3.84a) that the high frequency expansion of $G^{(\underline{h}\neq o)}$ starts with $\omega^{-4}$ for those atoms which are in the range of $\Phi^{(\underline{h})}$. Roughly speaking, the expansion starts with $\omega^{-6}$ for atoms within twice the range, and so forth. Consequently, $G^{(\underline{h})}(\omega^2)$ for large $\underline{h}$ must decrease more strongly than any power of $\omega^{-2}$. This hints to an exponential decrease, which is demonstrated in (3.71b) for the one-dimensional Green's function.

Equation (3.84) is for $\underline{h} = 0$ more general than (3.81) which is only valid for

---

[34] For an isotopic defect with a small mass $M - \bar{m} \ll M$ one obtains from (3.80), $(M\ddot{\underline{s}}^o \to (M - \bar{m})\ddot{\underline{s}}^o)$, for the localized frequency $\omega_L$: $\omega_L^2 = \frac{M}{M - \bar{m}}\Omega_E^2 \gg \Omega_E^2, \Omega_{max}^2$.

cubic crystals; instead of $\Omega_E^2$ in (3.83) we would have $\Phi^{(o)}/M$ which is isotropic only in cubic crystals. In general, (3.81) can be written as $\mathrm{tr}\left\{G^{(o)}(\lambda)\right\}/3 = (1/M)\cdot\int d\lambda'\, Z(\lambda')/(\lambda' - \lambda)$, and therefore we obtain by comparison

$$\lambda^{(\mu)} = \frac{1}{3}\,\mathrm{tr}\left\{\left[(\Phi/M)^{\mu}\right]^{(o)}\right\} = \frac{1}{3}\left[(\Phi/M)^{\mu}\right]^{oo}_{ii} ,$$

$$\lambda^{(1)} = \lambda_E = \langle\Omega^2\rangle = \langle\Omega_E^2\rangle = \frac{\Phi^{(o)}_{ii}}{3M}, \quad \lambda^{(2)} = \langle\Omega^4\rangle = \frac{\Phi^{(\underline{m})}_{il}\Phi^{(\underline{m})}_{li}}{3M^2}. \qquad (3.85)$$

These equations can also be verified for the linear chain (Sec. 3.3.1).

The low order moments are easily calculated, once the coupling is known. A simple example is the 1st neighbour longitudinal coupling in a fcc lattice (Fig. 3.15c), where

$$\Phi^{(o)}_{il} = 4f\,\delta_{il} , \quad \Phi^{(\underline{h})}_{il} = -f\hat{X}^{\underline{h}}_i\hat{X}^{\underline{h}}_l \quad \text{for the 12 first neighbours,}$$

$$\lambda^{(1)} = \langle\lambda\rangle = \lambda_E = \frac{4f}{M} = \frac{\lambda_{max}}{2} ; \quad \lambda^{(2)} = \langle\lambda^2\rangle = \frac{5}{4}\langle\lambda\rangle^2 . \qquad (3.86)$$

These moments give only integral information about the spectral distribution. They can be used to construct an approximate spectrum[35]. The average $\langle\lambda\rangle$ gives a very rough idea about the location of the spectrum, and $(\langle\lambda^2\rangle - \langle\lambda\rangle^2)^{1/2}$ about its extension such that $\lambda_{max} \cong \langle\lambda\rangle + (\langle\lambda^2\rangle - \langle\lambda\rangle^2)^{1/2}$. The low order moments do not give much detailed information about the spectrum, in particular none about the singular behaviour at the lower and upper limit. They can only be used to implement fitting data in approximate spectra in which the singularities are already contained.

The most general (local) spectra are, $\not{X}$ ,

$$Z^m(\omega^2) = \left[\delta(\Omega^2 - \omega^2)\right]^{mm} = \mathrm{Im}\left[\frac{\sqrt{M}\,G(\omega)\,\sqrt{M}}{\pi\ \mathrm{sgn}\ \omega}\right]^{mm} = \frac{M^m}{\pi\ \mathrm{sgn}\ \omega}\,\mathrm{Im}\left\{G^{mm}\right\},$$

$$\int d\omega^2\, Z^m(\omega^2) = \left[1\right]^{mm} = 1 ;$$

here $\Omega^2 = M^{-1/2}\Phi M^{-1/2}$ is the dynamical matrix, $M = (M^m\delta^{mn})$ is the mass matrix and the index m contains 1 atomic coordinate and 3 spatial directions. According to (2.33):

$$M^m \langle[s^m(0)]^2\rangle_{th} = \int d\omega^2\, Z^m(\omega^2)\,\varepsilon_{th}(\omega)/\omega^2 .$$

The low order moments of the local spectra become

$$\langle\omega^{2\mu}\rangle_m = \int d\omega^2\, Z^m\omega^{2\mu} = [\Omega^{2\mu}]^{mm} = \left[\left(\frac{1}{\sqrt{M}}\Phi\frac{1}{\sqrt{M}}\right)^{\mu}\right]^{mm} ,$$

---

35 From all moments one can construct the exact Z (/2.3/).

e.g. $\langle\omega^2\rangle_m = \dfrac{\Phi^{mm}}{M^m}$ , $\langle\omega^4\rangle_m = \dfrac{1}{M^m}\sum_n \Phi^{mn}\dfrac{1}{M^n}\Phi^{nm}$ ,

which for Bravais lattices pass into (3.85); for instance $\langle\omega^4\rangle_{m=o,x} = \sum_{\underline{n},l}\Phi^{o\underline{n}}_{xl}\Phi^{\underline{n}o}_{lx}/M^2$. For the local spectrum of the isotopic defect at 0, mass $M_d$, one obtains

$$\langle\omega^2\rangle = \frac{\Phi^{oo}_{xx}}{M_d} = \frac{M}{M_d}\Omega_E^2 \; , \qquad \langle\omega^4\rangle = \frac{1}{M_d^2}\sum_l (\Phi^{oo}_{xl})^2 + \frac{1}{M_d M}\sum_{\underline{n}\neq o}\sum_l (\Phi^{o\underline{n}}_{xl})^2 \; ,$$

and for the coupling of Fig. 3.15c,

$$\langle\omega^4\rangle = \frac{M^2}{M_d^2}\Omega_E^4 + \frac{M^2}{M_d M}\frac{\Omega_E^4}{4} = \left(\frac{M^2}{M_d^2} + \frac{M}{4M_d}\right)\Omega_E^4 \; ;$$

$\Omega_E\sqrt{M/M_d}$ is a first guess for the resonant or localized frequency.

c) *The behaviour of $G^{(o)}(\omega)$ for small $\omega$, $\omega^2 = \lambda \ll \lambda_{max}$* (cubic crystals), is different for Im $\{G^{(o)}\}$ than for Re $\{G^{(o)}\}$. It has been mentioned that the expansion of $G_2^{(o)}(\lambda) = \pi Z(\lambda)/M$ for small $\lambda$ begins with a term proportional to $\sqrt{\lambda}$. Small $\lambda$ also mean small k, $\Omega^2(\underline{k}\sigma) \cong c_\sigma^2(\hat{\underline{k}})k^2$. In this regime: $Z(\lambda) \cong a\sqrt{\lambda}$ [36], where a is determined by elastic continuum theory (elastic data). From (3.78) we have

$$Z(\lambda) = a\sqrt{\lambda} = \frac{1}{3V_B}\sum_\sigma \int \underbrace{k^2 dk}_{kdk^2/2} \, d\Omega_{\underline{k}} \; \delta\left(c_\sigma^2(\hat{\underline{k}})k^2 - \lambda\right) = \frac{1}{3V_B}\sum_\sigma \int d\Omega_{\underline{k}} \frac{\sqrt{\lambda}}{2c_\sigma^3}$$
$$= \frac{2\pi}{3V_B}\sum_\sigma \left\langle \frac{1}{c_\sigma^3(\hat{\underline{k}})}\right\rangle^{\Omega_{\underline{k}}} \sqrt{\lambda} \; , \qquad (3.87)$$

where the average is taken over the solid angle $\Omega_{\underline{k}}$ in $\underline{k}$ space ($\langle\ldots\rangle = (1/4\pi)\int d\Omega_{\underline{k}}\ldots$). For elastic isotropy one has one longitudinal and two equal transversal sound velocities, independent of $\hat{\underline{k}}$:

$$Z(\lambda) = a\sqrt{\lambda} \; , \quad a = \frac{2\pi}{3V_B}\left(\frac{1}{c_l^3} + \frac{2}{c_t^3}\right) = \frac{V_c}{12\pi^2}\left(\frac{1}{c_l^3} + \frac{2}{c_t^3}\right) \; , \quad z(\omega) = 2a\omega^2 \; . \qquad (3.87a)$$

Details and connections with the elastic data will be treated in Chapter 4. Further expansion, $\Omega^2(\underline{k}\sigma) = \ldots k^2 + \ldots k^4 + \ldots k^6 + \ldots$, will result in

---

[36] The kind of singularity at $\lambda = 0$ depends on dimension d:
d = 3: $Z(\lambda) \sim \int k^2 dk\, \delta(\lambda - c^2k^2) = \sqrt{\lambda}/(2c^{3/2})$;
d = 2: $Z(\lambda) \sim \int k\, dk\, \delta(\lambda - c^2k^2) = 1/(2c^2)$, step;
d = 1: $Z(\lambda) \sim \int dk\, \delta(\lambda - c^2k^2) = 1/(2c\sqrt{\lambda})$.

$$Z(\lambda) = \sqrt{\lambda}\,(a + b\lambda + c\lambda^2 + \dots) \tag{3.88}$$

such that $Z(\lambda)/\sqrt{\lambda}$ can be expanded in a Taylor series for small $\lambda$. Except for isotropic or nearly isotropic crystals the coefficients a,b,... have to be calculated numerically.

The real part $G_1^{(o)}(\lambda) = (1/M)\int d\lambda' Z(\lambda') \frac{P}{\lambda' - \lambda}$ is obtained from Z via the Kramers-Kronig-relations; for small $\lambda$ it will have an expansion (comp. Table 3.4)

$$G_1^{(o)}(\lambda) = \alpha + \beta\lambda + \gamma\lambda^2 + \dots = G_1^{(o)}(0) + \lambda\cdot\partial_\lambda G_1^{(o)}(\lambda)\Big|_{\lambda \to +0} + \dots \; . \tag{3.89}$$

In the value for $\lambda = 0$,

$$\alpha = \frac{1}{M}\int \frac{d\lambda' Z(\lambda')}{\lambda'} \; . \tag{3.89a}$$

Cauchy's principal value, P in (3.69c), need not be taken because the integral does not diverge (in *three* dimensions). But in the definition of $\beta$ by the derivative in (3.89) the limit $\eta \to +0$ (for P) and the limit $\lambda \to +0$ must be taken in this order. However, as is shown in Appendix H, for spectra without divergencies one has

$$\beta = 2\int \frac{d\lambda}{\sqrt{\lambda}}\left[\partial_\lambda \frac{Z(\lambda)}{\sqrt{\lambda}} = b + c\lambda + \dots\right] . \tag{3.89b}$$

One ought to realize that $Z(\lambda)$ for small $\lambda$ is completely determined by the behaviour of $\Omega(\underline{k}\sigma)$ at *small* $\underline{k}$, whereas to $G_1^{(o)}(\lambda)$ for small $\lambda$ *all* $\underline{k}$-values in $V_B$ contribute appreciably. The static response is given by $G^{(o)}(0) = G_1^{(o)}(0) = \alpha$; therefore $1/\alpha$ is an effective spring, $f_e$, with which the surrounding lattice binds one atom to its lattice site; because of relaxation, $f_e$ must be smaller than the Einstein spring $f_E$.

d) *The slow motion of an isotopic defect* can now be treated easily. If in (3.76), $\left[-m\omega^2 + 1/G^{(o)}(\omega)\right]\underline{s}^o = \underline{F}^o$, one employs the low-$\omega$ expansion of $G^{(o)}$,

$$G^{(o)}(\omega) = \alpha + i\underbrace{\frac{\pi a}{M}\sqrt{\omega^2}\,\mathrm{sgn}\,\omega}_{\omega} + \beta\omega^2 \; , \tag{3.90}$$

$$\frac{1}{G^{(o)}(\omega)} = \underbrace{\frac{1}{\alpha}}_{f_e} - i\omega\underbrace{\frac{\pi a}{M\alpha^2}}_{2\eta_e M_e} - \omega^2\underbrace{\left(\frac{\pi^2 a^2}{M^2\alpha^3} + \frac{\beta}{\alpha^2}\right)}_{M_e} = M_e\left(\partial_t^2 + 2\eta_e\partial_t + \underbrace{\Omega_e^2}_{f_e/M_e}\right) , \tag{3.90a}$$

one obtains the equation of slow motion of an ideal lattice atom,

$$M_e\left(\partial_t^2 + 2\eta_e\partial_t + \Omega_e^2\right)\underline{s}^o = \underline{F}^o \; , \tag{3.91}$$

which is that of an harmonic oscillator, (2.15), with a velocity-proportional damping caused by the emission of lattice waves. The force-free solutions are with

$\Omega_\pm = -i\eta_e \pm \sqrt{\Omega_e^2 - \eta_e^2}$ damped oscillations $\propto \exp(-i\Omega_\pm t) = \exp\left[(-\eta_e \pm i\sqrt{\Omega_e^2 - \eta_e^2})t\right]$.
To get an idea about the order of magnitude we employ the first spectrum of Table 3.4, where with

$$Z(\lambda) = \frac{8}{\pi\lambda_{max}}\left[\frac{\lambda}{\lambda_{max}}(1 - \frac{\lambda}{\lambda_{max}})\right]^{1/2} \quad ; \quad \alpha = \frac{4}{M\lambda_{max}}; \quad \beta = -\frac{8}{M\lambda_{max}^2};$$

$$f_E = M\langle\lambda\rangle = \frac{M\lambda_{max}}{2} = M\Omega_E^2 , \tag{3.91a}$$

we have

$$f_e = \frac{f_E}{2}, \quad \Omega_e = \frac{\Omega_{max}}{\sqrt{2}}, \quad M_e = \frac{M}{2}; \quad \eta_e = \frac{\Omega_{max}}{2}; \quad \Omega_\pm = -i\frac{\Omega_{max}}{2} \pm \frac{\Omega_{max}}{\sqrt{2}}. \tag{3.91b}$$

These results[37] represent a strongly damped motion; the oscillation with frequency $\Omega_{max}/\sqrt{2}$ decreases by a factor of 100 after one period, $\exp\left[-2\pi\eta_e/(\Omega_{max}/\sqrt{2})\right] \cong 10^{-2}$. For the motions of one lattice atom the characteristic frequencies, $\Omega_e$ and $\eta_e$, are rather high, and the expansions (3.90) are certainly not good approximations.

However, for a heavy isotopic defect ($m > M$) the characteristic frequencies are lowered appreciably such that the low-$\omega$ expansion becomes valid. Instead of (3.91) we have

$$\left[(M_e + m)\partial_t^2 + 2M_e\eta_e\partial_t + M_e\Omega_e^2\right]\underline{s}^o = M_d^e(\partial_t^2 + 2\eta_d\partial_t + \Omega_d^2)\underline{s}^o = \underline{F}^o \tag{3.92}$$

with

$$M_d^e = M_d + M_e - M = M_e + m , \quad \eta_d = \frac{M_e}{M_d^e}\eta_e , \quad \Omega_d^2 = \frac{M_e}{M_d^e}\Omega_e^2 ,$$

and the force-free solutions are proportional to $\exp(-i\Omega_\pm t)$ with $\Omega_\pm = -i\eta_d \pm \sqrt{\Omega_d^2 - \eta_d^2}$. For a very heavy isotopic defect ($m \gg M$, $\sqrt{M_e/M_d} = \varepsilon \ll 1$), one has $\eta_d = \varepsilon\eta_e$, $\Omega_d = \sqrt{\varepsilon}\Omega_e$, $\Omega_\pm = \varepsilon\eta_e \pm i\sqrt{\varepsilon}\Omega_e$, and the damping factor, $\exp(-\sqrt{\varepsilon}\, 2\pi\eta_e/\Omega_e)$, is small. The oscillation is only weakly damped and needs almost no force to be kept stationary. It resembles the localized mode because the (kinetic) energy is concentrated in the defect atom. Such a mode, which is almost stationary in the above sense, is called a resonant mode. Soundwaves with the frequency of the resonant modes are scattered most strongly; these resonances behave like the resonances in quantum theory, where they represent slowly decaying "almost bound" states. The heavy isotope also illustrates nicely the local spectrum which is proportional to the imaginary part of the isotope's response

$$Z_d \propto \mathrm{Im}\left\{\frac{1}{\Omega_d^2 - 2i\omega\eta_d - \omega^2}\right\} = \frac{2\omega\eta_d}{(\Omega_d^2 - \omega^2)^2 + 4\eta_d^2\omega^2} \cong \frac{2\Omega_d\eta_d}{(\Omega_d^2 - \omega^2)^2 + 4\eta_d^2\Omega_d^2}$$

---

[37] The Debye spectrum yields values quite similar to those of (3.91b). In fact, the features discussed here should be quite realistic in general.

(for $\eta_d \ll \Omega_d$ the $\omega$-factors of $\eta_d$ can be replaced by $\Omega_d$); therefore, $Z_d(\lambda) \propto 1/\left[(\lambda_d - \lambda)^2 + 4\eta_d^2\lambda_d\right]$ is a Lorentz curve about $\lambda_d$ with width $2\eta_d\Omega_d$. Because the relative width is small, $2\eta_d\Omega_d/\lambda_d = 2\eta_d/\Omega_d \ll 1$, $Z_d$ is a sharply peaked spectrum "similar" to that of a simple harmonic oscillator, where the spectrum is a $\delta$-function. The width here indicates the coupling to the environment (emission of radiation). Even though the maximum achievable values for isotopic defects[38] are at best of the order $\varepsilon^2 \cong 10$ ($m \cong 5M$, $M_e \cong M/2$) the results are roughly the same as above.

e) *The expansion of* $G^{(\underline{h})}(\omega)$ *for small* $\omega$ follows the procedure for $G^{(o)}$. One recognizes from (3.69b),

$$G_2^{(\underline{h})}(\lambda) = \frac{\pi}{M}\int\limits_{V_B}\frac{d\underline{k}}{V_B}\cos\underline{k}\underline{R}^{\underline{h}}\;\delta(\Omega^2(\underline{k}) - \lambda) = \frac{\pi}{M}\,Z^{(\underline{h})}(\lambda)\;,$$

that the "spectrum" $Z^{(\underline{h})}$ is weighted with $\cos\underline{k}\underline{R}^{\underline{h}}$, containing contributions of different sign in contrast to $Z^{(o)} \geqslant 0$. Indeed $\int d\lambda\; Z^{(\underline{h}\neq o)}(\lambda) = 0$. For small $\lambda$ (and $\underline{k}$) one can expand $\cos\underline{k}\underline{R}^{\underline{h}} = 1 - (\underline{k}\underline{R}^{\underline{h}})^2/2 + \dots$ and neglect the higher orders in the leading term (each $k^2$ means a factor $\lambda$). Consequently, the leading term for small $\lambda$ is independent of $\underline{h}$, $G_2^{(\underline{h})}(\lambda) \cong G_2^{(o)}(\lambda) \propto \sqrt{\lambda}$. The real part, $G_1^{(\underline{h})}$, is obtained via the Kramers-Kronig relations, and can be discussed as before. In cubic crystals $G_2^{(o)}(\lambda)$ is scalar. Consequently, only the diagonal elements of $G_2^{(\underline{h})}(\lambda)$ start with $\sqrt{\lambda}$, whereas the nondiagonal elements carry the factor $(\underline{k}\underline{R}^{\underline{h}})^2$ and must start with $\lambda^{3/2}$.

### 3.5.5 Asymptotic Expansion of $G^{(\underline{h})}(\omega)$, $(h \gg 1)$

Asymptotic expansions are rather complicated and tricky, in particular if the "large" quantity is an integer and one integrates over a periodic volume. Here, we cannot give a thorough review of the matter, but we will try to give an introduction using a few examples and handwaving, but plausible arguments. The main conclusions for $G^{(\underline{h})}(\omega)$ will be

1) *outside* the allowed frequencies, $\omega^2 > \Omega^2_{max}$, $G^{(\underline{h})}$ decreases exponentially with $R^{\underline{h}}$;

2) *inside* the allowed frequencies, $\omega^2 \leqslant \Omega^2_{max}$, $G^{\underline{h}}(\omega)e^{-i\omega t}$ behaves like an (outgoing) spherical wave ($e^{i\underline{k}\underline{R}^{\underline{h}}-i\omega t}/R^{\underline{h}}$) emitted from $R^{\underline{h}} = 0$;

3) for small $\omega$, $\omega^2 \ll \Omega^2_{max}$, only small $\underline{k}$ enter and $G^{(\underline{h})}(\omega)$ passes into the Green's function $G(\underline{R}^{\underline{h}},\omega)$ of elastic continuum theory.

[38] "Heavy isotopic" defect means here heavy mass as compared with host mass and only small spring changes, e.g., Ag in aluminium host, $M_d \cong 4M$, $m \cong 3M$ (comp. Sec. 6.3.1).

a) *The linear chain with 1st neighbour spring f,* (3.70-72), where

$$G^{(h)}(\omega) = \frac{i}{2f}\frac{e^{iz_p(\omega)|h|}}{\sin z_p(\omega)}, \quad \sin z_p(\omega) = 2\,\mathrm{sgn}\,\omega\sqrt{\frac{\omega^2}{\Omega^2_{max}}\left(1-\frac{\omega^2}{\Omega^2_{max}}\right)} \quad \text{for } \frac{|\omega|}{\Omega_{max}} \leq 1, \tag{3.93a}$$

$$G^{(h)}(\omega) = \frac{-1}{2f}\frac{(-1)^{|h|}e^{-\kappa_p(\omega)|h|}}{\sinh[\kappa_p(\omega)]}, \quad \sinh[\kappa_p(\omega)] = 2\sqrt{\frac{\omega^2}{\Omega^2_{max}}\left(\frac{\omega^2}{\Omega^2_{max}}-1\right)} \quad \text{for } \frac{|\omega|}{\Omega_{max}} \geq 1, \tag{3.93b}$$

can serve for illustration. For $|\omega| \leq \Omega_{max}$ the quantity $G^{(h)}(\omega)e^{-i\omega t}$ needed for $g^{(h)}(t)$ consists of waves emitted from the site h = 0, and outside the spectrum the amplitudes decrease exponentially with $|h|$. This remains true in 3 dimensions: exponential decrease outside the spectrum and outgoing waves inside. Of course, in one dimension the asymptotic expansion agrees with the exact result. Nevertheless, one can check with this example the methods used in deriving asymptotic expansions. In a more general model, with springs to further neighbours, we have more than one relevant pole given by $\Omega(k_p) = \omega$, $k_p a = z_p$, and G is a sum of pole contributions. If $\omega > \Omega_{max}$ all $k_p$ must be complex, $G^{(h)}$ must decrease exponentially, and asymptotically only the term with the smallest $\kappa_p$ survives. For $\omega < \Omega_{max}$ one must have at least *one* real $k_p$, the contributions from complex $k_p$'s die out exponentially and the wavelike contributions establish the asymptotic expansion.

b) *Simple examples of asymptotic expansions* are presented in this subsection so that the reader can become familiar with the standard methods. First we consider integrals, $\int \ldots dz$, which contain either a very rapidly increasing factor, exp(hz), or a very rapidly oscillating factor, exp(ihz), in the integrand ($h \gg 1$).

Let us consider the simple integral

$$I_1(h) = \int_{-\infty}^{\infty} dz\, \exp(hz - z^4/4) = \int_{-\infty}^{\infty} dz\, \exp[f_1(z)]\,. \tag{3.94}$$

The integrand has one maximum at $z_s = h^{1/3}$. To obtain a reliable representation of the integrand near the maximum one better employs a logarithmic expansion of the integrand, i.e.

$$\begin{aligned} f_1(z) &= f_1(z_s) + \frac{f_1''(z_s)}{2}\zeta^2 + \frac{f_1'''(z_s)}{3!}\zeta^3 + \frac{1}{4!}f_1^{(IV)}(z_s)\,\zeta^4 \\ &= \frac{3h^{4/3}}{4} - \frac{3h^{2/3}}{2}\zeta^2 - h^{1/3}\zeta^3 + \ldots, \quad \zeta = z - z_s\,. \end{aligned} \tag{3.94a}$$

If $h \gg 1$ the integrand falls off so fast from its maximum value due to the 2nd term in (3.94a), that higher order terms can be dropped in the region relevant for the integral: $f_1(z) \cong 3h^{3/4}/4 - (3h^{2/3}/2)\zeta^2$. In this approximation $e^{f_1}$ is represented by a Gaussian about $z_s$ with width $\propto h^{-1/3}$. For the integral only an interval of some widths about $z_s$ is essential, and there the 3rd term, $h^{1/3}\zeta^3$ (of order $h^{-2/3}$), can be neglected. As a consequence the maximum for large h is so strongly peaked, that $f_1(z)$ can be replaced by $f_1(z_s) + [f_1''(z_s)/2]\zeta^2$, and then

$$I_1(h) \sim \exp[f_1(z_s)] \int_{-\infty}^{\infty} d\zeta \exp\left[\frac{f_1''(z_s)}{2}\zeta^2\right] = \exp[f_1(z_s)] \cdot \left[\frac{2\pi}{|f_1''(z_s)|}\right]^{1/2} . \quad (3.95)$$

This asymptotic result is independent of the lower limit if this is outside the width of the Gaussian, e.g., $\int_0^\infty dz \exp(f_1)$ asymptotically agrees with $I_1$.

If h is replaced by ih, the integral

$$I_2(h) = \int_{-\infty}^{\infty} dz \exp(ihz - z^4/4) = \int_{-\infty}^{\infty} dz \exp[f_2(z)] \quad (3.96)$$

resembles more closely those needed for the Green's function. Whereas $I_1$ is large, $I_2$ must be small because of the rapidly oscillating factor exp(ihz). Here one expects only contributions from "stationary points" $z_s$ where the "phase" $\varphi_2(z) = f_2(z)/i$ does not rapidly vary with z, i.e., $f_2'(z_s) = 0$. Here the $z_s$ are complex (Fig. 3.32):

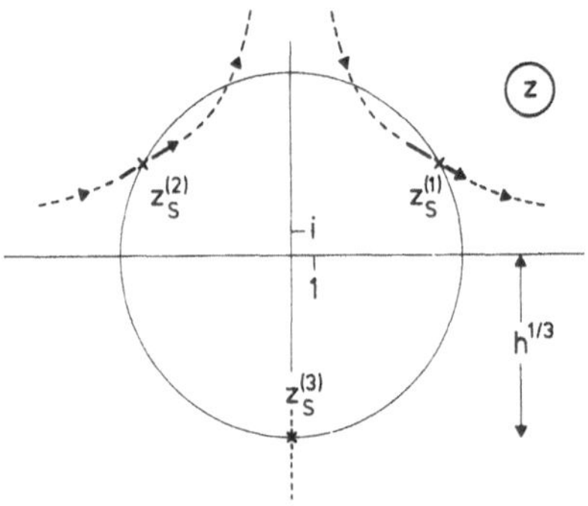


Fig. 3.32. The three stationary points of $f_2(z)$: $z_s^{(1,2,3)} = h^{1/3}i^{1/3}$. The paths of steepest descent are indicated by dashed lines. The path along the real axis for $I_2(h)$ can be deformed into the two paths through $z_s^{(2)}$ and $z_s^{(1)}$. The indicated directions at $z_s^{(2)}$ and $z_s^{(1)}$ are proportional to $-1/z_s^{(2)}$ and $1/z_s^{(1)}$

$$z_s^3 = ih , \quad z_s^{(1)} = h^{1/3}\frac{\sqrt{3}+i}{2}, \quad z_s^{(2)} = h^{1/3}\frac{-\sqrt{3}+i}{2}, \quad z_s^{(3)} = h^{1/3}(-i). \quad (3.96a)$$

The expansion about the stationary points is as above,

$$f_2(z) = f_2(z_s) + \frac{f''(z_s)}{2}\zeta^2 + \frac{f'''(z_s)}{3!}\zeta^3 + \ldots ,$$

but now in the complex z-plane. One must realize that one always has a saddlepoint[39] at $z_s$. This is most easily seen if one takes either $\zeta = \tilde{\zeta}/\sqrt{f''(z_s)}$, $f_2(z) \cong f_2(z_s) + \tilde{\zeta}^2/2$ or $\zeta = \tilde{\zeta}i/\sqrt{f_2''(z_s)}$, $f_2(z) = f_2(z_s) - \tilde{\zeta}^2/2$, $\tilde{\zeta}$ real. Consequently, $\exp(f_2) \cong \exp[f_2(z_s)]\cdot\exp(-\tilde{\zeta}^2/2)$ if one leaves $z_s$ in direction $i[f_2''(z_s)]^{-1/2} = 1/(z_s\sqrt{3})$, i.e., the absolute value of $\exp(f_2)$ decreases, whereas it increases in the perpendicular direction $[f_2''(z_s)]^{-1/2} = 1/(iz_s\sqrt{3})$. The original path along the real axis can be changed such that it passes through $z_s^{(1,2)}$ (Fig. 3.32). In particular, one can select the deformed path such that one has a maximum at $z_s$, direction $i[f_2''(z_s)]^{-1/2}$: at $z_s^{(2)}$, direction $-1/(z_s^{(2)}\sqrt{3})$, $\zeta = -\zeta/(z_s^{(2)}\sqrt{3})$, and at $z_s^{(1)}$, $\zeta = \tilde{\zeta}/(z_s^{(2)}\sqrt{3})$. This maximum is strongly peaked because $f_2''(z_s) \propto h^{2/3}$ is very large. Only the saddlepoint region contributes to $I_2$ if one chooses the path such that the decrease of the absolute value of the integrand is a maximum, the "path of steepest descent", parallel to the gradient of $\mathrm{Re}\{f_2(z)\}$. These paths of steepest descent must end either in infinity or pass through another saddle. The paths of steepest descent are indicated in Fig. 3.32. The result is then

$$\begin{aligned} I_2(h) &\sim \exp[f_2(z_s^{(2)})]\cdot\int_{-\infty}^{\infty}\frac{-d\tilde{\zeta}}{z_s^{(2)}\sqrt{3}}e^{-\tilde{\zeta}^2/2} + \exp[f_2(z_s^{(1)})]\cdot\int_{-\infty}^{\infty}\frac{d\tilde{\zeta}}{z_s^{(1)}\sqrt{3}}e^{-\tilde{\zeta}^2/2} \\ &= \sqrt{\frac{2\pi}{3}}\left[\frac{\exp(3ih^{1/3}z_s^{(1)}/4)}{z_s^{(1)}} - \frac{\exp(3ih^{1/3}z_s^{(2)}/4)}{z_s^{(2)}}\right] \\ &= \frac{2}{h^{1/3}}\sqrt{\frac{2\pi}{3}}\left(\sqrt{3}\cos\frac{3^{3/2}h^{1/3}}{8} + \sin\frac{3^{3/2}h^{1/3}}{8}\right)\exp(-3h^{1/3}/8)\ . \end{aligned} \tag{3.97}$$

If one looks back at the evaluation of $I_1$ in the complex z-plane, one realizes that the real axis there is also a path of steepest descent.

The method described above is called the method of the stationary phase or the saddlepoint method. It is important to realize the assumption that the original path can be deformed into paths of steepest descent. If this is not true, an uncritical application of the method gives nonsense. As an example let us consider

$$\begin{aligned} I_3(h) &= \int_0^{\infty} dz\, e^{-z^4/4}\underbrace{e^{izh}}_{\frac{1}{ih}\partial_z e^{izh}} = -\frac{1}{ih} - \frac{1}{ih}\int_0^{\infty} dz\,\underbrace{e^{izh}}_{\frac{1}{ih}\partial_z e^{izh}}\partial_z e^{-z^4/4} \\ &= -\frac{1}{ih} - \frac{1}{h^2}\int_0^{\infty} dz\, e^{izh}\,\partial_z^2\, e^{-z^4/4} \sim -\frac{1}{ih}\ , \end{aligned} \tag{3.98}$$

where the asymptotic expansion is obtained via repeated integration by parts. Instead

---

[39] Stationary points are always saddlepoints with respect to the absolute value.

of the exponential decrease of $I_2$ due to the saddlepoint contributions, the asymptotic behaviour of $I_3$ is rather given by $f_2(z)$ near $z = 0$. If one uses the method of integration by parts for $I_2$, the 1/h-term vanishes because the integrand becomes zero in infinity; eventually one can show by repeated integration by parts, that $I_2$ vanishes faster than any power of 1/h, i.e., exponentially.

c) *The asymptotic expansion of the linear* $G^{(h)}(\lambda > \lambda_{max})$, $\lambda = \omega^2$, $\lambda_{max} = \Omega^2_{max}$, will be treated first because it is simpler. We have[40] from (3.70)

$$2fG_1^{(h)}(\lambda) = \int_{-\pi}^{\pi} \frac{dz}{2\pi} \frac{e^{izh}}{1 - 2\lambda/\lambda_{max} - \cos z} = -(-1)^h \int_{-\pi}^{\pi} \frac{dz}{2\pi} \frac{e^{izh}}{\underbrace{2\lambda/\lambda_{max} - 1}_{\gamma} - \cos z} = (-1)^{h+1} I(h). \tag{3.99a}$$

The exact value of I(h) is

$$I(h > 0) = \frac{e^{-\kappa_p(\lambda)h}}{\sinh \kappa_p} = \frac{e^{-\kappa_p h}}{\sqrt{\gamma^2 - 1}}, \quad \cosh \kappa_p = \gamma > 1, \quad \sinh \kappa_p = \sqrt{\gamma^2 - 1} = 2[\lambda/\lambda_{max}(\lambda/\lambda_{max} - 1)]^{1/2}. \tag{3.99b}$$

The integrand of

$$I(h) = \int_{-\pi}^{\pi} \frac{dz}{2\pi} \frac{e^{izh}}{\gamma - \cos z} \tag{3.99c}$$

has poles at $z = i\kappa_p$, $\cos i\kappa_p = \cosh \kappa_p = \gamma > 1$ and is $2\pi$-periodic (Fig. 3.33). One

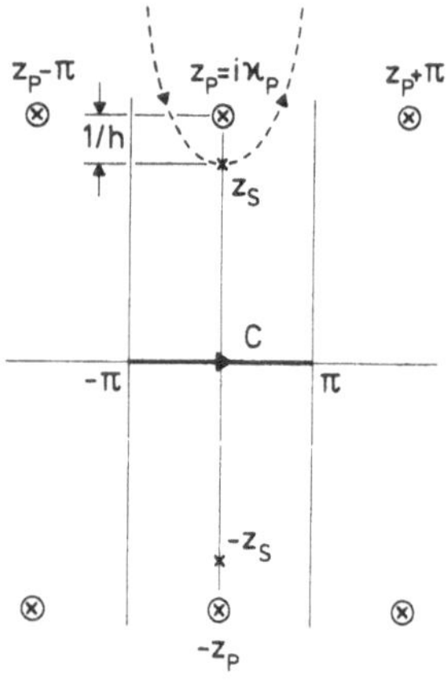


Fig. 3.33. Poles $z_p$ and saddlepoints $z_s$ of the integrand (3.99c). The original path C and the path of steepest descent (dashed line) are shown.

---

[40] Substitution $z \to z - \pi$, $\cos(z - \pi) = -\cos z$; the interval of integration can be any length $2\pi$.

ought to realize that the result (3.99b) is valid only for integer h. Indeed for non-integer h one obtains, besides the residuum at the pole, a contribution from the two paths from $-\pi \to i\infty$ and $i\infty \to +\pi$ (Fig. 3.27),

$$\int_0^\infty \frac{id\kappa}{2\pi} \frac{e^{-\kappa h}}{\gamma + \cosh \kappa} (e^{-ih\pi} - e^{ih\pi}) = \frac{\sin h\pi}{\pi} \int_0^\infty d\kappa \frac{e^{-\kappa h}}{\gamma + \cosh \kappa} \sim \frac{\sin h\pi}{\pi} \frac{1}{h(\gamma + 1)} ,$$

which becomes for large h $\left(\cosh \kappa \cong 1 \text{ for } \kappa \lesssim 1/h \ll 1\right)$ proportional to 1/h. Only for integer h is the exponential contribution the leading term in the asymptotic expansion. For all other intermediate h the exponential contribution can be neglected asymptotically. This indicates already that the asymptotic expansion not only requires large h, but that the constraint of periodicity (integer h) is important.

If one applies the saddlepoint method directly[41],

$$f(z) = izh - \ln(\gamma - \cos z) , \quad ih = \frac{\sin z_S}{\gamma - \cos z_S} = i \frac{\sinh \kappa_S}{\gamma - \cosh \kappa_S} \text{ with } z_S = i\kappa_S ,$$

one sees that for large h the denominator $(\gamma - \cosh \kappa_S)$ must be small and that therefore $\kappa_S$ must be close to $\kappa_P$; expanding $\sinh \kappa_S$, $\cosh \kappa_S$ about $\kappa_P$ and keeping only the leading term $\left(\gamma - \cosh \kappa_S \cong -(\kappa_S - \kappa_P) \sinh \kappa_P\right)$ we obtain

$$h \cong \frac{\sinh \kappa_P}{-(\kappa_S - \kappa_P) \sinh \kappa_P} \quad \text{or} \quad \kappa_S \cong \kappa_P - \frac{1}{h} .$$

The path of steepest descent is indicated in Fig. 3.33. It can be reached from the original path C because the two integrals from $\pm\pi$ to $i\infty$ cancel each other. The expansion of f(z) is

$$f(z) = f(z_S) + \underbrace{f''(z_S)}_{\sim -h^2} \zeta^2/2 + \underbrace{f'''(z_S)}_{\sim 2ih^3} \zeta^3/3 + \dots \quad . \tag{3.100}$$

Uncritical application of the saddlepoint method results in

$$I(h) \sim \frac{e^{-h\kappa_S}}{[2\pi(\gamma \cosh \kappa_S - 1)]^{1/2}} \sim \frac{e}{\sqrt{2\pi}} \cdot \frac{e^{-h\kappa_P}}{\sqrt{\gamma^2 - 1}} \cong 1.08\, I_{\text{exact}} , \tag{3.101}$$

which is comfortably close to the exact value, but nevertheless quite obviously wrong. Indeed, one cannot use the arguments used for $I_{1,2}$ to justify the saddlepoint method; here the range of the quadratic term is 1/h, but for $\zeta \cong 1/h$ the 3rd order term $h^3\zeta^3$ is of order unity and cannot be dropped. This is caused by the fact, that saddlepoint and pole are too close. Consequently, one should somehow remove the

41 Logarithmic expansion of the integrand is supposed to hold well in the stationary region.

pole, and this is done by rewriting

$$I(h) = \int_{-\pi}^{\pi} \frac{dz}{2\pi} \int_{0}^{\infty} d\theta\ e^{ihz-\theta(\gamma-\cos z)} = \int_{0}^{2\pi} \frac{dz}{2\pi} \int_{0}^{\infty} d\theta\ e^{F(z,\theta)} \ ; \tag{3.102}$$

now one can try to employ the method of stationary phase to the *two* variables z, θ:

$$\partial_z F = 0:\quad ih = \theta_S \sin z_S\ ,\quad h = \theta_S \sinh \kappa_S\ ;\qquad \partial_\theta F = 0:\quad \gamma = \cos z_S\ ,\ \gamma = \cosh \kappa_S$$

$$\text{or } \kappa_S = \kappa_P > 0 \quad \text{and} \quad \theta_S = h/\sqrt{\gamma^2 - 1} > 0\ . \tag{3.103}$$

An expansion of $F(z_S + \zeta, \theta_S + \tau)$ yields

$$F(z,\theta) = -h\kappa_P - \underbrace{\frac{\theta_S \cos z_S}{2}}_{\frac{h\gamma}{2\sqrt{\gamma^2-1}}}\zeta^2 - \underbrace{\sin z_S}_{i\sqrt{\gamma^2-1}} \cdot \zeta\tau + \underbrace{\frac{\theta_S \sin z_S}{3!}}_{\frac{ih}{3!}}\zeta^3 - \underbrace{\frac{\cos z_S}{2}}_{\frac{\gamma}{2}}\zeta^2\tau + \dots\ . \tag{3.104}$$

The range of $\zeta$ is $h^{-1/2}$, and the order of the 4th and 5th terms is $h^{-1/2}$ and $h^{-1}\tau$. The last term cannot be argued away dependably, because the integration interval for $\tau$ is: $-\theta_S = -h/\sqrt{\gamma^2 - 1} \leqslant \tau \leqslant \infty$. If one, nevertheless, includes only the quadratic terms in the expansion (3.104) and extends the $\tau$-interval from $-\infty$ to $+\infty$ because $\theta_S \to \infty$, one obtains asymptotically

$$\begin{aligned} I(h) &\sim \int_{-\infty}^{\infty}\int_{-\infty}^{\infty} \frac{d\zeta\, d\tau}{2\pi} \exp\left(-h\kappa_P - \frac{h\gamma}{2\sqrt{\gamma^2-1}}\zeta^2 - i\sqrt{\gamma^2-1}\,\zeta\tau\right) \\ &= e^{-h\kappa_P} \int_{-\infty}^{\infty} d\zeta \exp\left(-\frac{h\gamma}{2\sqrt{\gamma^2-1}}\zeta^2\right) \delta\left(\sqrt{\gamma^2-1}\,\zeta\right) = \frac{e^{-h\kappa_P}}{\sqrt{\gamma^2-1}}\ , \end{aligned} \tag{3.105}$$

the correct result for unconvincing reasons (note again that this result is valid only for integer h). The above *recipe* requires: find the saddlepoint, include only 2nd order terms and integrate from $-\infty$ to $+\infty$ in all the variables appropriately.

However, one can justify the above result by treating the integral (3.102) with the double saddlepoint method somewhat more carefully. First one applies the saddlepoint method to z, i.e., one determines from $\partial F(z,\theta)/\partial z = 0$ a $\theta$-dependent stationary point $z_S(\theta) = i\kappa_S(\theta)$ given by $h = \theta \sinh \kappa_S(\theta)$ or $\exp \kappa_S(\theta) = (h + \sqrt{h^2 + \theta^2})/\theta$ and expands $F(z,\theta)$ about $z_S$:

$$F(z,\theta) = -h\kappa_S - \theta(\gamma - \underbrace{\cos z_S}_{\sqrt{h^2+\theta^2}/\theta} + \frac{\zeta^2}{2}\cos z_S - \frac{\zeta^3}{3!}\underbrace{\sin z_S}_{ih/\theta} + \dots)\ .$$

The range of $\zeta$ is $(h^2 + \theta^2)^{-1/4}$, the last term can be dropped, and the saddlepoint method is applicable for z with the result,

$$I(h) \sim \int_0^\infty \frac{d\theta}{2\pi} \int_{-\infty}^\infty d\zeta \, \exp\left(-h\kappa_S(\theta) - \theta\gamma + \sqrt{h^2+\theta^2} - \frac{\sqrt{h^2+\theta^2}}{2}\zeta^2\right)$$

$$= \int_0^\infty \frac{d\theta}{\sqrt{2\pi}} \exp\left(-h\kappa_S(\theta) - \theta\gamma + \sqrt{h^2+\theta^2}\right) \cdot (h^2 + \theta^2)^{-1/4} \; .$$

The integrand has a strongly peaked maximum at the same $\theta_S$ as defined by (3.103); the factor $(h^2 + \theta^2)^{-1/4}$ can be disregarded when determining $\theta_S$ and can be replaced by $(h^2 + \theta_S^2)^{-1/4}$. From expansion in the exponent,

$$\underbrace{-h\kappa_S(\theta_S)}_{=\kappa_P} \;\; \underbrace{- \theta_S\gamma + \sqrt{h^2 + \theta_S^2}}_{=0} - \frac{(\gamma^2-1)^{3/2}}{h\gamma}\frac{\tau^2}{2} + \frac{(\gamma^2-1)^2}{h^2\gamma^3}(1 + 2\gamma^2)\frac{\tau^3}{3!} + \dots ,$$

one sees that the range of $\tau$ is "large", proportional to $\sqrt{h}$, but with respect to $\theta_S \sim h$ it is relatively small, so small that the 3rd order term (and by implication the higher order terms) are unimportant within the range of the Gaussian. Consequently,

$$I(h) \cong e^{-h\kappa_P} \int \frac{d\tau}{\sqrt{2\pi}} \frac{\exp -\left[\frac{(\gamma^2-1)^{3/2}}{h\gamma}\frac{\tau^2}{2}\right]}{(h^2 + \theta_S^2)^{1/4}} = e^{-h\kappa_P} \sqrt{\frac{h\gamma}{(\gamma^2-1)^{3/2}}} \frac{1}{(h^2+\theta^2)^{1/4}}$$

$$= \frac{e^{-h\kappa_P}}{\sqrt{\gamma^2 - 1}} ,$$

which is the exact result, this time derived with a better conscience. This discussion makes it plausible that one can neglect the higher order terms in (3.104). The range of $\zeta$ is $h^{-1/2}$, and the range of $\tau$ is $h^{1/2}$; the last two terms in (3.104) are both or order $h^{-1/2}$ and can be dropped. Even though the arguments are not entirely compelling, we will use the recipe leading from (3.102) to (3.105) from now on. The results are correct (also for $\gamma < 1$), and one can also find explicit three-dimensional examples.

d) *The expansion of the linear* $G^{(h)}$ *(*$0 \leq \omega \leq \Omega_{max}$*)* is obtained from

$$2f\, G^{(h)} = \int_{-\pi}^{\pi} \frac{dz}{2\pi} \frac{e^{ihz}}{1 - 2\lambda/\lambda_{max} - \cos z - i\eta}, \qquad \eta \to +0 \quad \text{for } \omega > 0$$

$$= \int_{-\pi}^{\pi} \frac{dz}{2\pi} \int_0^\infty i d\theta \, \exp[ihz - i(\underbrace{1 - 2\lambda/\lambda_{max}}_{\bar{\gamma}} - \cos z - i\eta)\theta] , \qquad (3.106)$$

where the integrand has already been written with the double saddlepoint method in mind. The integrand contains the factor $\exp(-\eta\theta)$, which makes the $\theta$-integral definite, but since $\eta \to +0$ it can be left out in the determination of the stationary values,

$$h = \theta_s \sin z_s\ , \quad 1 - \cos z_s = 2\lambda/\lambda_{max}\ , \quad 0 \leqslant z_s = k_s a \leqslant \pi\ , \quad h > 0. \tag{3.106a}$$

The double saddlepoint method can be performed according to the recipe given above with the correct result[42].

e) *The expansion of the three-dimensional* $G^{(\underline{h})}(\omega)$ follows the same lines; each branch can be treated separately (Table 3.3). Within the spectrum, the contribution of branch $\sigma$ is

$$\frac{1}{MV_B}\iint d\underline{k}\ d\theta\ e_i(\underline{k}\sigma)\ e_l(\underline{k}\sigma)\ e^{F_\sigma(\underline{k},\theta)} \quad \text{with}$$

$$F_\sigma(\underline{k},\theta) = i\underline{k}\underline{R}^{\underline{h}} - i[\Omega^2(\underline{k}\sigma) - (\omega + i\eta)^2]\theta\ .$$

The saddlepoint $\underline{k}_s$ is given by (if one chooses $\hat{\underline{R}}^{\underline{h}}$ as the x-axis in $\underline{k}$-space)

$$\underline{R}^{\underline{h}} = \theta\partial_{k_x}\Omega^2(\underline{k}\sigma)\ ; \quad \partial_{k_{y,z}}\Omega^2(\underline{k}\sigma) = 0\ , \quad \Omega^2(\underline{k}\sigma) = \omega^2\ . \tag{3.107}$$

The usual procedure gives then

$$G_{il}^{(\underline{h})}(\omega) \sim \frac{i}{MV_B}\sum_\sigma e_i(\underline{k}_s\sigma)\ e_l(\underline{k}_s\sigma)\ \frac{\exp[-i\frac{\pi}{4}(\mathrm{sgn}\ A_{yy}^\sigma + \mathrm{sgn}\ A_{zz}^\sigma)]}{(|A_{yy}^\sigma A_{zz}^\sigma|)^{1/2}} \cdot \frac{e^{i\underline{k}_s\underline{R}^{\underline{h}}}}{|\underline{R}^{\underline{h}}|}\ , \tag{3.108}$$

$$\text{with}\quad A_{yy}^\sigma = \partial_{k_y}^2\Omega^2(\underline{k}\sigma)\Big|_{\underline{k}_s}\ , \quad A_{zz}^\sigma = \partial_{k_z}^2\Omega^2(\underline{k}\sigma)\Big|_{\underline{k}_s}\ .$$

This can be seen as follows. In the integral

$$I = \int d\underline{k}\int d\theta\ e^{i\varphi(\underline{k},\theta)} \quad \text{with}\quad \varphi(\underline{k},\theta) = \underline{k}\underline{R}^{\underline{h}} - \theta[\Lambda(\underline{k}) - \lambda]\ , \quad \Lambda = \Omega^2\ , \quad \lambda = \omega^2\ ,$$

---

42 Actually one must be more careful because, for example, the expansion in powers of z is

$$F(z,\theta) = i[hz_s - \theta(\bar\gamma - \cos z_s)] - \underbrace{\theta\cos z_s}_{\sqrt{\theta^2 - h^2}} \cdot \zeta^2/2 + \dots\ , \quad h = \theta\sin z_s\ ,$$

which shows that the saddlepoint method cannot be applied if $\theta \cong h$, but it must be kept in mind that $\theta_s \cos z_s = h\bar\gamma/\sqrt{1-\bar\gamma^2} \gg 1$.

the phase $\varphi(\underline{k},\theta)$ can be expanded:

$$\varphi = \varphi_s - \frac{\theta_s}{2}\zeta_j\zeta_l\Lambda_{jl} - \zeta_1\tau\partial_{k_1}\Lambda + \dots , \qquad \text{where } \Lambda_{jl} = \partial_{k_j}\partial_{k_l}\Lambda ;$$

the 1st order terms vanish because of the saddlepoint conditions (3.107). Integration over $\tau$ yields

$$I = e^{i\varphi_s}\int d\underline{\zeta}\, \exp\left(-\frac{\theta_s}{2}\zeta_j\zeta_l\Lambda_{jl}\right)2\pi\delta(\zeta_1\partial_{k_1}\Lambda) = e^{i\varphi_s}\underbrace{\frac{2\pi}{|\partial_{k_1}\Lambda|}}_{2\pi\theta_s/R^{\underline{h}}} \cdot \int d\zeta_2\int d\zeta_3\, \exp\Bigl(-\frac{\theta_s}{2}\sum_{j,l(\neq 1)}\zeta_j\zeta_l\Lambda_{jl}\Bigr).$$

If one chooses the axes 2 and 3 $\perp \underline{R}^{\underline{h}}$ as the principal axes of $\Lambda_{jl}$ $(j,l \neq 1)$, then (Fig. 3.34)

$$I = e^{i\varphi_s}\cdot\frac{2\pi\theta_s}{R^{\underline{h}}}\int d\zeta_2\int d\zeta_3\, \exp\left[-i\frac{\theta_s}{2}(\zeta_2^2\Lambda_{22} + \zeta_3^2\Lambda_{33})\right]$$

$$= e^{i\underline{k}_s\underline{R}^{\underline{h}}}\,\frac{(2\pi)^2\exp\left[-i\frac{\pi}{4}(\operatorname{sgn}\Lambda_{22} + \operatorname{sgn}\Lambda_{33})\right]}{(\Lambda_{22}\Lambda_{33})^{1/2}\cdot R^{\underline{h}}} .$$

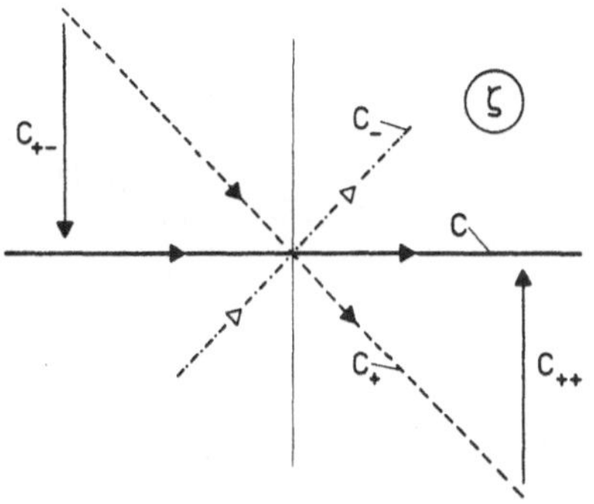


Fig. 3.34. Calculation of the "Fresnel-integral" $I = \int_{-\infty}^{\infty} d\zeta\, e^{-i\theta\zeta^2/2}$. For $\theta > 0$, the path of steepest descent is $C_+$; $C$ can be deformed into $C_+$ because the contributions of $C_{++}$ and $C_{+-}$ vanish exponentially in infinity. On $C_+$: $\zeta = \tilde{\zeta}\, e^{-i\pi/4}$; $I_+ = \int_{-\infty}^{\infty} e^{-i\pi/4}\, d\tilde{\zeta}\, \exp(-\theta\tilde{\zeta}^2/2) = e^{-i\pi/4}\sqrt{2\pi/\theta}$. For $\theta < 0$, the path of steepest descent is $C_-$ and one obtains analogously $I_- = e^{i\pi/4}\sqrt{2\pi/|\theta|}$. Consequently $I = \exp(-i\pi/4 \operatorname{sgn}\theta)\cdot\sqrt{2\pi/|\theta|}$

Note that, according to (3.107) (with $\theta > 0$), one has to choose $\underline{k}_s$ among equivalent $\underline{k}$'s (with $\Omega^2(\underline{k}\sigma) = \Omega^2(\underline{k}_s\sigma)$) such that $\left(\underline{R}^{\underline{h}}, \partial_{\underline{k}}\Omega^2(\underline{k}\sigma)\right) \geq 0$, or, in the coordinate system chosen above, $\partial_{k_x}\Omega^2(\underline{k}\sigma) \geq 0$; because $\partial_{\underline{k}}\Omega(\underline{k}\sigma)$ gives the group velocity of the wave $\exp[i\underline{k}\underline{R}^{\underline{h}} - i\Omega(\underline{k}\sigma)t]$, this means that (3.108) in fact represents an *outgoing* wave. If $\omega^2 > \Omega^2_{max}$, $\underline{k}_s$ is complex and $\exp(i\underline{k}_s\underline{R}^{\underline{h}}) = \exp(i\underline{k}_s\hat{\underline{R}}^{\underline{h}}\cdot R^{\underline{h}})$ exponentially decreases with increasing $R^{\underline{h}}$, $\mathrm{Im}\{k_{\underline{S}}\hat{R}^{\underline{h}}\} > 0$.

A case of particular interest is that of small $\omega^2$, where k must also be small. For small k one can expand $\Omega^2(\underline{k}\sigma)$ in powers of $\underline{k}$ and stop with the 1st nonvanishing term, which is of 2nd order in $k_j$ (comp. Sec. 3.4.3), $\Omega^2_{il}(\underline{k}\sigma) = e_i(\hat{\underline{k}}\sigma)e_l(\hat{\underline{k}}\sigma)c^2_\sigma(\hat{\underline{k}})k^2$. At the same time one has to replace $V_B$ by infinity[43], and one obtains asymptotically,

$$G^{(\underline{h})}_{il}(\omega) \sim G_{il}(\underline{R}^{\underline{h}},\omega) = \frac{\sum_\sigma}{MV_B}\int_{-\infty}^{\infty} d\underline{k}\, \frac{e_i(\hat{\underline{k}}\sigma)e_l\, e^{i\underline{k}\underline{R}^{\underline{h}}}}{c^2_\sigma k^2 - (\omega + i\eta)^2}\ , \tag{3.109}$$

the result of the corresponding continuum theory, which will be discussed in Chapter 4; even for cubic crystals $G(\underline{R},\omega)$ is not available analytically. For vanishing $\omega$ (and $\eta$) one obtains the static Green's function. Note, that the static Green's function diverges in one ($d\underline{k} \to dk$) and two ($d\underline{k} \to kdk$) dimensions, whereas in three dimensions ($d\underline{k} \to k^2dk$) the denominator $k^2$ is just cancelled.

The easiest model to obtain $G(\underline{R},\omega)$ is that of Fig. 3.15d, where $c_\sigma(\underline{k}) = c$ depends neither on direction nor on polarization,

$$G_{il}(\underline{R},\omega) = \delta_{il}\frac{1}{MV_B}\int \frac{d\underline{k}\ e^{i\underline{k}\underline{R}}}{c^2k^2 - (\omega + i\eta)^2} = \frac{\delta_{il}}{MV_B}\frac{2\pi^2\exp(i\omega R/c)}{c^2R}\quad ; \tag{3.110}$$

here also the corresponding g is simple,

$$g_{il}(\underline{R},t) = \int\frac{d\omega}{2\pi}G_{il}(\underline{R},\omega)e^{-i\omega t} = \frac{\delta_{il}}{MV_B}\frac{2\pi^2}{c^2R}\,\delta(\frac{R}{c} - t). \tag{3.110a}$$

For isotropy one has longitudinal and transversal soundvelocities, independent of $\hat{\underline{k}}$:

$$\begin{aligned} G_{il}(\underline{R},\omega) &= \frac{1}{MV_B}\int d\underline{k}\ e^{i\underline{k}\underline{R}}\left[\frac{\hat{k}_i\hat{k}_l}{c_l^2k^2 - (\omega + i\eta)^2} + \frac{\delta_{il} - \hat{k}_i\hat{k}_l}{c_t^2k^2 - (\omega + i\eta)^2}\right] \\ &= \frac{2\pi^2}{MV_B}\left[\frac{\exp(i\omega R/c_t)}{c_t^2R}\delta_{il} + \frac{1}{\omega^2}\partial_{X_i}\partial_{X_l}\left(\frac{\exp(i\omega R/c_t) - \exp(i\omega R/c_l)}{R}\right)\right], \end{aligned} \tag{3.111}$$

$$\begin{aligned} G_{il}(\underline{R},\omega = 0) &= \frac{2\pi^2}{MV_B}\left[\frac{1}{c_t^2R}\delta_{il} + \frac{1}{2}(\frac{1}{c_l^2} - \frac{1}{c_t^2})\,\partial_{X_i}\partial_{X_l}R\right] \\ &= \frac{\pi^2}{MV_B}\left[(\frac{1}{c_l^2} + \frac{1}{c_t^2})\,\delta_{il} + (\frac{1}{c_l^2} - \frac{1}{c_t^2})\frac{X_iX_l}{R^2}\right]\cdot\frac{1}{R}\ , \end{aligned} \tag{3.111a}$$

of which (3.110) is a special case.

---

[43] The argument is somewhat involved; essential is the periodicity (in $\underline{k}$) of the integrand (comp. /3.11/ for the static Green's function).

These two models give already a rather good overview of the continuum Green's function with the following main points: a sudden force spreads out with "the velocity of sound", $G \exp(-i\omega t) \sim \exp[i\omega(R/c - t)]/R$, it represents an outgoing spherical wave, and the static $G$ is proportional to $1/R$. Mostly the agreement of $G^{(\underline{h})}(\omega)$ and $G(\underline{R}^{\underline{h}},\omega)$ is surprisingly good, even for small lattice distances $R^{\underline{h}}$. Examples are given below.

### 3.5.6 Numerical Values for $G^{(\underline{h})}(\omega)$

In Figs. 3.35-3.38 and in Tables 3.5 and 3.6 we give some numerical results for $G^{(\underline{h})}(\omega)$ in order to illustrate the foregoing discussion. In the following the atomic positions are expressed by integer multiples of a/2: $G^{(101)}$ means $\underline{R}^{(101)} = a/2\ (1,0,1)$, $G^{(200)}$ means $\underline{R}^{(200)} = a/2\ (2,0,0)$, etc. $G^{(o)}(\omega)$ for the three models of Fig. 3.15b,c,d has been illustrated in Fig. 3.29. From these data one can extract the expansion coefficients $\alpha,\beta$, a of (3.88,89), the effective values (3.90a) for slow motion and the critical mass from (3.80a); they are summarized in Table 3.5.

Table 3.5. Expansion coefficients and critical mass of the first neighbour models corresponding to Fig. 3.15b,c,d: $G_1^{(o)}(\lambda) = \alpha + \beta\lambda + \ldots$, $Z(\lambda) = a\sqrt{\lambda} + \ldots$ ; $\frac{\bar{m}_c}{M} = -1/[M\lambda_{max}\, G_1^{(o)}(\lambda_{max})]$

| Model | $a\Omega^3_{max}$ | $\alpha M\Omega^2_{max}$ | $\beta M\Omega^4_{max}$ | $\frac{f_e}{f_E}$ | $\frac{M_e}{M}$ | $\frac{\Omega_e}{\Omega_{max}}$ | $\frac{\eta_e}{\Omega_{max}}$ | $\frac{\bar{m}_c}{M}$ |
|---|---|---|---|---|---|---|---|---|
| 15b | 2.20 | 4.25 | 0 | 0.52 | 0.62 | 0.62 | 0.31 | 0.33 |
| 15c | 1.32 | 3.36 | 1.79 | 0.60 | 0.61 | 0.70 | 0.30 | 0.25 |
| 15d | 0.41 | 1.79 | 1.00 | 0.74 | 0.60 | 0.97 | 0.34 | 0 |
| Regular spectrum of Table 3.4 | $\frac{8}{\pi} = 2.55$ | 4 | -8 | 1/2 | 1/2 | $\frac{1}{\sqrt{2}} = 0.72$ | 1/2 | 1/4 |

Values of $G^{(101)}$ for these three models are shown in Fig. 3.35. One recognizes:

1) the values are, in general, much smaller than those of $G^{(o)}$;
2) the models Fig. 3.15b,c give very similar results;
3) the imaginary parts of the diagonal elements start $\propto \sqrt{\omega^2}$, whereas the off-diagonal elements start with $(\omega^2)^{3/2}$;
4) the imaginary parts contain contributions of different sign (such that the integral vanishes).

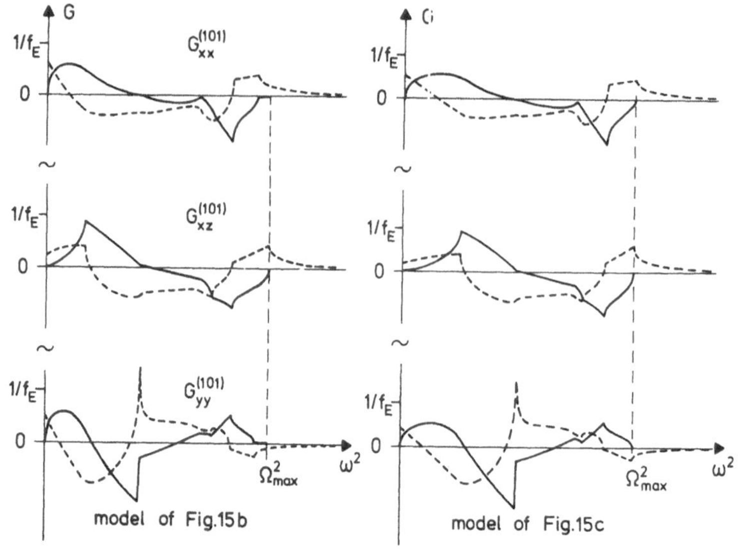


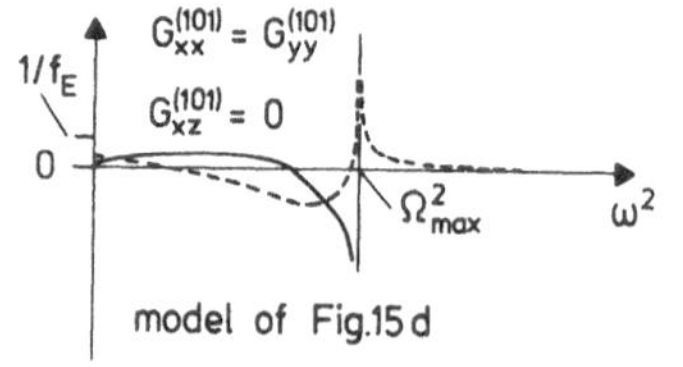


Fig. 3.35. $G^{(101)}$ for the three models of Fig. 3.15b,c,d; Re --- , Im —— ; $G_l = G_{xx} + G_{xz}$ , $G_{t'} = G_{xx} - G_{xz}$ , $G_t = G_{yy}$

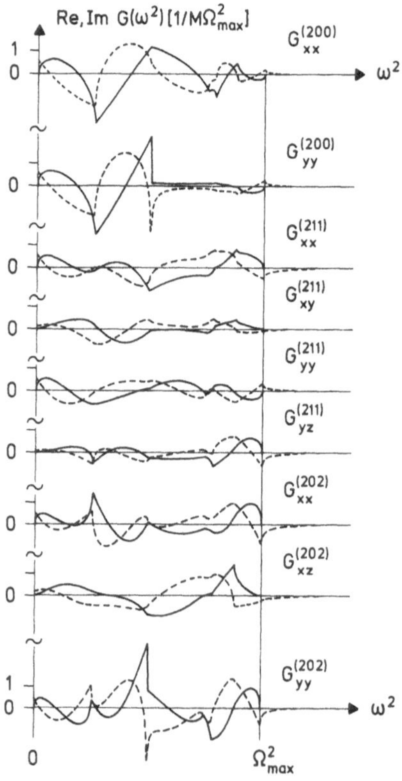


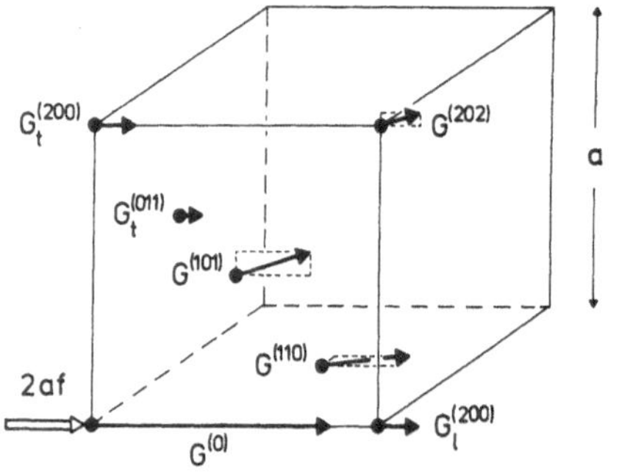


Fig. 3.37. Static G (model Fig. 3.15c). Shown are: the displacements (to scale) produced by a force 2af (1,0,0)

◁ Fig. 3.36. $G^{(200)}$ , $G^{(211)}$ and $G^{(202)}$ for the model of Fig. 3.15c; Re --- , Im ——

One can further check the sign above $\Omega_{max}$, because according to (3.84a), $G^{(101)}(\omega^2 \gg \Omega^2_{max}) \propto -\Phi^{(101)}/(M^2\Omega^4)$.

Values for larger $\underline{h}$ are shown in Fig. 3.36 for the model of Fig. 3.15c (longitudinal first neighbour spring $f_1$). The $G(\omega)$ are even more complicated than those of Fig. 3.35, and one realizes that analytical approximations covering the whole range of $\omega$ are not possible. Only for small and large frequencies analytical statements can be made, similar to those of Fig. 3.35.

Static Green's function is illustrated in Fig. 3.37 for the model Fig. 3.15c; the numerical values are given in Table 3.6, in addition the values of $G^{(o)}(0)$ and $G^{(101)}(0)$ obtained by variational methods (Sec. 2.5.4) are listed. If in the variation only atom 0 is allowed to move, its response $G^{(o)}$ is $1/f_E$, $fG^{(o)} = 0.25$; if, in addition, one permits also the first neighbours to be displaced, one obtains the values of Table 3.6.

Table 3.6. Static Green's functions for the model of Fig. 3.15c (purely longitudinal nearest neighbour coupling) in units of 1/f

| $\underline{R}^{\underline{h}}/(a/2)$ | $G^{(\underline{h})}_{xx}$ | $G^{(\underline{h})}_{yy}$ | $G^{(\underline{h})}_{zz}$ | $G^{(\underline{h})}_{xy}$ | $G^{(\underline{h})}_{yz}$ | $G^{(\underline{h})}_{zx}$ |
|---|---|---|---|---|---|---|
| 0 | 0.420 | 0.420 | 0.420 | 0 | 0 | 0 |
| variation | 0.330 | 0.330 | 0.330 | 0 | 0 | 0 |
| (1,1,0) | 0.130 | 0.130 | 0.104 | 0.041 | 0 | 0 |
| variation | 0.047 | 0.047 | 0.024 | 0.033 | 0 | 0 |
| (2,0,0) | 0.074 | 0.074 | 0.073 | 0 | 0 | 0 |
| (2,1,1) | 0.073 | 0.065 | 0.065 | 0.014 | 0.008 | 0.014 |
| (2,2,0) | 0.067 | 0.067 | 0.052 | 0 | 0 | 0.022 |

In Fig. 3.38 the static lattice Green's functions in $\langle 100\rangle$ and $\langle 110\rangle$ directions are compared with the elastic G of continuum theory (Sec. 4.8), which decreases proportional to the reciprocal distance. One sees that the agreement is rather good (except for $G^{(o)}$, of course). The spring model used for the lattice G is close to that of Fig. 3.15c, for details comp. /3.12/.

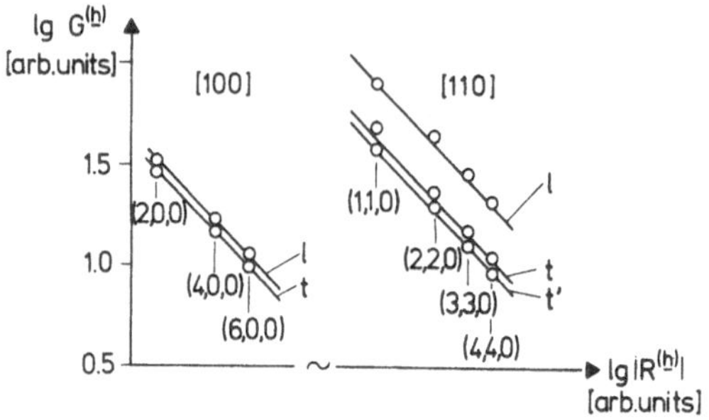


Fig. 3.38. Static Green function of lattice (0) and continuum (full line) in ⟨100⟩ and ⟨110⟩ directions /3.12/

### 3.5.7 Displacement-Displacement Correlation Functions

The correlation functions $l(t)$, $L(\omega)$ (2.32) can be compiled from Sections 2.3.3 and 2.4.4, e.g., for $\omega \geqslant 0$

$$\int dt\, e^{i\omega t} \langle s_i^{\underline{m}}(0) s_k^{\underline{n}}(t)\rangle_{th} = 2\hbar n_{th}(\omega)\ \mathrm{Im}\left\{G_{ik}^{\underline{mn}}(\omega)\right\} = 2\hbar n_{th}(\omega)\ \mathrm{Im}\{G_{ik}^{(\underline{m}-\underline{n})}(\omega)\}. \quad (3.112)$$

According to (2.32d), the equal time correlation in the classical limit, where $\varepsilon_{th}(\Omega) = kT$, can be expressed by the static G:

$$\langle s_i^{\underline{m}}(0) s_k^{\underline{n}}(0)\rangle_{th} = kTG_{ik}^{\underline{mn}}(0) = kTG_{ik}^{(\underline{m}-\underline{n})}(0)\ . \quad (3.113)$$

(Compare Chap. 7, where (3.112) and (3.113) are needed for the discussion of neutron and X-ray scattering.)

# 4. Continuum Theory

/4.1-4/

In continuum theory one views a material body as a system with continuous physical properties at every point $\underline{r}$ inside, in contrast to the discontinuous, discrete structure of a lattice. The basic quantity is the displacement field, $\underline{s}(\underline{r},t)$, the displacement of a mark at $\underline{r}$ in equilibrium. This field depends on the continuous variable $\underline{r}$, in contrast to the discrete displacements $\underline{s}^{\mathrm{m}}(t)$ of lattice theory. The atomistic description contains the continuum theory as a limiting case: if $\underline{s}^{\mathrm{m}}(t)$ changes only slowly from one atom to its neighbours, $\underline{s}^{\mathrm{m}}(t)$ can be replaced by a field $\underline{s}(\underline{r} = \underline{R}^{\mathrm{m}},t)$ and the difference equations of lattice theory pass into differential equations for the displacement field. This is equivalent to stating that phenomena in which only long lattice waves are involved can be treated as well in continuum theory. Another simple example is the static Green's function of the lattice, which is for large distances slowly varying and then passes into the corresponding Green's function of continuous elasticity theory.

In defect physics one must use microscopic lattice theory near the defect, but further out continuum theory is applicable and convenient to describe macroscopic features. Macroscopic mechanical properties of crystals, their change by defects and even the mechanical behaviour of single defects are expressed in terms of continuum theory. For this reason we give in the following an introduction to the continuum theory of crystals from basic principles. We concentrate on "harmonic" theory, where the derivatives of $\underline{s}(\underline{r})$ (the strains) are small, and for simplicity we consider only situations where the displacement field is small also.

In some ways continuum theory is simpler and easier to handle than lattice theory. It possesses fewer physical parameters, e.g., all the possible springs between atoms are condensed into a few elastic data. It neglects microscopic details which are prohibitively difficult to assess in lattice theory, e.g., the surface of a crystal is just a sharp cut, in contrast to the lattice description, where even the definition of a surface is a problem, in addition to the problem of changed springs in the surface region. These difficulties of the lattice description can be conveniently disregarded when discussing bulk properties. On the other hand, continuum elastic theory is conceptually more difficult. Besides the displacement field the basic quantities are two tensors of 2nd rank, strain and stress, which are connected in harmonic theory by a tensor of 4th rank, the elastic data. We have tried here to make this introduction as clear as possible, in particular by employing many illustrative drawings. For simplicity we concentrate mainly on cubic crystals in applications, though the general case is no more difficult but is less illustrative.

The last section is on static response. It contains in particular the response to internal force distributions, which is essential in defect theory. The most important new concept is that of the double force tensor of a force distribution. It characterizes the leading terms of an asymptotic expansion (analogous to the total charge of a charge distribution); it is the only macroscopically relevant property. The displacements caused by a single defect can be represented by forces acting in the ideal lattice. Macroscopically, then, a single defect will be completely described by its double force tensor.

## 4.1 The Strain Tensor

### 4.1.1 General Strain as a Symmetrical Tensor of 2nd Rank

Let us first discuss simple displacement fields, $\underline{s} = \left(s_i(\underline{r})\right)$. The simplest is certainly a common translation, $\underline{s}(\underline{r}) = \underline{T}$ independent or $\underline{r}$, which is uninteresting since it causes no real change of the material. The same would be true of a common (small) rotation, $\underline{s}(\underline{r}) = \varphi\hat{\underline{D}}\times\underline{r}$ or $s_i(\underline{r}) = \omega_{ik}x_k$ with antisymmetric $\omega$; here $\underline{s}$ is linear in $\underline{r}$. The most general linear field is

$$\underline{s} = v\underline{r}\ , \quad s_i = v_{ik}x_k\ , \quad \partial_k s_i = v_{ik} = \text{constant}\ , \tag{4.1}$$

called a homogeneous deformation v. Eq. (4.1) implies that v is constant over the whole body. However, the concept can be applied as well locally: eq. (4.1) can represent, for instance, an expansion of $\underline{s}$ about $\underline{r} = 0$ and v is then the deformation at $\underline{r} = 0$. This can be applied to any position $\underline{r}$ and

$$v_{ik}(\underline{r}) = \partial_k s_i(\underline{r}) \tag{4.2}$$

are then the local deformations. The conclusions drawn in the following for homogeneous deformations apply to local v's as well.

The elastic state of the material is not changed by a rotation. Consequently, only part of v causes a genuine change, a strain. This part is obtained by removing the rotational contribution to v. If v is separated into symmetrical and antisymmetrical parts, $\varepsilon$ and $\omega$,

$$v_{ik} = \frac{v_{ik} + v_{ki}}{2} + \frac{v_{ik} - v_{ki}}{2} = \varepsilon_{ik} + \omega_{ik}\ , \qquad v = \varepsilon + \omega, \tag{4.3}$$

it is obvious that the strain is given by $\varepsilon$, the strain tensor

$$\varepsilon_{ik} = \frac{v_{ik} + v_{ki}}{2} = \frac{\partial_k s_i + \partial_i s_k}{2} = \begin{bmatrix} \varepsilon_{11} & \varepsilon_{12} & \varepsilon_{13} \\ \varepsilon_{12} & \varepsilon_{22} & \varepsilon_{23} \\ \varepsilon_{13} & \varepsilon_{23} & \varepsilon_{33} \end{bmatrix} , \tag{4.4}$$

which is symmetric and possesses six independent components.

There are many versions of how to define six independent components. It is important to choose the most lucid ones which are simplest to grasp because later on we will have to deal with a linear relation between two tensors, the strain tensor and the stress tensor $\sigma$: $\sigma = C\varepsilon$. Here C is a tensor of 4th rank with 4 subcripts, the properties of which can be made transparent only if one makes a good choice. One obvious choice is to express $\varepsilon$ in terms of its eigenvectors and eigenvalues

$$\varepsilon = \sum_{j=1}^{3} |j\rangle \overset{j}{\varepsilon} \langle j| \ . \tag{4.5}$$

The meaning of (4.5) is illustrated in Fig. 4.1. The strain can be represented by the ellipsoid into which a sphere of radius 1 deforms.

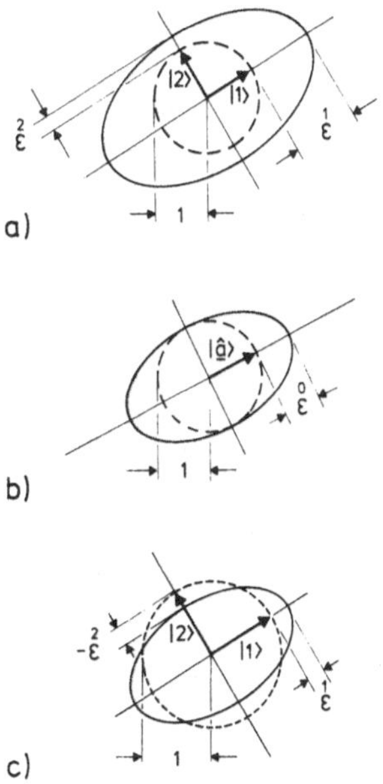


Fig. 4.1a-c. Representation of $\varepsilon$ by its eigenvectors and eigenvalues. A sphere of radius 1 undergoing a strain a deforms into an ellipsoid. The (principal) axes are the eigenvectors of $\varepsilon$, the eigenvalues determine the lengths, $1 + \overset{j}{\varepsilon}$ .

a) general situation $\varepsilon = \sum_j |j\rangle \overset{j}{\varepsilon} \langle j|$ ,

b) uniaxial strain $v = \varepsilon = |\hat{\underline{a}}\rangle \overset{o}{\varepsilon} \langle\hat{\underline{a}}|$ ,

c) shear strain ellipsoid, $\overset{2}{\varepsilon} = -\overset{1}{\varepsilon}$ , $\overset{3}{\varepsilon} = 0$

### 4.1.2 Voigt's Notation

It has become customary in crystal elasticity to use Voigt's notation,

$$\varepsilon_{ik} = \begin{bmatrix} \varepsilon_1 & \varepsilon_6 & \varepsilon_5 \\ \varepsilon_6 & \varepsilon_2 & \varepsilon_4 \\ \varepsilon_5 & \varepsilon_4 & \varepsilon_3 \end{bmatrix} = \varepsilon_1 \underbrace{\begin{bmatrix} 1 & 0 & 0 \\ 0 & 0 & 0 \\ 0 & 0 & 0 \end{bmatrix}}_{V(1)} + \ldots + \varepsilon_5 \underbrace{\begin{bmatrix} 0 & 0 & 1 \\ 0 & 0 & 0 \\ 1 & 0 & 0 \end{bmatrix}}_{V(5)} + \varepsilon_6 \underbrace{\begin{bmatrix} 0 & 1 & 0 \\ 1 & 0 & 0 \\ 0 & 0 & 0 \end{bmatrix}}_{V(6)} \tag{4.6}$$

$$= \sum_{\alpha=1}^{6} \varepsilon_\alpha V_{ik}(\alpha)$$

This can be viewed as introducing a kind of six-dimensional vector $\underline{\varepsilon} = (\varepsilon_{\alpha=1,\ldots,6})$. We will illustrate Voigt's representation by the deformation of a unit cube (with one corner in the origin which is simpler to draw). The first three terms V(1,2,3) in (4.6) are called uniaxial strains. Fig. 4.2 shows as an example $\varepsilon_1 V(1)$, the uniaxial strain in x-direction, together with the deformation of the inscribed sphere. In general, a uniaxial strain in direction $\underline{a}$ is given by (Fig. 4.1b):

$$\varepsilon = \overset{\circ}{\varepsilon}\, |\hat{\underline{a}}\rangle \langle \hat{\underline{a}}| \ . \tag{4.7}$$

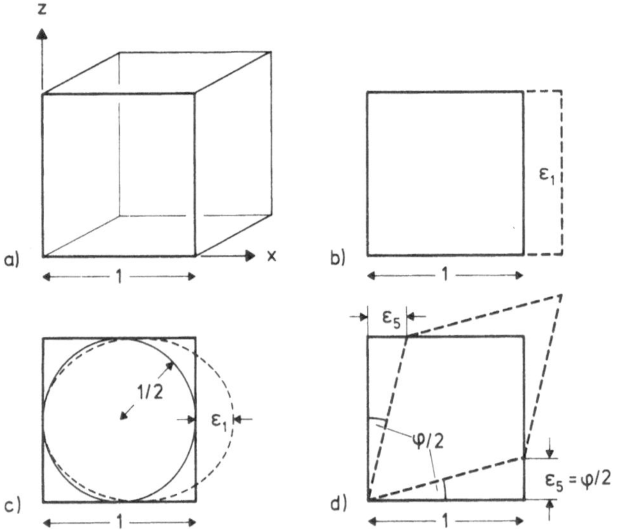


Fig. 4.2a-d. Voigt's deformations. a) unit cube (undergoing deformation), b) $\varepsilon = \varepsilon_1 V(1) = \varepsilon_1 |x\rangle \langle x|$ , uniaxial strain in x-direction, $\Delta V = \varepsilon_1$, c) $\varepsilon = \varepsilon_1 V(1)$, deformation of inscribed sphere into ellipsoid, comp. Fig. 4.1b, d) $\varepsilon = \varepsilon_5 V(5)$, xz-shear, shear angle $\varphi$, $\Delta V = 0$

The last three terms V(4,5,6) in (4.6) are called shear strains. As Fig. 4.2d demonstrates for V(5), the volume of the cube does not change, which is the general definition of a shear strain.

A simple shear can best be viewed as shearing planes $\underline{\hat{b}}$ (a deck of cards) in direction $\underline{\hat{a}}$ by an angle $\varphi$ (Fig. 4.3a),

$$\underline{s}^{(a)} = \varphi\ |\underline{\hat{a}}\rangle \langle\underline{\hat{b}}|\underline{r}\rangle\ , \quad V = \varphi\ |\underline{\hat{a}}\rangle \langle\underline{\hat{b}}|\ , \quad |\underline{\hat{a}}\rangle \perp |\underline{\hat{b}}\rangle\ , \quad V_{ik} = \varphi \hat{a}_i \hat{b}_k\ . \tag{4.8a}$$

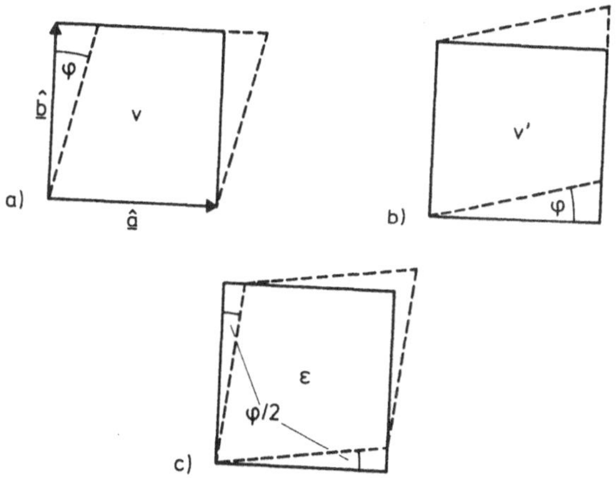


Fig. 4.3a-c. Shears and shear strain. The first two shears are connected by a rotation with angle $\varphi$; they are linked to the third one by a rotation with angle $\varphi/2$.

a) shear $\underline{\hat{a}},\underline{\hat{b}},\varphi$ ; $V = \varphi|\underline{\hat{a}}\rangle \langle\underline{\hat{b}}|$ , $V_{ik} = \varphi\hat{a}_i\hat{b}_k$ ,

b) shear $\underline{\hat{b}},\underline{\hat{a}},\varphi$ ; $V' = \varphi|\underline{\hat{b}}\rangle \langle\underline{\hat{a}}|$ , $V'_{ik} = \varphi\hat{b}_i\hat{a}_k$ ,

c) shear strain $\varepsilon = \dfrac{V + V'}{2}$ , $\varepsilon_{ik} = \varphi\,\dfrac{\hat{a}_i\hat{b}_k + \hat{b}_i\hat{a}_k}{2}$

Interchanging $\underline{\hat{a}}$ and $\underline{\hat{b}}$ results in another simple shear (Fig. 4.3b),

$$\underline{s}^{(b)} = \varphi\ |\underline{\hat{b}}\rangle \langle\underline{\hat{a}}|\underline{r}\rangle\ , \quad V' = \varphi\ |\underline{\hat{b}}\rangle \langle\underline{\hat{a}}|\ , \quad V'_{ik} = \varphi\hat{b}_i\hat{a}_k = V_{ki}\ . \tag{4.8b}$$

The difference, $\underline{s}^{(b)} - \underline{s}^{(a)} = \varphi[\underline{\hat{b}}(\underline{\hat{a}},\underline{r}) - \underline{\hat{a}}(\underline{\hat{b}},\underline{r})] = \varphi(\underline{\hat{a}}\times\underline{\hat{b}})\times\underline{r}$, is a (small) rotation by $\varphi$ about the axis $\underline{\hat{a}}\times\underline{\hat{b}}$. The *strain* resulting from the simple shears (4.8a) and (4.8b) is the same,

$$\underline{s}^{(c)} = \frac{\underline{s}^{(a)} + \underline{s}^{(b)}}{2} = \frac{V + V'}{2} = \varepsilon\underline{r}\ , \quad \varepsilon = \varphi\,\frac{|\underline{\hat{a}}\rangle \langle\underline{\hat{b}}| + |\underline{\hat{b}}\rangle \langle\underline{\hat{a}}|}{2}\ . \tag{4.8c}$$

The strain of (4.8c) can be written in a form which makes its eigenvectors and eigenvalues evident

$$\varepsilon = \varphi \frac{|\hat{\underline{a}}\rangle\langle\hat{\underline{b}}| + |\hat{\underline{b}}\rangle\langle\hat{\underline{a}}|}{2} = \frac{\varphi}{2}\left(\left|\frac{\hat{\underline{a}}+\hat{\underline{b}}}{\sqrt{2}}\right\rangle\left\langle\frac{\hat{\underline{a}}+\hat{\underline{b}}}{\sqrt{2}}\right| - \left|\frac{\hat{\underline{a}}-\hat{\underline{b}}}{\sqrt{2}}\right\rangle\left\langle\frac{\hat{\underline{a}}-\hat{\underline{b}}}{\sqrt{2}}\right|\right). \tag{4.8d}$$

Obviously one has

$$\text{the eigenvectors } |\underline{a}_\pm\rangle = \frac{\hat{\underline{a}} \pm \hat{\underline{b}}}{\sqrt{2}}, \quad \hat{\underline{a}}\times\hat{\underline{b}} \text{ with eigenvalues } \pm 1, 0. \tag{4.8e}$$

This is illustrated in Fig. 4.4 for the xz-shear V(5). A simple shear then can be viewed as a pair of two perpendicular uniaxial strains of opposite sign. The most general shear consists of two such pairs.

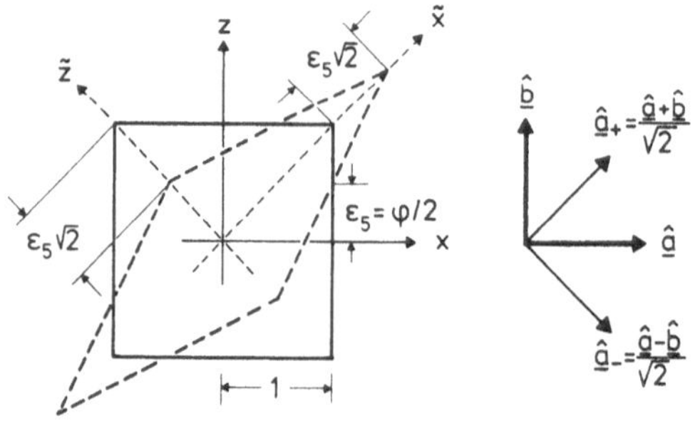


Fig. 4.4. The shear strain $\varepsilon_5 V(5) = \frac{\varphi}{2} V(5)$, $V(5) = |x\rangle\langle z| + |z\rangle\langle x| = |\hat{\underline{a}}_+\rangle\langle\hat{\underline{a}}_+| + |\hat{\underline{a}}_-\rangle\langle\hat{\underline{a}}_-| = |\hat{\underline{a}}\rangle\langle\hat{\underline{b}}| + |\hat{\underline{b}}\rangle\langle\hat{\underline{a}}|$ corresponds to shear strain $\frac{\varphi}{2}\begin{bmatrix}0&0&1\\0&0&0\\1&0&0\end{bmatrix}$ in the xz-system or $\frac{\varphi}{2}\begin{bmatrix}1&0&0\\0&0&0\\0&0&-1\end{bmatrix}$ in the $\tilde{x}\tilde{z}$-system.

Because the three shears V(4,5,6) do not change the volume, the volume change $\Delta V$ is determined by the uniaxial strains alone. It is additive because $\varepsilon$ is small $[(1+\varepsilon_1)(1+\varepsilon_2)(1+\varepsilon_3) \cong 1 + \varepsilon_1 + \varepsilon_2 + \varepsilon_3]$, and it is the relative volume change because we have considered a unit cube:

$$\frac{\Delta V}{V} = \varepsilon_1 + \varepsilon_2 + \varepsilon_3 = \varepsilon_{11} + \varepsilon_{22} + \varepsilon_{33} = \varepsilon_{ii} = \mathrm{tr}\{\varepsilon\}. \tag{4.9}$$

The physical meaning of Voigt's parameters is then: relative length changes $\varepsilon_{1,2,3}$ for three perpendicular uniaxial strains, and shear angles $\varphi_{4,5,6} = 2\varepsilon_{4,5,6}$ for three independent simple shears. One recognizes that the strains represented by 1,2,3 still contain two independent simple shears, such as V(1) - V(3) for which $\Delta V = \mathrm{tr}\{V(1) - V(3)\} = 0$.

Voigt's notation is used almost exclusively in crystal elasticity. However, it has its drawbacks; for instance the "scalar product" $\varepsilon_\alpha \varepsilon_\alpha$ of the "six-vector" changes if one rotates the coordinate system, because only

$$\mathrm{tr}\left\{\varepsilon^2\right\} = \varepsilon_{ik}\varepsilon_{ki} = (\varepsilon_1^2 + \varepsilon_2^2 + \varepsilon_3^2) + 2(\varepsilon_4^2 + \varepsilon_5^2 + \varepsilon_6^2) = \sum_{\alpha=1}^{6} \varepsilon_\alpha g^\alpha \varepsilon_\alpha$$

with $g^\alpha = 1$ for $\alpha = 1,2,3$, $g^\alpha = 2$ for $\alpha = 4,5,6$ is invariant against rotations. Therefore, to make rotational invariance more evident, one ought to introduce $\sqrt{g^\alpha}\,\varepsilon^\alpha$ as the proper and most simple quantities although their introduction will be difficult after so many years of use.

### 4.1.3 Representation by Six Orthonormal Basis Tensors (Adapted to Cubic Symmetry)

Nevertheless we introduce another notation that is very practical, at least for cubic symmetry (including isotropy). For this purpose we define first an (invariant) scalar product for two tensors $\alpha,\beta$,

$$(\alpha,\beta) = \langle\alpha|\beta\rangle = \alpha_{ik}\beta_{ki} = \mathrm{tr}\left\{\alpha\beta\right\} = \mathrm{tr}\left\{\beta\alpha\right\}. \tag{4.10}$$

Secondly, we search for a convenient basis consisting of six "orthonormal" symmetrical tensors $T(\nu = 1\ldots6)$

$$\langle T(\mu)|T(\nu)\rangle = T_{ik}(\mu)T_{ki}(\nu) = \delta^{\mu\nu}\ ; \tag{4.11}$$

an arbitrary strain $\varepsilon$ can then be written as

$$\varepsilon = |T(\nu)\rangle\ \langle T(\nu)|\varepsilon\rangle = |T(\nu)\rangle\ \varepsilon^\nu\ , \tag{4.11a}$$

where $\varepsilon^\nu = \langle T(\nu)|\varepsilon\rangle = T_{ik}(\nu)\varepsilon_{ki}$ is the component of $\varepsilon$ with respect to the basis tensor $T(\nu)$, and for the scalar product of two symmetrical tensors $\alpha,\beta$ one obtains

$$\langle\alpha|\beta\rangle = \alpha^\nu\beta^\nu\ . \tag{4.10a}$$

One such basis could be

$$\hat{V}(\nu) = \frac{1}{\sqrt{g^\nu}}V(\nu)\ ,\quad \varepsilon^\nu = \sqrt{g^\nu}\,\varepsilon_\nu\ ,\ \not\Sigma\ , \tag{4.11b}$$

i.e., a modified and only appropriately normalized Voigt-basis. Yet this is not the most convenient choice. It is natural to try to separate $\varepsilon$ into five independent shear strains with vanishing trace and one so-called "uniform" dilatation which takes care of $\mathrm{tr}\,\{\varepsilon\}$. It is easy to see that the dilatation is given by

$$T(1) = \frac{1}{\sqrt{3}}\begin{bmatrix}1 & 0 & 0\\ 0 & 1 & 0\\ 0 & 0 & 1\end{bmatrix} = \frac{1}{\sqrt{3}}[V(1) + V(2) + V(3)]\ ,\quad T_{ik}(1) = \frac{\delta_{ik}}{\sqrt{3}}\ , \tag{4.12}$$

$$\langle T(1)|T(1)\rangle = 1 \ , \quad \langle T(1)|\varepsilon\rangle = \varepsilon^1 = \frac{1}{\sqrt{3}}\,\mathrm{tr}\{\varepsilon\} \ , \quad \varepsilon^1 T_{ik}(1) = \frac{\mathrm{tr}\,\{\varepsilon\}}{3}\,\delta_{ik} \ . \quad (4.12)$$

If the other vectors of the basis are orthogonal to T(1) then they must have vanishing trace and must, therefore, represent shears. The dilatation is shown in Fig. 4.5a for a cube with its center in the origin and edge length 2. An arbitrary body is deformed in a geometrically similar way, the relative length changes being $1/\sqrt{3}$.

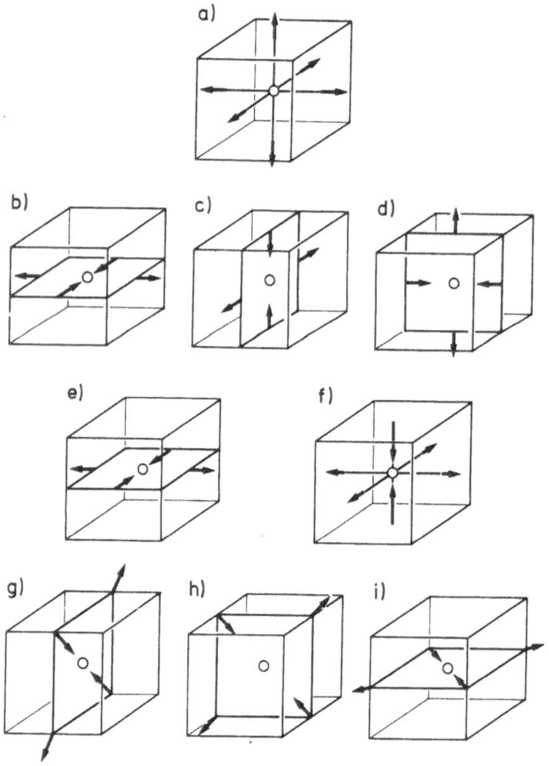


Fig. 4.5a-i. Basis tensors T(ν); the displacements are not to scale; their absolute values are also given. a) The dilatation T(1); $1/\sqrt{3}$. b-d) The (101)-shears T(2,2',2''); $1/\sqrt{2}$. e,f) The complete orthonormal set of two (101)-shears; T(2), $1/\sqrt{2}$; T(3), $2/\sqrt{6}$, $1/\sqrt{6}$. g-i) The (100)-shears T(4,5,6);1

Next we consider simple shears with diagonal elements, for instance (Fig. 4.5b)

$$T(2) = \frac{1}{\sqrt{2}}[-V(2) + V(1)] = \frac{1}{\sqrt{2}}\begin{bmatrix} 1 & 0 & 0 \\ 0 & \bar{1} & 0 \\ 0 & 0 & 0 \end{bmatrix} \ . \quad (4.13)$$

There are two more equivalent diagonal shears,

$$T(2') = \frac{1}{\sqrt{2}}[-V(3) + V(2)] \; , \quad T(2'') = \frac{1}{\sqrt{2}}[-V(1) + V(3)] \; , \tag{4.13a}$$

pictured in Fig. 4.5c,d. These three shears are not independent because T(2) + T(2') + T(2'') = 0. Therefore, we must select two independent and orthogonal shears from these. If one keeps T(2), one can choose

$$T(3) = \frac{1}{\sqrt{3}}[T(2'') - T(2')] = \frac{1}{\sqrt{6}}[-V(1) - V(2) + 2V(3)] = \frac{1}{\sqrt{6}}\begin{bmatrix} \bar{1} & 0 & 0 \\ 0 & \bar{1} & 0 \\ 0 & 0 & 2 \end{bmatrix} \tag{4.13b}$$

as the second independent shear (Fig. 4.5f). Finally, we choose

$$T(4,5,6) = \frac{1}{\sqrt{2}}V(4,5,6) = \hat{V}(4,5,6) \; , \quad \text{e.g.,} \quad T(5) = \frac{1}{\sqrt{2}}\begin{bmatrix} 0 & 0 & 1 \\ 0 & 0 & 0 \\ 1 & 0 & 0 \end{bmatrix} , \tag{4.14}$$

shown in Fig. 4.5g,h,i as pairs of uniaxial strains as in Fig. 4.4.

A simple shear is defined by shear direction and shear plane (Fig. 4.4). In the shears T(2,2',2'') both directions are of the ⟨101⟩ type (face diagonal). We call them (101)-shears, also T(3) which is not a simple shear but a composite of 2 simple (101)-shears. The shears T(4,5,6) are all simple and of (100) type (cube edge). Fig. 4.6 shows the cube before and after the basic deformations.

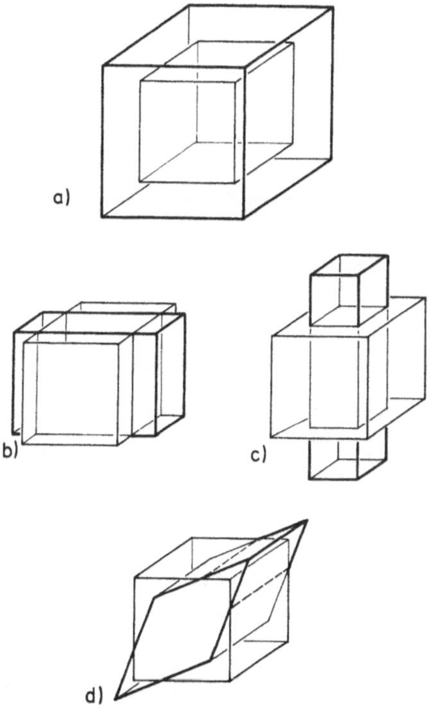


Fig. 4.6a-d. The deformation of a cube due to basis tensors.
a) T(1): cube transforms into a dilatated cube, b) T(2): cube transforms into a brick, c) T(3): cube transforms into a square column, d) T(5): cube transforms into a rhombohedron

Of particular interest is the behaviour of these basis strains under rotation. Obviously T(1) is invariant against any rotation whereas the shears transform among each other, e.g.,

rotation about z-axis by $90^0$:
$T(2) \to -T(2)$ , $T(3) \to T(3)$ , $T(4) \to -T(5)$ , $T(5) \to T(4)$ , $T(6) \to -T(6)$ ,

rotation about z-axis by $45^0$: $T(6) \to -T(2)$ etc. .

With respect to rotations the strains can be separated into a one-dimensional dilatational subspace with projector $P^d$,

$$P^d = |T(1)\rangle \langle T(1)| = (\delta_{ik}\delta_{mn}/3) = (P^d_{ik,mn}) ,$$
$$P^d\varepsilon = |T(1)\rangle \langle T(1)|\varepsilon\rangle = (P^d_{ik,mn}\varepsilon_{mn}) = (\delta_{ik}\ tr\{\varepsilon\}/3) , \tag{4.15a}$$
$$\langle T(1)|T(1)\rangle = P^d_{ik,ki} = 1 \quad \text{(one-dimensional subspace)} ,$$

and a five-dimensional shear subspace with projector

$$P^s = \sum_{\nu=2}^{6} |T(\nu)\rangle \langle T(\nu)| = 1 - |T(1)\rangle \langle T(1)| \tag{4.15b}$$

$$P^s_{ik,mn} = \frac{\delta_{im}\delta_{kn} + \delta_{in}\delta_{km}}{2} - \frac{\delta_{ik}\delta_{mn}}{3} , \quad P^s_{ik,ki} = 6 - 1 = 5 \ \text{(five-dimensional subspace)};$$

$P^d$ and $P^s$ are invariant under rotations.

If we now consider cubic crystals and identify the axes (which so far were arbitrary) with the cube edges, then we recognize another separation if we only consider the admitted rotations (3.23) of cubic symmetry (which leave the cube invariant). It is immediately obvious from Fig. 4.5 that now the (101) shears do not mix with the (100) shears and that with respect to cubic symmetry we can distinguish three subspaces:

$$P^d = |T(1)\rangle \langle T(1)| \text{ , dilatational subspace, one-dimensional ,} \tag{4.16a}$$

$$P^{s(101)} = |T(2)\rangle \langle T(2)| + |T(3)\rangle \langle T(3)| \text{ , the two-dimensional subspace of (101) shears, and} \tag{4.16b}$$

$$P^{s(100)} = 1 - P^d - P^{s(101)}, \text{ the three-dimensional subspace of (100) shears.} \tag{4.16c}$$

This representation will enable us to treat cubic crystals in a very concentrated and nevertheless transparent way. Furthermore the displacement patterns of Fig. 4.5 will enter in a natural way when we discuss defects of cubic symmetry.

## 4.2 The Stress Tensor

### 4.2.1 Stress Tensor and Surface Forces

Experimentally one deforms an elastic body by applying forces to its surface S. For small surface elements $d\underline{S}$ the forces $d\underline{F}$ are proportional to the surface area,

$$d\underline{F} = \sigma d\underline{S} \ , \quad dF_i = \sigma_{ik} dS_k \ , \tag{4.17}$$

where $\sigma$(force/unit area) is the *stress tensor* determining the surface forces (Fig. 4.7a).

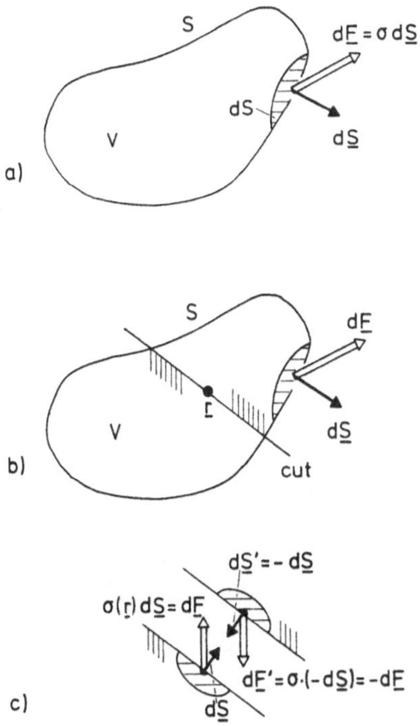


Fig. 4.7a-c. The stress tensor $\sigma$.
a) $\sigma$ determines the surface forces,
b) Cut through the volume V,
c) Equilibrium at each side of the cut is maintained by surface forces determined by $\sigma(\underline{r})$

This concept can also be used to define the stress, $\sigma(\underline{r})$, in the interior where the forces are transferred via the tensions of microscopic (short range) springs. Imagine a cut through the elastic body (Fig. 4.7b) which cuts through many springs under tension. If one were to separate the two sides of the cut (Fig. 4.7c), external forces according to (4.17) would have to applied to maintain the original state in order to compensate for the cut springs (comp. Sec. 5.5). By varying the cut plane one can then in principle obtain the complete $\sigma_{ik} = \partial F_i/\partial S_k$.

Besides "surface-forces", represented by stresses, "volume-forces" $\underline{f}(\underline{r})$, force densities or forces per unit volume, have to be considered, for example gravitational forces.

### 4.2.2 Conservation of Linear and Angular Momentum, Equation of Motion

Now it is easy to obtain the equation of motion from momentum conservation of an arbitrary volume V (surface S) inside an elastic body (Fig. 4.8). The momentum density is the product of mass density[1] (of the initial state) $\rho_o(\underline{r})$ and velocity $\dot{\underline{s}}$. Therefore[2],

$$\underbrace{\partial_t \int_V d\underline{r}\, \rho_o \dot{\underline{s}}}_{\text{change of momentum in V}} = \int_V d\underline{r}\, \rho_o \ddot{\underline{s}} = \underbrace{\underbrace{\int_S \sigma\, d\underline{S}}_{\text{surface forces}} + \underbrace{\int_V d\underline{r}\, \underline{f}}_{\text{volume forces}}}_{\text{force on material in V}} \quad \text{(integral version of momentum conservation)}, \tag{4.18}$$

in components,

$$\int d\underline{r}\, \rho_o \ddot{s}_i = \int_S \sigma_{ik} dS_k + \int_V d\underline{r}\, f_i = \int_V d\underline{r}(\partial_k \sigma_{ik} + f_i) , \tag{4.18a}$$

or, because (4.18a) is valid for arbitrary V,

$$\rho_o \ddot{s}_i(\underline{r},t) = \partial_k \sigma_{ik}(\underline{r},t) + f_i(\underline{r},t) \quad \text{(differential version of momentum conservation)}. \tag{4.18b}$$

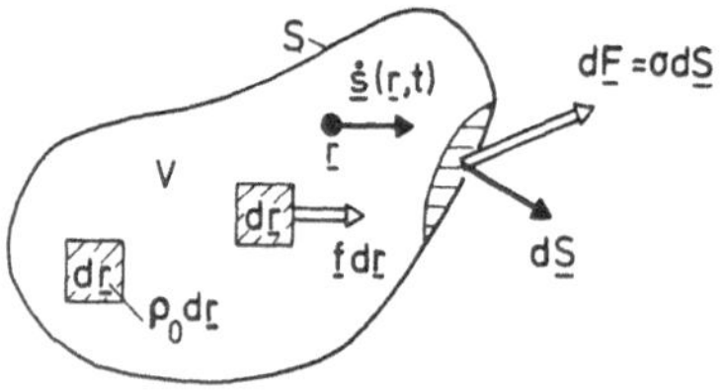


Fig. 4.8. Momenta and forces. With the velocity $\dot{\underline{s}}(\underline{r},t)$ and the mass density $\rho_o(\underline{r})$ the momentum-density is $\rho_o\dot{\underline{s}}$ ; $\rho_o\dot{\underline{s}}\, d\underline{r}$ is the momentum of the volume element $d\underline{r}$. The forces on V are given by the integrated volume force densities, $\int_V d\underline{r}\, \underline{f}$, and by the integrated surface forces, $\int_S \sigma\, d\underline{S}$

---

[1] For a homogeneous crystal $\rho_o$ is independent of $\underline{r}$; in any case it does *not* depend on t.

[2] It is assumed that the displacements are small such that the stresses can be taken at $\underline{r}$ rather than at $\underline{r} + \underline{s}(\underline{r})$.

In eq. (4.18a) we have used Gauss' theorem

$$\int_S A_i(\underline{r})\, dS_i = \int_V d\underline{r}\; \partial_i A_i(\underline{r}) \quad , \tag{4.19}$$

which connects surface and volume integrals[3].

Conservation of angular momentum,

$$\partial_t \int_V d\underline{r}\; \underline{r}\times\rho_o\dot{\underline{s}} = \int_V d\underline{r}\; \underline{r}\times\rho_o\ddot{\underline{s}} = \int_S \underline{r}\times\sigma\; d\underline{S} + \int_V d\underline{r}\; \underline{r}\times\underline{f} \quad , \tag{4.20}$$

leads, upon inserting (4.18), to

$$\int_V d\underline{r}\; \varepsilon_{ikl}\sigma_{lk} = 0 \;,\quad \varepsilon_{ikl}\sigma_{lk} = 0 \;,\quad \text{for } \varepsilon_{ikl} \text{ comp. Appendix A,} \tag{4.20a}$$

which implies that

$$\sigma_{lk} = \sigma_{kl} \;, \text{ the stress tensor is symmetrical.} \tag{4.20b}$$

The symmetry of $\sigma$ holds, if no other contributions to the angular momentum, e.g., magnetic contributions, are present. We will assume symmetrical stress from now on.

### 4.2.3 Representation of the Stress by the Basis Tensors of Section 4.1.3

The symmetrical stress tensor can be expanded in complete analogy to the symmetrical strain,

$$\sigma = \sigma_\alpha V(\alpha) = \sigma^\nu T(\nu) \;. \tag{4.21}$$

Some surface force patterns $T(\nu)d\underline{S}$ are shown in Fig. 4.9. The T(1) contribution is shown in Fig. 4.9a. Because $T(1)d\underline{S} = d\underline{S}/\sqrt{3}$, the corresponding forces are always perpendicular to the surface. This corresponds to a pressure p, (Fig. 4.9b), where

$$\sigma = -p \;,\quad \sigma_{ik} = -p\delta_{ik} \;,\quad \sigma^1 = \frac{\mathrm{tr}\{\sigma\}}{\sqrt{3}} = -\sqrt{3}\, p \tag{4.22}$$

or $p = -\mathrm{tr}\{\sigma\}/3$ as the general definition of the pressure for arbitrary stress. The other stresses are called shear stresses for which $\mathrm{tr}\{\sigma\} = 0$.

---

3 As a simple application consider the change $\Delta V$ of a volume V by (surface) displacements $\underline{s}(\underline{r})$

$$\Delta V = \int_S (\underline{s}, d\underline{S}) = \int_V d\underline{r}\; \partial_i s_i(r) = \int_V d\underline{r}\; \mathrm{tr}\{\varepsilon(\underline{r})\} \quad (= V\,\mathrm{tr}\{\varepsilon\} \text{ for homogeneous } \varepsilon).$$

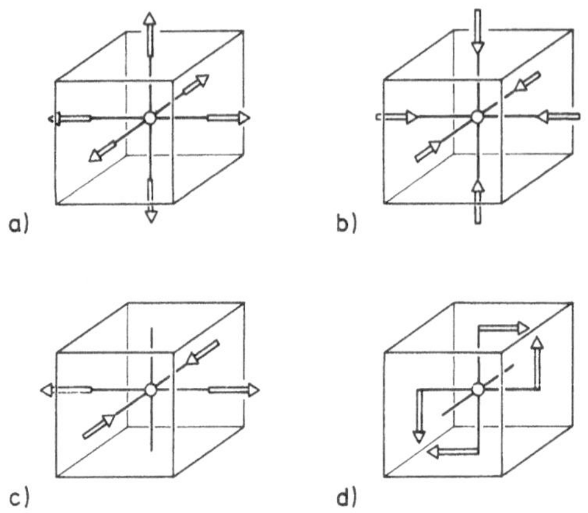

Fig. 4.9a-d. Forces on the surface of a unit cube for some basic stresses.
a) $\sigma = T(1)$, all forces are normal to the surface,
b) $\sigma = -p$, pressure p, $\sigma = \sigma^1 T(1) = -\sqrt{3}pT(1) = -p$,
c) $\sigma = T(2)$, (101) shear stress,
d) $\sigma = T(5)$, (100) shear stress

### 4.2.4 Statics

In the static limit the equation of motion (4.18b) becomes

$$0 = \partial_k \sigma_{ik} + f_i \ . \tag{4.23}$$

One particularly simple situation is a homogeneous $\sigma_{ik}$, not dependent on $\underline{r}$, and vanishing volume forces $\underline{f}$ which is a solution of (4.23). This implies surface forces $d\underline{F} = \sigma d\underline{S}$, and these forces have neither a total force,

$$\int_S d\underline{F} = \int_S \sigma \, d\underline{S} = \sigma \int_S d\underline{S} = 0 \ , \tag{4.24}$$

nor a total torque[4],

$$\int_S \underline{r}\times d\underline{F}\Big|_i = \int_S \underline{r}\times\sigma \, d\underline{S}\Big|_i = \varepsilon_{ikm}\int_S x_k \sigma_{ml} dS_l = \varepsilon_{ikm}\int_V d\underline{r} \, \underbrace{\partial_l x_k \sigma_{ml}}_{\delta_{lk}\sigma_{ml} = \sigma_{mk}} = 0 \tag{4.24a}$$

because of (4.20). Consequently translational-rotational modes are excluded and static equilibrium with these surface forces is possible (comp. Sec. 2.5).

---

[4] This is also illustrated in Fig. 4.9d, where $\sqrt{2}\sigma = |x\rangle\langle z| + |z\rangle\langle x|$. The torque produced by $|x\rangle\langle z|$ on the two z-surfaces is cancelled by the contribution $|z\rangle\langle x|$ on the two x-faces.

## 4.3 Harmonic Theory

### 4.3.1 Linear Stress-Strain Relations

In lattice theory the harmonic approximation is characterized by a linear relation between forces and displacements (or rather, differences of displacements). In continuum theory this means a linear relation between stresses and strains (Hooke's law):

$$\sigma = C\varepsilon \ , \quad \sigma_{ik} = C_{ik,mn}\varepsilon_{mn} \ , \quad C = \text{elastic moduli or stiffnesses} \tag{4.25}$$

$$\varepsilon = S\sigma \ , \quad \varepsilon_{ik} = S_{ik,mn}\sigma_{mn} \ , \quad S = \text{elastic coefficients or compliances.} \tag{4.25a}$$

The symmetries of $\varepsilon,\sigma$ imply that in C, S the two first and the two last subscripts can be interchanged

$$C_{ik,mn} = C_{ki,mn} = C_{ik,nm} \ ; \quad \text{ditto for S.} \tag{4.26}$$

### 4.3.2 The Elastic (Potential) Energy

The elastic (potential) energy U (same symbol as in lattice theory) and its density u (potential energy/unit volume),

$$U = \int_V d\underline{r}\ u \ , \tag{4.27}$$

can be expressed as follows: Consider a *static* situation with $\underline{f} = 0$ and surface forces $d\underline{F} = \sigma d\underline{S}$ corresponding to forces $\underline{F}$ in lattice theory. In lattice theory the total internal energy U is $(\underline{s},\underline{F})/2 = (\underline{s},\Phi\underline{s})/2$, where the factor 1/2 is due to the linear relation between displacements and forces. In complete analogy one has in continuum theory

$$U = \frac{1}{2}\int_S \underline{s}\ d\underline{F} = \frac{1}{2}\int_S s_i\sigma_{ik}dS_k \ . \tag{4.27a}$$

Consequently ($\partial_k\sigma_{ik} = 0$ in statics !),

$$U = \int_V d\underline{r}\,\frac{1}{2}\,\partial_k(s_i\sigma_{ik}) = \int_V d\underline{r}\,\frac{1}{2}\,\varepsilon_{ki}\sigma_{ik} = \int d\underline{r}\,\frac{(\varepsilon,\sigma)}{2} = \int d\underline{r}\ u \ ; \tag{4.27b}$$

the energy density u is given by

$$2u = (\varepsilon,\sigma) = (\varepsilon,C\varepsilon) = (\sigma,S\sigma) = \varepsilon_{ki}C_{ik,mn}\varepsilon_{nm} = \sigma_{ki}S_{ik,mn}\sigma_{nm} \ . \tag{4.28}$$

Since one can start, as in lattice theory, with the potential energy as the primary quantity, one recognizes from (4.28) that one can assume symmetry in the 1st and 2nd pairs of subscripts,

$$C_{ik,mn} = C_{mn,ik} \; ; \quad \text{ditto for S.} \tag{4.26a}$$

Only the corresponding symmetrized version of C determines the energy density and hence the physics of elasticity. Elastic stability requires $u > 0$ for non-vanishing $\varepsilon$ or $\sigma$ and puts restraints on the values of stiffnesses or compliances.

### 4.3.3 The Elastic Moduli in Voigt's Notation

Because of (4.26) one can use Voigt's notation as in (4.6), $C_{ik,mn} \to c_{\alpha\beta}$,

| $C_{ik} \diagdown_{mn}$ | 11 | 22 | 33 | 23 | 31 | 12 |
|---|---|---|---|---|---|---|
| 11 | $c_{11}$ | $c_{12}$ | $c_{13}$ | $c_{14}$ | $c_{15}$ | $c_{16}$ |
| 22 | $c_{21}$ | $c_{22}$ | $c_{23}$ | $c_{24}$ | $c_{25}$ | $c_{26}$ |
| 33 | $c_{31}$ | $c_{32}$ | $c_{33}$ | $c_{34}$ | $c_{35}$ | $c_{36}$ |
| 23 | $c_{41}$ | $c_{42}$ | $c_{43}$ | $c_{44}$ | $c_{45}$ | $c_{46}$ |
| 31 | $c_{51}$ | $c_{52}$ | $c_{53}$ | $c_{54}$ | $c_{55}$ | $c_{56}$ |
| 12 | $c_{61}$ | $c_{62}$ | $c_{63}$ | $c_{64}$ | $c_{65}$ | $c_{66}$ |

(4.29)

and from (4.26a) one has

$$c_{\alpha\beta} = c_{\beta\alpha} \; . \tag{4.29a}$$

One recognizes that the 6×6 matrix c is symmetrical and contains 21 independent elements. Then ( $\not\Sigma$ in (4.30))

$$\varepsilon = \sum_{\alpha} \varepsilon_{\alpha} V(\alpha) \; ; \quad \sigma = \sum_{\alpha} \sigma_{\alpha} V(\alpha) \; ; \quad \varepsilon_{\alpha} = \frac{1}{g^{\alpha}} \Big( V(\alpha), \varepsilon \Big) ; \quad \begin{array}{l} g^{\alpha} = 1 \text{ for } \alpha = 1,2,3 \\ g^{\alpha} = 2 \text{ for } \alpha = 4,5,6 \; , \end{array} \tag{4.30a}$$

$$\sigma_{\alpha} = \sum_{\beta} c_{\alpha\beta} g^{\beta} \varepsilon_{\beta} \; , \quad c_{\alpha\beta} = \frac{\Big( V(\alpha), CV(\beta) \Big)}{g^{\alpha} g^{\beta}} = \frac{s^{-1}_{\alpha\beta}}{g^{\alpha} g^{\beta}} \; , \; C = \sum_{\alpha,\beta} V(\alpha) c_{\alpha\beta} V(\beta) \; , \tag{4.30b}$$

$$\varepsilon_{\alpha} = \sum_{\beta} s_{\alpha\beta} g^{\beta} \varepsilon_{\beta} \; , \quad s_{\alpha\beta} = \frac{\Big( V(\alpha), SV(\beta) \Big)}{g^{\alpha} g^{\beta}} = \frac{c^{-1}_{\alpha\beta}}{g^{\alpha} g^{\beta}} \; , \; S = \sum_{\alpha,\beta} V(\alpha) s_{\alpha\beta} V(\beta) . \tag{4.30c}$$

The somewhat complicated relations (4.30) can be traced back to the fact that the $V(\alpha)$ are not normalized: $\left(V(\alpha),V(\alpha)\right) = g^{\alpha}$. If one uses instead normalized $\hat{V}(\alpha) = V(\alpha)/\sqrt{g^{\alpha}}$ one obtains more transparent relations:

$$\varepsilon = \hat{\varepsilon}_{\alpha}\hat{V}(\alpha) , \quad \sigma = \hat{\sigma}_{\alpha}V(\alpha) , \quad \hat{\varepsilon}_{\alpha} = \left(\hat{V}(\alpha),\varepsilon\right) = \sqrt{g^{\alpha}}\,\varepsilon_{\alpha} \left(\not\Sigma\right), \tag{4.31a}$$

$$\hat{\sigma}_{\alpha} = \hat{c}_{\alpha\beta}\hat{\varepsilon}_{\beta} , \quad \hat{c}_{\alpha\beta} = \left(\hat{V}(\alpha),C\hat{V}(\beta)\right) = \hat{s}^{-1}_{\alpha\beta} , \quad C = \hat{V}(\alpha)\hat{c}_{\alpha\beta}\hat{V}(\beta) , \tag{4.31b}$$

$$\hat{\varepsilon}_{\alpha} = \hat{s}_{\alpha\beta}\hat{\sigma}_{\beta} , \quad \hat{s}_{\alpha\beta} = \left(V(\alpha),SV(\beta)\right) = \hat{c}^{-1}_{\alpha\beta} , \quad S = \hat{V}(\alpha)\hat{s}_{\alpha\beta}\hat{V}(\beta) . \tag{4.31c}$$

Here the quantities $\varepsilon_{\alpha}$, $c_{\alpha\beta}$ are direct components of the tensors $\varepsilon$, $C$ whereas $\hat{\varepsilon}_{\alpha}$, $\hat{c}_{\alpha\beta}$ differ by factors $\sqrt{g^{\alpha}}$, $\sqrt{g^{\alpha}g^{\beta}}$.

Voigt's notation is used in tables, where one finds values for $c_{\alpha\beta}$[dyn/cm$^2$ = erg/cm$^3$ or eV/Å$^3$] according to (4.29, 30b), for given orientation with respect to certain crystal axes. The $s_{\alpha\beta}$'s found in tables are mostly not defined by (4.30c) analogously to $c_{\alpha\beta}$; the table values of $s_{\alpha\beta}$ are, as a rule, identical with $c^{-1}_{\alpha\beta}$. Again, this lack of symmetry has historical reasons.

The condition for elastic stability ($u > 0$) is

$$2u = (v,Cv) = (\varepsilon,C\varepsilon) = \hat{\varepsilon}_{\alpha}\hat{c}_{\alpha\beta}\hat{\varepsilon}_{\beta} = \sum_{\alpha,\beta} \varepsilon_{\alpha}g^{\alpha}c_{\alpha\beta}g^{\beta}\varepsilon_{\beta} > 0 \quad \text{if } \varepsilon \neq 0 , \tag{4.32}$$

which means that $c_{\alpha\beta}$ or $\hat{c}_{\alpha\beta}$ are positive definite matrices (positive eigenvalues).

### 4.3.4 The Equation of Motion

Equation (4.18b) can be written in terms of $\underline{s}$ alone,

$$\rho_0(\underline{r})\ddot{s}_i(\underline{r},t) = \underbrace{\partial_k\sigma_{ik}}_{\text{internal}} + \underbrace{f_i(\underline{r},t)}_{\text{external}} = \partial_k C_{ik,mn}(\underline{r})\partial_m s_n + f_i ; \tag{4.33}$$

force densities

this differential equation for $\underline{s}(\underline{r},t)$ has to be supplemented by boundary conditions on S, e.g., specifying displacements, $\underline{s}$, or surface forces, $\sigma d\underline{S}$, on all points of S. On interior surfaces S, e.g., grain boundaries or inclusions, $\underline{s}$ *and* $\sigma d\underline{S}$ must be continuous, otherwise the differentiations in (4.33) diverge. The symmetrical positive matrix D of lattice theory is replaced by the three-dimensional matrix differential operator

$$D_{in} = -(1/\sqrt{\rho_0})\ \partial_k C_{ik,mn}\partial_m(1/\sqrt{\rho_0})$$

which has positive eigenvalues because of elastic stability (App. I).

Eq. (4.33) represents the "conservation of momentum": $\rho_o \dot{s}_i$ is the density of momentum in direction i; $-\sigma_{ik}$ is the corresponding current density which changes the momentum within a volume by flowing through its surface; $f_i$ represents the change of momentum by forces. As in classical mechanics one obtains conservation of energy by multiplying the equation of motion with the velocity: $\partial_t(u_{kin} + u) - \mathrm{div}\{\sigma\dot{\underline{s}}\} = (\underline{f},\dot{\underline{s}})$, where $u_{kin} = \rho_o\dot{\underline{s}}^2/2$ is the density of kinetic energy, $u_{kin} + u$ the total energy density, $-\sigma\dot{\underline{s}}$ the density of the energy current and $(\underline{f},\dot{\underline{s}})$ the density of the work done by the forces. Since the theory is more or less symmetrical with respect to time derivatives and spatial derivatives one obtains corresponding conservation equations upon multiplying (4.33) with $\partial_m s_i$:

$$\partial_t(-\rho_o\dot{s}_i\partial_m s_i) - \partial_j[\delta_{jm}(u - u_{kin}) - \sigma_{ij}\partial_m s_i] = -f_i\partial_m s_i = \partial_t\Pi_m - \partial_j\Pi_{jm} ;$$

similar conservation equations are obtained in all field theories, e.g., for the electromagnetic field; without forces these equations can be traced back to the invariance of the field theory with respect to shifts in time and space; the quantity $\Pi_m$ is called the field momentum or the quasimomentum; in normal fields, where $\underline{r}$ and its shift refer to a position in space (and not to a mark as in elastic theory), $\Pi_m$ is the actual momentum density and the $-\Pi_{jm}$ are the corresponding currents; in elastic theory the quasimomentum conservation requires homogeneity of the material (where $\rho_o$ and C are independent of $\underline{r}$) and is therefore not universal; nonetheless it has its uses, e.g., the static quasi stress tensor $\Pi_{mj}$,

$$\Pi_{mj} = u\delta_{jm} - \sigma_{ij}\partial_m s_i , \quad \partial_j\Pi_{mj} = f_i\partial_m s_i ,$$

can be used to represent the forces on defects (/4.13/), comp. Sec. 4.8.8. Similar considerations apply to angular momentum.

In homogeneous crystals, $\partial_{\underline{r}} C = 0$, $\partial_{\underline{r}}\rho_o = 0$, one realizes that not C itself enters directly into (4.33),

$$\rho_o\ddot{s}_i = C_{ik,mn}\partial_k\partial_m s_n + f_i = H_{in,km}\partial_k\partial_m s_n + f_i , \tag{4.33a}$$

but rather a symmetrized version (symmetrized with respect to k and m), the Huang tensor:

$$H_{in,km} = \frac{1}{2}(C_{ik,nm} + C_{im,nk}) ,$$

$$H_{ik,mn} = \frac{1}{2}(C_{in,mk} + C_{im,nk}) , \tag{4.34}$$

$$h_{11} = c_{11} , \quad h_{12} = c_{66} , \quad h_{14} = c_{56} , \quad h_{15} = c_{15} , \quad h_{66} = \frac{1}{2}(c_{12} + c_{66}), \text{ etc.}$$

The Huang tensor has the same symmetries as C and therefore the same number of independent components. The relation between H and C is linear and can be described by an operator L,

$$H = LC \, . \tag{4.34a}$$

If one repeats the operation of L on H one sees immediately that

$$LH = \frac{1}{2}(C + H) = \frac{1}{2}(1 + L)C = L^2C \quad \text{or} \quad 2L^2 = 1 + L \, , \tag{4.34b}$$

and that the reciprocal to (4.34a) becomes

$$\begin{gathered} C = 2LH - H \, , \quad C_{ik,mn} = H_{im,kn} + H_{in,mk} - H_{ik,mn} \, , \quad \text{e.g.,} \\ c_{11} = h_{11} \, , \quad c_{12} = 2h_{66} - h_{12} \, , \quad c_{14} = 2h_{56} - h_{14} \, , \quad c_{15} = h_{15} \, , \quad c_{66} = h_{12}, \text{etc.} \end{gathered} \tag{4.34c}$$

The introduction of the Huang tensor is essential for a comparison of lattice with continuum theory in the proper limit (Chap. 5). For infinite crystals, it turns out that in the limit of slowly varying displacements or long waves one obtains directly H and has then to determine C via (4.34c). If one tries to calculate the elastic data by considering the elastic energy of finite crystals under homogeneous deformations, one also has to calculate H first. The direct expressions for C are surface sensitive, whereas the expressions for H are not.

### 4.3.5 Cauchy Relations

A special case is a totally symmetric C, where one can interchange all subscripts,

$$C_{ik,mn} = C_{im,kn} = C_{in,mk} \, , \text{etc.} \, , \quad C = H \, . \tag{4.35}$$

The $c_{\alpha\beta} = c_{\beta\alpha}$ are no longer independent but obey the 6 relations

$$\begin{gathered} C_{11,22} = c_{12} = c_{66} = C_{12,12} \, , \quad c_{13} = c_{55} \, , \quad c_{23} = c_{44} \, , \\ c_{14} = c_{56} \, , \quad c_{25} = c_{46} \, , \quad c_{36} = c_{45} \, , \end{gathered} \tag{4.35a}$$

which are called Cauchy's relations. We will see later that they are valid for Bravais lattices with two-body interactions.

One can always separate C into a symmetrized tensor T with total symmetry and a tensor $\Gamma$ representing the deviations from (4.35a). One recognizes from (4.34) that C + 2H is totally symmetric, therefore

$$C = \frac{1}{3}(C + 2H) + \frac{2}{3}(C - H) = T + \Gamma \, , \quad \text{i.e.,} \tag{4.35b}$$

$$\begin{gathered} t_{11} = c_{11} \, , \quad \gamma_{11} = 0 \, ; \quad t_{12} = t_{66} = \frac{1}{3}(c_{12} + 2c_{66}) \, , \\ \gamma_{12} = \frac{2}{3}(c_{12} - c_{66}) = -2\gamma_{66} \, , \quad \text{etc.} \, , \end{gathered}$$

where $t_{\alpha\beta}$ , $\gamma_{\alpha\beta}$ are the components of T , $\Gamma$ in Voigt's notation. The tensor T has 15 independent components[5] and the tensor $\Gamma$ has 6. The separation into T and $\Gamma$ is conserved under rotation, i.e., the rotated tensor T is also totally symmetric. The Cauchy tensor, $\Gamma$, has 6 independent components, which transform among themselves under rotations and which can be represented by a symmetrical tensor of 2nd rank, $\Gamma^{(2)}$. If one introduces the tensor $E_{ik,mn,st} = \varepsilon_{ims}\varepsilon_{knt}$, symmetrized with respect to ik and mn, which is invariant with respect to interchange of pairs and interchange within one pair, one obtains

$$\Gamma^{(2)}_{st} = -\frac{1}{6}E_{ik,mn,st}\Gamma_{ik,mn} = -\frac{1}{6}\varepsilon_{ims}\varepsilon_{knt}\Gamma_{ik,mn} = -\frac{1}{6}\varepsilon_{ims}\varepsilon_{knt}C_{ik,mn}, \quad (4.35c)$$

$$\Gamma_{ik,mn} = -2E_{ik,mn,st}\Gamma^{(2)}_{st} .$$

## 4.4 Cubic Symmetry

### 4.4.1 Elastic Moduli, C, for Cubic Symmetry

If one starts from the relations

$$\sigma = C\varepsilon , \quad \sigma_{ik} = C_{ik,mn}\varepsilon_{mn} \quad \text{or} \quad \sigma^{\mu} = \Big(T(\mu),CT(\nu)\Big)\varepsilon^{\nu} = C^{\mu\nu}\varepsilon^{\nu} ,$$

and rotates the crystal, the relation between $\sigma$ and $\varepsilon$ is changed unless the rotation is a symmetry operation, under which C transforms into itself. If the tilde denotes any of the 48 cubic rotations D (as an even rank tensor C is already invariant against inversion, and the 24 proper rotations are sufficient), one has for the transformed tensor $\tilde{C}$:

$$\tilde{C} = C , \quad \tilde{C}_{ik,mn} = D_{ii'}D_{kk'}D_{mm'}D_{nn'}C_{i'k',m'n'} ,$$
$$\Big(T(\mu),\tilde{C}T(\nu)\Big) = \Big(T(\mu),CT(\nu)\Big) \quad \text{or also} \quad \Big(\tilde{T}(\mu),C\tilde{T}(\nu)\Big) = \Big(T(\mu),CT(\nu)\Big) . \quad (4.36)$$

If one uses all the restrictions (4.26), which is not difficult[6], one is left with

[5] The transformation behaviour under rotations is discussed in Appendix J.

[6] Compare the remarks in Section 3.2.4. All subscripts x,y,z can be interchanged; elements with one single subscript x,y or z are zero. Then only components $C_{xx,xx} = C_{yy,yy} = \ldots = c_{11}$, $C_{xx,yy} = C_{xx,zz} = \ldots = c_{12}$, $C_{xy,xy} = C_{xz,xz} = \ldots = c_{44}$ are left.

$$C_{ik,mn} = c_{12}\delta_{ik}\delta_{mn} + c_{44}(\delta_{im}\delta_{kn} + \delta_{in}\delta_{km}) + (c_{11} - c_{12} - 2c_{44})\delta_{ikmn} ,$$
$$C = c_{12}A_1 + c_{44}A_2 + \underbrace{(c_{11} - c_{12} - 2c_{44})}_{c_a}A_3 \qquad (4.37)$$

where $\delta_{ikmn} = 1$ if all subscripts are equal and $\delta_{ikmn} = 0$ otherwise. In Voigt's notation

$$c_{\alpha\beta} = \begin{bmatrix} c_{11} & c_{12} & c_{12} & 0 & 0 & 0 \\ c_{12} & c_{11} & c_{12} & 0 & 0 & 0 \\ c_{12} & c_{12} & c_{11} & 0 & 0 & 0 \\ 0 & 0 & 0 & c_{44} & 0 & 0 \\ 0 & 0 & 0 & 0 & c_{44} & 0 \\ 0 & 0 & 0 & 0 & 0 & c_{44} \end{bmatrix} \qquad (4.37a)$$

The first two terms, $A_{1,2}$, in (4.37) are isotropic (invariant against all rotations), and the third term, $A_3$ , has cubic symmetry (invariant against all cubic rotations). If

$$c_a = c_{11} - c_{12} - 2c_{44} = 0 , \quad \text{isotropy-condition } (c_a \text{ is a measure of anisotropy}), \qquad (4.38)$$

the crystal is isotropic[7]. Cubic crystals have three independent elastic data, isotropic crystals have two. Elastic stability requires the eigenvalues of the matrix (4.37a) to be positive,

$$c_{11} + 2c_{12} > 0 , 1\times ; \quad c_{11} - c_{12} > 0 , 2\times ; \quad c_{44} > 0 , 3\times . \qquad (4.39)$$

### 4.4.2 Eigentensors and Eigenvalues of C

Voigt's scheme is not very transparent; in particular, the $c_{\alpha\beta}$ so far have no direct physical meaning. For our purpose the basis $T(\nu)$, (4.12-14), is most convenient and illustrative because it is adapted to cubic symmetry, and it turns out that the matrix $C^{\mu\nu}$ is diagonal:

[7] The condition for isotropy is the same for the $s_{\alpha\beta}$ defined by (4.30c), $s_{11} - s_{12} - 2s_{44} = 0$ but not for the $c^{-1}_{\alpha\beta}$ which are given in tables, $c^{-1}_{11} - c^{-1}_{12} - c^{-1}_{44}/2 = 0$.

$$C^{\mu\nu} = \overset{\nu}{C}\delta^{\mu\nu}\ , \ \forall \quad ; \quad \overset{1}{C}\ , \quad \overset{2}{C} = \overset{3}{C}\ , \quad \overset{4}{C} = \overset{5}{C} = \overset{6}{C}\ ;$$

$$C = T(\nu)C^{\nu\mu}T(\mu) = \sum_{\nu} T(\nu)\overset{\nu}{C}T(\nu) \quad ; \quad CT(\mu) = \overset{\mu}{C}T(\mu),\ \forall\ . \tag{4.40}$$

This can easily be seen from the behaviour of the states $T(\nu)$ under cubic rotations, some of which are listed in Table 4.1. These results can be obtained by inspection of Fig. 4.5. The 1st column shows that the states 1,2,2',2'',3,4 are eigenstates of the $\pi$-rotation about [100] with eigenvalue 1 and that states 5,6 have eigenvalue -1; therefore,

$$C^{15} = \Big(T(1),CT(5)\Big) = \Big(\tilde{T}(1),C\tilde{T}(5)\Big) = -\Big(T(1),CT(5)\Big) = -C^{15} = 0\ .$$

Table 4.1. Transformation of basis tensors $T(\nu)$ under cubic symmetry operations

| Axis of rotation / Angle | | [100] | | [010] | | [001] | |
|---|---|---|---|---|---|---|---|
| | | $\pi$ | $\pi/2$ | $\pi$ | $\pi/2$ | $\pi$ | $\pi/2$ |
| transformed (xyz) $\tilde{x}\,\tilde{y}\,\tilde{z}$ ($\bar{x} = -x$) | | $x\,\bar{y}\,\bar{z}$ | $x\,\bar{z}\,y$ | $\bar{x}\,y\,\bar{z}$ | $z\,y\,\bar{x}$ | $\bar{x}\,\bar{y}\,z$ | $\bar{y}\,x\,z$ |
| $T(\nu)$ changes into | $\nu$ = 1 | T(1) | T(1) | T(1) | T(1) | T(1) | T(1) |
| | 2 | T(2) | -T(2'') | T(2) | -T(2') | T(2) | -T(2) |
| | 2' | T(2') | -T(2') | T(2') | -T(2) | T(2') | -T(2") |
| | 2" | T(2") | -T(2) | T(2") | -T(2") | T(2") | -T(2') |
| | 3 | T(3) | -- | T(3) | -- | T(3) | T(3) |
| | 4 | T(4) | -T(4) | -T(4) | T(6) | -T(4) | -T(5) |
| | 5 | -T(5) | -T(6) | T(5) | -T(5) | -T(5) | T(4) |
| | 6 | -T(6) | T(5) | -T(6) | -T(4) | T(6) | -T(6) |

Going through all the columns, one recognizes that $C^{\mu\nu}$ is diagonal[8]. From the second column we have $\tilde{T}(6) = T(5)$ and therefore

$$\overset{6}{C} = C^{66} = \left(T(6), CT(6)\right) = \left(\tilde{T}(6), C\tilde{T}(6)\right) = C^{55} = \overset{5}{C} ,$$

and similarly $\overset{6}{C} = \overset{4}{C}$ ; further $C^{22} = C^{2'2'} = C^{2''2''}$ , $C^{22'} = C^{2'2''} = C^{2''2}$ . That $\overset{2}{C} = \overset{3}{C}$ can also be seen from $C^{33} = (C^{2''2''} + C^{2'2'} - C^{2''2'} - C^{2'2''})/3 = (2C^{22} - C^{22'} - C^{22''})/3 = C^{22}$, if one recognizes from $T(2') + T(2'') = -T(2)$ that $C^{22'} + C^{22''} = C^{2,2'+2''} = -C^{22}$.

Equation (4.40) shows then that the $T(\nu)$ are the eigenstates (tensors of 2nd rank) of the 4th rank tensor C with eigenvalues $\overset{\nu}{C}$ and, of course, the most convenient representation of C is by its eigenstates and eigenvalues. Eq. (4.40) states further that shear strains and shear stresses, dilatation and pressure are directly connected

$$\sigma^{\mu} = \overset{\mu}{C}\varepsilon^{\mu} , \not\Sigma \quad . \tag{4.40a}$$

It is obvious that the cubic C has three independent moduli, the three independent eigenvalues $\overset{1}{C},\overset{2}{C},\overset{5}{C}$. An isotropic C has two independent moduli, $\overset{2}{C} = \overset{5}{C}$, because the relation between shear stresses and strains cannot depend on the special type of shear. The elastic compliances become

$$S = \sum_{\nu} T(\nu) \frac{1}{\overset{\nu}{C}} T(\nu) \quad , \tag{4.40b}$$

and elastic stability means simply

$$\overset{\nu}{C} > 0 \ , \quad \text{comp. (4.39)} \ , \tag{4.41}$$

because $(\varepsilon, C\varepsilon) = \sum_{\nu} \varepsilon^{\nu}\overset{\nu}{C}\varepsilon^{\nu}$.

The physical meaning of the $\overset{\nu}{C}$ is contained in (4.40a) and demonstrated in Fig. 4.10. The T(1) pattern is considered in Fig. 4.10a; stress and strain patterns are the same, differing only be the factor $\overset{1}{C}$. Accordingly, we have $p = -\overset{1}{C}\,\Delta V/3V$ or $-Vp/\Delta V = -V\partial_V p = \overset{1}{C}/3 = K$, where K is called the compressional or bulk modulus[9]. Fig. 4.10b shows the shear pattern T(5). The ratio of shear stress and shear angle, here $\overset{5}{C}/2$, is called the shear modulus, $\mu$ , for (100) shears. For (110) shears the situation is analogous, the modulus, $\overset{2}{C}/2$, is denoted by $\mu'$. The connection between Voigt's moduli, $c_{\alpha\beta}$ , and the eigenvalues $\overset{\nu}{C}$ are obtained from

---

8 That $C^{13} = 0$ is not immediately evident. It follows from $C^{12'} = 0 = C^{12''}$, because T(3) is a linear combination of 2' and 2''.

9 The compressibility is 1/K.

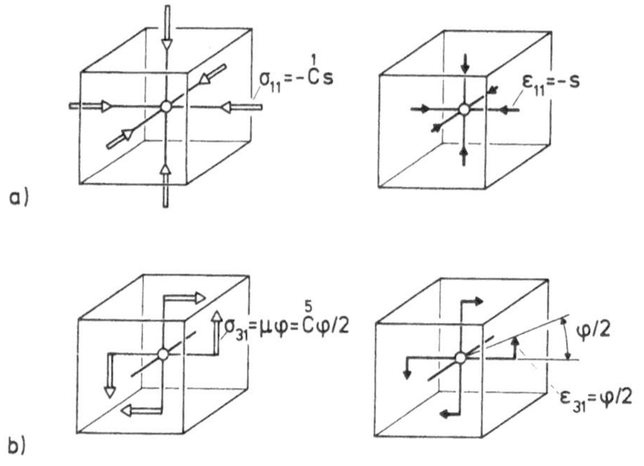


Fig. 4.10a and b. Physical meaning of the eigenvalues of C.
a) T(1)-pattern

$$\sigma = \underbrace{-\overset{1}{C}s}_{-p} \begin{bmatrix} 1 & 0 & 0 \\ 0 & 1 & 0 \\ 0 & 0 & 1 \end{bmatrix} , \quad \varepsilon = \underbrace{-s}_{\frac{\Delta V}{3V}} \begin{bmatrix} 1 & 0 & 0 \\ 0 & 1 & 0 \\ 0 & 0 & 1 \end{bmatrix} ; \quad \overset{1}{C} = -3V \frac{p}{\Delta V} = 3K$$

b) Shear pattern T(5)

$$\sigma = \overset{5}{C} \frac{\varphi}{2} \begin{bmatrix} 0 & 0 & 1 \\ 0 & 0 & 0 \\ 1 & 0 & 0 \end{bmatrix} , \quad \varepsilon = \frac{\varphi}{2} \begin{bmatrix} 0 & 0 & 1 \\ 0 & 0 & 0 \\ 1 & 0 & 0 \end{bmatrix} ; \quad \mu = \frac{\sigma}{\varphi} = \frac{\overset{5}{C}}{2}$$

$$\overset{\nu}{C} = \Big(T(\nu), CT(\nu)\Big), \not\Sigma ; \quad c_{\alpha\beta} = \Big(V(\alpha), CV(\beta)\Big)/g^{\alpha}g^{\beta} , \not\Sigma , \quad (4.30b) , \tag{4.42}$$

and (4.12-14) which give $T(\nu)$ in terms of $V(\alpha)$. Eventually,

$$\overset{1}{C} = 3K = c_{11} + 2c_{12} \qquad K \text{ bulk modulus} , \tag{4.43a}$$

$$\overset{2}{C} = \overset{3}{C} = 2\mu' = c_{11} - c_{12} \qquad \mu' \text{ shear modulus for (110) shears} , \tag{4.43b}$$

$$\overset{4}{C} = \overset{5}{C} = \overset{6}{C} = 2\mu = 2c_{44} \qquad \mu \text{ shear modulus for (100) shears} , \tag{4.43c}$$

$$c_{11} = K + 4\mu'/3 , \quad c_{12} = K - 2\mu'/3 , \quad c_{44} = \mu . \tag{4.43d}$$

If we introduce projectors $P^K$, $P^{\mu'}$, $P^{\mu}$ onto the states T(1), T(2) and T(3), T(4) and T(5) and T(6), where $P^K$ picks out the dilatation and $P^{\mu'}$, $P^{\mu}$ the corresponding shears, one can write

$$C = \overset{1}{C} P^K + \overset{2}{C} P^{\mu'} + \overset{5}{C} P^{\mu} = 3K P^K + 2\mu' P^{\mu'} + 2\mu P^{\mu} ,$$

and can by comparison with (4.37), $C = c_{12}A_1 + c_{44}A_2 + (c_{11} - c_{12} - 2c_{44})A_3$, establish the relations[10]

$$A_1 = 3P^K \;; \quad A_2 = 2 \;; \quad A_3 = 1 - P^\mu \;.$$

Consequently,

$$C = (c_{11} + 2c_{12})P^K + (c_{11} - c_{12})P^{\mu'} + 2c_{44}P^\mu \;,$$

$$S = \underbrace{(s_{11} + 2s_{12})}_{\frac{1}{c_{11}+2c_{12}}}P^K + \underbrace{(s_{11} - s_{12})}_{\frac{1}{c_{11}-c_{12}}}P^{\mu'} + \underbrace{2s_{44}}_{\frac{1}{2c_{44}}}P^\mu \;, \qquad (4.44a)$$

which establishes the relations between $s_{\alpha\beta}$ and $c_{\alpha\beta}$.

The above representation of C is very transparent; it is certainly much more transparent than Voigt's representation, and we will use it from now on, whenever we deal with cubic crystals. One useful application of the new notation is the calculation of the so-called Voigt averages. It consists of taking the average of C over all orientations of the crystal; the resulting $\bar{C}$ is isotropic and thought to be an approximation to the elastic moduli of polycrystals, which are isotropic. For cubic crystals this average can be calculated without difficulty from (4.44a). Since $P^K$ is *invariant*: $\overline{P^K} = P^K$. Further it is clear that for the shear components $\overline{T_{ik}(\nu)T_{mn}(\nu)}$ must be independent of $\nu$ and equal $(1 - P^K)/5$ where $(1 - P^K) = P^s$ picks out all shears. Therefore, $\overline{P^{\mu'}} = 2P^s/5$, $\overline{P^\mu} = 3P^s/5$ and

$$\bar{C} = 3KP^K + 2\bar{\mu}P^s \;, \text{ where} \qquad (4.45)$$

$$\bar{\mu} = \frac{2\mu' + 3\mu}{5} \qquad (4.45a)$$

is a weighted average of the two cubic shear moduli; the weights 2 and 3 are the dimensions of the two shear spaces $P^{\mu'}$, $P^\mu$. That K must be equal for single- and poly-crystal is easy to show. But Voigt's average $\bar{\mu}$ is only an approximation to the shear modulus of an isotropic polycrystal, $\mu^{eff}$. Actually $\bar{\mu}$ is an upper limit. A lower limit is established by the so-called Reuss averages, where $\bar{S}$ is calculated, $\bar{S} = (1/3K)P^K + (1/2\mu^{Reuss})P^s$: $\mu^{Reuss} = \left(\overline{\frac{1}{\mu}}\right)^{-1} < \mu^{eff} < \bar{\mu} = \mu^{Voigt}$. These limits will be discussed in more detail in Section 4.8.7.

---

10 $A_2/2$ represents unity only for symmetrical tensors of 2nd rank. $A_3\varepsilon$ picks out the diagonal elements of $\varepsilon$; $(1 - A_3)\varepsilon$ removes the diagonal elements and leaves only the off diagonal elements which represent (100) shears.

### 4.4.3 The Equation of Motion

Employing the Huang tensor (4.34),

$$H_{ik,mn} = \underbrace{h_{12}}_{c_{44}}\delta_{ik}\delta_{mn} + \underbrace{h_{44}}_{\frac{c_{12}+c_{44}}{2}}(\delta_{im}\delta_{kn} + \delta_{in}\delta_{km}) + \underbrace{(h_{11} - h_{12} - 2h_{44})}_{c_{11} - c_{12} - 2c_{44}}\delta_{ikmn} \ , \quad (4.46)$$

the equation of motion becomes

$$\rho_o \ddot{s}_i = H_{ik,mn}\partial_n\partial_m s_k + f_i \ , \quad (4.46a)$$

$$\text{e.g.,} \quad \rho_o \ddot{s}_x = h_{12}\Delta s_x + (h_{11} - h_{12} - 2h_{44})\partial_x^2 s_x + 2h_{44}\partial_x\partial_k s_k + f_x \ . \quad (4.46b)$$

For $2h_{44} = c_{12} + c_{44} = 0$ (4.46a,b) are separated in the three components $s_{x,y,z}$; if one further requires isotropy one has

$$H_{ik,mn} = h_{12}\delta_{ik}\delta_{mn} = c_{44}\delta_{ik}\delta_{mn} \ , \quad (4.47)$$

corresponding to $c_{11} = c_{44}$, $c_{12} = -c_{44}$, i.e.,

$$C_{ik,mn} = c_{44}(-\delta_{ik}\delta_{mn} + \delta_{im}\delta_{kn} + \delta_{in}\delta_{km}) \ , \quad C = c_{44}(-P^K + 2P^S) \ . \quad (4.47a)$$

This represents a very simple model substance with only *one* elastic modulus, which unfortunately is *statically unstable* because $3K = -c_{44} < 0$ if $c_{44} > 0$. The lattice analogue is the spring model used in Fig. 3.15d, which is *dynamically stable* and which can be employed to show the structure of dynamical quantities, for instance $\tilde{G}(\underline{k},\omega)$.

The equation of motion for isotropy is in vector notation

$$\rho_o \ddot{\underline{s}} = c_{44}\Delta s + (c_{12} + c_{44}) \operatorname{grad} \operatorname{div}\{\underline{s} + \underline{f}\} \ , \quad (4.46c)$$

and for the simple model (4.47),

$$\rho_o \ddot{\underline{s}} = c_{44}\Delta\underline{s} + \underline{f} \ ,$$

we obtain a simple and easily solvable wave equation for the three components of $\underline{s}$.

### 4.4.4 A Survey of the Elastic Data

Tables 4.2 and 4.3 contain the elastic data at room temperature (if not otherwise stated) for most fcc and bcc metals. The moduli have the dimension of an energy density; they are in the neighbourhood of $1\ \text{eV/Å}^3 = 1.6 \cdot 10^{12}\ \text{dyn/cm}^2$, an appropriate unit. Fig. 4.11 shows a comparison of the bulk moduli: the normal bcc metals are stiffest, the fcc metals follow and the bcc alkalis are least stiff. Fig. 4.12

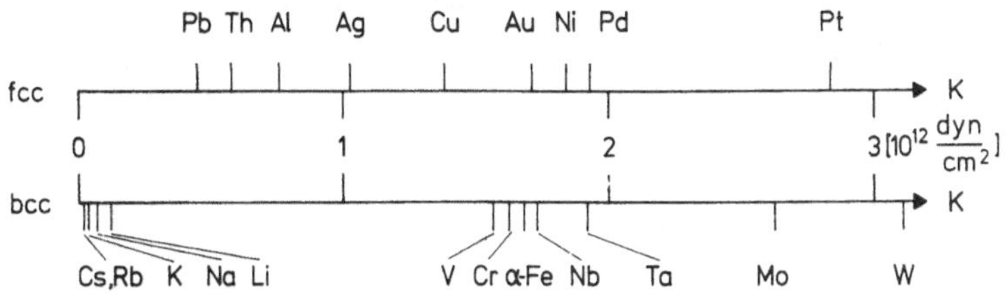


Fig. 4.11. Bulk moduli of cubic metals, $K = \overset{1}{C}/3 = (c_{11} + 2c_{12})/3$. (Ir with the large value 3.73 is not shown)

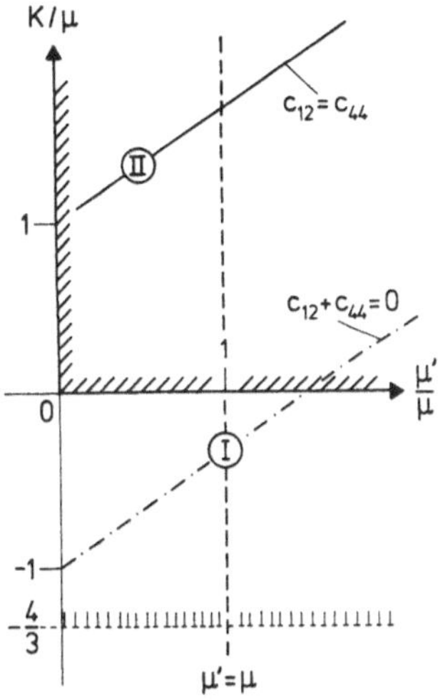


Fig. 4.12. Static stability limits of cubic elasticity. The region of stability is the quadrant $K/\mu > 0$, $\mu'/\mu > 0$. The limits of dynamical stability $K/\mu > -4/3$ , $\mu'/\mu > 0$ are indicated (comp. Sec. 4.6.3). The relations $\mu' = \mu$ (isotropy), $c_{12} = c_{44}$ (Cauchy), $c_{12} = -c_{44}$ (separation of equation of motion in the three components) are entered. Point I corresponds to the statically unstable model (4.47) where $K/\mu = -1/3$, $\mu'/\mu = 1$; point II corresponds to the data of an fcc crystal with purely longitudinal 1st neighbour spring where $K/\mu = 2/3$, $\mu'/\mu = 1/2$ (comp. Chapter 5)

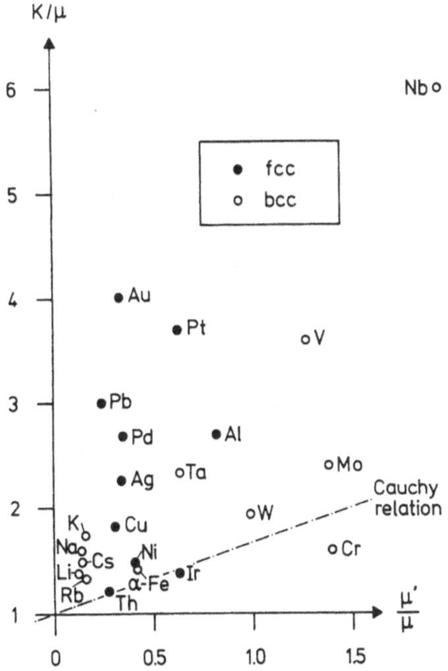


Fig. 4.13. Elastic data of cubic metals

shows in a coordinate system $K/\mu$, $\mu'/\mu$ the stability limits and other relations of interest. In the same coordinate system the data of most metals in Tables 4.2 and 4.3 are entered in Fig. 4.13; there we see some metals which are almost isotropic, such as W and Al and some which deviate appreciably from isotropy such as the alkalis.

Table 4.2. Elastic moduli [$10^{12}$ dyn/cm$^2$] at room temperature, lattice distance a [Å] and density $\rho_o$[g/cm$^3$] for fcc metals; $\mu = c_{44} = \overset{5}{C}/2$; $\mu' = (c_{11} - c_{12})/2 = \overset{2}{C}/2$; $K = (c_{11} + 2c_{12})/3 = \overset{1}{C}/3$

| Metal | $\rho_o$ | a | $c_{11}$ | $c_{12}$ | $c_{44}$ | $\mu'$ | K | Ref. |
|---|---|---|---|---|---|---|---|---|
| Ag | 10.5 | 4.09 | 1.23 | 0.92 | 0.453 | 0.16 | 1.02 | α |
| Al | 2.7 | 4.05 | 1.07 | 0.607 | 0.282 | 0.23 | 0.76 | β |
| Au | 19.3 | 4.08 | 1.90 | 1.61 | 0.423 | 0.15 | 1.71 | α |
| Cu | 8.92 | 3.61 | 1.69 | 1.22 | 0.755 | 0.24 | 1.38 | α |
| Ir | 22.5 | 3.84 | 6.0 | 2.6 | 2.7 | 1.7 | 3.73 | β |
| Ni | 8.9 | 3.52 | 2.51 | 1.50 | 1.24 | 0.51 | 1.84 | β |
| Pb | 11.3 | 4.95 | 0.495 | 0.423 | 0.149 | 0.04 | 0.45 | β |
| Pd | 12.0 | 3.89 | 2.26 | 1.76 | 0.717 | 0.25 | 1.93 | β |
| Pt | 21.5 | 3.92 | 3.47 | 2.51 | 0.765 | 0.48 | 2.83 | α |
| Th | 11.7 | 5.08 | 0.755 | 0.489 | 0.478 | 0.13 | 0.58 | α |

α) Landolt-Börnstein, New Series Vol. III/2, (Berlin-Heidelberg-New York Springer 1969).

β) S. Allard (ed.): Int. Tables of Selected Constants Vol. 16. (Oxford Pergamon Press 1969).

γ) G. Simmons, H. Wang: Single Crystal Elastic Constants (Cambridge/Mass. MIT Press 1971).

δ) F.W. Vahldiek, S.A. Mersol (eds): Anisotropy in Single Crystal Refractory Compounds (New York-London Plenum Press 1968).

Table 4.3. Elastic moduli, lattice distance and density for bcc metals; for the references compare Table 4.2.

| Metal | $\rho_o$ | a | $c_{11}$ | $c_{12}$ | $c_{44}$ | $\mu'$ | K | Ref. |
|---|---|---|---|---|---|---|---|---|
| Cr | 7.2 | 2.89 | 3.5 | 0.678 | 1.01 | 1.41 | 1.62 | β |
| Cs | 1.98 | 6.14 | 0.0247 | 0.0209 | 0.0148 | 0.005 | 0.0165 | γ, 78°K |
| α-Fe | 7.86 | 2.87 | 2.33 | 1.35 | 1.18 | 0.49 | 1.68 | β |
| K | 0.85 | 5.32 | 0.037 | 0.031 | 0.019 | 0.003 | 0.03 | β |
| Li | 0.53 | 3.51 | 0.135 | 0.114 | 0.088 | 0.01 | 0.12 | γ |
| Mo | 10.2 | 3.15 | 4.63 | 1.61 | 1.09 | 1.51 | 2.62 | β |
| Na | 0.97 | 4.29 | 0.0768 | 0.0645 | 0.0434 | 0.006 | 0.0686 | α |
| Nb | 8.58 | 3.30 | 2.47 | 1.35 | 0.287 | 0.56 | 1.73 | β |
| Rb | 1.53 | 5.70 | 0.0241 | 0.021 | 0.0095 | 0.0016 | 0.622 | δ |
| Ta | 16.7 | 3.30 | 3.61 | 1.57 | 0.818 | 0.52 | 1.92 | β |
| V | 6.02 | 3.62 | 2.29 | 1.19 | 0.432 | 0.55 | 1.56 | β |
| W | 19.3 | 3.62 | 5.23 | 2.05 | 1.61 | 1.59 | 3.11 | β |

## 4.5 Hexagonal Symmetry

### 4.5.1 Simple Hexagonal Lattices

Simple lattices with hexagonal symmetry are the so-called "close packed" hexagonal structures (hcp). To build a close packed structure one starts with a close packed α-plane using spheres of diameter a, Fig. 4.14. One can put another plane on top of the 1st plane and has two choices for a close packed arrangement, β and γ. The distance between the two planes α-γ or α-β is $\sqrt{2}\,a/\sqrt{3}$. Any sequence of such planes is a close packed arrangement. The periodic sequence ...αβγαβγ... is identical with the fcc lattice; the planes are (111) planes. Hexagonal close packed lattices have the periodic sequence ...αβαβαβ..., and they are non-Bravais lattices (the basis vectors and the elementary cell are shown in Fig. 4.14). Inside the elementary cell, spanned by α-positions, is one more atom, in β position. The lattice data are given by a and by $|\underline{a}^{(3)}| = c$. For the ideal packing: $c/a = 2\sqrt{2}/\sqrt{3} = 1.633$. In general the interplanar distance is not ideal but differs slightly from 1.633, Table 4.4. The $\underline{a}^{(3)}$ direction is called the hexagonal axis, and the planes perpendicular to it

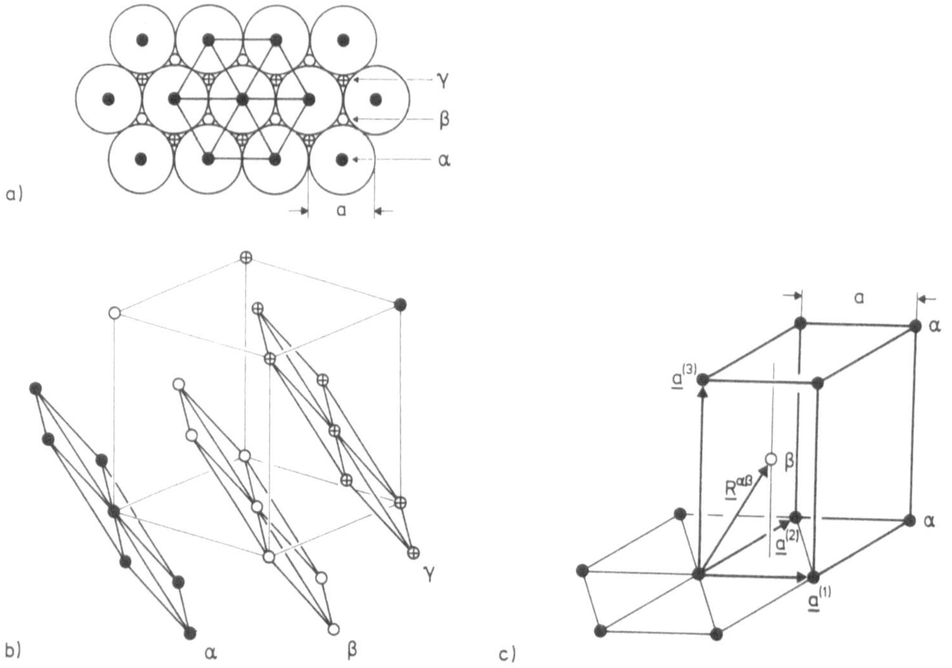


Fig. 4.14a-c. Hexagonal close packed lattice
a) Close packed plane of spheres with diameter a: α-positions (full circles), β and γ positions out of Plane ( o and ⊕ ).
b) A sequence αβγαβγ... forms a fcc lattice.
c) A sequence αβαβ... forms an hcp lattice. The basis vectors are $\underline{a}^{(1)} = a(1,0,0)$, $\underline{a}^{(2)} = a/2\ (1,\sqrt{3},0)$, $\underline{a}^{(3)} = 2\sqrt{2}a\,/\sqrt{3}\ (0,0,1)$; within one elementary cell there are two atoms: one in α-position, $\underline{R}^{\alpha} = 0$, and one in β-position, $\underline{R}^{\alpha\beta} = (\underline{a}^{(1)} + \underline{a}^{(2)})/3 + \underline{a}^{(3)}/2$

are called the basal planes.

Whereas the close packed plane has a sixfold axis ∥ $\underline{a}^{(3)}$ as a symmetry operation, the hcp lattice has not because a rotation by 2π/6 about an α-atom transforms α into α but β into γ. However, an inversion also transforms β into γ; therefore, a rotation by 2π/6 followed by an inversion is a symmetry operation of the hcp lattice. Since an inversion does not influence C (as a 4th rank tensor), it is invariant under a 2π/6 rotation about $\underline{a}^{(3)}$. This means isotropy[11] about $\underline{a}^{(3)}$. The remaining

[11] If a tensor of rank 1 is invariant under rotations with angle 2π/n the tensor is invariant under any rotation (isotropy) if n > 1. The proof is given in Appendix K.

Table 4.4. Elastic moduli [$10^{12}$ dyn/cm$^2$], lattice distances a and c [Å], and density $\rho_o$ [g/cm$^3$] of hcp metals; for the references β and γ comp. Tables 4.2 and 4.3

| Metal | $\rho_o$ | a | c | $c_{11}$ | $c_{12}$ | $c_{13}$ | $c_{33}$ | $c_{44}$ | Ref. |
|---|---|---|---|---|---|---|---|---|---|
| Be | 1.85 | 2.29 | 3.58 | 2.92 | 0.267 | 0.140 | 3.36 | 1.63 | γ |
| Ce | 8.64 | 2.98 | 5.62 | 1.15 | 0.395 | 0.399 | 0.509 | 0.199 | γ |
| Co | 8.84 | 2.51 | 4.07 | 3.07 | 1.65 | 1.03 | 3.58 | 0.755 | β |
| Hf | 12.7 | 3.19 | 5.05 | 1.81 | 0.772 | 0.661 | 1.97 | 0.557 | β |
| Mg | 1.74 | 3.21 | 5.21 | 0.597 | 0.262 | 0.217 | 0.617 | 0.164 | β |
| Re | 21.0 | 2.76 | 4.46 | 6.13 | 2.70 | 2.06 | 6.83 | 1.63 | β |
| Ru | 12.4 | 2.71 | 4.28 | 5.63 | 1.88 | 1.68 | 6.24 | 1.81 | γ |
| Ti | 4.51 | 2.95 | 4.68 | 1.62 | 0.920 | 0.690 | 1.81 | 0.467 | β |
| Tl | 11.6 | 3.46 | 5.53 | 0.408 | 0.354 | 0.29 | 0.528 | 0.073 | β |
| Y | 4.47 | 3.65 | 5.73 | 0.779 | 0.285 | 0.21 | 0.769 | 0.243 | β |
| Zn | 7.14 | 2.66 | 4.95 | 1.64 | 0.364 | 0.530 | 0.635 | 0.388 | β |
| α-Zr | 6.51 | 3.23 | 5.15 | 1.43 | 0.728 | 0.653 | 1.65 | 0.320 | β |

symmetry operations are the rotation about the center of a column with a regular hexagon as cross section, and the form of C compatible with that symmetry can be worked out easily.

### 4.5.2 The Elastic Data of Hexagonal Metals

For hexagonal symmetry the elastic data contain five independent parameters; in Voigt's notation one has:

$$C = \begin{bmatrix} c_{11} & c_{12} & c_{13} & 0 & 0 & 0 \\ c_{12} & c_{11} & c_{13} & 0 & 0 & 0 \\ c_{13} & c_{13} & c_{33} & 0 & 0 & 0 \\ 0 & 0 & 0 & c_{44} & 0 & 0 \\ 0 & 0 & 0 & 0 & c_{44} & 0 \\ 0 & 0 & 0 & 0 & 0 & c_{66} \end{bmatrix} \qquad c_{66} = \frac{c_{11} - c_{12}}{2} \tag{4.48}$$

Table 4.4 gives the data for many hexagonal metals. The $s_{\alpha\beta}$ follow the same scheme. For the four shear moduli connected with the eigentensors T(2,4,5,6), one obtains as in the cubic case

$$2s_{44} = \frac{1}{2c_{44}} \; , \quad s_{11} - s_{12} = 2s_{66} = \frac{1}{c_{11} - c_{12}} \; . \tag{4.48a}$$

For the other $s_{\alpha\beta}$ one must calculate the reciprocal of the upper left 3×3 matrix in (4.48), with the result:

$$\left.\begin{matrix} s_{11} + s_{12} \\ s_{13} \\ s_{33} \end{matrix}\right\} = \frac{1}{(c_{11} + c_{12})c_{33} - 2c_{13}^2} \left\{\begin{matrix} c_{33} \\ -c_{13} \\ c_{11} + c_{12} \end{matrix}\right. \; . \tag{4.48b}$$

## 4.5.3 The Eigenstates of C

Four eigenstates of C are already contained in the cubic patterns of Fig. 4.5. Obviously the two states T(4,5) are eigenstates with eigenvalues $2c_{44}$; e.g., CT(4) = $CV(4)/\sqrt{2} = 2c_{44}T(4)$. Because of the degeneracy of T(4) and T(5), the shear modulus $c_{44}$ applies to all shears with shear plane or direction parallel to the hexagonal axis; this establishes the isotropy about that axis for this kind of shear. Also T(6) and T(2) are eigenstates, eigenvalue $c_{11} - c_{12}$, shear modulus $(c_{11} - c_{12})/2$ for shears with shear plane and direction perpendicular to the hexagonal axis[12]. In the remaining two states, T(1) and T(3), C is not diagonal in general. The five independent moduli so far are the two shear moduli, $c_{44}$ and $(c_{11} - c_{12})/2$, and the three matrix elements $C^{11}$, $C^{13}$, $C^{33}$. One would have to solve a quadratic equation to find the remaining two eigenvalues and to see how the eigenstates look in terms of T(1) and T(3).

In any case, T(1) is no longer an eigenstate and, therefore, a pressure does not produce a geometrically similar dilatation. From

$$\varepsilon_\alpha = \sum_\beta s_{\alpha\beta} g^\beta \sigma_\beta \; , \tag{4.49}$$

one obtains for $\sigma = -p$, i.e., $\sigma_1 = \sigma_2 = \sigma_3 = -p$,

$$\varepsilon_\alpha = -p \sum_{\beta=1}^{3} s_{\alpha\beta} \; ; \quad \varepsilon_1 = \varepsilon_2 = -p(s_{11} + s_{12} + s_{13}) \; , \quad \varepsilon_3 = -p(2s_{13} + s_{33}) . \tag{4.49a}$$

The compressibility is

---

12 Due to the asymmetry of Voigt's notation, the eigenvalues of c with components in T(4,5,6) are half the eigenvalues of C.

$$1/K = -\frac{\Delta V}{pV} = -\frac{\varepsilon_1 + \varepsilon_2 + \varepsilon_3}{p} = 2(s_{11} + s_{12} + s_{13}) + (2s_{13} + s_{33}). \tag{4.50}$$

The strains parallel ($\varepsilon_3$) and perpendicular ($\varepsilon_1$) to the hexagonal axis differ. In terms of the $c_{\alpha\beta}$ it is

$$\left.\begin{aligned} -\frac{\varepsilon_1}{p} &= s_{11} + s_{12} + s_{13} \\ -\frac{\varepsilon_3}{p} &= 2s_{13} + s_{33} \end{aligned}\right\} = \frac{1}{(c_{11} + c_{12})c_{33} - 2c_{13}^2} \cdot \left\{\begin{aligned} &c_{33} - c_{13} \\ &c_{11} + c_{12} - 2c_{13} \end{aligned}\right. , \tag{4.50a}$$

e.g., for Zn: $-\frac{\varepsilon_1}{p} = 0.15$, $-\frac{\varepsilon_3}{p} = 1.33$ $[10^{-12}\mathrm{cm}^2/\mathrm{dyn}]$, hence $1/K = 1.62$ $[10^{-12}\mathrm{cm}^2/\mathrm{dyn}]$, $K = 0.62$ $[10^{12}\mathrm{dyn/cm}^2]$.

Elastic stability requires all eigenvalues of C (or c) to be positive. The eigenvalues are $2c_{44}$, $c_{11} - c_{12}$ and $C_\pm = \frac{1}{2}[c_{11} + c_{12} + c_{33} \pm \sqrt{(c_{11} + c_{12} - c_{33})^2 + 8c_{13}^2}]$; therefore

$$c_{44}\ ,\quad c_{11} - c_{12}\ ,\ c_{11} + c_{12} + c_{33}\ ,\quad (c_{11} + c_{12})c_{33} - 2c_{13}^2 > 0\ .$$

The first two conditions guarantee positive shear moduli; the last two conditions are sufficient for the two remaining eigenvalues of C to be positive.

## 4.6 Dynamics in Infinite Crystals, Elastic Waves

### 4.6.1 The Equations of Motion

The equation of motion (4.33a) for a homogeneous elastic continuum, (comp. App I),

$$\rho_o \ddot{s}_i(\underline{r},t) = H_{ik,mn}\partial_m\partial_n s_k + f_i(\underline{r},t)\ ; \qquad \rho_o\ddot{\underline{s}} = -\rho_o D\underline{s} + \underline{f}\ , \tag{4.51}$$

has to be contrasted with that of the homogeneous lattice

$$M\ddot{s}_i^{\underline{m}}(t) = -\ \Phi_{ik}^{(\underline{m}-\underline{n})} s_k^{\underline{n}} + F_i^{\underline{m}}\ . \tag{4.52}$$

One recognizes that

$\Phi/M$ , $\underline{F}/M$ correspond to D, $\underline{f}/\rho_o$ or, because $\rho_o = M/V_c$ ,

$\Phi/V_c$ , $\underline{F}/V_c$ correspond to $\rho_o D$, $\underline{f}$.

The symmetrical and positive matrix $\Phi/V_c$ corresponds to the hermitean and positive differential operator $\rho_o D$.

### 4.6.2 Elastic Waves

The general treatment of (4.51) is completely analogous to that in lattice theory. The eigenfunctions to D are again plane waves

$$<_i^{\underline{r}}|\underline{k}\sigma> = e_i(\underline{k}\sigma)\,\frac{e^{i\underline{k}\underline{r}}}{(2\pi)^{3/2}}\,,$$
$$[D\underline{e}(\underline{k}\sigma)e^{i\underline{k}\underline{r}}/(2\pi)^{3/2}]_i = \underbrace{\frac{1}{\rho_o}H_{ik,mn}k_m k_n}_{\tilde{D}_{ik}(\underline{k})} e_k(\underline{k}\sigma)\,\frac{e^{i\underline{k}\underline{r}}}{(2\pi)^{3/2}}\,, \tag{4.53}$$

and the polarizations $\underline{e}(\underline{k}\sigma)$ are eigenvectors of $\tilde{D}(\underline{k})$, which corresponds to $\tilde{\Phi}(\underline{k})/M$ of the lattice,

$$\tilde{D}(\underline{k})\underline{e} = \tilde{D}(\underline{k}\sigma)\underline{e}(\underline{k}\sigma) \quad \text{with the eigenvalue} \tag{4.53a}$$

$$\tilde{D}(\underline{k}\sigma) = \left(e(\underline{k}\sigma),\tilde{D}(\underline{k})e(\underline{k}\sigma)\right) = \frac{1}{\rho_o}e_i e_k H_{ik,mn}k_m k_n = \frac{1}{\rho_o}e_i e_k C_{im,kn}k_m k_n > 0. \tag{4.53b}$$

It follows from elastic stability that $\tilde{D}(\underline{k}\sigma) > 0$, if one uses $v_{im} = e_i k_m$ in (4.32).

The difference to lattice theory is that $\underline{k}$ extends over infinite space and that $\tilde{D}(\underline{k})$ is quadratic in $\underline{k}$:

$$\tilde{D}(\underline{k}) = k^2\tilde{D}(\hat{\underline{k}}) = k^2\hat{\tilde{D}}\,, \quad \tilde{D}(\underline{k}\sigma) = k^2\tilde{D}(\hat{\underline{k}}\sigma)\,; \tag{4.53c}$$

indeed, $\tilde{D}(\underline{k})$ must equal the lowest order of the expansion of $\Phi(\underline{k})/M = \Omega^2(\underline{k})$ in powers of $\underline{k}$ (comp. Chap. 5). The solutions of the force-free equations are again waves:

$$\underline{s}(\underline{r},t) \propto \underline{e}(\underline{k}\sigma)\,e^{i[\underline{k}\underline{r} - \Omega(\underline{k}\sigma)t]}\,; \tag{4.54}$$

if $\underline{r}$ is replaced by $\underline{R}^{\underline{m}}$ eq. (4.54) agrees with (3.29a). Because of (4.53c), the frequency is proportional to k and one obtains for the velocity of sound

$$c_\sigma(\hat{\underline{k}}) = \frac{\Omega(\underline{k}\sigma)}{k} = \Omega(\hat{\underline{k}}\sigma)\,,$$

which does only depend on the direction of $\hat{\underline{k}}$. The group velocity,

$$\underline{v}_\sigma(\hat{\underline{k}}) = \partial_{\underline{k}}\Omega(\underline{k}\sigma)\,,$$

is, in general, different from $c_\sigma$.

### 4.6.3 Cubic Crystals

For cubic symmetry $c_\sigma(\hat{\underline{k}})$ agrees with $|\underline{v}_\sigma(\hat{\underline{k}})|$ if $\hat{\underline{k}}$ is parallel to one of the main symmetry directions; this case will be discussed now as the simplest example. Here eqs. (4.53) yield

$$\rho_o\hat{\tilde{D}}_{ik} = \rho_o\frac{\tilde{D}_{ik}(\underline{k})}{k^2} = \underbrace{h_{12}}_{c_{44}}\delta_{ik} + \underbrace{2h_{44}}_{c_{12}+c_{44}}\hat{k}_i\hat{k}_k + \underbrace{(h_{11}-h_{12}-2h_{44})}_{c_{11}-c_{12}-2c_{44}=c_a}\begin{bmatrix}\hat{k}_x^2 & 0 & 0\\ 0 & \hat{k}_y^2 & 0\\ 0 & 0 & \hat{k}_z^2\end{bmatrix}. \quad (4.56)$$

For isotropy, $c_a = 0$ or $c_{11} = c_{12} + 2c_{44}$, hence

$$\rho_o\hat{\tilde{D}}_{ik} = c_{44}\delta_{ik} + (c_{12} + c_{44})\hat{k}_i\hat{k}_k \ , \quad (4.57)$$

one obviously has longitudinal polarization,

$$\underline{\overset{1}{e}} = \hat{\underline{k}} \ , \quad \rho_o\hat{\tilde{D}}\hat{\underline{k}} = (c_{12} + 2c_{44})\hat{\underline{k}} \ , \quad \rho_o c_1^2 = c_{12} + 2c_{44} = c_{11} \ , \quad c_1^2 = \frac{c_{11}}{\rho_o} \ , \quad (4.57a)$$

and transversal (two-fold degenerate) polarization

$$\underline{\overset{t}{e}} \perp \hat{\underline{k}} \ , \quad \rho_o\hat{\tilde{D}}\underline{\overset{t}{e}} = c_{44}\underline{\overset{t}{e}} \ , \quad \rho_o c_t^2 = c_{44} \ , \quad c_t^2 = \frac{c_{44}}{\rho_o} \ . \quad (4.57b)$$

In the main symmetry directions also the 3rd (anisotropic) term in (4.56) has only longitudinal and transversal eigenvectors, e.g., for $\hat{\underline{k}} = (1,0,0)$

$$c_a\begin{bmatrix}1 & 0 & 0\\ 0 & 0 & 0\\ 0 & 0 & 0\end{bmatrix} \ , \quad \begin{matrix}\text{longitudinal eigenvalue } c_a\\ \text{transversal eigenvalue } 0\end{matrix} \ .$$

The results are given in Fig. 4.15 and Table 4.5. For each branch the value of $\rho_o c^2$ is indicated. From measuring three independent velocities one obtains the three elastic moduli of cubic crystals. For isotropic material $c_1$ and $c_t$ determine the two elastic moculi. Elastic stability $c_{11} + 2c_{12} = 3c_{11} - 4c_{44} > 0$ requires here $c_t^2 > (4/3)c_1^2$.

In the model (4.47) one has $c_1^2 = c_t^2 = c_{44}/\rho$ . This model is unstable in elastostatics because the compressional modulus $3K = -c_{44}$ is negative, if $c_{44} > 0$. One does not realize this fact, if one discusses only elastic waves in infinite crystals, because $\Omega^2 = D$ is positive for all $\hat{\underline{k}}$ as for the corresponding spring model (Fig. 3.15d). Consequently, from a dynamical point of view, one can employ this model for purposes of simple demonstration. One realizes from Fig. 4.15 that for isotropy ($c_a = 0$) $c_{11}$ and $c_{44}$, but not K, enter the sound velocities; and therefore, negative values of $K > -4c_{44}$ would still be admissible dynamically. The same inequality is obtained if one asks for the dynamical stability of an arbitrary cubic material:

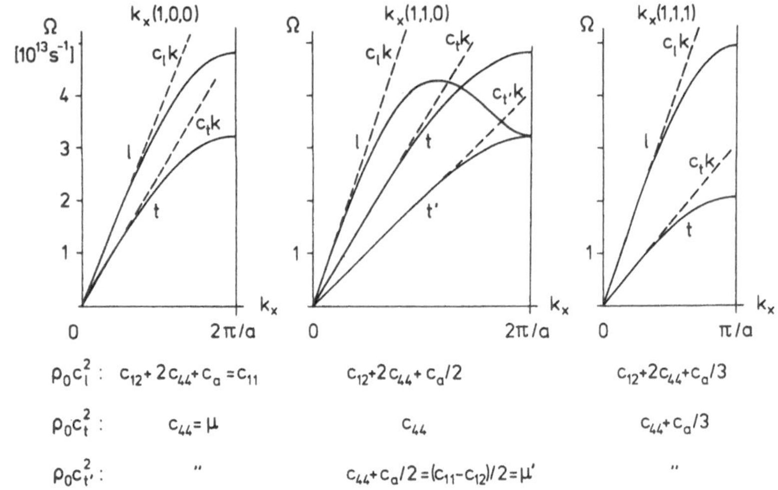


Fig. 4.15. Sound velocities in the main symmetry directions of cubic crystals. The figure shows the lattice dispersion curves for Cu (Fig. 3.15b) in the main symmetry directions. The slopes for small k are the velocities of sound, c. For each branch the value of $\rho_0 c^2$ is indicated. In continuum theory the linear behaviour, $\Omega = ck$, is valid for arbitrarily large k

Table 4.5. Eigenvalues and eigenvectors of the cubic term in equation (4.56). For $\hat{\underline{k}} = (1,\bar{1},0)/\sqrt{2}$ the transversal polarizations are $\underline{e}^{t'} = (1,1,0)/\sqrt{2}$ and $\underline{e}^{t} = (0,0,1)$.

| Direction | Cubic term | Eigenvalues to indicated polarizations | | |
|---|---|---|---|---|
| $\hat{\underline{k}}$ | $\delta_{ikmn}\hat{k}_m\hat{k}_n$ | l | t' | t |
| $(1,0,0)$ | $c_a\begin{bmatrix}1&0&0\\0&0&0\\0&0&0\end{bmatrix}$ | $c_a$ | 0 | 0 |
| $\frac{1}{\sqrt{2}}(1,\bar{1},0)$ | $\frac{c_a}{2}\begin{bmatrix}1&0&0\\0&1&0\\0&0&0\end{bmatrix}$ | $c_a/2$ | $c_a/2$ | 0 |
| $\frac{1}{\sqrt{3}}(1,1,1)$ | $\frac{c_a}{3}\begin{bmatrix}1&0&0\\0&1&0\\0&0&1\end{bmatrix}$ | $c_a/3$ | $c_a/3$ | $c_a/3$ |

here one can easily show that this condition must be valid for $\hat{\underline{k}}$ in the main symmetry directions, and then one can extend the proof to general $\hat{\underline{k}}$. The areas of static and dynamical stability have been already shown in Fig. 4.12. One must keep in mind, however, that for a finite crystal with a free surface, static and dynamical stability must be equivalent because here a homogeneous deformation can be constructed from the eigensolutions of the finite system. The requirement of positive u leads then to the static stability conditions.

## 4.7 Green's Functions

### 4.7.1 The Representation of Green's Functions

The Green's functions of continuum theory can be represented in complete analogy to those of lattice theory (comp. Sec. 3.5); instead of the lattice Green's function (comp. Tab. 3.3)

$$g^{(\underline{h})}(t) = \Theta(t)\,\frac{1}{MV_B}\int d\underline{k}\,\frac{\sin\sqrt{\tilde{\Phi}/M}\,t}{\sqrt{\tilde{\Phi}/M}}\,e^{i\underline{k}\underline{R}^{\underline{h}}}\;,\qquad MV_B = (2\pi)^3\rho_o$$

one has in elasticity theory

$$g(\underline{r},t) = \Theta(t)\,\frac{1}{(2\pi)^3\rho_o}\int_\infty d\underline{k}\,\frac{\sin\sqrt{\tilde{D}(\underline{k})}\,t}{\sqrt{\tilde{D}(\underline{k})}}\,e^{i\underline{k}\underline{r}}\;, \tag{4.58}$$

which is the solution of (4.51) for vanishing displacement at $t < 0$ and for momentum transfer 1 to the point $\underline{r} = 0$ at $t = 0$:

$$\int d\underline{r}\;\rho_o\dot{g}(\underline{r},t > 0) = \int d\underline{r}\;\rho_o\dot{g}(\underline{r},t = +0) = \Theta(t > 0) = 1\;. \tag{4.58a}$$

The displacement field is

$$g(\underline{r},t)\underline{\kappa} = (g_{il}\kappa_l)\ \text{for a force density}\ \underline{f}(\underline{r},t) = \underline{\kappa}\delta(\underline{r})\delta(t)\;, \tag{4.58b}$$

and the general retarded solution becomes

$$\underline{s}(\underline{r},t) = \iint dt'd\underline{r}'\;g(\underline{r}-\underline{r}',t-t')\;\underline{f}(\underline{r}',t')\;. \tag{4.58c}$$

The time Fourier transform is

$$g(\underline{r},t) = \int\frac{d\omega}{2\pi}\,e^{-i\omega t}\,G(\underline{r},\omega)\;,\qquad G(\underline{r},\omega) = \int dt\;e^{i\omega t}\,g(\underline{r},t)\;, \tag{4.59}$$

where

$$G(\underline{r},\omega) = \int\limits_{\infty} d\underline{k}\, \frac{1}{(2\pi)^3} \underbrace{\frac{1}{\rho_o[\tilde{D}(\underline{k}) - (\omega + i\eta)^2]}}_{\tilde{G}(-\underline{k},\omega) \,=\, \tilde{G}(\underline{k},\omega) \text{ of continuum theory}} e^{i\underline{k}\underline{r}} \; , \tag{4.60}$$

instead of

$$G^{(\underline{h})}(\omega) = \int\limits_{V_B} \frac{d\underline{k}}{V_B} \underbrace{\frac{1}{M[\Omega^2(\underline{k}) - (\omega + i\eta)^2]}}_{\tilde{G}(-\underline{k},\omega) \,=\, \tilde{G}(\underline{k},\omega) \text{ of lattice theory}} e^{i\underline{k}\underline{R}^{\underline{h}}} \; . \tag{4.60a}$$

### 4.7.2 The Physical Meaning of Green's Functions

Table 4.6, which corresponds directly to Table 3.3 for lattice Green's functions, summarizes the physical meaning of Green's functions in continuum theory. In the proper limit these Green's functions agree with those of the lattice, i.e., $g(\underline{r},t)$, $G(\underline{r},\omega)$ agree with $g^{(\underline{m})}(t)$, $G^{(\underline{m})}(\omega)$ for large $\underline{r} = \underline{R}^{\underline{m}}$ if one uses the small-$\underline{k}$ expansion of the lattice functions and extends the $\underline{k}$ integration over the Brillouin-Zone to infinite volume. However, the corresponding $\tilde{G}(k,\omega)$ differ for small $\underline{k}$: in the

Table 4.6. Green's functions and force densities

| Force density $\underline{f}(\underline{r},t)$ | Response $\underline{s}(\underline{r},t)$ |
|---|---|
| $\underline{\kappa}\delta(\underline{r})\delta(t)$ | $g(\underline{r},t)\underline{\kappa}$ ; $g(\underline{r},t) = \int_{-\infty}^{\infty} \frac{d\omega}{2\pi} e^{-i\omega t} G(\underline{r},\omega)$ |
| $\underline{\kappa}\delta(\underline{r})e^{-i\omega t}$ | $G(\underline{r},\omega)\underline{\kappa}e^{-i\omega t}$ ; $G(\underline{r},\omega) = \int\limits_{\infty} \frac{d\underline{k}}{(2\pi)^3} e^{i\underline{k}\underline{r}} \tilde{G}(\underline{k},\omega)$ |
| $\underline{\kappa}e^{i(\underline{k}\underline{r}-\omega t)}$ | $\tilde{G}(\underline{k},\omega)\underline{\kappa}e^{i(\underline{k}\underline{r}-\omega t)}$ ; $\tilde{G}(\underline{k},\omega) = \frac{1}{\rho_o[\tilde{D}(\underline{k}) - (\omega + i\eta)^2]}$ |
| $\underline{e}(\underline{k}\sigma)e^{i(\underline{k}\underline{r}-\omega t)}$ | $\tilde{G}(\underline{k}\sigma,\omega)\underline{e}(\underline{k}\sigma)e^{i(\underline{k}\underline{r}-\omega t)}$ ; $\tilde{G}(\underline{k}\sigma,\omega) = \frac{1}{\rho_o[\tilde{D}(\underline{k}\sigma) - (\omega + i\eta)^2]}$ |
| $\underline{\kappa}\delta(\underline{r})$ | $G(\underline{r},\omega=0)$ ; $G(\underline{r},0) = \int\limits_{\infty} \frac{d\underline{k}}{(2\pi)^3} e^{i\underline{k}\underline{r}} \frac{1}{\rho_o\tilde{D}(\underline{k})}$ |

lattice, $\underline{\kappa}$ is the force per atom, and in the continuum, $\underline{\kappa}$ in Table 4.6 denotes the force per unit volume; equivalent displacements would be caused by equivalent forces, i.e., $\underline{\kappa}_{cont} = \underline{\kappa}_{latt}/V_c$. Consequently $\tilde{G}_{latt}$ passes for small $\underline{k}$ into

$$\frac{1}{M[\tilde{D}(\underline{k}) - (\omega + i\eta)^2]} = \frac{\rho_o}{M}\tilde{G}_{cont} = \tilde{G}_{cont}/V_c \ . \tag{4.60b}$$

### 4.7.3 The Simplest Models

A particularly simple model is that of eq. (4.47); it can be derived from the lattice model of Fig. 3.15d and has been discussed already in Section 3.5.5, eqs. (3.110, 110a). For the Green's functions one obtains ($\tilde{D}$,G,g $\propto$ unit matrix):

$$\rho_o\tilde{D}(\underline{k}) = c_{44}k^2 = \mu k^2 = \rho_o c^2 k^2 \ ,$$

$$G(\underline{r},\omega) = \frac{\exp(i\frac{\omega r}{c})}{4\pi\mu r} = \frac{\exp(i\frac{\omega r}{c})}{4\pi\rho_o c^2 r} \ ; \quad G_1 = \frac{\cos\frac{\omega r}{c}}{4\pi\mu r} \ ; \quad \mathrm{sgn}\,\omega \ G_2 = \frac{\sin\frac{\omega r}{c}}{4\pi\mu r} \ , \tag{4.61}$$

$$G(\underline{r},0) = \frac{1}{4\pi\mu r} \ ; \quad g(\underline{r},t) = \frac{\delta(\frac{r}{c} - t)}{4\pi\mu r} = \frac{\delta(r - ct)}{4\pi\rho_o c r} = \frac{1}{4\pi\rho_o}\frac{t}{r^2}\delta(r - ct) \ .$$

Note that for small $\omega$ $\mathrm{Im}\{G(\underline{r},\omega)\} \to \omega/4\pi\mu c$ independent of r, just as $\mathrm{Im}\{G^{(\underline{h})}(\omega)\}$ becomes independent of $\underline{h}$, according to Section 3.5.4 (p. 86).

Isotropy has also been treated in Section 3.5.5, eqs. (3.111,111a),

$$G_{ij}(\underline{r},\omega) = \frac{1}{4\pi\rho_o}\left(\frac{\delta_{ij}\exp(i\omega r/c_t)}{c_t^2 r} + \partial_i\partial_j\frac{\exp(i\omega r/c_t) - \exp(i\omega r/c_l)}{\omega^2 r}\right);$$

$$\mathrm{Re}|G_{ij}(\underline{r},\omega)| = \frac{1}{4\pi\rho_o}\left(\frac{\delta_{ij}\cos(\omega r/c_t)}{c_t^2 r} + \partial_i\partial_j\frac{\cos(\omega r/c_t) - \cos(\omega r/c_l)}{\omega^2 r}\right). \tag{4.62}$$

The static result is obtained by taking the proper limit $\omega \to 0$:

$$G_{ij}(\underline{r},0) = \frac{1}{4\pi\rho_o}\left(\frac{\delta_{ij}}{c_t^2 r} + \partial_i\partial_j\frac{r}{2}\left(\frac{1}{c_l^2} - \frac{1}{c_t^2}\right)\right)$$

$$= \frac{1}{4\pi\rho_o r}\left(\frac{\hat{x}_i\hat{x}_j}{c_t^2} + \frac{1}{2}\left(\frac{1}{c_l^2} + \frac{1}{c_t^2}\right)(\delta_{ij} - \hat{x}_i\hat{x}_j)\right), \tag{4.63}$$

and one recognizes a longitudinal and a transversal contribution[13],

---

[13] $G(\underline{r},\omega)$ can also be split into longitudinal and transversal contributions.

$$G_l = \frac{1}{4\pi\rho_o r}\,\frac{1}{c_t^2}\,, \qquad G_t = \frac{1}{4\pi\rho_o r}\,\frac{1}{2}\left(\frac{1}{c_l^2} + \frac{1}{c_t^2}\right). \tag{4.63a}$$

With (4.62) one obtains $g(\underline{r},t)$ via (4.59). The factor $G(\underline{r},\omega)$ has no singularity at $\omega = 0$; the path of integration can be shifted into the lower half of the complex $\omega$-plane. The integration for the three summands can be performed separately

$$\begin{aligned} 4\pi\rho_o g_{ij}(\underline{r},t) &= \frac{\delta_{ij}}{c_t^2 r}\,\delta\!\left(\frac{r}{c_t} - t\right) + \partial_i\partial_j\,\frac{1}{r}\Big[\left(\frac{r}{c_l} - t\right)\underbrace{\Theta\!\left(\frac{r}{c_l} - t\right)}_{\Theta_l} - \left(\frac{r}{c_t} - t\right)\underbrace{\Theta\!\left(\frac{r}{c_t} - t\right)}_{\Theta_t}\Big] \\ &= \delta_{ij}\,\frac{t}{r^3}\Big[r\delta(r - c_t t) + \Theta_l - \Theta_t\Big] + \hat{x}_i\hat{x}_j\,\frac{t}{r^2}\Big[\delta(\underline{r} - c_l t) \\ &\qquad - \delta(r - c_t t) - \frac{3}{r}\,(\Theta_l - \Theta_t)\Big]\,, \end{aligned} \tag{4.64}$$

and again split into longitudinal and transversal components:

$$4\pi\rho_o g_l = \frac{t}{r^2}\,\delta(r - c_l t) - \frac{2t}{r^3}\,(\Theta_l - \Theta_t)\,, \quad 4\pi\rho_o g_t = \frac{t}{r^2}\,\delta(r - c_t t) + \frac{t}{r^3}\,(\Theta_l - \Theta_t)\,. \tag{4.64a}$$

A sketch of g is shown in Fig. 4.16. In the model (4.47) with *one* sound velocity the sudden force at t = 0 produces a spherical outgoing displacement field confined to the surface of a sphere with radius ct. For general isotropy with *two* sound velocities the displacement is confined to a spherical shell between $c_t t$ and $c_l t$, i.e.,

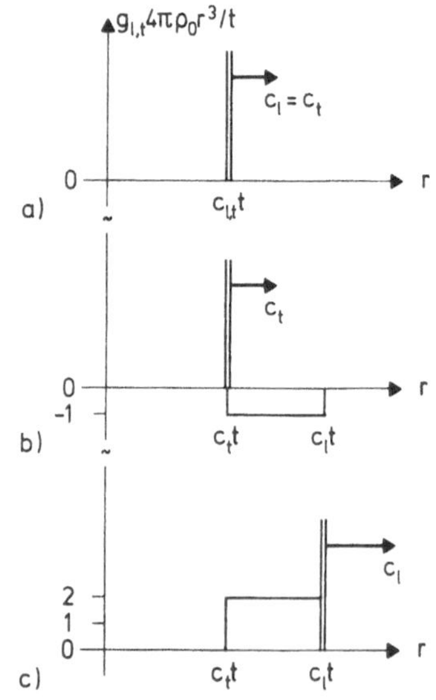


Fig. 4.16a-c. $g(\underline{r},t)$ for isotropy (schematic)

a) $\dfrac{4\pi\rho_o r^3}{t}\,g_l = \dfrac{4\pi\rho_o r^3}{t}\,g_t = r\delta(r - c_t t)$ according to the model (4.47), $c_l = c_t$ ,

b) $\dfrac{4\pi\rho_o r^3}{t}\,g_t = r\delta(r - c_t t) - \big[\Theta(r - c_t t) - \Theta(r - c_l t)\big]$, $c_l > c_t$ ,

c) $\dfrac{4\pi\rho_o r^3}{t}\,g_l = r\delta(r - c_l t) + 2\big[\Theta(r - c_t t) - \Theta(r - c_l t)\big]$

to a region which can be reached from the origin by sound waves with these velocities. This is also the general behaviour, namely that g is confined to a region defined by a maximum and minimum velocity which must be selected from the many (group) velocities[14] in the anisotropic case.

For cubic crystals one simple solution exists if $c_{12} + c_{44} = 0$. According to (4.46b) the equation of motion is separated, e.g.,

$$\rho_o \ddot{s}_x = \{c_{11}\partial_x^2 + c_{44}(\partial_y^2 + \partial_z^2)\} s_x + f_x \ . \tag{4.65}$$

Green's function is diagonal as in model (4.47), and (4.65) can be easily transformed by scaling ($\tilde{x} = \sqrt{c_{44}/c_{11}}\, x$ , $\tilde{y} = y$, $\tilde{z} = z$) into an equation of that model. Consequently,

$$g_{xx} = \frac{\delta(\tilde{r} - ct)}{4\pi\rho_o c\tilde{r}\sqrt{c_{11}/c_{44}}} \ , \qquad \tilde{r} = \left(\frac{c_{44}}{c_{11}} x^2 + y^2 + z^2\right)^{1/2} \ , \qquad \rho_o c^2 = c_{44} \ , \tag{4.65a}$$

and the displacement field is restricted to the surface of a rotational ellipsoid,

$$\tilde{r}^2 = c^2 t^2 \ , \qquad \frac{x^2}{c_l^2 t^2} + \frac{y^2 + z^2}{c_t^2 t^2} = 1 \ , \qquad c_t = c \ , \qquad c_l = \sqrt{c_{11}/\rho_o} \ ,$$

with principal distances given by the longitudinal and transversal sound velocities in [100].

Analytical solutions for general cubic symmetry are not known. Approximations can be employed by expansions, for instance near isotropy. However, the general behaviour is given well enough by the examples discussed above.

## 4.8 The Static Response $G(\underline{r},0)$, /4.5,6/

In this section we will denote the static response simply by G.

---

[14] For an elastic wave $\underline{s}(\underline{r},t) = \underline{e}(\underline{k}\sigma)\cos[\underline{k}\underline{r} - \Omega(\underline{k}\sigma)t]$ the energy current density $\underline{j} = -\sigma\underline{s}$ (comp. Sec. 4.3.4) is in the time average (denoted by a bar), where $\overline{\sin^2 \ldots} = \overline{\cos^2 \ldots} = 1/2$ , $\overline{j_i} = (\Omega/2)C_{ik,mn}e_k e_n k_m = (\Omega/2)H_{im,kn}e_k e_n k_m$ ; the average total energy density is $\overline{u_{tot}} = \rho_o\Omega^2/2$. The velocity $v_i = \overline{j_i}/\overline{u_{tot}} = H_{im,kn}e_k e_n k_m/(\rho_o\Omega)$ is identical with the group velocity $v_i = \partial\Omega/\partial k_i = (\partial\Omega^2/\partial k_i)/(2\Omega)$ because $\rho_o\Omega^2 = H_{im,kn}k_i k_m e_k e_n$ ; it is always $c^2 \leqslant v^2$.

## 4.8.1 The Static Response in an Infinite, Homogeneous Medium

The static Green's function of an infinite, homogeneous continuum is

$$G = \int \frac{d\underline{k}}{(2\pi)^3\rho_o}\,\frac{\cos \underline{k}\underline{r}}{k^2\tilde{D}(\hat{\underline{k}})} = \frac{1}{(2\pi)^3\rho_o}\int d\Omega_{\underline{k}} \int_0^\infty dk\,\frac{\cos kr(\hat{\underline{k}}\hat{\underline{r}})}{\tilde{D}(\hat{\underline{k}})}$$

$$= \frac{1}{8\pi^2 r}\int d\Omega_{\underline{k}}\,\frac{\delta(\hat{\underline{k}}\hat{\underline{r}})}{\rho_o\tilde{D}(\hat{\underline{k}})} = \frac{1}{8\pi^2 r}\hat{G}\ , \qquad (4.66)$$

where $\int_0^\infty dk \cos k\alpha = \pi\delta(\alpha)$ has been used. The integration extends now over the circle $\hat{\underline{k}} \perp \hat{\underline{r}}$ on the unit sphere in $\underline{k}$ space. For isotropy,

$$\rho_o\tilde{D}(\hat{\underline{k}}) = c_{11}P_l(\hat{\underline{k}}) + c_{44}P_t(\hat{\underline{k}})\ , \qquad \frac{1}{\rho_o\tilde{D}(\hat{\underline{k}})} = \frac{P_l(\hat{\underline{k}})}{c_{11}} + \frac{P_t(\hat{\underline{k}})}{c_{44}}$$

where $P_{l,t}(\hat{\underline{k}})$ are the longitudinal and transversal projectors referring to $\hat{\underline{k}}$. Because of

$$\int d\Omega_{\underline{k}}\ \delta(\hat{\underline{k}}\hat{\underline{r}}) \left\{\begin{matrix} 1 \\ P_l(\hat{\underline{k}}) \\ P_t(\hat{\underline{k}}) \end{matrix}\right. = \left\{\begin{matrix} 2\pi \\ \pi P_t(\hat{\underline{r}}) \\ \pi[1 + P_l(\hat{\underline{r}})] \end{matrix}\right. \ ,$$

(4.66) can be easily evaluated; the result is

$$G = \frac{1}{4\pi r}\left[\frac{1}{c_{44}}P_l(\hat{\underline{r}}) + \frac{1}{2}\left(\frac{1}{c_{11}} + \frac{1}{c_{44}}\right)P_t(\hat{\underline{r}})\right]\ , \qquad (4.67)$$

which agrees with (4.63) and is illustrated in Fig. 4.17.

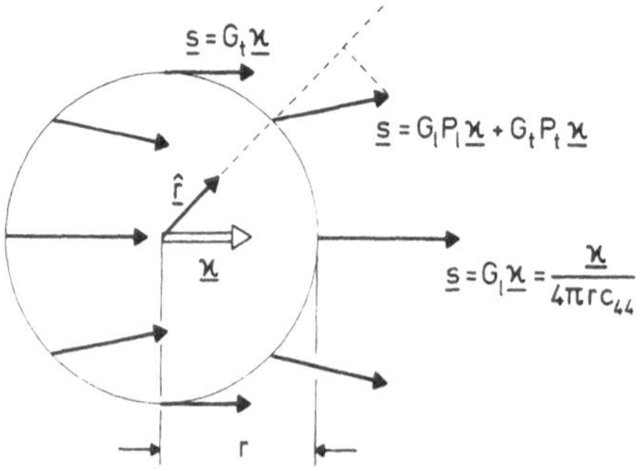


Fig. 4.17. Isotropic $G = \frac{1}{4\pi r}\left[\frac{1}{c_{44}}P_l(\hat{\underline{r}}) + \frac{1}{2}\left(\frac{1}{c_{44}} + \frac{1}{c_{11}}\right)P_t(\hat{\underline{r}})\right] = G_lP_l + G_tP_t$. A single force $\underline{\kappa}$ in the origin produces the displacement $G\underline{\kappa} = G_lP_l\underline{\kappa} + G_tP_t\underline{\kappa}$. Here $G_t/G_l = 0.65$ for W is used; for (4.47), where $c_{11} = c_{44}$, one has: $G\underline{\kappa} = \frac{1}{4\pi r c_{44}}\underline{\kappa}$

### 4.8.2 Cubic Symmetry[15]

For $\hat{\underline{r}}$ parallel to the main symmetry directions of cubic crystals we can evaluate (4.66) analytically; here G or $\hat{G}$ must have the longitudinal-transversal structure of the corresponding coupling matrices $\Phi(\hat{\underline{r}})$ in Sec. 3.2.4. We will demonstrate the procedure for $\hat{\underline{r}} = (1,0,0)$, where $G_l = \int d\Omega_{\underline{k}}\ \delta(\hat{k}_x)[\tilde{D}^{-1}(\hat{\underline{k}})]_{xx}/\rho_o$. Because

$$\rho_o\tilde{D}(\hat{\underline{k}}) = [(c_{12} + c_{44})\hat{k}_i\hat{k}_k] + c_{44}[\delta_{ik}] + c_a\begin{bmatrix} \hat{k}_x^2 & 0 & 0 \\ 0 & \hat{k}_y^2 & 0 \\ 0 & 0 & \hat{k}_z^2 \end{bmatrix}$$

$$= \begin{bmatrix} c_{44} & 0 & 0 \\ 0 & \rho_o\tilde{D}_{22} & \rho_o\tilde{D}_{23} \\ 0 & \rho_o\tilde{D}_{23} & \rho_o\tilde{D}_{33} \end{bmatrix} \quad \text{for } \hat{k}_x = 0 ,$$

one obtains $\hat{G}_l = \int d\Omega_{\underline{k}}\ (\hat{k}_x)$. Also for $\hat{G}_t$ the determinant (det $\tilde{D}(\hat{\underline{k}})$, $\hat{k}_x = 0$) needed for the calculation of $\tilde{D}^{-1}$ factorizes conveniently, and the result is

$$\hat{G}_t = 2\pi \frac{c_{11} + c_{44}}{[c_{11}c_{44}(4c_{11}c_{44} + c_a(c_{11} + c_{12}))]^{1/2}} \ ; \quad \hat{G}_l = \frac{2\pi}{c_{44}} \quad \text{for } \hat{\underline{r}} = \langle 100\rangle. \tag{4.68a}$$

The expression for $G_t$ is still simple but not very transparent.

For $\hat{\underline{r}} = \langle 111\rangle$ one has again one longitudinal and one degenerate transverse contribution. The resulting expressions,

$$\frac{c_{44}}{2\pi}[4c_{11}c_{44} + c_a(c_{11} + c_{12})]\sqrt{\alpha} \begin{Bmatrix} \hat{G}_l \\ \hat{G}_t \end{Bmatrix} = \frac{1}{3}c_a(c_{11} + c_{12} + 8c_{44}) + \begin{Bmatrix} 4c_{44}(c_{12} + 2c_{44}) \\ 2c_{44}(c_{12} + 3c_{44}) \end{Bmatrix}$$

$$\alpha = 1 + \frac{2}{27}\frac{c_a^2(c_{11} + 2c_{12} + c_{44})}{c_{44}(c_{11} - c_{12})(c_{11} + c_{12} + 2c_{44})} \quad \text{for } \hat{\underline{r}} = \langle 111\rangle , \tag{4.68b}$$

are very clumsy. Nevertheless, they are useful, for instance when one investigates the behaviour of the response near elastic instability, e.g., for $c_{44} \to 0$, where $\hat{G}_l$ for $\langle 100\rangle$ diverges most strongly.

For $\hat{\underline{r}} = \langle 110\rangle$ the longitudinal, $\hat{G}_l$, and the transversal, $\hat{G}_t$, $\hat{G}_{t'}$, contributions can be calculated analytically too. Only $\tilde{G}_l$ is simple,

$$\hat{G}_l = \frac{2}{c_{44}}\frac{1}{\sqrt{1 + \frac{c_a}{2c_{44}}}} , \tag{4.68c}$$

[15] For hexagonal symmetry G can be calculated analytically for any $\hat{\underline{r}}$, comp. /4.5/.

and instead of discussing the lengthy analytical results we rather summarize the results for some typical metals in Table 4.7.

Table 4.7. Static Green's functions in the main symmetry directions of cubic crystals, $G = \frac{1}{4\pi r c_{44}} \hat{g}(\hat{\underline{r}})$, calculated with the elastic data of Tables 4.2 and 4.3

| Metal | $\hat{g}_l(100)$ | $\hat{g}_t(100)$ | $\hat{g}_l(111)$ | $\hat{g}_t(111)$ | $\hat{g}_l(110)$ | $\hat{g}_{t'}(110)$ | $\hat{g}_t(110)$ |
|---|---|---|---|---|---|---|---|
| Ag | 1 | 1.05 | 2.07 | 0.88 | 1.71 | 0.83 | 0.92 |
| Al | 1 | 0.68 | 1.14 | 0.66 | 1.10 | 0.66 | 0.67 |
| Au | 1 | 0.98 | 2.05 | 0.82 | 1.71 | 0.78 | 0.84 |
| Cu | 1 | 1.13 | 2.21 | 0.93 | 1.80 | 0.86 | 0.99 |
| Ir | 1 | 0.85 | 1.37 | 0.78 | 1.26 | 0.76 | 0.82 |
| Ni | 1 | 1.03 | 1.85 | 0.88 | 1.57 | 0.83 | 0.95 |
| Pb | 1 | 1.19 | 2.58 | 0.96 | 2.03 | 0.86 | 0.96 |
| Pd | 1 | 1.02 | 2.04 | 0.86 | 1.69 | 0.81 | 0.89 |
| Pt | 1 | 0.74 | 1.37 | 0.69 | 1.26 | 0.68 | 0.71 |
| Th | 1 | 1.29 | 2.43 | 1.02 | 1.90 | 0.92 | 1.11 |
| Cr | 1 | 0.58 | 0.80 | 0.61 | 0.85 | 0.62 | 0.58 |
| Cs | 1 | 1.75 | 3.78 | 1.27 | 2.69 | 1.01 | 1.24 |
| α-Fe | 1 | 1.03 | 1.83 | 0.88 | 1.55 | 0.83 | 0.95 |
| K | 1 | 1.59 | 3.45 | 1.19 | 2.52 | 0.98 | 1.17 |
| Li | 1 | 1.91 | 4.16 | 1.36 | 2.90 | 1.04 | 1.30 |
| Mo | 1 | 0.55 | 0.81 | 0.58 | 0.85 | 0.58 | 0.56 |
| Na | 1 | 1.71 | 3.71 | 1.25 | 2.66 | 1.00 | 1.22 |
| Nb | 1 | 0.42 | 0.65 | 0.46 | 0.72 | 0.46 | 0.44 |
| Rb | 1 | 1.50 | 3.34 | 1.13 | 2.48 | 0.95 | 1.09 |
| Ta | 1 | 0.78 | 1.36 | 0.72 | 1.25 | 0.71 | 0.75 |
| V | 1 | 0.54 | 0.85 | 0.56 | 0.89 | 0.56 | 0.55 |
| W | 1 | 0.66 | 1.01 | 0.66 | 1.01 | 0.66 | 0.66 |

## 4.8.3 Double Force Tensor of a Force Distribution in an Infinite Medium

For an arbitrary force density the displacement field is given by

$$\underline{s}(\underline{r}) = \int d\underline{r}'\ G(\underline{r} - \underline{r}')\ f(\underline{r}') \ , \qquad s_i(\underline{r}) = \int d\underline{r}'\ G_{ik}(\underline{r} - \underline{r}')\ f_k(\underline{r}') \ . \tag{4.69}$$

We only will consider forces with vanishing total force and torque (comp. Sec. 4.2.2),

$$\int d\underline{r}'\ f_k(\underline{r}') = 0 \ , \tag{4.69a}$$

$$\varepsilon_{ijk} \int d\underline{r}\ x_j'\ f_k(r') = 0 \ , \tag{4.69b}$$

which also can be applied to a finite crystal without destroying static equilibrium.

We will now discuss the asymptotic behaviour of $\underline{s}(\underline{r})$, roughly speaking for distances $|\underline{r} - \underline{r}'|$ much larger than the extension of $\underline{f}$. If $\underline{f}$ is centered near the origin this expansion corresponds to an expansion of G in powers of the "relatively small" $\underline{r}'$

$$G_{ik}(\underline{r} - \underline{r}') = \underbrace{G_{ik}(\underline{r})}_{\propto \frac{1}{r}} - \underbrace{x_s' \partial_s G_{ik}(\underline{r})}_{\propto \frac{1}{r^2}} + \frac{1}{2}\, \underbrace{x_s' x_t' \partial_s \partial_t G_{ik}(\underline{r})}_{\propto \frac{1}{r^3}} + \dots \ , \tag{4.70}$$

or an expansion in powers of 1/r. If one inserts (4.70) into (4.69), the 1st term vanishes because of (4.69a), and the leading term of the expansion becomes

$$s_i(\underline{r}) \sim - P_{sk} \partial_s G_{ik}(\underline{r}) \tag{4.71}$$

where

$$P_{sk} = \int d\underline{r}'\ x_s'\ f_k(\underline{r}') \ , \qquad P_{sk} = P_{ks} \ \text{from (4.69b)} \ , \tag{4.71a}$$

is the *dipole* or *double force tensor* of the force distribution $\underline{f}(\underline{r})$. The double force tensor is an inherent property of $\underline{f}$; it does not change if $\underline{f}$ is translated, $\underline{f}(\underline{r}) \to \underline{f}(\underline{r} - \underline{T})$.

This expansion is analogous to the asymptotic expansions of stationary electromagnetic fields produced by charge-current densities. In the electric case the leading term corresponds to the total charge or, if this vanishes, the electric dipole moment of the charge distribution. Magnetically, the leading term for stationary currents is determined by an antisymmetrical tensor of 2nd rank, the magnetic dipole moment.

The double force tensor is, without exaggeration, the most important new concept needed in defect physics. The displacements produced by a point defect in the surrounding lattice can be represented by effective forces, so-called Kanzaki forces /4.7/, which would produce the *actual* displacements in the *unperturbed* crystal. In the framework of continuum theory one can use effective force densities, which must obey (4.69a,b). The effective range is microscopic; therefore the corresponding double force tensor alone is sufficient to characterize the macroscopic situation. This is also true for finite crystals; if the defect is in the bulk it can be represented by forces and a double force tensor independent of the defect location. Moreover, somewhat farther from the defect but still in the bulk, the above expansion would hold, still with the G of the infinite crystal.

If one deals with forces $F^{\underline{m}}$ in the lattice the displacements are

$$\underline{s}^{\underline{m}} = G^{(\underline{m}-\underline{m}')}\,\underline{F}^{\underline{m}'} , \tag{4.72}$$

and asymptotically[16]

$$s_i^{\underline{m}} = s_i(\underline{R}^{\underline{m}} = \underline{r}) \sim G_{ik}(\underline{r} - \underline{R}^{\underline{m}'})\,F_k^{\underline{m}'} \sim - P_{sk}\partial_s G_{ik} , \tag{4.72a}$$

where

$$P_{sk} = X_s^{\underline{m}} F_k^{\underline{m}} , \tag{4.72b}$$

which corresponds to a force density

$$f(\underline{r}) = \underline{F}^{\underline{m}}\,\delta(\underline{r} - \underline{R}^{\underline{m}}) , \tag{4.72c}$$

which must obey (4.69a,b). A force density, for which the asymptotic expansion (4.71) holds exactly, is

$$f_s(\underline{r}) = - P_{sl}\partial_l\,\delta(\underline{r}) ; \tag{4.73}$$

it can be obtained from (4.72b,c) by performing the limit $\underline{F}^{\underline{m}} \to \infty$, $\underline{R}^{\underline{m}} \to 0$ such that $P_{sk}$ stays constant. This is illustrated in Fig. 4.18 for a "double force" and a "dilatation center".

---

[16] For large distances $|\underline{R}^{\underline{m}} - \underline{R}^{\underline{m}'}|$, the lattice Green's function $G^{(\underline{m}-\underline{m}')}$ passes into the Green's function $G(\underline{R}^{\underline{m}} - \underline{R}^{\underline{m}'})$ of continuum theory.

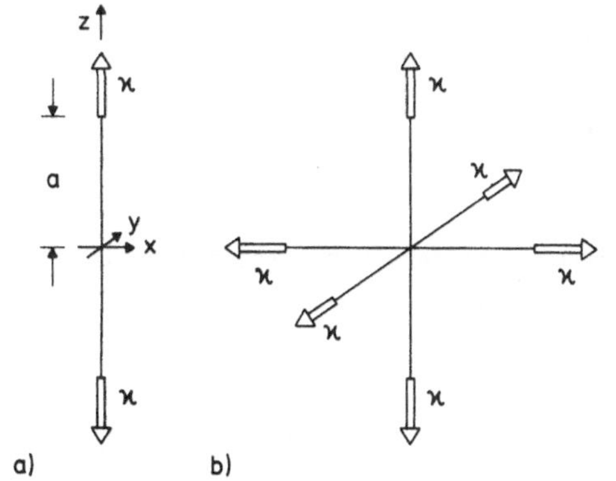


Fig. 4.18a and b. Double force tensors
a) Double force in z-direction: the force density $f_z(\underline{r}) = \kappa[\delta(z-a) - \delta(z+a)]$ $\delta(x)\delta(y)$ has the double force tensor

$$P = 2\kappa a\ |z\rangle\langle z|\quad ,\quad P_{sk} = 2\kappa a \begin{bmatrix} 0 & 0 & 0 \\ 0 & 0 & 0 \\ 0 & 0 & 1 \end{bmatrix}\ ;$$

by taking the limes $a \to 0$, $\kappa \to \infty$ with fixed $\kappa a$, one obtains the "concentrated" force density $f_z(\underline{r}) = -2\kappa a \partial_z\ \delta(\underline{r})$ or $f_s(\underline{r}) = -P_{sl}\partial_l\ \delta(\underline{r})$ with the same double force tensor.
b) Dilatation center: three orthogonal double forces of equal strength. The force density $\underline{f}(\underline{r}) = (f_x, f_y, f_z)$ with $f_z$ as above and $f_x = \kappa[\delta(x-a) - \delta(x+a)]\ \delta(y)\delta(z)$, $f_y = [\kappa\ \delta(y-a) - \delta(y+a)]\ \delta(z)\delta(x)$ has the double force tensor $P = 2\kappa a(|x\rangle\langle x| + |y\rangle\langle y| + |z\rangle\langle z|) = 2\kappa a$, $P_{sk} = 2\kappa a\delta_{sk}$; the corresponding "concentrated" density is $f_s = -2\kappa a\partial_s\delta(\underline{r})$

### 4.8.4 The Volume Change ΔV of a Finite Crystal

If a (static) force density $f(\underline{r})$ inside a finite crystal with volume V and surface S produces a displacement field $s(\underline{r})$, the corresponding volume change ΔV can be expressed by a surface or volume integral:

$$\Delta V = \int_S d\underline{S}\ \underline{s} = \int_V d\underline{r}\ \mathrm{div}\{\underline{s}\} = \int_V d\underline{r}\ \mathrm{tr}\{\varepsilon\}\ . \tag{4.74}$$

To calculate ΔV we start from the original static equation,

$$\partial_k\sigma_{ik} + f_i = 0\ , \tag{4.75}$$

multiply with $x_s$ and integrate over V:

$$\int_V d\underline{r}\,(\underbrace{x_s\partial_k\sigma_{ik}}_{\partial_k x_s\sigma_{ik}-\sigma_{is}} + x_s f_i) = \int_S x_s\sigma_{ik}\,dS_k - \int_V d\underline{r}\,\sigma_{is} + \underbrace{\int_V d\underline{r}\,x_s f_i}_{P_{si}} = 0 \qquad (4.75a)$$

The force must be translational-rotational invariant; then P has the properties discussed in the foregoing subsection. For a *free surface*, where $\sigma_{ik}dS_k = 0$, eq. (4.75a) becomes

$$\int_V d\underline{r}\,\sigma = P \quad , \qquad (4.76)$$

which does not depend on the special location of the force pattern[17]; for a homogeneous crystal, where C and S are independent of $\underline{r}$ (in V), one has

$$C\int_V d\underline{r}\,\varepsilon = P \; , \qquad \int_V d\underline{r}\,\varepsilon = SP \; , \qquad (4.77)$$

$$\Delta V = \int_V d\underline{r}\,\mathrm{tr}\{\varepsilon\} = \mathrm{tr}\{SP\} = S_{ii,mn}P_{mn} \; . \qquad (4.78)$$

Eq. (4.78) shows that $\Delta V$ depends on the forces only through the dipole force tensor and on the crystal properties only through the compliances S; it does not depend on the location of the forces nor on the extension of the crystal. This result confirms, as has been stated in Section 4.8.3, that, macroscopically, a force pattern can solely be characterized by its double force tensor. Quite crudely one can argue as follows: Imagine a sphere with macroscopic radius R and $\underline{f}$ in its center; near the surface the asymptotic expansion (4.71) can be applied; the displacement due to P is proportional to $R^{-2}$ and yields a $\Delta V$, which is independent of R (because $dS \propto R^2$) and proportional to P; the higher order terms of the expansion (4.71) would lead to a $\Delta V$ contribution of order $R^{-1}$, which can be dropped.

For *cubic crystals*, (4.78) can be simplified further; from the representation of S by the basis tensors, (4.40b), and from eqs. (4.12-14) one recognizes that because of $\mathrm{tr}\{T(\nu)\} = \sqrt{3}\,\delta_{\nu 1}$ only T(1) remains if the trace with respect to the first two subscripts is taken:

---

17 According to (4.75a), eq. (4.76) is also valid if P includes the contribution of surface forces $\int_S x_s df_i$.

$$S_{ii,mn} = \sum_{\nu} \mathrm{tr}\{T(\nu)\}(1/\overset{\nu}{C})T_{mn}(\nu) = \sqrt{3}\,(1/\overset{1}{C})T_{mn}(1) = (1/\overset{1}{C})\delta_{mn} \,, \qquad (4.78a)$$

$$\overset{1}{C} = 3K = c_{11} + 2c_{12} \,,$$

therefore $\Delta V = P_{mm}/(3K) = \mathrm{tr}\{P\}/(3K)$. This result is still valid for arbitrary symmetry of the crystal if P is isotropic, i.e., if $P_{mn} = P_o\delta_{mn} = \delta_{mn}\mathrm{tr}\{P\}/3$ and therefore $\Delta V = S_{ii,mm}\mathrm{tr}\{P\}/3$, because $S_{ii,mm} = K^{-1}$ for arbitrary symmetry (see Sec. 4.8.6).

Equation (4.76) can be discussed in another way, which turns out to be important for crystals with many defects, where the physics can be represented by averages. The force densities are given by $f(\underline{r} - \underline{R})$, where $\underline{R}$ denotes the location of the force pattern. The double force tensor does not depend on $\underline{R}$, but $\sigma(\underline{r},\underline{R})$ does. Eq. (4.76) states for the volume average of $\sigma$,

$$\langle\sigma\rangle^{\underline{r}} = \frac{1}{V}\int_V d\underline{r}\; \sigma(\underline{r},\underline{R}) = \frac{P}{V} \,, \text{ independent of } \underline{R}, \text{ force density } f(\underline{r} - \underline{R}). \qquad (4.79)$$

On the other hand, one can take the average over location $\underline{R}$. This average becomes meaningful for many equivalent patterns $f(\underline{r} - \underline{R})$, which represent defects and are distributed homogeneously in $\underline{R}$. We will show below that the two averages are identical:

$$\langle\sigma\rangle^{\underline{R}} = \frac{1}{V}\int_V d\underline{R}\; \sigma(\underline{r},\underline{R}) = \frac{P}{V} \,, \quad \text{independent of } \underline{r}, \qquad (4.79a)$$

$$\text{force density } \frac{1}{V}\int_V d\underline{R}\; \underline{f}(\underline{r} - \underline{R}) = \langle\underline{f}\rangle^{\underline{R}} \,.$$

A homogeneous distribution of force patterns causes a homogeneous stress inside the crystal, or, in other words, the averaged force density must correspond to surface forces $Pd\underline{S}/V$. This is easy to show for a pure[18] double force

$$f_s = -P_{sk}\partial_k\delta(\underline{r} - \underline{R}) \,, \qquad (4.80)$$

corresponding to

$$\langle f_s\rangle^{\underline{R}} = -\frac{P_{sk}}{V}\partial_k\int_V d\underline{R}\; \delta(\underline{r} - \underline{R}) = -\frac{P_{sk}}{V}\partial_k\Theta_V(\underline{r}) \,, \quad \Theta_V(\underline{r}) = \begin{cases} 1 \text{ if } \underline{r} \text{ inside V} \\ 0 \text{ if } \underline{r} \text{ outside V} \end{cases} \qquad (4.80a)$$

which vanishes except on the surface. The dipole forces cancel each other in the bulk and leave only $\delta$-like force densities in the surface, which are identical with surface forces $Pd\underline{S}/V$ (Fig. 4.19).

If each $\underline{f}$ represents a single defect and if N defects are homogeneously distri-

---

18 This amounts to neglecting microscopic contributions, which are due to the extension of the force distribution and arise near the surface.

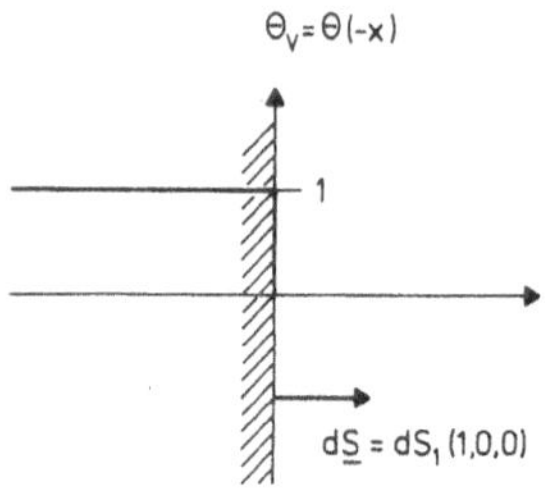


Fig. 4.19. Average force densities. For simplicity the surface is $x = 0$; then $\Theta_V(\underline{r}) = \Theta(-x)$, $\partial_k\Theta_V = -\delta_{k1}\delta(x)$ and $\langle f_s\rangle^{\underline{R}} = \frac{P_{s1}}{V}\delta(x)$ corresponds to surface forces $\frac{P_{s1}}{V}\, dS_1 = \frac{P_{sk}}{V}\, dS_k$

buted in the crystal, one expects

$$\text{an average stress} \quad \langle\sigma\rangle = \frac{N}{V} P \ , \tag{4.81a}$$

$$\text{an average strain} \quad \langle\varepsilon\rangle = \frac{N}{V} SP \ , \tag{4.81b}$$

$$\text{and a volume change} \quad \Delta V = N\,\mathrm{tr}\{SP\} \ , \quad \frac{\Delta V}{V} = \frac{N}{V}\,\mathrm{tr}\{SP\} \ . \tag{4.81c}$$

### 4.8.5 Local, Microscopic Displacements Derived from the Macroscopic Volume Change $\Delta V$

Macroscopically, a defect is characterized alone by its double force tensor P, which can be determined experimentally. Near the defect, the displacements have to be described by lattice theory, and one must consider the distribution of the Kanzaki forces on the lattice atoms around the defect site. A crude approximation is to use continuum theory for a concentrated force density at the defect site determined by P alone, $\underline{f} = -P\underline{\partial}\delta(\underline{r})$ for a defect at $\underline{r} = 0$. In the bulk, the displacements near the defect ought to be those in an *infinite* crystal:

$$\overset{\infty}{s}_i = -P_{sk}\partial_s G_{ik}(\underline{r}) \ , \quad \text{where G refers to the infinite crystal.} \tag{4.82}$$

However, the displacement field $\overset{\infty}{\underline{s}}$ will not meet the boundary conditions. For instance, $\overset{\infty}{\underline{s}}$ will be compatible only with surface forces $\overset{\infty}{\sigma}d\underline{S}$; if the surface is free, $\sigma d\underline{S} = 0$, one has to superimpose a displacement field $\overset{I}{\underline{s}}$ produced by surface forces $-\overset{\infty}{\sigma}d\underline{S}$ to obtain the required force-free surface. The complete solution is then

$$\underline{s} = \overset{\infty}{\underline{s}} + \overset{I}{\underline{s}} \ . \tag{4.82a}$$

The superscript I means "image" because $\overset{I}{\underline{s}}$ can be produced by forces $\underline{f}^I$ outside in the infinite crystal and in many cases the $\underline{f}^I$ are mirror images of the original $\underline{f}$ (comp. Sec. 4.8.9). The volume change of the whole crystal can be separated analogously

$$\Delta V = \int_S (\overset{\infty}{\underline{s}} + \overset{I}{\underline{s}})\, d\underline{S} = \Delta V^\infty + \Delta V^I = (1+\gamma)\Delta V^\infty \ , \quad \gamma = \frac{\Delta V^I}{\Delta V^\infty} \ , \tag{4.82b}$$

where $\gamma$, the Eshelby factor, /4.8/, indicates the influence of the surface; $\gamma$ is of the order 1 for most of the metals in Tables 4.2 and 4.3.

For purposes of illustration let us consider the simplest example, a dilatation center, $\underline{f} = -P_o \underline{\partial}\delta(\underline{r})$, in the center of a sphere of isotropic material, where with (4.63,67)

$$\overset{\infty}{s}_i = -P_o \partial_k G_{ik}(\underline{r}) = \frac{P_o}{4\pi c_{11}} \frac{x_i}{r^3} ; \quad \overset{\infty}{\underline{s}} = \frac{P_o}{4\pi c_{11}} \frac{\hat{r}}{r^2} = \overset{\infty}{s}_r \hat{\underline{r}} ; \tag{4.83a}$$

$$\overset{\infty}{\varepsilon}_{ik} = \frac{P_o}{4\pi c_{11} r^3}(\delta_{ik} - 3\hat{x}_i\hat{x}_k) \text{ , pure shear because } \overset{\infty}{\varepsilon}_{ii} = \mathrm{div}\{\overset{\infty}{\underline{s}}\} = 0 \tag{4.83b}$$

$$\overset{\infty}{\sigma} = 2\mu\overset{\infty}{\varepsilon} = 2c_{44}\overset{\infty}{\varepsilon} ; \quad \overset{\infty}{\sigma}\hat{\underline{r}} = -\frac{4c_{44}P_o}{4\pi c_{11} r^3}\hat{\underline{r}} = -p(r)\hat{\underline{r}} . \tag{4.83c}$$

Together with an external pressure, p(R), the displacement $\overset{\infty}{\underline{s}}$ is a correct solution (Fig. 4.20):

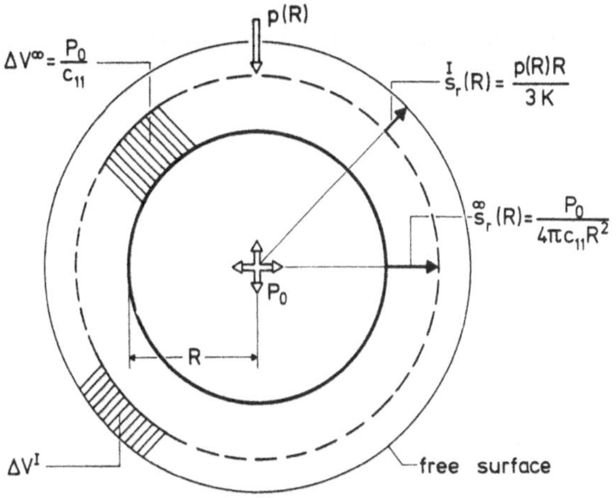


Fig. 4.20. Dilatation center $P_o$ in an elastically isotropic sphere. The displacement field $\overset{\infty}{\underline{s}} = \overset{\infty}{s}_r\hat{\underline{r}}$ is a radial field; together with a pressure, p(R), it is a correct static solution. The volume change by this field is $\Delta V^\infty = 4\pi R^2 \overset{\infty}{s}_r(R)$; if the pressure p(R) is removed (free surface), one obtains an additive "image" displacement, $\overset{I}{\underline{s}} = \overset{I}{s}_r\hat{\underline{r}}$, which produces an additive change in volume.

$$\overset{I}{\underline{s}} = p(R)\underline{r}/(3K) = \underline{r}\, c_{44}P_o/(3\pi c_{11} K R^3) , \tag{4.84}$$

$$\Delta V = \underbrace{\frac{P_o}{c_{11}}}_{\Delta V^\infty} + \underbrace{P_o\frac{4c_{44}}{3c_{11}K}}_{\Delta V^I} = \frac{P_o}{c_{11}}\Big(1 + \underbrace{\frac{4c_{44}}{c_{11}+2c_{12}}}_{\gamma}\Big) = \frac{P_o}{K} , \tag{4.85}$$

$\gamma \cong 0.69$ and 0.49 for nearly isotropic W and Al.

One recognizes that $\overset{I}{s}_r/\overset{\infty}{s}_r$ is of the order $r^3/R^3$ and that, therefore, $\overset{I}{\underline{s}}$ can be disregarded near the defect, but that at the surface, where $\Delta V$ is observed, both displacements are of equal order.[19] This example demonstrates clearly that the image effects do not influence the local behaviour, which is determined by $\overset{\infty}{\underline{s}}$ alone.

Actually div $\overset{\infty}{\underline{s}}$, the volume change per unit volume, does not vanish; more correctly, one has

$$\mathrm{div}\{\overset{\infty}{\underline{s}}\} = \frac{P_o}{c_{11}}\,\delta(\underline{r})\ , \text{ consistent with } \Delta V^\infty = \int_V d\underline{r}\ \mathrm{div}\{\overset{\infty}{\underline{s}}\} = \frac{P_o}{c_{11}}\ , \tag{4.83d}$$

i.e., the displacement field springs from the source in the origin. This means that any volume containing the origin, e.g., a sphere, changes by the *same* amount $\Delta V^\infty$ and the volumes not containing the origin, e.g., spherical shells, do not change their volume at all. An arbitrarily small volume would produce a void $\Delta V^\infty$ in the center, for negative $\Delta V^\infty$ one would have to cut out a void before applying the displacement. Here lattice theory gives a hint how to handle that situation. Fig. 4.21 shows a substitutional atom in the origin of a fcc lattice; this defect has cubic symmetry and $P$ must correspond to a dilatation center. Suppose $\Delta V$ has been measured and $\Delta V^\infty$ has been calculated from (4.85). One can then obtain a crude picture of the

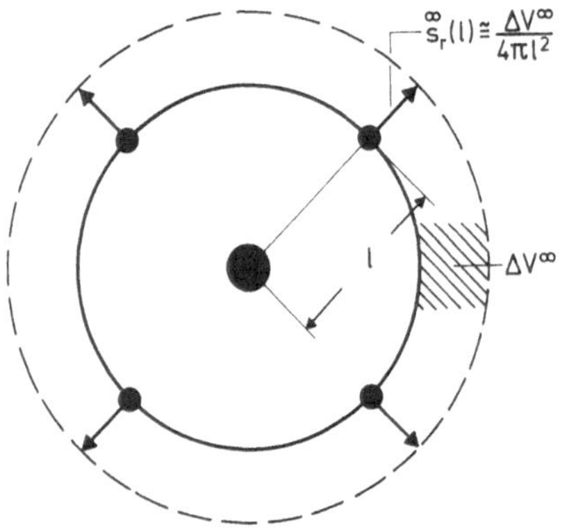


Fig. 4.21. Local lattice displacements around a dilatation center in a fcc lattice. If $\Delta V^\infty = V_c$, the displacement of 1st neighbours is about 6% of their distance: $\overset{\infty}{s}_r(l)/l = 1/4\pi\sqrt{2} \cong 0.06$

[19] The same holds for all surface conditions, e.g., for fixed surface one has simply $\underline{s} = P_o/(4\pi c_{11})\underline{r}(1/r^3 - 1/R^3)$, $\overset{I}{s}_r/\overset{\infty}{s}_r = r^3/R^3$.

local displacements by requiring that the 1st neighbour displacements[20] lead to the calculated volume change, $\Delta V^{\infty} \cong 4\pi l^2 \overset{\infty}{s}_r(l)$ (Fig. 4.21). Consequently, $\Delta V^{\infty}$ gives some information about the local changes near the defect.

The volume change, $\Delta V = \mathrm{tr}\{SP\}$, does not depend on form and size of the crystal. However, the single distributions $\Delta V^{\infty}$ and $\Delta V^{I}$ depend on form and size because $\mathrm{div}\{\overset{\infty}{\underline{s}}\}$ is, in general, not concentrated at the defect site as in the example of the dilatation center in an isotropic crystal. With $\overset{\infty}{s}_i = -P_{sk}\partial_s G_{ik}(\underline{r})$, $\mathrm{div}\{\overset{\infty}{\underline{s}}\} = -P_{sk}\partial_i\partial_s \cdot G_{ik}(\underline{r})$ one obtains from (4.66)

$$\mathrm{div}\{\overset{\infty}{\underline{s}}\} = P_{sk}\int \frac{d\underline{k}}{(2\pi)^3\rho_o} k_i k_s \left[\tilde{D}^{-1}(\underline{k})\right]_{ik} e^{i\underline{k}\underline{r}} , \tag{4.86}$$

where $\hat{k}_i\hat{k}_s[\tilde{D}^{-1}(\underline{k})]_{ik}$ depends only on $\hat{\underline{k}}$, but not on k. One recognizes that the average of $\mathrm{div}\{\overset{\infty}{\underline{s}}\}$ over a spherical surface arround the defect vanishes and that therefore $\langle\mathrm{div}\{\overset{\infty}{\underline{s}}\}\rangle$ is concentrated at the defect site. This can be seen as follows: The average of $\exp(i\underline{k}\underline{r})$ over all directions becomes $\langle\exp(i\underline{k}\underline{r})\rangle = \sin(kr)/kr$, which depends only on k. Therefore, the directional average $\langle\ldots\rangle^{\underline{k}}$ in $\underline{k}$ space of the factor in (4.86) can be taken out of the integral,

$$\langle\mathrm{div}\{\overset{\infty}{\underline{s}}\}\rangle = \underbrace{P_{sk}\left\langle \hat{k}_i\hat{k}_k\left[\tilde{D}^{-1}(\hat{\underline{k}})/\rho_o\right]_{ik}\right\rangle^{\hat{\underline{k}}}}_{\Delta V^{\infty}} \underbrace{\int \frac{d\underline{k}}{(2\pi)^3} e^{i\underline{k}\underline{r}}}_{\delta(\underline{r})} , \tag{4.87}$$

where $\Delta V^{\infty}$ refers to a sphere with the defect at its center. For cubic crystals the factor of $P_{sk}$ in (4.87) is a tensor of 2nd rank which must be isotropic. Consequently,

$$\Delta V^{\infty} = \frac{\mathrm{tr}\{P\}}{3}\left\langle \hat{k}_i\left[\tilde{D}^{-1}(\hat{\underline{k}})/\rho_o\right]_{ik}\hat{k}_k\right\rangle^{\hat{\underline{k}}} = \frac{\mathrm{tr}\{P\}}{3}\left\langle\left(\hat{\underline{k}},\tilde{G}(\hat{\underline{k}})\hat{\underline{k}}\right)\right\rangle^{\hat{\underline{k}}} = \frac{\mathrm{tr}\{P\}}{3}\left\langle\left(\underline{k},\tilde{G}(\underline{k})\underline{k}\right)\right\rangle^{\hat{\underline{k}}} \quad \text{for cubic crystals,} \tag{4.87a}$$

$$\Delta V^{\infty} = \frac{\mathrm{tr}\{P\}}{3}\tilde{G}_1 = \frac{\mathrm{tr}\{P\}}{3}\frac{1}{c_{11}} \quad \text{for isotropy.} \tag{4.87b}$$

In cubic crystals $\Delta V^{\infty}/\Delta V$ is obtained from (4.78a) and (4.87a). The integral (4.87a) can be expanded in powers of the anisotropy $c_a$. The choice of averaged isotropic data $\bar{c}_{\alpha\beta}$ (with $\bar{c}_{11} - \bar{c}_{12} - 2\bar{c}_{44} = 0$), about which this expansion is made, is not unique; it turns out that Voigt's averages

---

[20] On the other hand, one can try to describe P by radial forces $\kappa$ acting on the 1st neighbours, $P_o = 4\kappa l = K\Delta V$, and one obtains $\overset{\infty}{s}_r$ from the lattice G's if they are available. Both methods are equally crude. A comparison is made in Appendix M.

$$\bar{c}_{11} = c_{11} - 2c_a/5 \; , \quad \bar{c}_{12} = c_{12} + c_a/5 \; , \quad \bar{c}_{44} = c_{44} + c_a/5$$

are distinguished, because then the linear term in $c_a$ vanishes and the expansion starts with a quadratic term:

$$\frac{\Delta V^{\infty}}{\Delta V} = \frac{K}{\bar{c}_{11}} \left[ 1 + \left( \frac{c_a}{\bar{c}_{44}} \right)^2 \int \frac{d\Omega_{\hat{\underline{k}}}}{4\pi} \cdots \right] .$$

This integral[21] is evaluated numerically; the results are given in Table 4.8, which contains also the values of $\overline{\Delta V^{\infty}}/\Delta V = K/\bar{c}_{11}$ , calculated with Voigt's averages. One recognizes that $\overline{\Delta V^{\infty}}$ is, as a rule, already a good approximation for $\Delta V^{\infty}$, but that it is always smaller than the exact value. The reason for this inequality is given in Sec. 4.8.7, where $\overline{\Delta V^{\infty}}$ is obtained from a variational principle which implies $\overline{\Delta V^{\infty}} < \Delta V^{\infty}$.

Table 4.8. Values of $\Delta V^{\infty}/\Delta V = 1/(1 + \gamma)$ for cubic crystals; variational $(K/\bar{c}_{11})$ and numerical values obtained with the elastic data of Tables 4.2 and 4.3

| | Var. | Num. | | Var. | Num. |
|---|---|---|---|---|---|
| Ag | 0.697 | 0.725 | Cr | 0.509 | 0.513 |
| Al | 0.686 | 0.687 | Cs | 0.629 | 0.718 |
| Au | 0.804 | 0.824 | α-Fe | 0.582 | 0.606 |
| Cu | 0.654 | 0.690 | K | 0.663 | 0.737 |
| Ir | 0.549 | 0.556 | Li | 0.614 | 0.715 |
| Ni | 0.592 | 0.618 | Mo | 0.609 | 0.613 |
| Pb | 0.764 | 0.800 | Na | 0.644 | 0.728 |
| Pd | 0.732 | 0.756 | Nb | 0.765 | 0.776 |
| Pt | 0.765 | 0.770 | Rb | 0.723 | 0.785 |
| Th | 0.560 | 0.611 | Ta | 0.673 | 0.679 |
| | | | V | 0.709 | 0.711 |
| | | | W | 0.593 | 0.593 |

21 The explicit expression for the integrand is rather clumsy; it is obtained by straightforward expansion of $\tilde{D}^{-1}(\hat{\underline{k}})$ in powers of $c_a/\bar{c}_{44}$. The linear term does not vanish in the integrand but only in the directional average, $\int d\Omega_{\hat{\underline{k}}} \ldots$ .

To obtain approximate core displacements one determines $\Delta V^\infty$ from $\Delta V$ for the defect in the center of a sphere and approximates the core displacements as in Fig. 4.21. Details are discussed in Appendix M.

### 4.8.6 Homogeneous Stresses

In static experiments it is usually simpler to apply a stress, $\sigma$, and measure $\varepsilon$ to determine the elastic compliances. We will discuss only the very simplest examples.

1) Homogeneous pressure p:

$$\sigma_{ik} = -p\delta_{ik}, \quad \varepsilon_{ik} = -pS_{ik,mm}, \tag{4.88a}$$

$$\frac{\Delta V}{V} = \varepsilon_{ii} = -pS_{ii,mm} = -\frac{P}{K}, \quad \frac{1}{K} = S_{ii,mm}, \tag{4.88b}$$

$$\frac{1}{K} = \begin{cases} 3\overset{1}{S} = \dfrac{3}{\overset{1}{C}} = \dfrac{3}{c_{11} + 2c_{12}} & \text{for cubic crystals} \\[2ex] \dfrac{c_{11} + 2c_{33} + c_{12} - 4c_{13}}{(c_{11} + c_{12})c_{33} - 2c_{13}^2} & \text{for hexagonal crystals.} \end{cases} \tag{4.88c}$$

2) Uniaxial stress $\sigma_o$ in direction $\hat{\underline{a}}$:

$$\sigma_{ik} = \sigma_o \hat{a}_i \hat{a}_k, \quad \varepsilon_{ik} = \sigma_o S_{ik,mn} \hat{a}_m \hat{a}_n; \tag{4.89a}$$

relative length change in direction $\hat{\underline{a}}$:

$$(\hat{\underline{a}}, \varepsilon\hat{\underline{a}}) = \sigma_o \hat{a}_i \hat{a}_k S_{ik,mn} \hat{a}_m \hat{a}_n = \sigma_o / E(\hat{\underline{a}}), \quad E(\hat{\underline{a}}) \text{ Young's modulus.} \tag{4.89b}$$

In cubic crystals, where

$$\frac{1}{E(\hat{\underline{a}})} = (s_{12} + 2s_{44}) + (s_{11} - s_{12} - 2s_{44})(\hat{a}_x^4 + \hat{a}_y^4 + \hat{a}_z^4),$$

$$s_{12} + 2s_{44} = \frac{1}{3(c_{11} + 2c_{12})} - \frac{1}{3(c_{11} - c_{12})} + \frac{1}{2c_{44}}, \tag{4.89c}$$

$$s_{11} - s_{12} - 2s_{44} = \frac{1}{c_{11} - c_{12}} - \frac{1}{2c_{44}},$$

$E(\hat{\underline{a}})$ and K determine the three cubic moduli.

3) Uniaxial stress in cube edge direction, $\hat{\underline{a}} = (1,0,0)$:

$$\sigma = \sigma_o \begin{bmatrix} 1 & 0 & 0 \\ 0 & 0 & 0 \\ 0 & 0 & 0 \end{bmatrix} = \frac{\sigma_o}{3} \left\{ \underbrace{\begin{bmatrix} 1 & 0 & 0 \\ 0 & 1 & 0 \\ 0 & 0 & 1 \end{bmatrix}}_{\text{dilatation}} + \underbrace{\begin{bmatrix} 2 & 0 & 0 \\ 0 & \bar{1} & 0 \\ 0 & 0 & \bar{1} \end{bmatrix}}_{\text{(011) shear}} \right\} ; \tag{4.90a}$$

$$\varepsilon_{11} = \frac{\sigma_o}{3} (1/\overset{1}{C} + 2/\overset{2}{C}) = \sigma_o \frac{c_{11} + c_{12}}{(c_{11} + 2c_{12})(c_{11} - c_{12})} = \frac{\sigma_o}{E(100)} ,$$

(4.90b)

$$\varepsilon_{22} = \varepsilon_{33} = \frac{\sigma_o}{3} (1/\overset{1}{C} - 1/\overset{2}{C}) = \sigma_o \frac{c_{12}}{(c_{11} + 2c_{12})(c_{11} - c_{12})} , \text{ isotropic about } \hat{\underline{a}};$$

the ratio of transversal contraction, $-\varepsilon_{22}$, to longitudinal expansion, $\varepsilon_{11}$, is Poisson's ratio,

$$\nu = \frac{c_{12}}{c_{11} + c_{12}} . \tag{4.90c}$$

In isotropic crystals $\nu$ and E do not depend on direction and can be used as independent elastic data. Table 4.9 summarizes the various combinations of moduli which are used for isotropic media.

Table 4.9. Isotropic elastic moduli

| | |
|---|---|
| $\nu > 0$<br>$K > 0$ | shear modulus, eigenvalue of C: $2\mu$<br>compressional modulus, eigenvalue of C: $3K$ |
| $c_{44} = \mu > 0$<br>$c_{11} = K + \frac{4\mu}{3} > \frac{4c_{44}}{3}$ | transversal sound velocity: $\sqrt{c_{44}/\rho_o}$<br>longitudinal sound velocity: $\sqrt{c_{11}/\rho_o}$ |
| $\mu = c_{44} > 0$<br>$\lambda = c_{12} = K - \frac{2\mu}{3} > -\frac{2\mu}{3}$ | Lamé's constants |
| $E = \frac{3K + \mu}{9K\mu} > 0$<br>$-1 < \nu = \frac{K - 2\mu/3}{2K + 2\mu/3} < 0.5$ | Young's modulus — uniaxial stress: longitudinal modulus $E$<br>Poisson's ratio — uniaxial stress: transversal modulus $-\nu E$ |

### 4.8.7 Variational Methods

As in lattice theory (see Sec. 2.5.4), one can employ variational methods. The crucial quantity is the total energy L which includes all volume forces, $\underline{f}$, and surface forces, $\Sigma d\underline{S} = d\underline{F}$, and which is given by

$$L[\underline{s}(\underline{r})] = \int_V d\underline{r}\, \frac{(\varepsilon(s), C\varepsilon)}{2} - \int_V d\underline{r}\, (\underline{s},\underline{f}) - \int_S \underline{s}\Sigma\, d\underline{S} \quad , \tag{4.91}$$

in analogy to (2.44). The 1st term is the internal elastic energy, the 2nd represents the potential energy of volume forces and the 3rd term describes the potential energy of the surface forces.

If $\underline{s}(\underline{r})$ is the correct solution and $\underline{\eta}(\underline{r})$ a deviation, then $[\sigma = C\varepsilon(\underline{s})]$

$$L[\underline{s} + \underline{\eta}] = L[\underline{s}]$$

correct total energy: $L[\underline{s}] = L_{min} = -U[\underline{s}]$

$$- \int_V d\underline{r}\, \eta_i(\partial_k\sigma_{ik} + f_i) + \int_S \underline{\eta}(\sigma - \Sigma)\, d\underline{S}$$

1st order term: vanishes for all $\underline{\eta}$ if $\underline{s}$ is the correct solution

$$+ \int_V d\underline{r}\, \frac{\big(\varepsilon(\eta), C\varepsilon(\eta)\big)}{2}$$

2nd order term: $U[\underline{\eta}]$, positive for all $\underline{\eta} \neq 0$ because of elastic stability.

One recognizes that the exact solution is determined by requiring the linear term to vanish, which leads to

$$\partial_k\sigma_{ik} + f_i = 0 \ , \tag{4.92a}$$

$$(\sigma - \Sigma)\, d\underline{S} = 0 \quad \text{on the surface} \ , \tag{4.92b}$$

the correct equation of motion and the correct boundary condition. That $L[\underline{s}]$ is a minimum is due to the positive 2nd order term.[22]

This property of L to be a minimum for the correct solution can be used as in Section 2.5.4 to determine optimal parameters in variational solutions, $\underline{s}_v$ , the

[22] That the equations (4.92a,b) lead to a unique solution can be argued as follows: Suppose $\underline{s}$ and $\tilde{\underline{s}} = \underline{s} + \underline{\eta}$ are both solutions then $L[\tilde{\underline{s}}] = L[\underline{s} + \underline{\eta}] = L[\underline{s}] + U[\underline{\eta}]$, $L[\underline{s}] = L[\tilde{\underline{s}} - \underline{\eta}] = L[\tilde{\underline{s}}] + U[\underline{\eta}]$. This implies $U[\underline{\eta}] = 0$ which means $\underline{\eta} = 0$ except for translations and/or rotations.

lower the value of $L[\underline{s}_v = \underline{s} + \underline{\eta}] = L_v = -U[\underline{s}_v] = -U_v$ the better the approximate $\underline{s}_v$. We will discuss two simple but useful applications.

First we derive a variational principle for $\Delta V^\infty$ using a radially symmetric force density[23] in an infinite crystal (vanishing surface term in (4.91)):

$$f_i(\underline{r}) = -P_o\partial_i h(r) = -P_o\hat{x}_i h'(r) , \qquad \int d\underline{r}\; h(r) = 1 , \tag{4.93a}$$

$$P_{si} = \int d\underline{r}\; x_s f_i(\underline{r}) = P_o\delta_{si}\int d\underline{r}\; h(r) = P_o\delta_{si} . \tag{4.93b}$$

The minimum value of L becomes

$$\begin{aligned} L_{min} &= -\frac{1}{2}\int d\underline{r}\;(f,\underline{s}) = -\frac{1}{2}\iint d\underline{r}\, d\underline{r}'\,\Big(\underline{f}(\underline{r}),G(\underline{r}-\underline{r}')f(\underline{r}')\Big) \\ &= -\frac{P_o^2}{2}\iint d\underline{r}\, d\underline{r}'[\partial_i h(r)]G_{ik}(\underline{r}-\underline{r}')[\partial_k' h(r')] \\ &= -\frac{P_o^2}{2}\iint d\underline{r}\, d\underline{r}'\; h(r)\, h(r')\int d\underline{k}\,\Big(\underline{k},\tilde{G}(\underline{k})\underline{k}\Big)\,\frac{e^{i\underline{k}(\underline{r}-\underline{r}')}}{(2\pi)^3} , \end{aligned} \tag{4.94}$$

where the last integral is obtained by integration by parts and by use of the Fourier representation of $G(\underline{r})$. From (4.86,87) one recognizes that the $\underline{k}$-integral yields $(\Delta V^\infty/P_o)\;\delta(\underline{r}-\underline{r}')$, and therefore

$$L_{min} = -\frac{P_o\Delta V^\infty}{2}\int d\underline{r}\; h^2(r) . \tag{4.94a}$$

Consequently, a trial $L_{v,min} > L_{min}$ will yield an approximate and optimal $\Delta V_v^\infty < \Delta V^\infty$. For a radially symmetric trial displacement, $\underline{s}_v = -\,\mathrm{grad}\;\alpha(r)$, $\Big(\alpha = P_o/(4\pi c_{11} r)$ for isotropy outside the range of $h\Big)$, one obtains

$$L[\underline{s}_v] = \frac{C_{ik,mn}}{2}\int d\underline{r}\;\partial_i\partial_k\alpha(r)\cdot\partial_m\partial_n\alpha(r) - P_o\int d\underline{r}\;\partial_i\alpha(r)\cdot\partial_i h(r) ;$$

the factor of C is rotationally invariant and therefore C can be replaced by its average over all orientations, i.e., Voigt's average $\bar{C}$. The variational solution must be the isotropic solution with $\bar{C}$ and the approximate $\Delta V_v^\infty = \overline{\Delta V^\infty} = P_o/\bar{c}_{11}$ must be smaller than $\Delta V^\infty$ (comp. Table 4.8).

---

23 For $h(\underline{r}) = \delta(\underline{r})$, the usual center of dilatation, $L_{min}$ diverges. To avoid this we take a somewhat smeared-out density, for which $\int d\underline{r}\; h^2(\underline{r})$ stays finite.

In a second application we consider a finite crystal (volume V, surface S) with given surface forces $\Sigma d\underline{S}$ (homogeneous $\Sigma$, constant on S) and vanishing volume forces $\underline{f}$. For homogeneous material, where C does not depend on $\underline{r}$, the deformation is homogeneous, $\underline{s} = \varepsilon\underline{r}$, $\varepsilon = S\Sigma$ and $L_{min} = -\frac{1}{2}\int_S \underline{s}\Sigma\, d\underline{s} = \frac{V}{2}(\Sigma,S\Sigma)$. We consider now a crystal where $C(\underline{r})$ depends on $\underline{r}$ in a statistical way, for example a polycrystal where C depends on the orientation of the grains, or an isotropic medium, $C_1$, with isotropic inclusions, $C_2$. For each situation one has $L_{min} = \frac{1}{2}\int_S \underline{s}\Sigma\, d\underline{S}$, where $\underline{s}$ is the correct and very complicated solution depending, for instance, on the location, extension and orientation of grains. One expects that macroscopically the crystal which contains many grains or inclusions will behave like a homogeneous crystal with *effective* elastic data, $C^{eff} = (S^{eff})^{-1}$. Hopefully[24], the averages over the statistical distributions will represent the macroscopic behaviour. Then $\langle L_{min}\rangle = -\frac{1}{2}\int \langle\underline{s}\rangle\Sigma\, d\underline{S}$, and, under the circumstances considered here, $\langle\underline{s}\rangle$ must be a homogeneous deformation on the surface: $\langle\underline{s}\rangle = \langle\varepsilon\rangle\,\underline{r}$ , $\langle\varepsilon\rangle = S^{eff}\Sigma$ , $\langle L_{min}\rangle = -V/2(\Sigma,S^{eff}\Sigma)$. If one uses trial displacements, $\underline{s}_v$ , then $\langle L_{v,min}\rangle \geqslant \langle L_{min}\rangle$. If one chooses a homogeneous $\underline{s}_v = \varepsilon_v\underline{r}$, where the ($\underline{r}$-independent) $\varepsilon_v$ is still subject to variation, one recognizes that $L_v$ via $2U_v = (\varepsilon_v, \int d\underline{r}C\varepsilon_v)$ contains only the volume average of C, $\bar{C} = \int d\underline{r}\, C(\underline{r})/V$. Consequently $L_{v,min} = -\frac{V}{2}(\Sigma,\bar{C}^{-1}\Sigma) > -\frac{V}{2}(\Sigma,S^{eff}\Sigma)$ for all $\Sigma$. In both the above examples the effective and averaged elastic data are isotropic and we have for the eigenvalues $S^{eff} > \bar{C}^{-1}$, $\bar{C} > C^{eff}$. For the polycrystal, $\bar{C}$ is identical with Voigt's average and we have $K^{eff} = K$, $\bar{\mu} > \mu^{eff}$ in cubic crystals (comp. Sec. 4.2); note that $\bar{\mu}$ is an upper bound independent of grain sizes and correlations between orientations. The only assumption needed is that no orientation is preferred.

For given displacements and no forces otherwise, the elastic energy U is a minimum. If the surface displacements are given by $\varepsilon_s\underline{r}$ with homogeneous (constant) $\varepsilon_s$ and if the medium is homogeneous, it is $U_{min} = \frac{V}{2}(\varepsilon_s,C\varepsilon_s) = \frac{1}{2}\int(\varepsilon_s\underline{r},\sigma d\underline{S})$. In a statistical situation, $\sigma$ is subject to statistics and one has to calculate $\langle\sigma\rangle = C^{eff}\varepsilon_s$ on the surface: $\langle U_{min}\rangle = \frac{V}{2}(\varepsilon_s,C^{eff}\varepsilon_s)$. If one chooses $\underline{s}_v = \varepsilon_s\underline{r}$ throughout V, then $U_v = \frac{V}{2}(\varepsilon_s,\bar{C}\varepsilon_s) > \frac{V}{2}(\varepsilon_s,C^{eff}\varepsilon_s)$, and again one has $\bar{C} > C^{eff}$.

Estimates of this kind can also be obtained starting from elastic stability $2U = (\varepsilon,C\varepsilon) = (\sigma,S\sigma) > 0$. Then

$$0 < \int d\underline{r}\,(\varepsilon-\varepsilon_s)C(\underline{r})(\varepsilon-\varepsilon_s) = \int d\underline{r}\left[(\varepsilon,C\varepsilon) + (\varepsilon_s,C\varepsilon_s)\right] - 2\int \underbrace{\varepsilon_s\underline{r}}_{\underline{s}\text{ on the surface}}\cdot\sigma d\underline{S}$$

$$= V(\varepsilon_s,\bar{C}\varepsilon_s) - 2U ,$$

where $\underline{s}$, $\varepsilon$ , $\sigma$ are the exact microscopic fields for given surface displacement $\varepsilon_s\underline{r}$.

---

[24] Comp. Chapter 8.

Upon averaging one obtains

$$- 2 \langle U\rangle + V(\varepsilon_s,\bar{C}\varepsilon_s) = V\left[-(\varepsilon_s,C^{eff}\varepsilon_s) + (\varepsilon_s,\bar{C}\varepsilon_s)\right] > 0 \ , \quad \text{or } \bar{C} > C^{eff} \ .$$

For given $\Sigma d\underline{S} = \sigma d\underline{S}$ one averages

$$0 < \int d\underline{r}\ (\sigma - \Sigma)S(\sigma - \Sigma) = \int d\underline{r}\ \left[(\sigma,s\sigma) + (\underbrace{\Sigma,S\Sigma)}_{\bar{S}}\right] - 2\int \underbrace{\underline{s}\Sigma d\underline{S}}_{\sigma d\underline{S}\text{ on surface}} = V(\Sigma,\bar{S}\Sigma) - \int \underline{s}\Sigma d\underline{S}$$

and obtains

$$\int \langle\underline{s}\rangle\, \Sigma\, d\underline{S} = V(\Sigma,S^{eff}\Sigma) < V(\Sigma,\bar{S}\Sigma) \quad \text{or} \quad \bar{S} > S^{eff} \ , \qquad (S^{eff})^{-1} = C^{eff} > \bar{S}^{-1} \ .$$

Consequently, Voigt's averages[25] $\bar{C}$ establish *upper* bounds and the Reuss averages $\bar{S}$ *lower* bounds for $C^{eff}$. Take for example Cu where in units of $10^{11}$ dyn cm$^{-2}$: $\mu' = 2.4 < (\overline{1/\mu})^{-1} = 4.1 < \mu^{eff} < 5.5 = \bar{\mu} < \mu = 7.5$; the best guess for the shear modulus of a polycrystal should be the average (Sec. 4.4.8) between the bounds. Of course, one must realize that $\mu^{eff}$ finally depends on microscopic details, whereas the bounds do not.

### 4.8.8 Interaction Energies

a) *The total potential energy for given forces* is ($L_{min} = L$)

$$L = U[\underline{s}] = -\frac{1}{2}\int d\underline{r}\ f_i(\underline{r})s_i(\underline{r}) = -\frac{1}{2}\iint d\underline{r}\ d\underline{r}'\ f_i(\underline{r})G_{ik}(\underline{r},\underline{r}')f_k(\underline{r}') \ , \tag{4.95}$$

where $\underline{s}$ is the exact solution for given $\underline{f}$; surface forces can be included in $\underline{f}$. If $\underline{f} = \underline{f}^a + \underline{f}^b$ can be separated into two physically meaningful parts, the energy splits into three parts:

$$L = -\int d\underline{r}\ \frac{1}{2}\ (\underline{f}^a\underline{s}^a + \underline{f}^b\underline{s}^b + \underline{f}^a\underline{s}^b + \underline{f}^b\underline{s}^a) = L^{aa} + L^{bb} + L^{ab} \ . \tag{4.96}$$

Here $L^{ab} = W$ is the "interaction energy" whereas $L^{aa}$, $L^{bb}$ are the energies corresponding to the single force densities. The interaction energy can be expressed in various ways by using the static equilibrium equation combined with integration by

---

[25] For cubic crystals: $\bar{c}_{11} = c_{11} + 4(\mu - \mu')/5$, $\bar{c}_{12} = c_{12} - 2(\mu - \mu')/5$, $\bar{c}_{44} = c_{44} - 2(\mu - \mu')/5$. For noncubic crystals $\bar{C}$ can be easily determined from C by requiring that the two scalars (the two rotational invariants) contained in $\bar{C}$ and C be identical:

$$C_{ii,mm} = \bar{C}_{ii,mm} = 9\bar{c}_{12} + 6\bar{c}_{44} \ , \quad C_{ik,ik} = \bar{C}_{ik,ik} = 3\bar{c}_{12} + 12\bar{c}_{44} \ .$$

parts[26],

$$- W = \int_V d\underline{r}\ \underline{f}^a \underline{s}^b = \int_V d\underline{r}\ \sigma^a \varepsilon^b = \int_V d\underline{r}\ \varepsilon^a \sigma^b = \int_V d\underline{r}\ \underline{f}^b \underline{s}^a \ . \tag{4.97}$$

The lattice analogue is obviously

$$- W = (\underline{F}^{\underline{m},a}, \underline{s}^{\underline{m},b}) = (\underline{F}^a, \underline{s}^b) = (\underline{s}^a, \underline{F}^b) \ . \tag{4.97a}$$

The separation can refer to one defect a,

$$f^a_i(\underline{r}) = - P^a_{ij} \partial_j \delta(\underline{r} - \underline{R}^a) \quad \text{with double force tensor } P^a \text{ at } \underline{R}^a \ , \tag{4.98}$$

and to its interaction with surface forces or another defect.

b) *The interaction of a defect (P at $\underline{R}$) with external (surface) forces* is

$$W = \int d\underline{r}\ P_{ij} \partial_j \delta(\underline{r} - \underline{R})\ s_i(\underline{r}) = - P_{ij} \varepsilon_{ji}(\underline{R}) \ , \tag{4.99}$$

where $\underline{s}$ is the displacement field due to the external forces. The same 1st order term results if one expands $\underline{s}(\underline{r})$ about the location of the defect $\underline{R}$ assuming that the Kanzaki forces are rather concentrated. For slowly varying external fields higher order terms can certainly be neglected.

For a dilatation center, $W = - P_o \varepsilon_{ii}(\underline{R}) = - P_o \operatorname{div}\{\underline{s}(\underline{R})\}$, the energy is lowest where the local volume change is largest ($P_o > 0$), and the defect tends to *diffuse* to regions where most space is available.

For a simple double force in direction $\hat{\underline{a}}$, $P = P_o\ |\hat{\underline{a}}\rangle \langle\hat{\underline{a}}|$ , $W = - P_o\ \langle\hat{\underline{a}}|\varepsilon|\hat{\underline{a}}\rangle$ the energy is lowest if $\hat{\underline{a}}$ points in the direction of maximum strain. The defect will tend to *change its orientation* $\hat{\underline{a}}$ until $\langle\hat{\underline{a}}|\varepsilon|\hat{\underline{a}}\rangle = \hat{a}_i \varepsilon_{ik} \hat{a}_k$ becomes largest ($P_o > 0$).

If one represents P and $\varepsilon$ by the basis tensors $T(\nu)$ of Section 4.1.3, one sees that $- W = P^\nu \varepsilon^\nu$. As an example consider a tetragonal double force tensor, oriented parallel to the z-axis:

$$P = \begin{bmatrix} B & 0 & 0 \\ 0 & B & 0 \\ 0 & 0 & A \end{bmatrix}, \quad - W = P^1 \varepsilon^1 + P^3 \varepsilon^3 \text{ with } P^1 = \frac{2B + A}{\sqrt{3}}, \quad P^3 = 2\frac{A - B}{\sqrt{6}}.$$

One realizes that diffusion takes place only in strain fields of type T(1) (dilata-

26 The equalities (4.97) can most easily be obtained by assuming a free surface $\sigma^a d\underline{S} = \sigma^b d\underline{S} = 0$. This seems to exclude surface forces but actually does not. One can imagine the surface forces to be applied a bit below the actual surface which does not change the bulk situation.

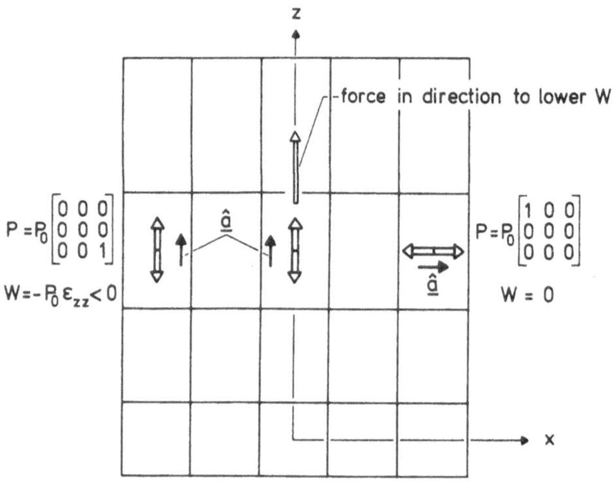


Fig. 4.22. Uniaxial defect, $P = P_o\,|\hat{a}\rangle\langle\hat{a}|$, in a uniaxial strain field,

$$\varepsilon = \begin{bmatrix}0 & 0 & 0\\ 0 & 0 & 0\\ 0 & 0 & \varepsilon_{zz}\end{bmatrix}, \quad \varepsilon_{zz} = \alpha z$$

tion) and of type T(3) (101 shear), and that only $\varepsilon^{2,3}$ strains lead to changes of orientations parallel to the x,y-axes, Fig. 4.22.

For a homogeneous external strain $\varepsilon$, only change of orientation can occur whereas for inhomogeneous strains also diffusion takes place.[27] These changes tend toward thermal equilibrium, where the density of defects is proportional to $\exp(-W/kT)$. It must be mentioned, however, that the above description of a defect in an external field is not complete. We will discuss in Section 6.6 the "polarizability" of defects by an external field, where $P(\varepsilon) = P(0) + \alpha\varepsilon$ depends on the external field.

c) *Two defects in an infinite medium* at sites $\underline{R}^a$ and $\underline{R}^b$ have an interaction energy

$$W(\underline{R}) = +P^a_{ij}P^b_{mn}\partial_j\partial_n G_{im}(\underline{r} = \underline{R}) = -P^a_{ij}P^b_{mn}\int \frac{d\underline{k}\,e^{i\underline{k}\underline{R}}}{(2\pi)^3}\hat{k}_j\hat{k}_n\tilde{G}_{im}(\hat{\underline{k}}) \;, \tag{4.100}$$

$$\underline{R} = \underline{R}^b - \underline{R}^a \;,$$

where $G(\underline{r} - \underline{r}')$ is the Green's function of an infinite medium (with translational symmetry). In an expansion of the interaction energy of two extended (not concentrated) force distributions in powers of $1/R$, eq. (4.100) gives the leading term ($\propto R^{-3}$). One can easily see that the average of $W(\underline{R})$ over all directions $\hat{\underline{R}}$ vanishes[28]. Therefore $W(R)$ contains attractive and repulsive parts of equal weight.

---

27 The change of orientation in a homogeneous strain is the "Snoek-effect", /4.9/; the change of defect distribution in inhomogeneous strains is the "Gorski--effect", /4.10/.

28 $\langle\exp(i\underline{k}\underline{R})\rangle^{\hat{R}}$ depends only on k. After averaging over $\hat{\underline{k}}$, the $\hat{\underline{k}}$ dependent factor in the integrand can be taken out of the integral, and one obtains $\langle W\rangle^{\hat{R}} \propto \delta(\underline{R})$; the argument is the same as that for the evaluation of (4.86.87). It is simple to extend that proof to higher order terms in the expansion of W.

The simplest example is one dilatation center $P_o^a$ and one general $P^b$ in an isotropic medium,

$$W = \frac{P_o^a}{4\pi c_{11} R^3} [\mathrm{tr}\{P^b\} - 3(\hat{\underline{R}}, P^b \hat{\underline{R}})] \quad , \tag{4.100a}$$

where one can easily verify that $\langle W \rangle^{\hat{\underline{R}}} = 0$. This shows also that the interaction of two dilatation centers vanishes, unless the medium is anisotropic:

$$W = P_o^a P_o^b \partial_i \partial_m G_{im} = - P_o^a P_o^b \delta(\underline{R})/c_{11} \quad \text{for isotropy} . \tag{4.100b}$$

In anisotropic cubic crystals an approximate result is obtained by starting from the isotropic crystal as defined by Voigt's average, $\bar{C}$, and then expanding in powers of the anisotropy $c_a$ (see App. N)[29]; the linear term, first derived by ESHELBY,/4.8/, is

$$W(\underline{R}) \cong - \underbrace{\frac{15}{8\pi R^3} \frac{P_o^a P_o^b}{\bar{c}_{11}^2}}_{\Delta V^a \Delta V^b (K/\bar{c}_{11})^2} c_a A(\hat{\underline{R}}) \; , \; A(\hat{\underline{R}}) = \frac{3}{5} - \sum_j \hat{X}_j^4 = \begin{cases} -\frac{2}{5} & \text{for } \langle 100 \rangle \\ \frac{1}{10} & \text{for } \langle 101 \rangle, \; \langle A \rangle^{\hat{\underline{R}}} = 0 \\ \frac{3}{15} & \text{for } \langle 111 \rangle \end{cases} \tag{4.100c}$$

In the main symmetry directions one can obtain analytical expressions for W. For like dilatation centers, attractive and repulsive directions depend on the sign of $c_a$. For all the fcc metals in Tab. 4.2, the anisotropy $c_a < 0$, and one finds attraction in $\langle 100 \rangle$ and repulsion in $\langle 101 \rangle$ and $\langle 111 \rangle$.

d) *The force on a defect*, say the force $\underline{F}^b$ on defect b, is defined by the change of total energy $L_{min}$ due to a change of the position $\underline{R}$ of defect b:

$$\underline{F}^b = - \partial_{\underline{R}} L_{min}(\underline{R}) \; .$$

The defect position $\underline{R}$ enters $L_{min}$ via the force density $\underline{f}^b$ (and via the corresponding displacement field $\underline{s}^b$); if the force density for the defect at 0 is $\underline{f}^b(\underline{r})$, then a translation of the defect by $\underline{R}$ will change the force density into $\underline{f}^b(\underline{r} - \underline{R})$. Therefore, one obtains from (4.95,96), with $\partial_{\underline{R}} \underline{f}^b(\underline{r} - \underline{R})\big|_{\underline{R}=0} = - \partial_{\underline{r}} \underline{f}^b(\underline{r})$, for the force on defect b at $\underline{R} = 0$:

$$F_m^b = - \int d\underline{r} \; s_i \partial_m f_i^b = \int d\underline{r} \; f_i^b \partial_m s_i \; ,$$

---

29 The method used in Appendix N to evaluate the integrals in (4.100), yields $8\pi W = P_{ji}^a P_{mn}^b \left[ \frac{1}{c_{44}} \delta_{im} \partial_j \partial_n + \left(\frac{1}{c_{11}} - \frac{1}{c_{44}}\right) \partial_i \partial_j \partial_m \partial_n \right] R$.

where $s_i = s_i^a + s_i^b$ is the total displacement field including the displacement $s_i^a$ due to a second defect a (or surface forces etc.). The contribution by $s_i^a$ alone gives the force exerted by defect a, it can be obtained directly from W according to (4.97); the contribution by $s_i^b$ gives the force due to the surface (boundary conditions, images), it vanishes in infinite material. The integrand can be expressed by the static quasi stress tensor of Sec. 4.3.4 for which $\partial_j \Pi_{mj} = f_i \partial_m s_i$. Consequently,

$$F_m^b = \int_{V_b} d\underline{r}\, f_i \partial_m s_i = \int_{S_b} \Pi_{mj}\, dS = \int_{S_b} u\, dS_m - \int_{S_b} (\partial_m s_i)\sigma_{ij}\, dS_j \ ,$$

provided that the volume $V_b$ covers only defect b or that $S_b$ (the surface enclosing $V_b$) separates defect b from all other defects. Therefore, one has now a wide choice of surfaces to obtain $\underline{F}^b$, which can simplify calculations. The surface $S_b$, surrounding the defect, is in "good" material, i.e., free of defects. This expression for the force is quite general and holds also for inhomogeneities in $C(\underline{r})$, elastic inclusions, which cannot be represented by force densities /4.8,13/.

### 4.8.9 Surface Effects, Images

In the following we will discuss some very simple examples to demonstrate the influence of a surface, and to see under what conditions the surface contributions can be neglected. The general behaviour will be the following: the problem of a finite crystal with surface conditions and given inside force densities f can be considered as the problem of an infinite crystal with force f inside and "image" forces outside the original volume; the image forces can be chosen such that on a cut along the original surface the boundary conditions are obeyed.

Because surface effects even in isotropic[30] media are difficult to handle and lead to extremely clumsy expressions, we employ the simplest possible model substance, a cubic substance with $c_{12} + c_{44} = 0$, where the equation of motion (4.46b) is separated in the three components of $\underline{s}$ (comp. also Sec. 4.7.3):

$$c_{12} + c_{44} = 0\ ; \quad c_{11} - 2c_{44} > 0\ , \quad c_{44} > 0$$

$$[c_{11}\partial_x^2 + c_{44}(\partial_y^2 + \partial_z^2)]s_x + f_x = 0\ , \text{ correspondingly for } s_y \text{ and } s_z\ , \qquad (4.101)$$

$$\sigma_{xx} = c_{11}\varepsilon_{xx} - c_{44}(\varepsilon_{yy} + \varepsilon_{zz}) \text{ etc.}\ , \quad \sigma_{xy} = 2c_{44}\varepsilon_{xy}\ ,$$

$$\overset{\infty}{G}_{xx}(r) = \frac{1}{4\pi\sqrt{c_{11}}\, c_{44}} \frac{1}{\sqrt{x^2/c_{11} + (y^2 + z^2)/c_{44}}}\ , \text{ correspondingly for } \overset{\infty}{G}_{yy} \text{ and } \overset{\infty}{G}_{zz} \text{ (comp. Sec. 4.7.3).}$$

[30] Plane surfaces of an isotropic medium are treated in /4.11/.

In the unstable isotropic model:

$$c_{11} = c_{44} = -c_{12} \ ; \quad c_{44}\Delta s_x + f_x = 0 \ ; \quad \sigma_{xx} = c_{44}(2\varepsilon_{xx} - \varepsilon_{mm}) \ , \quad \sigma_{xy} = 2c_{44}\varepsilon_{xy} ;$$

$$\overset{\infty}{G}_{xx}(\underline{r},\underline{R}) = \frac{1}{4\pi c_{44}} \frac{1}{|\underline{r} - \underline{R}|} \quad \text{for } f(\underline{r}) = \delta(\underline{r} - \underline{R}) \text{ , infinite crystal.} \tag{4.102}$$

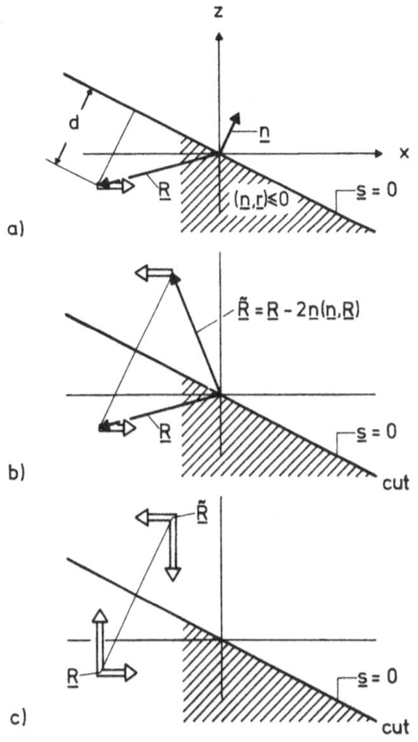


Fig. 4.23a-c. Image forces.
a) Original situation: Force in x-direction at $\underline{R}$ inside, boundary condition $\underline{s} = 0$ in $(\underline{n},\underline{r}) = 0$.
b) Can be replaced by infinite crystal with force at $\underline{R}$, opposite image force at $\tilde{\underline{R}}$ and cut along $(\underline{n},\underline{r}) = 0$ where $\underline{s} = 0$.
c) Arbitrary force at $\underline{R}$ leads to opposite image force at $\tilde{\underline{R}}$

We consider a semi-infinite crystal bounded by a plane through the origin with normal $\underline{n}$ (Fig. 4.23a), and ask for Green's function, $G_{xx}(\underline{r};\underline{R})$, for $f_x = \delta(\underline{r} - \underline{R})$ and $\underline{s} = 0$ in the boundary. This problem can be solved by considering the infinite crystal and adding the opposite force at the mirror image $\tilde{\underline{R}}$ of $\underline{R}$ (Fig. 4.23b,c):

$$\overset{\infty}{G}_{xx}(\underline{r};\underline{R}) = \overset{\infty}{G}_{xx}(\underline{R}) - \overset{\infty}{G}_{xx}(\tilde{\underline{R}}) = \frac{1}{4\pi c_{44}} \left(\frac{1}{|\underline{r} - \underline{R}|} - \frac{1}{|\underline{r} - \tilde{\underline{R}}|}\right) , \tag{4.103}$$

which obviously solves the original problem for $(\underline{n},\underline{r}) \leqslant 0$. One sees that:

1) near $\underline{R}$, $|\underline{r} - \underline{R}| \ll d = |(\underline{n},\underline{R})|$ = distance from the surface, the response is that of the infinite crystal;

2) far from $\underline{R}$, $|\underline{r} - \underline{R}| \gg d$, the response is that of a double force, proportional to $|\underline{r} - \underline{R}|^{-2}$ rather than to $|\underline{r} - \underline{R}|^{-1}$ as in the infinite crystal.

This also represents the general behaviour even though the "images" become more complicated. In this unstable model a free surface solution is not unique (App. O).

Therefore, it is useful to demonstrate the mirror procedure employing the stable but anisotropic model (4.101), where $G^{\infty}_{xx}(\underline{r})$ is not constant on a sphere but on an ellipsoid, $(\underline{r},A\underline{r})$ = constant, $A_{xx} = 1/c_{11}$, $A_{yy} = A_{zz} = 1/c_{44}$. The "image" point $\tilde{\underline{R}}$ is not the mirror image, but given by $\sqrt{A}(\underline{r} - \underline{R}) = \sqrt{A}(\underline{r} - \tilde{\underline{R}})$ with $\underline{r}$ on the surface, $(\underline{n},\underline{r}) = 0$. The result is

$$\tilde{\underline{R}} = \underline{R} - 2\,\frac{A^{-1}\underline{n}}{(\underline{n},A^{-1}\underline{n})}\,(\underline{n},\underline{R})\ ; \qquad (\underline{n},\tilde{\underline{R}}) = -\,(\underline{n},\underline{R}), \tag{4.104}$$

e.g., for $\underline{n} = \frac{1}{\sqrt{2}}(101)$:

$$\tilde{X} = X - \frac{2c_{11}}{c_{11} + c_{44}}(X + Z)\ ,\quad \tilde{Z} = Z - \frac{2c_{44}}{c_{11} + c_{44}}(X + Z)\ ,\quad \tilde{Y} = Y, \tag{4.104a}$$

(Fig. 4.24a).

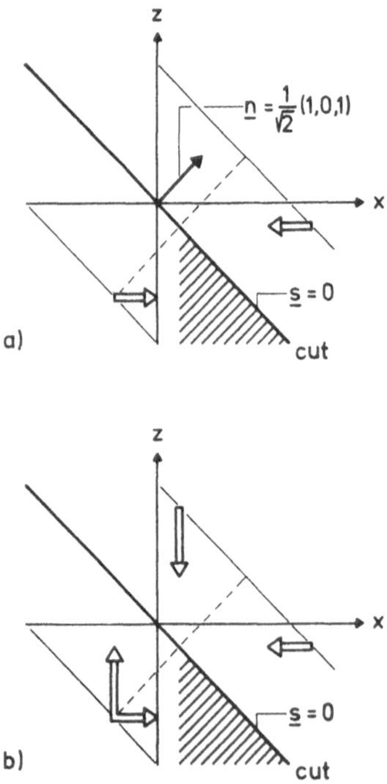


Fig. 4.24a and b. Images for (101)-planes, $c_{12} + c_{44} = 0$, $c_{11} = 3c_{44}$.
a) only x components,
b) x and z components

For pure z-components one must interchange $c_{11}$ and $c_{44}$ and obtains another image point shown in Fig. 4.24b. Except for $\underline{n} = \langle 100 \rangle$ one always obtains several images. The general behaviour is unchanged. This should be sufficient to demonstrate the influence of surfaces which roughly can be represented by image forces for qualitative estimates.

## 4.9 Anharmonicity

In the harmonic approximation the energy density,

$$u = \frac{1}{2}(\varepsilon, C\varepsilon) = \frac{1}{2}\varepsilon_{ik}C_{ik,mn}\varepsilon_{mn} , \tag{4.105}$$

is quadratic in the strains $\varepsilon = (v + v')/2$, $\varepsilon_{ik} = (v_{ik} + v_{ki})/2$, which vanish if the deformation represents a small rotation. Eq. (4.105) has to be considered as the first term in an expansion of u in powers of "strains". The $\varepsilon$'s are not the appropriate "strains" for a higher order expansion of u; they only vanish for *small* rotations. What one needs for a general expansion are generalized strains, E(v), which vanish for *arbitrary* rotations. They are easy to obtain by considering the change of $(\underline{r},\underline{r})$ under a deformation v,

$$\Big((1 + v)\underline{r},(1 + v)\underline{r}\Big) - (\underline{r},\underline{r}) = \Big(\underline{r},(v + v' + v'v)\underline{r}\Big) = 2(\underline{r},E\underline{r})$$

$$\text{where}\quad E = \frac{v + v'}{2} + \frac{v'v}{2}, \quad E_{ik} = \frac{v_{ik} + v_{ki} + v_{li}v_{lk}}{2} = E_{ki} \tag{4.106}$$

obviously vanishes, if v represents an arbitrary rotation. For small deformations, the quadratic terms in v can be dropped and E agrees with $\varepsilon$.

The proper expansion of the energy density becomes

$$u = \frac{CE^2}{2!} + \frac{DE^3}{3!} + \ldots = \frac{1}{2!}C_{ik,mn}E_{ik}E_{mn} + \frac{1}{3!}D_{ik,mn,st}E_{ik}E_{mn}E_{st} + \ldots . \tag{4.107}$$

The symmetry of the expansion coefficients is evident. A small deformation, $\delta E$, of a prestrained ($\mathring{E}$) material ($E = \mathring{E} + \delta E$) leads to

$$u = \frac{C\mathring{E}^2}{2} + \frac{D\mathring{E}^3}{3!} + (C\mathring{E} + \frac{D\mathring{E}^2}{2})\delta E + (C + D\mathring{E})\frac{(\delta E)^2}{2} + \ldots , \tag{4.107a}$$

i.e., $C\mathring{E} + D\mathring{E}^2/2$ represents the stress tensor in the prestrained state, and $C + D\mathring{E}$ are the elastic moduli of the prestrained material. Therefore, D determines the change of the elastic moduli by small deformations $\mathring{\varepsilon}$:

$$\delta C_{ik,mn} = D_{ik,mn,st}\mathring{\varepsilon}_{st} . \tag{4.108}$$

In cubic crystals D contains only six independent elastic moduli of 3rd order[31], in Voigt's notation

$$c_{111} = c_{222} = c_{333} = D_{11,11,11} \; ; \quad c_{112} = c_{113} = \ldots = D_{11,11,22} \; ;$$
$$c_{123} = D_{11,22,33} \; ; \quad c_{144} = c_{255} = \ldots = D_{11,23,23} \; ; \quad c_{166} = c_{266} = \ldots \tag{4.109}$$
$$= D_{11,12,12} \; ; \; c_{654} = \ldots = D_{12,23,31} \; .$$

Table 4.10. Third order elastic moduli, $c_{\alpha\beta\gamma}$, $[10^{12}\ \mathrm{dyn/cm^2}]$, /4.12/

| $\alpha\ \beta\ \gamma$ | 111 | 112 | 123 | 144 | 166 | 654 | $\partial 3K/\partial p$* | $\partial 2\mu'/\partial p$ | $\partial 2\mu/\partial p$ |
|---|---|---|---|---|---|---|---|---|---|
| Al | - 9 | -2.7 | 0.8 | 0.8 | -3.8 | 0.2 | 10.34 | 4.29 | 5.95 |
| Cu | -12.7 | -8.1 | -0.5 | -0.03 | -7.8 | -0.95 | 15.1 | 2.95 | 7.55 |
| Ag | - 8.4 | -5.3 | 1.9 | 0.56 | -6.4 | 0.83 | 11.85 | 3.34 | 7.97 |
| Au | -17.3 | -9.2 | -2.3 | -0.13 | -6.5 | -0.12 | 15.06 | 2.93 | 5.13 |

* For 1st neighbour two body interactions assuming a $\mu > \nu$-Lennard-Jones potential (comp. Sec. 3.4) one obtains $\partial 3K/\partial p = 6 + (\mu + \nu)$. For a 12-6 potential $\partial 3K/\partial p = 24$ has the right order of magnitude. In general, two body potential models tend to exaggerate anharmonicity because the "effective powers" in metals are smaller.

Table 4.10 contains the values for several fcc metals. D is, at most, a factor of 10 larger than C. The change $\delta C$ and C become comparable for $\varepsilon \cong 10^{-1}$, which is a very large strain as compared with the usually applied strains of about $10^{-6}$. We will need these data later to calculate the trivial change of C due to the volume expansion by defects, in order to separate out the effect of genuine spring changes. The order of magnitude of the change $\delta C$ due to expansion of the crystal by defects can be roughly estimated as follows: if the volume change per defect is $V_c$, the strain is given by the atomic concentration c of defects, $\delta C \cong Dc$, $\delta C/cC \cong D/C \cong -10$, which corresponds to a relatively large change $\delta C/C = -10^{-1}$ for 1 at% defects.

It is advantageous to use the basis tensors $T(\nu)$ of Section 4.1.3, because then

31 In the non-vanishing elements of D the cartesian indices can only occur in pairs; all elements, which can be obtained by interchange of cartesian indices, are equal.

the physical meaning of the 3rd order moduli becomes clearer. For $\overset{\circ}{\varepsilon} = T(\lambda)$ one has from

$$D = \sum_{\mu\nu\lambda} D^{\mu\nu\lambda} T(\mu)T(\nu)T(\lambda) , \quad D^{\mu\nu\lambda} = T_{ik}(\mu)T_{mn}(\nu)T_{st}(\lambda)D_{ik,mn,st} , \tag{4.110}$$

$$\delta C^{(\lambda)} = DT(\lambda) = \sum_{\mu\nu} D^{\mu\nu\lambda} T(\mu)T(\nu)T(\lambda) , \quad \not\sum_{\lambda} ,$$

$$\delta C^{(\lambda)}_{ik,mn} = D_{ik,mn,st}T_{st}(\lambda) = \sum_{\mu\nu} D^{\mu\nu\lambda} T_{ik}(\mu)T_{mn}(\nu) , \quad \text{fixed } \lambda. \tag{4.110a}$$

The change of C by a dilatation $\varepsilon^1 T(1)$ is most important; it corresponds to $\lambda = 1$, $T_{st}(1) = \delta_{st}/\sqrt{3}$. The change $\delta C^{(1)}_{ik,mn} = D_{ik,mn,ss}/\sqrt{3}$ has full cubic symmetry. Therefore the change of the eigenvalues $\overset{\mu}{C}$ by the dilatation $\varepsilon^1 T(1)$ is

$$\delta \overset{\mu}{C}{}^{(1)} = T_{ik}(\mu)T_{mn}(\mu)D_{ik,mn,st}T_{st}(1)\varepsilon^1 = D^{\mu\mu 1}\varepsilon^1 , \quad \not\sum_{\mu} , \tag{4.111}$$

where

$$\sqrt{3}\, D^{111} = c_{111} + 6c_{112} + 2c_{123} , \quad \sqrt{3}\, D^{221} = c_{111} - c_{123} , \quad \sqrt{3}\, D^{551} = 2c_{144} + 4c_{166} .$$

If we take again the above example of an atomic concentration c of defects with volume change $\delta_1 V$ per defect then $\overset{\circ}{\varepsilon}_{st} = c(\delta_1 V/3V_c)\delta_{st}$, $\varepsilon^1 = c(\delta_1 V/3V_c)\sqrt{3}$, and one obtains

$$\delta \overset{\mu}{C} = D^{\mu\mu 1} \frac{c}{\sqrt{3}} \text{ for } \delta_1 V = V_c , \quad \delta \overset{1}{C}/\overset{1}{C} = \delta K/K = \frac{c}{3}\, \frac{c_{111} + 6c_{112} + 2c_{122}}{c_{11} + 2c_{12}},$$

$$\delta \overset{2}{C}/\overset{2}{C} = \delta\mu'/\mu' = \frac{c}{3}\, \frac{c_{111} - c_{123}}{c_{11} - c_{12}} , \quad \delta \overset{5}{C}/\overset{5}{C} = \delta\mu/\mu = \frac{c}{3}\, \frac{c_{144} + 2c_{166}}{c_{44}},$$

e.g., for Cu: $\delta \overset{1}{C}/\overset{1}{C} \cong -2c$, $\delta \overset{2}{C}/\overset{2}{C} \cong -7c$ and $\delta \overset{5}{C}/\overset{5}{C} = -2c$. It is more illustrative to express the changes (4.111) in terms of an applied pressure p, $\varepsilon = -Sp\sqrt{3}\, T(1) = -\sqrt{3}\, pT(1)/3K$, whereupon[32]

$$\delta \overset{\mu}{C} = - D^{\mu\mu 1} \frac{p}{\sqrt{3}\, K} ; \quad \frac{\partial 3K}{\partial p} = - \frac{D^{111}}{\sqrt{3}\, K} , \quad \frac{\partial 2\mu'}{\partial p} = - \frac{D^{221}}{\sqrt{3}\, K} , \quad \frac{\partial 2\mu}{\partial p} = - \frac{D^{551}}{\sqrt{3}\, K} \tag{4.111a}$$

Of the remaining three (independent) elements of $D^{\mu\nu\lambda}$,

$$D^{442} = \sqrt{2}\,(c_{441} - c_{442}) , \quad D^{223} = (c_{111} - 3c_{112} + 2c_{123})/\sqrt{6} , \quad D^{645} = 2\sqrt{2}\, c_{654} ,$$

---

32 The order of magnitude is $\delta K/K \cong 10\ p/K = 10^{-8}\ p[\text{atm}]/K[10^{12}\ \text{dyn/cm}^2]$.

the first two are directly connected with the change of elastic moduli by shears: the change of $\overset{5}{C} = 2\mu$ by shears T(2) is determined by $D^{442}$, the change of $\overset{2}{C}$ by shears of type T(3) can be expressed by $D^{223}$, whereas $D^{645}$ has no simple meaning in terms of moduli adapted to cubic symmetry.

# 5. Transition From Lattice to Continuum Theory

It has been pointed out before that lattice theory must pass into continuum theory for very slowly varying displacements. Under this assumption the equation of motion of the *infinite* lattice passes into the equation of motion of the continuum (Sec. 5.1); the Huang tensor, H, is expressed by the springs and the elastic moduli are obtained from H. This method cannot easily be used directly in defect crystals where one rather investigates the dynamical behaviour via the response to appropriate conditions. A homogeneous deformation is certainly the most slowly-varying displacement. It can be used to determine C via the elastic energy U of a finite crystal (Sec. 5.2); this is also the most convenient procedure for defect crystals. It turns out that C is determined by a surface sum and seems to be surface sensitive; actually H, and with it C, can be expressed by a bulk sum. The two methods described above can be used for the most general springs compatible with crystal symmetry, without any assumption about the interactions. The calculation of elastic moduli from simple Born -von Karman models is discussed in Section 5.3.

It would be most lucid and illustrative if one would be able to evaluate the local stress $\sigma$ (or the energy density which is equivalent) for an infinite lattice under a homogeneous deformation v. The connection between $\sigma$ and $\varepsilon$ would then directly define the elastic data. One could, for instance, cut all the springs at the surface of one Wigner-Seitz-Cell or an agglomerate of Wigner-Seitz-Cells, apply appropriate tensions and try to define $\sigma$ by the force per unit area. This procedure, however, cannot work for general interactions. This becomes clear if one realizes that the physical meaning of $\sigma$ is a momentum current density: if a spring possesses two body character it is at least plausible, that the momentum connected with it flows along the connection between the two atoms linked by the spring; this can no longer be argued for *many* body forces where momentum can no longer flow along the connection of *two* atoms. Consequently, detailed knowledge about the character of the interaction is necessary to define local stresses, and the same problem arises for local energy densities. On the other hand, one needs the same information, when discussing defects with necessarily local spring changes. For this reason two body models, including shell and bond charge models, have been discussed before in Section 3.4. They are again discussed in Section 5.4 and 5.5, and they can be directly employed in simple defect models.

## 5.1 Infinite Lattice

In the lattice equation,

$$M\ddot{s}_i^{\underline{m}}(t) = M\ddot{s}_i(\underline{r} = \underline{R}^{\underline{m}},t) = -\sum_{\underline{n}} \Phi_{ij}^{(\underline{m}-\underline{n})} s_j(\underline{r} + \underline{R}^{\underline{n}-\underline{m}}) + F_i(\underline{r} = \underline{R}^{\underline{m}}) , \tag{5.1}$$

one can consider $s_i^{\underline{m}}(t) = s_i(\underline{r} = \underline{R}^{\underline{m}},t)$ as a smooth, continuous and differentiable function of $\underline{r}$, if the displacements vary slowly. On the right-hand side one can expand in powers of $\underline{R}^{\underline{n}-\underline{m}}$, and keep only the lowest, nonvanishing term, if the changes of $\underline{s}$ within the range of the coupling are small:[1]

$$M\ddot{s}_i(\underline{r},t) = -\frac{1}{2} \sum_{\underline{h}} \Phi_{ij}^{(\underline{h})} X_l^{\underline{h}} X_k^{\underline{h}} \partial_k \partial_l s_j(\underline{r},t) + F_i(\underline{r},t) \tag{5.2}$$

which agrees with (4.33a),

$$\rho_o \ddot{s}_i = H_{ij,kl} \partial_k \partial_l s_j + f_i , \tag{5.2a}$$

provided that

$$H_{ij,kl} = -\frac{1}{2V_c} \sum_{\underline{h}} \Phi_{ij}^{(\underline{h})} X_k^{\underline{h}} X_l^{\underline{h}} ; \quad f_i(\underline{r}) = \frac{F_i(\underline{r} = \underline{R}^{\underline{m}})}{V_c} , \quad \rho_o = \frac{M}{V_c} . \tag{5.2b}$$

The expressions for $\underline{f}$ and $\rho_o$ are obvious (slowly varying $\underline{F}^{\underline{m}}$ is required). The symmetries of H are also obvious, except for the interchange between the two pairs of subscripts, which is needed to convert H into C. One can show that this symmetry is a consequence of the equilibrium conditions (following section).

A completely equivalent method is to compare the spatial Fourier transforms of continuum, $-H_{ij,kl} k_k k_l$, and lattice theory for small $\underline{k}$, $-\tilde{\Phi}_{ij}(\underline{k})/V_c \cong \sum \Phi_{ij}^{(\underline{h})} X_k^{\underline{h}} X_l^{\underline{h}} \cdot k_k k_l / 2V_c$. This is the method of long waves which requires the wavelength, $2\pi/k$, to be much larger than the range of the coupling.

## 5.2 Finite Crystals

For a homogeneous deformation,

$$\underline{s} = v\underline{R} , \quad \underline{s}^{\underline{m}} = v\underline{R}^{\underline{m}} , \quad s_i^{\underline{m}} = v_{ik} X_k^{\underline{m}} , \tag{5.3}$$

the elastic energy U of a finite crystal is

---

[1] The first two terms of the expansion vanish because of translational invariance, $\sum_{\underline{h}} \Phi^{(\underline{h})} = 0$, and inversion symmetry, $\sum_{\underline{h}} \Phi_{ij}^{(\underline{h})} X_l^{\underline{h}} = 0$.

$$2U = (v\underline{R},\Phi v\underline{R}) = v_{ik}X_k^{\underline{m}}\Phi_{ij}^{\underline{mn}}X_l^{\underline{n}}v_{jl} \; ; \qquad (5.3a)$$

comparison with the corresponding expression in continuum theory,

$$2U = Vu = V(v,Cv) = V(\varepsilon,C\varepsilon) = Vv_{ik}C_{ik,jl}v_{jl} \; , \qquad (5.3b)$$

leads to

$$VC_{ik,jl} = X_k^{\underline{m}}\Phi_{ij}^{\underline{mn}}X_l^{\underline{n}} \; . \qquad (5.4)$$

The symmetries of $\Phi_{ij}^{\underline{mn}}$ discussed in Chapter 2 guarantee the needed symmetries of C: (2.5a) allows interchange of the pairs (i,k) and (j,l), and (2.5c) allows interchange within one pair (i.e., the lattice theoretical expression for rotational invariance guarantees that the continuum theoretical energy density u depends only on $\varepsilon$); translational invariance, (2.5b), says only that C does not change under a translation of the crystal, $\underline{R}^{\underline{m}} \to \underline{R}^{\underline{m}} + \underline{T}$. The forces $\underline{F} = \Phi v\underline{R}$, necessary to maintain the deformation, are surface forces, because

$$F_i^{\underline{m}} = \Phi_{ij}^{\underline{mn}}X_l^{\underline{n}}v_{jl} = \sum_{\underline{n}} \Phi_{ij}^{\underline{mn}}(X_l^{\underline{n}} - X_l^{\underline{m}})v_{jl} \; , \quad \not{\sum_{\underline{m}}} \; , \qquad (5.5)$$

vanishes, if $\underline{m}$ is in the bulk where coupling and sites are those of the infinite lattice:

$$F_i^{\underline{m}} = \sum_{\underline{n}} \Phi_{ij}^{(\underline{m}-\underline{n})}X_l^{\underline{n}-\underline{m}}v_{jl} = \sum_{\underline{h}} \Phi_{ij}^{(\underline{h})}X_l^{\underline{h}}v_{jl} = 0 \quad \text{(inversion symmetry)}. \qquad (5.5a)$$

Therefore, the sums in (5.3a) extend only over the surface and C seems to be sensitive to surface properties such as structure and change of coupling near the surface. By comparison one recognizes that the (surface) stresses $\Sigma_{ik}$ are

$$\Sigma_{ik} = X_k^{\underline{m}}F_i^{\underline{m}}/V = X_k^{\underline{m}}\Phi_{ij}^{\underline{mn}}X_l^{\underline{n}}v_{jl}/V \; . \qquad (5.5b)$$

On the other hand, C must be a bulk property and, consequently, the surface sum (5.4) must be equivalent to a bulk sum. This can be seen by calculating the Kun Huang tensor H, (4.34),

$$2VH_{ij,kl} = V(C_{ik,lj} + C_{il,kj}) = X_k^{\underline{m}}\Phi_{ij}^{\underline{mn}}X_l^{\underline{n}} + X_l^{\underline{m}}\Phi_{ij}^{\underline{mn}}X_k^{\underline{n}} \; . \qquad (5.6)$$

The correct symmetries of C imply the right symmetries of H defined by (5.6). Eq. (5.6) can be written as

$$-2VH_{ij,kl} = \sum_{\underline{m}}\sum_{\underline{n}} (X_k^{\underline{m}} - X_k^{\underline{n}})\Phi_{ij}^{\underline{mn}}(X_l^{\underline{m}} - X_l^{\underline{n}}) \; , \qquad (5.6a)$$

since terms like $\sum_{\underline{n}} X_k^{\underline{m}}\Phi_{ij}^{\underline{mn}}X_l^{\underline{m}}$ vanish because of translational invariance. Obviously the sum (5.6a) is a bulk sum: if $\underline{m}$ is in the bulk its contribution to $-2VH_{ij,kl}$ is $\sum_{\underline{n}} X_k^{\underline{m}-\underline{n}} \Phi_{ij}^{(\underline{m}-\underline{n})} X_l^{\underline{m}-\underline{n}} = \sum_{\underline{h}} X_k^{\underline{h}} \Phi_{ij}^{(\underline{h})} X_l^{\underline{h}}$, independent of $\underline{m}$. Neglecting *surface terms* one has

$$-2VH_{ij,kl} = N \sum_{\underline{h}} X_k^{\underline{h}} \Phi_{ij}^{(\underline{h})} X_l^{\underline{h}} , \quad H_{ij,kl} = -\frac{1}{2V_c} \sum_{\underline{h}} X_k^{\underline{h}} \Phi_{ij}^{(\underline{h})} X_l^{\underline{h}} . \tag{5.6b}$$

which agrees with (5.2b), i.e., H defined by (5.2b) has the correct symmetries if one neglects surface terms.

Physically, one can only prescribe surface conditions, either forces or displacements. Because forces must be invariant, whereas displacements are not restricted by conditions, it is more convenient to prescribe surface displacements (comp. Sec. 2.5.3). If one chooses homogeneous surface displacements, $pv\underline{R}$, the energy according to (2.43) becomes ($p$ = projector onto the surface, $q = 1 - p$ = projector onto the bulk)

$$2U = \Big(v\underline{R}, \underbrace{(\Phi - \Phi G \Phi) v\underline{R}}_{0 \text{ in the bulk}}\Big) , \quad G = \frac{1}{q\Phi q} , \quad G = 0 \text{ on the surface.} \tag{5.7}$$

This is a general result, valid also for defect crystals to which it will be applied later. For perfect crystals one comes back to (5.3a): the 2nd term in (5.7) vanishes because $\Phi v\underline{R}$ vanishes in the bulk whereas G vanishes in the surface, $G\Phi v\underline{R} = 0$. In perfect crystals a homogeneous deformation of the surface leads to a homogeneous deformation of the bulk. In defect crystals this is not the case, and it is the deviation from homogeneous behaviour in the bulk which determines the change of elastic data in defect crystals (comp. Sec. 5.6).

## 5.3 Simple Born-von Karman Models in Cubic Crystals

In this section we demonstrate the connection between microscopic springs and macroscopic moduli employing simple Born -von Karman models in cubic crystals. The Kun Huang tensor is calculated from (5.6b). We restrict ourselves to coupling along the three main symmetry directions, where the form of the coupling matrices, compatible with symmetry, has already been given in Section 3.2.4; the occuring lattice sums are divided into sums over shells of (symmetrically) equivalent atoms. Finally, the elastic moduli can be calculated via (4.34c):[2]

$$c_{11} = h_{11} , \quad c_{12} = 2h_{44} - h_{12} , \quad c_{44} = h_{12} . \tag{5.8}$$

---

[2] The symmetry of H when exchanging the two pairs of subscripts is an additive restriction (besides lattice symmetries) for the most general coupling. In cubic crystals, however, H has cubic symmetry, which includes the general symmetries H must obey.

For longitudinal springs, $-\Phi_{ij}^{(\underline{h})} = f_1 d^2 \hat{X}_i^{\underline{h}} \hat{X}_j^{\underline{h}}$ ,eq. (5.6b) yields

$$H_{ij,km}^{(1)} = \frac{f_1 d^2}{2V_c} \sum_{\underline{h}}{}' \hat{X}_i^{\underline{h}} \hat{X}_j^{\underline{h}} \hat{X}_k^{\underline{h}} \hat{X}_m^{\underline{h}} = \frac{f_1 d^2}{2V_c} \hat{H}_{ij,km}^{(1)} , \tag{5.9}$$

where $d = R^{\underline{h}}$ and where the sum extends over one shell of equivalent sites. For isotropic transversal springs (degeneracy of the two components t and t')

$$\begin{aligned} H_{ij,km}^{(t)} + H_{ij,km}^{(t')} = H_{ij,km}^{(t+t')} &= \frac{f_t d^2}{2V_c} \sum_{\underline{h}}{}' (\delta_{ij} \hat{X}_k^{\underline{h}} \hat{X}_m^{\underline{h}} - \hat{X}_i^{\underline{h}} \hat{X}_j^{\underline{h}} \hat{X}_k^{\underline{h}} \hat{X}_m^{\underline{h}}) \\ &= \frac{f_t d^2}{2V_c} \sum_{\underline{h}}{}' (\delta_{ij} \delta_{km}/3 - \hat{X}_i^{\underline{h}} \hat{X}_j^{\underline{h}} \hat{X}_k^{\underline{h}} \hat{X}_m^{\underline{h}}) = \frac{f_t d^2}{2V_c} \hat{H}_{ij,km}^{(t+t')} . \end{aligned} \tag{5.10}$$

For coupling along ⟨100⟩ or ⟨111⟩ directions, (5.9) and (5.10) are already sufficient because the coupling is of l,t-type. For ⟨101⟩ coupling one has three different springs $f_1$ , $f_t$ , $f_{t'}$ ; therefore

$$H = \frac{d^2}{2V_c} \left[ (f_1 \hat{H}^{(1)} + f_t \hat{H}^{(t)} + f_{t'} \hat{H}^{(t')}) \right] \tag{5.11}$$

where, with a t-type transversal coupling $\Phi_{ij}^{(\underline{h})} = f_t \left[1 - 2(\hat{X}_i^{\underline{h}})^2\right] \left[1 - 2(\hat{X}_j^{\underline{h}})^2\right]$,

$$\hat{H}_{ij,km}^{(t)} = \sum_{\underline{h}}{}' \left[1 - 2(\hat{X}_i^{\underline{h}})^2\right] \left[1 - 2(\hat{X}_j^{\underline{h}})^2\right] \hat{X}_k^{\underline{h}} \hat{X}_m^{\underline{h}} \quad \text{for } \hat{\underline{X}}^{\underline{h}} = \langle 101 \rangle . \tag{5.11a}$$

Further,

$$\hat{H}^{(1+t+t')} = \hat{H}^{(1)} + \hat{H}^{(t)} + \hat{H}^{(t')} , \quad \hat{H}_{ij,km}^{(1+t+t')} = \sum_{\underline{h}} \delta_{ij} \hat{X}_k^{\underline{h}} \hat{X}_m^{\underline{h}} = \frac{Z_s}{3} \delta_{ij} \delta_{km} , \tag{5.12}$$

$Z_s$ = number of shell sites,

which would be needed if all springs were equal, $\Phi_{ij}^{(\underline{h})} = -f\delta_{ij}$, corresponding to the unstable model of Fig. 3.15d. The results for $\hat{H}$ and the corresponding $\hat{C}$ are summarized in Table 5.1. From this table the contribution of the single springs to the elastic data can be extracted, as shown in Table 5.2 for 1st neighbour coupling in the three cubic lattices.[3] On the other hand, one can express the springs by the elastic data, e.g., the three first neighbour springs of the fcc lattice by the three elastic moduli:

$$f_1 = \frac{a}{24} (4\overset{1}{C} + 2\overset{2}{C} + 3\overset{5}{C}) , \quad f_{t'} = \frac{a}{8} (2\overset{2}{C} - \overset{5}{C}) , \quad f_t = \frac{a}{12} (-\overset{1}{C} - 2\overset{2}{C} + 3\overset{5}{C}) \tag{5.13}$$

(Table 5.3).

3 For additive 2nd neighbour coupling in an fcc lattice one must add twice these values with f's referring to the ⟨100⟩ shell.

Table 5.1. Values of $\hat{H}$ and $\hat{C}$ for three cubic lattice shells

| Shell | Spring | $\hat{H}$-components | $\hat{h}_{11}$ | $\hat{h}_{12}$ | $\hat{h}_{44}$ | $\hat{c}_{12}$ | $\hat{c}_{11}+2\hat{c}_{12}$ | $\hat{c}_{11}-\hat{c}_{12}$ | $2\hat{c}_{44}$ |
|---|---|---|---|---|---|---|---|---|---|
| ⟨100⟩ | $f_l$ | 1 | 2 | 0 | 0 | 0 | 2 | 2 | 0 |
| | $f_t$ | t' + t | 0 | 2 | 0 | -2 | -4 | 2 | 4 |
| | $f_l = f_t$ | 1 + t' + t | 2 | 2 | 0 | -2 | -2 | 4 | 4 |
| ⟨111⟩ | $f_l$ | 1 | 8/9 | 8/9 | 8/9 | 8/9 | 8/3 | 0 | 16/9 |
| | $f_t$ | t' + t | 16/9 | 16/9 | -8/9 | -32/9 | -16/3 | 16/3 | 32/9 |
| | $f_l = f_t$ | 1 + t' + t | 8/3 | 8/3 | 0 | -8/3 | -8/3 | 16/3 | 16/3 |
| ⟨101⟩ | $f_l$ | 1 | 2 | 1 | 1 | 1 | 4 | 1 | 2 |
| | $f_{t'}$ | t' | 2 | 1 | -1 | -3 | -4 | 5 | 2 |
| | $f_t$ | t | 0 | 2 | 0 | -2 | -4 | 2 | 4 |
| | all f equal | 1 + t' + t | 4 | 4 | 0 | -4 | -4 | 8 | 8 |

Table 5.2. Elastic moduli from 1st neighbour coupling in the three cubic lattices

$\overset{1}{C} = c_{11} + 2c_{12}$, $\overset{2}{C} = c_{11} - c_{12}$, $\overset{5}{C} = 2c_{44}$, $c_{11} = (\overset{1}{C} + 2\overset{2}{C})/3$, $c_{12} = (\overset{1}{C} - \overset{2}{C})/3$

| Lattice | Shell | $d^2/2V_c$ | $\overset{1}{C} = 3K$ | $\overset{2}{C} = 2\mu'$ | $\overset{5}{C} = 2\mu$ |
|---|---|---|---|---|---|
| sc | ⟨100⟩ | $\frac{1}{2a}$ | $\frac{f_l - 2f_t}{a}$ | $\frac{f_l + f_t}{a}$ | $\frac{2f_t}{a}$ |
| bcc | ⟨111⟩ | $\frac{3}{4a}$ | $\frac{2f_l - 4f_t}{a}$ | $\frac{4f_t}{a}$ | $\frac{4f_l + 8f_t}{3a}$ |
| fcc | ⟨101⟩ | $\frac{1}{a}$ | $4\,\frac{f_l - f_{t'} - f_t}{a}$ | $\frac{f_l + 5f_{t'} + 2f_t}{a}$ | $2\,\frac{f_l + f_{t'} + 2f_t}{a}$ |

Table 5.3. Elastic moduli and first neighbour springs of fcc metals [$10^{12}$ dyn/cm$^2$]

| Metal | $\overset{1}{C}$ | $\overset{2}{C}$ | $\overset{5}{C}$ | $\frac{f_l}{a}$ | $\frac{f_{t'}}{a}$ | $\frac{f_t}{a}$ |
|---|---|---|---|---|---|---|
| Ag | 3.07 | 0.31 | 0.906 | 0.65 | -0.036 | -0.081 |
| Al | 2.284 | 0.463 | 0.564 | 0.49 | +0.045 | -0.126 |
| Au | 5.12 | 0.29 | 0.846 | 0.98 | -0.033 | -0.264 |
| Cu | 4.13 | 0.47 | 1.51 | 0.92 | -0.071 | -0.045 |
| Ir | 11.2 | 3.4 | 5.4 | 2.825 | +0.175 | -0.15 |
| Ni | 5.51 | 1.01 | 2.48 | 0.79 | -0.057 | -0.007 |
| Pb | 1.341 | 0.072 | 0.298 | 0.27 | -0.019 | -0.05 |
| Pd | 5.78 | 0.5 | 1.434 | 1.18 | -0.054 | -0.251 |
| Pt | 8.49 | 0.96 | 1.53 | 1.69 | +0.049 | -0.485 |
| Th | 1.731 | 0.264 | 0.956 | 0.43 | -0.054 | +0.05 |

This procedure has been used to obtain the dispersion curves of Fig. 3.15b from the elastic moduli of Cu. The springs obtained in this way are given in Table 5.3 for several fcc metals.

For the bcc lattice the two 1st neighbour springs cannot be uniquely expressed by the three moduli, but if one adds the second neighbour ⟨100⟩ shell with purely longitudinal coupling $f_l'$, one has three springs which can be fitted to the moduli:

$$\overset{1}{C}a = 2f_l - 4f_t + 2f_l' \;, \qquad \overset{2}{C}a = 4f_t + 2f_l' \;, \qquad \overset{5}{C}a = \frac{4f_l + 8f_t}{3} \;, \tag{5.14a}$$

$$f_l = \frac{a}{2}(c_{12} + 2c_{44}) \;, \qquad f_t = \frac{a}{4}(c_{44} - c_{12}) \;, \qquad f_l' = \frac{a}{2}(c_{11} - c_{44}) \;. \tag{5.14b}$$

In Table 5.4 the ratios $f_t/f_l$ , $f_l'/f_l$ for several bcc metals are listed. One sees, that, as in the fcc case, the transversal spring is negative and small, but that the two longitudinal springs can be of equal order.

From (5.14a) and also from Table 5.2 it appears that the compressional modulus depends on the longitudinal *and* transversal springs. This is surprising in view of the fact that in a homogeneous dilatation all springs are stretched only longitudinally, $\underline{s}^m - \underline{s}^n \parallel \underline{R}^m - \underline{R}^n$, and one would, therefore, expect that only longitudinal spring contributions enter $\overset{1}{C}$. This paradox is connected with the many body character of the transversal springs; without knowing details of the interaction, one cannot connect spring tensions and stress locally.

Table 5.4. Spring ratios in bcc metals

$$\frac{f_t}{f_l} = \frac{(c_{44} - c_{12})/2}{c_{12} + 2c_{44}} , \quad \frac{f_l'}{f_l} = \frac{c_{11} - c_{44}}{c_{12} + 2c_{44}}$$

| Metal | $-f_t/f_l$ | $f_l'/f_l$ |
|---|---|---|
| Mo | 0.07 | 0.89 |
| W | 0.03 | 0.44 |
| K | 0.06 | 0.23 |
| Na | 0.07 | 0.23 |
| Li | 0.02 | 0.12 |

## 5.4 Two Body Potentials

For two body potentials the connection between springs and elastic data is simple; in particular the above paradox does not occur. From Section 3.4.1 we know that the coupling is longitudinal and (isotropic) transversal:

$$-\Phi_{ij}^{(\underline{h})} = V''(R^{\underline{h}})\, \hat{X}_i^{\underline{h}}\hat{X}_j^{\underline{h}} + \frac{V'(R^{\underline{h}})}{R^{h}}\,(\delta_{ij} - \hat{X}_i^{\underline{h}}\hat{X}_j^{\underline{h}}) = f_l^{(\underline{h})}\hat{X}_i^{\underline{h}}\hat{X}_j^{\underline{h}} + f_t^{(\underline{h})}(\delta_{ij} - \hat{X}_i^{\underline{h}}\hat{X}_j^{\underline{h}}), \quad \underline{h} \neq 0. \tag{5.15}$$

The energy per atom is $E_c = (1/2) \sum_{\underline{h}\neq 0} V(R^{\underline{h}}) = (1/2) \sum_{\underline{h}\neq 0} V(|A\underline{h}|)$, and the equilibrium condition,

$$0 = \frac{\partial E_c}{\partial A_{mn}} = \frac{1}{2} \sum_{\underline{h}\neq 0} h_n X_m^{\underline{h}} \frac{V'(R^{\underline{h}})}{R^{\underline{h}}} ,$$

yields (after multiplication with $A_{kn}$)

$$\sum_{\underline{h}} X_k^{\underline{h}} X_m^{\underline{h}} f_t^{(\underline{h})} = 0 \tag{5.16}$$

a restriction on the transversal springs which for cubic lattices was discussed in Section 3.4.1. The Huang tensor

$$H_{ij,km} = -\frac{1}{2V_c} \sum_{\underline{h}} \Phi_{ij}^{(\underline{h})} X_k^{\underline{h}} X_m^{\underline{h}} = \frac{1}{2V_c} \sum_{\underline{h}} [f_l^{(\underline{h})} - f_t^{(\underline{h})}] \hat{X}_i^{\underline{h}} \hat{X}_j^{\underline{h}} X_k^{\underline{h}} X_m^{\underline{h}} ,$$

$$\text{where } \frac{\delta_{ij}}{2V_c} \sum_{\underline{h}} f_t^{(\underline{h})} X_k^{\underline{h}} X_m^{\underline{h}} = 0 \quad \text{because of (5.16)} , \tag{5.17}$$

becomes totally symmetric, i.e., Cauchy's relations[4] are fulfilled: H = C. A homogeneous dilatation, $\varepsilon_{ij} = \varepsilon_o\delta_{ij} = \varepsilon^1 T_{ij}(1)$, yields a stress

$$\sigma_{km} = C_{km,ii}\varepsilon_o = \frac{\varepsilon_o}{2V_c}\sum_{\underline{h}} f_1^{(\underline{h})} X_k^{\underline{h}} X_m^{\underline{h}} \tag{5.18}$$

where the $f_t$ contributions again drop out because of (5.16). Only the longitudinal springs contribute to the stress for a given dilatation and the paradox of Section 5.3 is thus resolved. Further, for the elastic energy we have

$$U = \frac{1}{4}\sum_{\underline{m},\underline{n},i,j} (s_i^{\underline{m}} - s_j^{\underline{n}})\Phi_{ij}^{\underline{mn}}(s_j^{\underline{n}} - s_j^{\underline{m}}).$$

For cubic crystals the atoms can be arranged in shells (comp. Sec. 3.4.1) and we have

$$\sum_\nu Z_\nu R_\nu^2 f^\nu = 0 \tag{5.16a}$$

$$C_{ik,mn} = \frac{1}{2V_c}\sum_\nu (f_1^\nu - f_t^\nu)R_\nu^2 \sum_{\underline{h}}{}^\nu \hat{X}_i^{\underline{h}}\hat{X}_k^{\underline{h}}\hat{X}_m^{\underline{h}}\hat{X}_n^{\underline{h}} = \frac{1}{2V_c}\sum_\nu (f_1^\nu - f_t^\nu)R_\nu^2\hat{H}_{ik,mn}^{(1,\nu)} \tag{5.17a}$$

where $\sum_{\underline{h}}^\nu$ means summing over all atoms in shell $\nu$. For some shells the values of H are given in Table 5.1. Purely 1st neighbour interaction corresponds to $f_1$ alone in Table 5.2.

In metals Cauchy's relations are not well obeyed (comp. Fig. 4.12 and Tables 4.2,3); this indicates the importance of (effective) many body interactions due to the metal electrons. Cauchy's relation should be well valid in the fcc rare gas lattices, e.g., argon, where (at 10K, /5.1/) $c_{11} = 4.24$, $c_{12} = 2.39$, $c_{44} = 2.25$ [$10^{10}$ dyn/cm$^2$]; however, the assumption of 1st neighbour interaction is not fulfilled, because $c_{11} + 2c_{12} = 9.02$, $c_{11} - c_{12} = 1.85$, $2c_{44} = 4.50$ [$10^{10}$ dyn/cm$^2$], which according to Table 5.2 ought to be in the ratio 4:1:2. For neon at 6K (in brackets data for 24K), /5.2/, where $c_{11} = 1.65$ (1.21), $c_{12} = 0.90$ (0.76), $c_{44}$ = 0.93 (0.63) [$10^{10}$ dyn/cm$^2$], the data are closer to a 1st neighbour two body interaction model: $c_{11} + 2c_{12} = 3.45$ (2.73), $c_{11} - c_{12} = 0.75$ (0.45), $2c_{44} = 1.86$ (1.26); but this has to be viewed with caution, because the light Ne atom has large zero point oscillations, and anharmonic effects are significant.

---

4 Cauchy's relations need not be valid under pressure and for non Bravais lattices (see e.g. /2.1/).

5.5 Remarks on the Problem of Microscopic Stresses

The most direct and lucid way to determine the elastic moduli would be to find the stress $\sigma$ to a given deformation v, and to extract C from $\sigma = Cv = C\varepsilon$. The stress would have to be defined by tensions of cut springs (forces) divided by appropriate surface areas. It has already been pointed out that the procedure is not possible for general springs because the compressional modulus depends also on transversal springs (Table 5.2). A dilatation does not utilize the transversal springs, the spring tension must be purely longitudinal, and therefore the compressibility cannot contain transversal contributions. Consequently, the simple picture of spring tensions cannot be employed; the reasons have been given in the introductory remarks to this chapter.

Two body forces represent the most favourable case for a definition of stresses by tensions of cut springs. Let us discuss first the simplest case, namely the sc lattice with 1st neighbour interaction (only $f_1$). Fig. 5.1 shows the situation for a T(1)-dilatation and a T(2)-shear; T(5) shears do not stretch the springs, $\overset{5}{C} = 0$. If one considers the Wigner-Seitz-Cell around one atom, cuts the springs along the surface, replaces the tension by forces, and defines the stresses by force/area, then one obviously obtains the correct result for all basis strains T($\nu$). The contributions of a longitudinal spring connecting 2nd neighbours in ⟨100⟩ must be four times

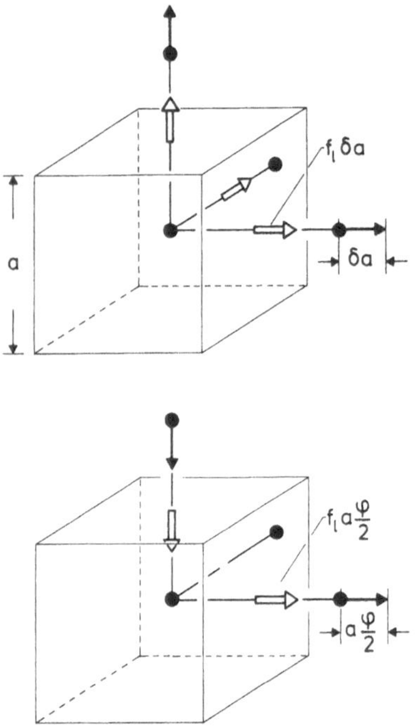


a)

b)

Fig. 5.1a and b. Stresses defined by spring tensions on the surface of a Wigner-Seitz-Cell (sc lattice with 1st neighbour longitudinal spring).
a) Dilatation, T(1): $\varepsilon = \delta a/a$; the forces $f_1\delta a$, normal to the six faces, can be represented by $\sigma = f_1\delta a/a^2 = (f_1/a)(\delta a/a) = \overset{1}{C}\varepsilon$.
b) (101) shear, -T(2'') (comp. Fig. 4.9d): the forces, $\pm f_1 a\varphi/2$, can be expressed by a stress $\sigma = \overset{2}{C}\varepsilon$ where $\overset{2}{C} = f_1/a$. If one were to include a transversal spring in this picture it would not enter $\overset{2}{C}$ because the displacement is longitudinal, in contrast to the general result of Tab. 5.2

as large: the spring tension is (for given $\varepsilon$) twice as large (the distance changes from a to 2a) and there are now two springs cut at each surface (e.g., one spring from 0 to 2a and one from -a to a, the latter crossing the Wigner-Seitz-Cell. The matter becomes complicated, however, if one considers longitudinal ⟨101⟩ springs to 2nd neighbours: the springs go right through an edge and one must divide up the tension properly between the adjoining faces.

As a second example let us consider the fcc lattice with 1st neighbour spring $f_1$ and its Wigner-Seitz-Cell, Fig. 5.2. The twelve faces have the area $A = a^2/4\sqrt{2}$. For a

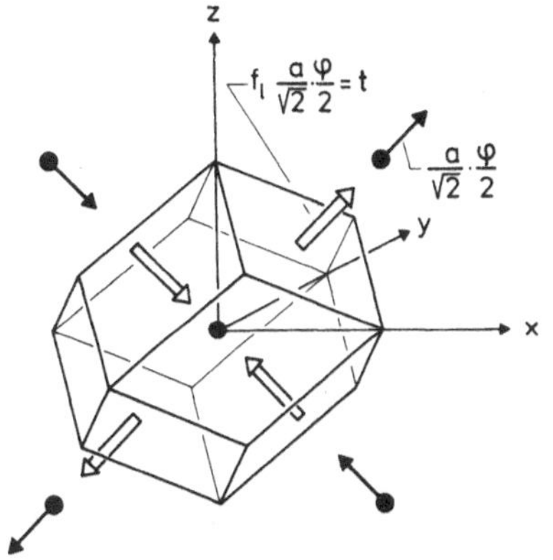


Fig. 5.2. T(5) shear in fcc lattice

$$\varepsilon = \frac{\varphi}{2}\begin{bmatrix}0&0&1\\0&0&0\\1&0&0\end{bmatrix}, \quad \varepsilon\frac{a}{2}\begin{pmatrix}1\\0\\1\end{pmatrix} = \frac{a\varphi}{4}\begin{pmatrix}1\\0\\1\end{pmatrix}, \quad A = \frac{a^2}{4\sqrt{2}}$$

Springs are stretched only for the four neighbours with $Y^{\underline{h}} = 0$. The forces on the four corresponding faces are $\pm f_1 a\varphi/2\sqrt{2}$

dilatation $\varepsilon$, the tension t is $f_1\varepsilon a/\sqrt{2}$, and the corresponding stress would be $\sigma = t/A = 4f_1\varepsilon/a$ leading to the correct $\overset{1}{C} = 4f_1/a$. The simple T(5) shear is most instructive. Stretched springs cross only the four indicated faces in Fig. 5.2. The stress should be $\sigma = \overset{5}{C}\varepsilon = 2f_1\varepsilon/a$. This would lead to forces $\pm\overset{5}{C}A/2 = \pm f_1 a/4\sqrt{2}$, half that of the spring tensions. For compensation the other 8 faces crossed by unstretched springs carry now tangential forces $\overset{5}{C}\varphi A/2$

These two examples show that a simple definition of a microscopic stress via "tension per area" is not practicable. In the above example, one must consider a macroscopic surface passing through many Wigner-Seitz-Cells or other elementary volumes in order for one to obtain the correct $\sigma$ by an averaging process, because a general plane will contain many nonparallel faces of Wigner-Seitz-Cells.

However, the result (5.17) lends itself to a very illustrative interpretation; because C and H are totally symmetric, $\sigma = Cv = Hv$, we have

$$\sigma_{ik} = H_{ik,jm}v_{jm} = H_{ij,km}v_{jm} = \frac{1}{V_c}\sum_{\underline{h}} \underbrace{-\Phi_{ij}^{(\underline{h})}v_{jm}X_m^{\underline{h}}}_{F_i^{(\underline{h})}}\cdot(X_k^{\underline{h}}/2) = \frac{1}{V_c}\underbrace{F_i^{(\underline{h})}X_k^{\underline{h}}/2}_{P_{ik}} , \qquad (5.19)$$

where $\underline{F}^{(\underline{h})} = -\Phi^{(\underline{h})}v\underline{R}^{\underline{h}} = -\Phi^{(\underline{h})}\underline{s}^{\underline{h}}$ are the forces exerted by the spring between 0 and

$\underline{R}^{h}$. The double force tensor P, as given by (5.19), is constructed from the spring tensions cut off at half the spring distance, $\underline{R}^{h}/2$, (comp. (4.72b)). For the shear of Fig. 5.2 the construction of P is sketched in Fig. 5.3. For a homogeneous strain every site carries the same forces by cut springs. By adding the double force pat-

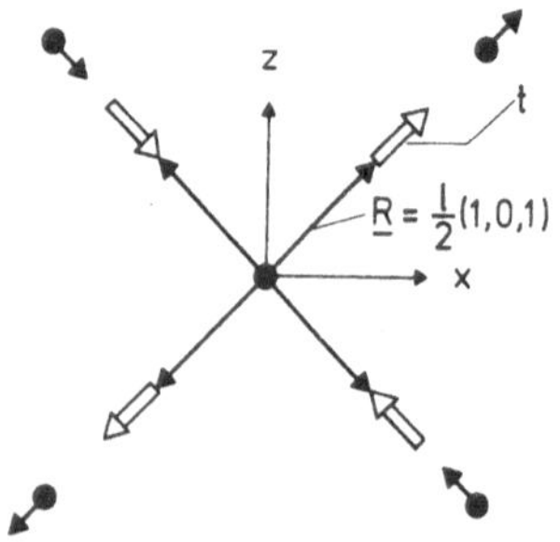


Fig. 5.3. Forces, acting at half spring distance, define a double force tensor P which determines the stress, $\sigma = P/V_c$. Sketch for T(5) shear of Fig. 5.2, $l = a/\sqrt{2}$, t = spring tension. Here

$$P = \frac{tl}{V_c}\begin{bmatrix}0 & 0 & 1\\ 0 & 0 & 0\\ 1 & 0 & 0\end{bmatrix} = \frac{f_1 al}{\sqrt{2}\,V_c}\,\frac{\varphi}{2}\begin{bmatrix}0 & 0 & 1\\ 0 & 0 & 0\\ 1 & 0 & 0\end{bmatrix} = \frac{f_1 al}{\sqrt{2}\,V_c}\,\varepsilon \ ;$$

$$\overset{5}{C} = \frac{f_1 al}{\sqrt{2}\,V_c} = \frac{2f_1}{a}$$

terns over all sites, the bulk forces disappear and only surface forces remain which are represented by a stress

$$\sigma = \frac{N}{V}\,P = \frac{1}{V_c}\,P \ . \tag{5.19a}$$

This has been discussed in Section 4.8.4; (5.19a) corresponds to (4.81a); the number of double forces N in (4.81a) is here the number of lattice sites. One, therefore, can try to define a local $\sigma$ (at site $\underline{m}$) via P by

$$\sigma^{\underline{m}}_{ik} = \sum_{\underline{n},j} -\frac{1}{2V_c}\,\Phi^{\underline{mn}}_{ij}(s^{\underline{n}}_j - s^{\underline{m}}_j)(X^{\underline{n}}_k - X^{\underline{m}}_k) \ , \tag{5.20}$$

which passes into (5.19) for homogeneous deformations. The stress from (5.20) is not necessarily symmetrical, but it could be made so by symmetrizing, $(\sigma^{\underline{m}}_{ik} + \sigma^{\underline{m}}_{ki})/2$. The consequences, though, of this assumption have not been investigated.

By cutting the spring at half distance, one makes sure that by adding the equivalent force patterns the forces by cut springs compensate exactly such that the accumulated pattern for any agglomerate is indeed given by pure surface forces. Fig. 5.4 illustrates that for 1st neighbour interaction in the (101) plane of a fcc lattice under the shear of Fig. 5.2. The same holds for the force pattern used in (5.20): forces $-\Phi^{\underline{mn}}\underline{s}^{\underline{n}}$ at $(\underline{R}^{\underline{m}} + \underline{R}^{\underline{n}})/2$. This point of view can be used to discuss the influence of bond charges, which we have used to represent many body forces and transversal springs. For illustration we use the example of tetrahedral bond charges (Sec.

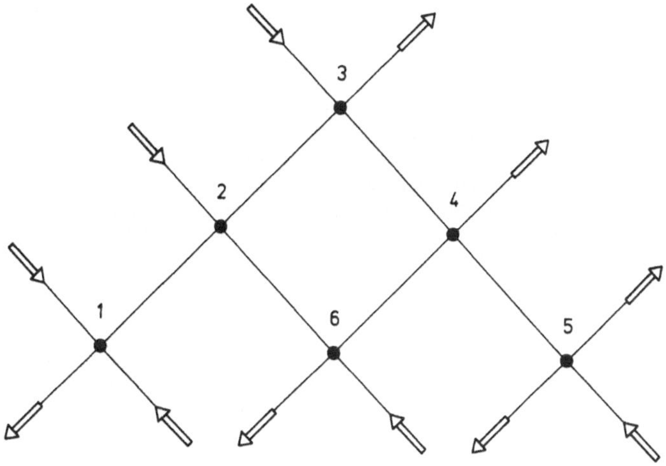


Fig. 5.4. Agglomerate of six atoms. The force patterns of Fig. 5.3 for the six atoms are added: only surface forces remain

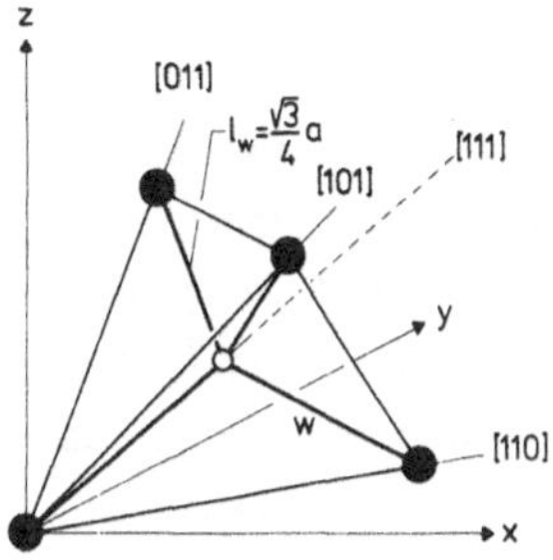


Fig. 5.5. Tetrahedral bond charge. The contributions to the elastic moduli are obtained from (3.64): $f_l = w/3$, $f_{t'} = 0$, $f_t = -w/6$, and from Table 5.2: $\overset{1}{C} = 4(f_l - f_t)/a = 2w/a$ , $\overset{2}{C} = (f_l + 2f_t)/a = 0$ , $\overset{5}{C} = 0$

3.4.3, p. 66, (3.64) and Fig. 3.23) bound by a longitudinal spring w to the four corners of its tetrahedron, Fig. 5.5. In this case, however, one must keep the whole length of the spring to avoid dangling forces when adding patterns.[5] Simplest is the dilatation where the bond charge must remain in the center of the dilatated tetrahedron. The double force tensor is isotropic: $P = (4w/3)\ l_w \varepsilon l_w$ (whole length $l_w$, not $l_w/2$). The stress is $\sigma = 2P/V_c$ because there are two bond charges per unit cell; therefore one obtains the expected result

$$\sigma = \frac{8wl_w^2}{3V_c}\varepsilon\ , \quad \overset{1}{C} = \frac{8wl_w^2}{3V_c} = \frac{2w}{a}\ . \tag{5.20a}$$

For (101) shears, the displacement of the corner atoms is perpendicular to the spring direction, the bond charge is not coupled to a (101) shear, and P and $\overset{2}{C}$ vanish. Also for (100) shears no contribution by the bond charge is to be expected, Fig. 5.6.

[5] For springs between bond charges one again must take half the length.

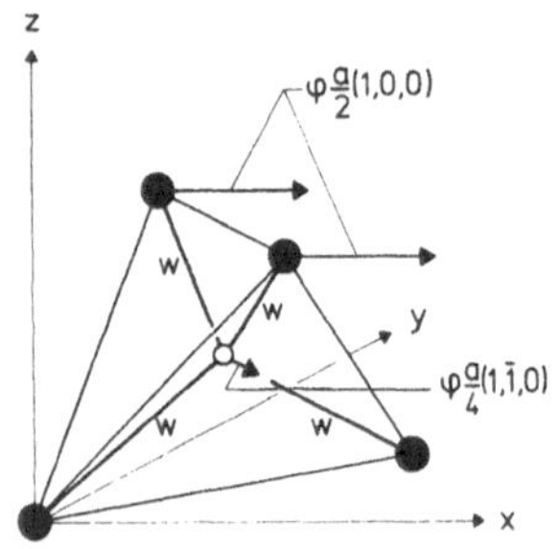


Fig. 5.6. Tetrahedral bond charge under T(5)-like simple shear, $s_x = \varphi z$. The bond charge does not join the shear which would lead to a displacement $\varphi a(1,0,0)/4$. It prefers to evade additively in direction $[0\ \bar{1}\ 0]$, displacement $\varphi a(1,\bar{1},0)$, into an equilibrium position, where none of the four w-springs is stretched. Note, that this displacement is perpendicular to the springs between the bond charge and the two atoms with $Z^{h} = 0$, which is not exhibited by the drawing

This illustrates also that, when calculating P's under homogeneous deformations, one must take into account the relative displacement of atomic vs. bond charge lattice which determines the tension of the connecting springs.[6]

## 5.6 The Change of Elastic Data by Single Tetrahedral Bond Charges

If one adds to an atomic lattice tetrahedral bond charges in small atomic concentration c, $0 \leqslant c \leqslant 2$ (per atom there are two bond charge sites available), the change of elastic data will be proportional to c. One can then calculate the change for a single "defect" and need only to add these simple changes. It is already obvious that a shear does not involve a single bond charge at all. The shear moculi $\overset{2}{C}$ and $\overset{5}{C}$ are not changed. Only $\overset{1}{C}$ is subject to change. Therefore, we consider a homogeneous dilatation $\varepsilon$; in the ideal lattice, where $\sigma = \overset{1}{C}\varepsilon$, all atoms in the bulk are force--free. However, around a single bond charge as defect the w-springs are stretched (tension $t = wl_w\varepsilon$), and in the dilatated state the four corner atoms experience forces $\underline{F}^{(\nu)} = t\underline{a}^{(\nu)}$, $\nu = 1...4$, Fig. 5.7. This is an invariant force pattern which is applied to the actual defect lattice and produces an additive displacement $\underline{s}_d$. The double force tensor constructed with the $\underline{F}$'s is not directly connected with the change of volume or the average stress as in Section 4.8.4. However, one can easily determine the "Kanzaki" forces $\underline{K}$ which produce the actual displacement, $\underline{s}_d$, in the perfect lattice[7] (comp. Secs. 4.8.3 and 6.2). The displacement is given by

$$\underline{s}_d = \frac{1}{\Phi + \psi}\underline{F} = \frac{1}{(1 + \psi G)\Phi}\underline{F} = G\frac{1}{1 + \psi G}\underline{F} = G\underline{K}\ , \quad G = \frac{1}{\Phi}\ , \tag{5.21}$$

---

6 Relative displacements of sublattices must be considered in all non-Bravais lattices of which an atomic with an additive bond charge lattice is an example. Only if every site is a center of symmetry (inversion center), such as in the alkali--halides, relative displacements do not occur and a homogeneous deformation is homogeneous throughout. In this example the bond charge sites are not centers of symmetry.

7 Because the $\underline{K}$'s are acting in the *perfect* lattice the methods of Section 4.8 can be employed.

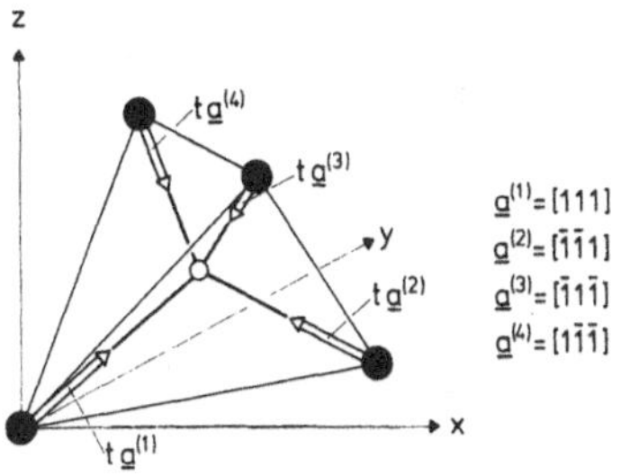


Fig. 5.7. Force pattern, $\underline{F}^{(\nu)} = t\underline{a}^{(\nu)}$, about a single bond charge in a dilatated fcc lattice

where $\Phi$ is the coupling of the perfect lattice, $G = 1/\Phi$ its Green's function and $\psi$ is the additive coupling due to the bond charge springs, proportional to w. The Kanzaki forces $\underline{K}$ are renormalized $\underline{F}$'s. This renormalization is of utmost importance if $1 + \psi G$ is small ("resonances" treated in Chap. 6). If $\psi$ or w is small one approximately can replace $\underline{K}$ by $\underline{F}$, "Born's approximation"; then the double force tensor

$$P^d_{ik} = - t l_w \sum_\nu a_i^{(\nu)} a_k^{(\nu)} \;, \quad P^d = -\frac{4 t l_w}{3} = -\frac{4 w l_w^2}{3} \varepsilon = \alpha\varepsilon \tag{5.22}$$

is isotropic and proportional to $\varepsilon$. The factor $\alpha$ is called polarizability; here it is only a scalar factor, but also in general $P^d = \alpha\varepsilon$ where $\alpha$ is a 4th rank tensor.[8] In the average over the possible defect sites, one defect produces a stress $P^d/V = \alpha\varepsilon/V$ which supports the external stress $\sigma$, and for $N_d$ defects in small atomic concentration $c = N_d/N$ one has approximately ($V = NV_c$)

$$\sigma + \frac{N_d}{V} \alpha\varepsilon = C\varepsilon \;, \quad \sigma = \left(C - \frac{c}{V_c}\alpha\right)\varepsilon = (C + \delta_1 C)\varepsilon \;, \quad \text{i.e.,}\ \delta_1 C = -\frac{c}{V_c}\varepsilon \;. \tag{5.23}$$

The change of C is directly proportional to the polarizability of the single defect. In our case of dilatation we obtain

$$\delta \overset{1}{C} = -\frac{c}{V_c}\alpha = \frac{c}{V_c}\frac{4 w l_w^2}{3} = c\,\frac{w}{a} \;. \tag{5.23a}$$

This result (Born's approximation) can easily be understood as a linear interpolation in concentration between the perfect lattice without bond charges (c = 0) and the perfect lattice with bond charges (c = 2) where the bond charge contribution to $\overset{1}{C}$ is 2w/a (comp. (5.20a)).

---

[8] Note, that according to (5.21) $P^d$ and $\varepsilon$ are always linearly connected, because $\underline{F}$ and $\underline{K}$ are proportional to $\varepsilon$.

# 6. Statics and Dynamics of Simple Single Point Defects

/6.1,2/

It is easy to describe the behaviour of one point defect in terms of perturbations of the matrices $\Phi$, M and of Green's functions referring to the perfect lattice. The reaction of defects to external sources such as externally produced lattice waves or strains, and incoming neutrons or X-rays is treated by scattering theory. After the extensive introduction into Green's functions, formal scattering theory is very simple. Still the details can be clumsy and we will treat only the simplest examples for purposes of illustration to elucidate the physics and to demonstrate the methods. In this chapter we treat only a single point defect. The essential quantity will be Green's function, G, of the defect lattice. It will be expressed by $\overset{o}{G}$, Green's function of the perfect lattice, and the "transition" (scattering) amplitude t, which can be expressed by $\overset{o}{G}$ and the changes of $\Phi$ and M. The results give directly:

1) the intensity of the Mößbauer line (Sec. 2.4.5), determined by $Z_d(\omega) = M^d \, \mathrm{Im}\{G^{dd}\}/\pi$, where d refers to the (defect) Mößbauer atom;
2) the scattering of lattice waves by one defect, determined by the "transition" matrix $t(\omega)$;
3) the polarizability, $\alpha$, of the defect (Sec. 5.6), determined essentially by $t(\omega = 0)$ ("static" scattering);
4) the change of thermal energy by one defect (Sec. 2.4.5), $\delta_1 E(T) = \int_0^\infty d\omega^2 \, \varepsilon_{th}(\omega) \cdot (Z_{ii}^{mm} - \overset{o}{Z}{}_{ii}^{mm})$;
5) the change of elastic moduli by one defect (Sec. 5.6), $\delta_1 C = -\alpha/V$.

The intensity of the Mößbauer line can be observed directly (comp.,e.g.,/6.3/). For small concentrations or for a "small" number of defects, $N_d$, the quantities $\delta_1 E$ and $\delta_1 C$ are approximately additive; consequently, $N_d \delta_1 E(T)$ determines the change of energy and specific heat, and $N_d \delta_1 C = -N_d \alpha/V = -c\alpha/V_c$ the change of moduli. The scattering cross sections are required for lattice transport phenomena, which are not treated here. The transition matrix t of a single defect determines (for small concentrations) the average G which in turn is measured in neutron and X-ray scattering.

## 6.1 Simple Defect Structures

Schematic structures are shown in a simple square lattice, the (001) plane of a simple cubic lattice, Fig. 6.1. The figures indicate only the structure of the defect; this structure will relax by displacements which at the surface will cause a volume change $\delta_1 V$; these displacements are not shown.

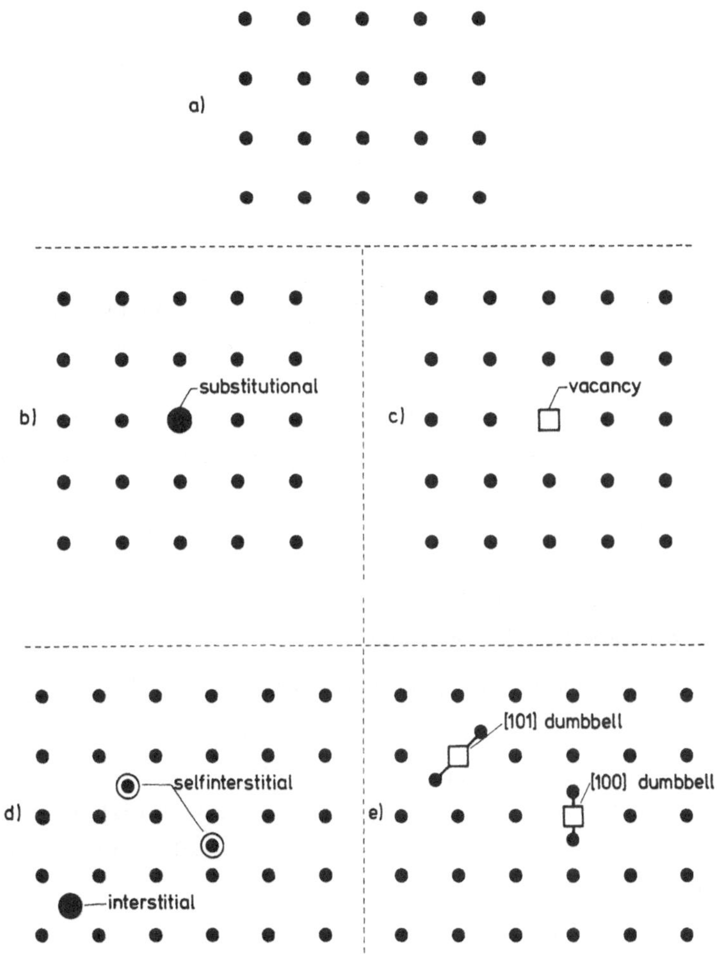


Fig. 6.1a-e. Simple defect structures (schematic for the (001) plane of a sc lattice, no relaxation). a) ideal lattice, b) substitutional impurity, c) vacancy, d) interstitial impurity, two self-interstitial structures, e) two self-interstitial dumbbell structures

The substitutional defect of Fig. 6.1b is the most simple defect. The number of atomic degrees of freedom is not changed, but one mass and a number of springs around the defect are. The simplest kind of a substitutional defect is the isotopic defect, where only one mass is changed and which has been treated preliminarily in Section 3.5.3.

The vacancy (Fig. 6.1c) is also a simple structure. One atomic coordinate is missing; this can approximately be represented by cutting the interactions of the central atom with the environment, such that the coordinate does not couple to the rest.

Fig. 6.1d shows various possibilities for interstitial defects, one interstitial impurity and two self-interstitial structures, where interstitial and host are like. There are many more possible structures for interstitials and self-interstitials. It is rather difficult to decide structures theoretically because the energy differences between various structures are small, and therefore even reasonable approximations made in the calculations can favour the wrong structure. Theoretical *and* experimental results have shown that in the simple metals with cubic symmetry the self-interstitial favours a dumbbell form indicated in Fig. 6.1. This can be described by one vacancy and two symmetrically located self-interstitials. In the fcc lattices (e.g., Al and Cu, /6.1/) the stable direction of the dumbbell axis is ⟨100⟩, in the bcc lattices ⟨101⟩ (e.g., Mo, /6.1/). The surplus atomic coordinate leads to minor complications in the theory.

Of course, there are many possible defect structures, e.g., various close pairs of defects such as the divacancy, the di-self-interstitial, the Frenkel pair (vacancy + self-interstitial) and the higher order agglomerates of vacancies and self-interstitials which finally pass into dislocation loops.[1] These agglomerates can hardly be considered as point defects. We will concern ourselves here only with the simplest structures of simple point defects.

Structures in the fcc lattice which will be treated are:

1) *the substitutional defect* (including the vacancy) which is easy to visualize;

2) *the octahedral interstitial* (Fig. 6.2) with full cubic symmetry in fcc lattices; in bcc lattices the octahedral interstitial has tetragonal symmetry, the ⟨001⟩ octahedral site at $\frac{a}{2}$ (0,0,1) is surrounded by eight atoms: two with distance a/2 and four in the cube centers with distance $a/\sqrt{2}$; the three possible [100] structures are not identical;

3) *the tetrahedral interstitial* (shown in Fig. 5.5).

---

[1] Comp. Appendix L.

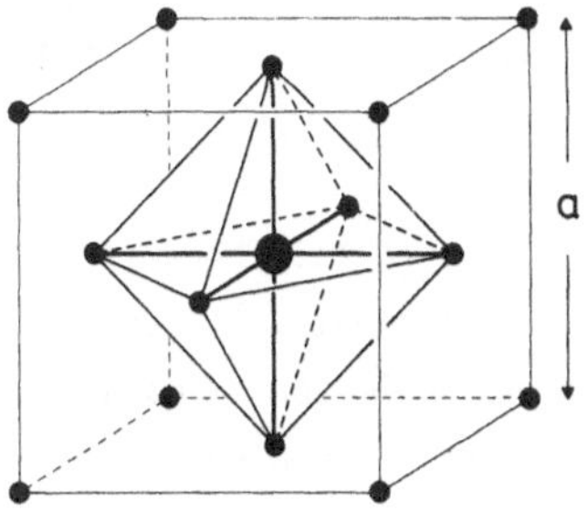


Fig. 6.2. Interstitial in octahedral position of fcc lattice. The octahedral position in the center of the elementary cube is shown together with the surrounding octahedron; this position corresponds also to the center between 2nd neighbours (in ⟨100⟩ directions)

In most cases the host will be represented by the simple 1st neighbour model with longitudinal coupling for which the needed Green's functions have been tabulated in Section 3.5.6.

## 6.2 Static Structure

The static structure of the defect is described by its schematic structure exhibiting the symmetry and by the "permanent" displacements $\underline{u}$ from the original ideal positions $\underline{R}$ of the infinite lattice. This is sketched in Fig. 6.3a for a substitutional impurity; in the new equilibrium positions, $\underline{R} + \underline{u}$, all atoms are force-free. A particularly useful representation of $\underline{u}$ is that by fictitious (Kanzaki) forces $\underline{K}$, first introduced by KANZAKI, /6.2/, which produce the same displacements $\underline{u}$ in the *perfect* harmonic crystal (Fig. 6.3b),

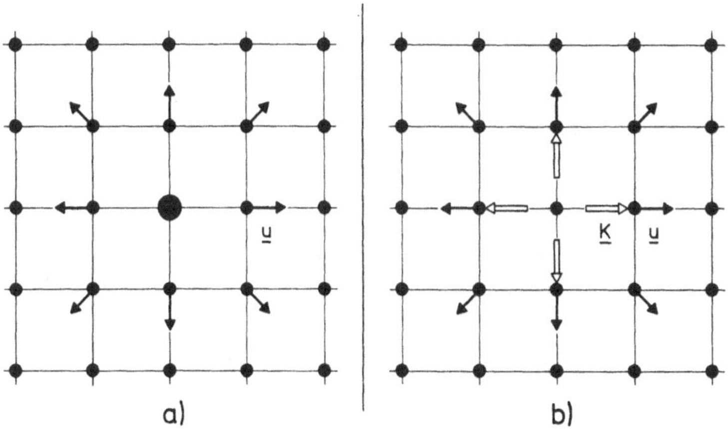


Fig. 6.3a and b. Displacements and Kanzaki forces. a) The displacement field $\underline{u}$ in the defect lattice caused by a substitutional impurity, b) the same displacements $\underline{u}$, but produced by the Kanzaki forces $\underline{K}$ in the *perfect* harmonic lattice

$$u = \overset{o}{G}\underline{K} = \frac{1}{\overset{o}{\Phi}}\underline{K}\,, \quad \text{where } \overset{o}{G} = \frac{1}{\overset{o}{\Phi}} \text{ refers to the perfect crystal.} \tag{6.1}$$

Of course, one can always define $\underline{K} = \overset{o}{\Phi}\underline{u}$ and then gets back $\underline{u}$ via (6.1). The introduction of $\underline{K}$ is just another way to express the permanent displacements. It must also be noted that (6.1) does not imply the validity of the harmonic approximation for actual forces $\underline{K}$. The real advantages of introducing $\underline{K}$ are the following:

1) Because of screening, the range of $\underline{K}$ in a metal will be small; therefore the long range $\underline{u}$ can be expressed (parametrized) by a few forces $\underline{K}$. Indeed the assumption of nearest neighbour $\underline{K}$'s, e.g., Fig. 6.4, usually yields very satisfactory agreement with measured $\underline{u}$'s. The Kanzaki force pattern must be invariant, i.e., the total force and torque must vanish.

2) The displacements $\underline{u}$ can still contain a translational-rotational contribution. One cannot superimpose the displacements of two defects which are far apart, but one can add the Kanzaki forces.

3) The long range displacements can be expressed by a double force tensor, $P_{ik} = X_i^m K_k^m$, constructed from the Kanzaki forces and the perfect lattice sites. Continuum theory can be used and the results of Sections 4.8.3,4 apply: $\delta_1 V = P_{ii}/3K = (\underline{R},\underline{K})/3K$, where K is the compressional modulus. This relation permits us to express $\underline{K}$ by $\delta_1 V$ in simple cases. If one assumes radial forces, $\kappa$, to the 1st neighbours of the defect one obtains for tr{P}: $12\kappa a\sqrt{2}$ for the substitutional, $3\kappa a$ for the octahedral interstitial, and $\sqrt{3}\kappa a$ for the tetrahedral interstitial (comp. App. M). The pattern, averaged over all defect positions in the lattice, leaves only surface forces corresponding to a stress P/V as in Section 4.8.4.

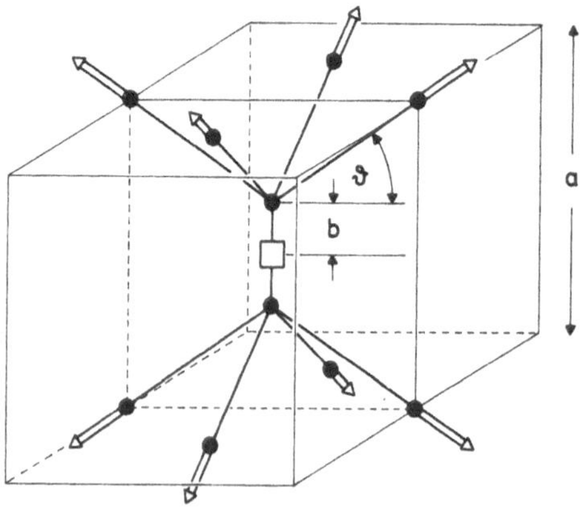


Fig. 6.4. [001] self-interstitial (fcc lattice). The dumbbell parameters are b or $\vartheta$ = arctg (1 - 2b/a). If one assumes radial forces $\kappa$ to the eight 1st neighbours of the dumbbell one obtains $P_\perp = 2\kappa a \cos\vartheta$, $P_\parallel = 4\kappa a \sin\vartheta$. For b = a/4 the dumbbell becomes elastically isotropic: $P_\perp = P_\parallel = 2\kappa a/\sqrt{5}$

It is evident that in an infinite crystal the $\underline{K}$-pattern moves with the position of the defect. The same $\underline{K}$-pattern can also be used in the bulk of a finite crystal (which is not so evident). Obviously it cannot be used near a surface, and this is clearly seen in Fig. 6.3b: if there the "defect" were in the surface, one of the forces would not even have a partner to act on. The same argument holds for the superposition of force patterns from two defects: superposition becomes meaningless if the two patterns overlap; but for interaction of two defects at large distances, or for that of one bulk defect with an external strain or with a surface, one can use the superposition of $\underline{K}$-patterns. The double force tensor defined microscopically is *the* link to macroscopic behaviour. For this reason the concept of the double force tensor is of central importance in the theory of defects.

Values of $\delta_1 V$ for substitutional defects in fcc lattices are given in Table 6.1. The substitutional defect has cubic symmetry, hence $P_{ik} = P_o \delta_{ik}$, $P_o = K\delta_1 V$. The values of $P_o$ are usually given in eV, and the order of magnitude is 10 eV for $\delta_1 V = V_c$, e.g., for Ag (Table 2.2): $K \cong 0.6\ \mathrm{eV/\AA^3}$, $a \cong 4\mathrm{\AA}$, $V_c = a^3/4$, $KV_c = 10$ eV.

Table 6.1. $10^2 \delta_1 V/V_c$ for some fcc metals. Taken from: H.W. KING: J. Mat. Sc. 1, 79 (1965)

| Substitute \ Host | Ag | Al | Au | Cu | Ni | Pd | Pt |
|---|---|---|---|---|---|---|---|
| Ag | | 0.1 | -0.6 | 44 | | 13 | 9 |
| Al | -9 | | -10 | 20 | 15 | 1 | -9 |
| Au | -1.8 | | | 48 | 64 | 16 | 11 |
| Cu | -28 | -38 | -28 | | 7 | | -20 |
| Ni | | | -22 | -8.5 | | | |
| Pd | -17 | | -14 | 28 | 41 | | -4 |
| Pt | -20 | | -13 | 31 | 46 | 1.5 | |

Another example is the self-interstitial dumbbell in ⟨100⟩ orientation in a fcc lattice (Fig. 6.4). In the [001] orientation shown there, the double force tensor must have the form

$$P = \begin{bmatrix} P_\perp & 0 & 0 \\ 0 & P_\perp & 0 \\ 0 & 0 & P_\parallel \end{bmatrix},$$

in the model of Fig. 6.4:

$$P_\perp = 2\kappa a \cos \vartheta, \quad P_\parallel = 4\kappa a \sin \vartheta,$$

$$\frac{P_\perp - P_\parallel}{\mathrm{tr}\{P\}} = \frac{\cos \vartheta - 2 \sin \vartheta}{2(\cos \vartheta + \sin \vartheta)} = \frac{b - a/4}{a - b} = \text{anisotropy}.$$

The actual values for Al are (comp. /6.1/): $P_\perp \cong 16$ eV, $P_\parallel \cong 15$ eV, $(P_\perp - P_\parallel)/\mathrm{tr}\{P\} \cong 0.02$, which corresponds in the model of Fig. 6.4 to $b/a \cong 0.27$. Of course, this dumbbell structure has three equivalent orientations parallel to the three cube edges. Under normal conditions, no orientation is preferred. By applying an external strain one can change the population and can, for instance, favour a particular dumbbell direction.

## 6.3 Simple Models for Defects

The theory, as a rule, will be kept quite general. We will demonstrate its results by the simplest models in order to avoid numerical complications. The lattice model employed most often will be the fcc lattice with 1st neighbour longitudinal spring f, which results from short range two body interaction, $\Omega_E^2 = 4f/\overset{\circ}{M} = \Omega_{max}^2/2$; its Green's functions are found in Section 3.5. In this section we denote the host coupling by $\overset{\circ}{\Phi}$ and the host mass by $\overset{\circ}{M}$. The actual quantities are denoted by

$$\Phi = \overset{\circ}{\Phi} + \psi , \qquad M = \overset{\circ}{M} + m , \tag{6.2}$$

where $\psi$, m describe the changes introduced by the defect. The defect models also will be stripped down to the essentials for simplicity. They are discussed below in detail.

### 6.3.1 The Isotopic Defect

The simplest point defect at all is the isotopic defect; it is a substitutional defect, and only the mass is changed at the defect site, $\underline{\mu}$,

$$\psi = 0 , \quad m_{ik}^{\underline{mn}} = (M_d - \overset{\circ}{M})\, \delta^{\underline{mn}} \delta_{ik} \delta^{\underline{m\mu}} = (M_d - \overset{\circ}{M})\, P^{\underline{\mu}} , \quad ⨳ , \tag{6.3}$$

where $P^{\underline{\mu}}$ is the projector onto site $\underline{\mu}$. The introduction of an isotope into an isotopically pure host lattice practically does not change the electronic behaviour, and therefore does not change the atomic interaction. Only for an isotopic mass $M_d$ very different from $\overset{\circ}{M}$ the effects are appreciable. For normal metals, however, the isotopic mass changes are minimal (order of 1%) and can be neglected. Consequently, one does not have to worry about normal metals which are actually rather impure isotopically; there it suffices to consider $\overset{\circ}{M}$ as the average mass.

However, any substitutional atom which causes only small changes in the interaction can be treated like an isotopic defect. An indication of the change in interaction is the volume change. It might be possible, therefore, to treat substitutional atoms with "small" $\delta_1 V$ as isotopic defects. Examples are (only cases $|\delta_1 V/V_c| \leqslant 0.1$ are listed): Al in Ag, $\delta_1 V/V_c = -0.09$, $M_d/\overset{o}{M} = 0.25$; Au in Ag, -0.02, 1.83; Mn in Ag, 0.001, 0.51; Mg in Ag, 0.07, 0.23; Al in Au, -0.1, 0.14; Mn in Au, -0.05, 0.28; Ag in Au, -0.006, 0.55; Ag in Al, 0.001, 4. Cases with large $M_d/\overset{o}{M}$ are candidates for resonances; small $M_d/\overset{o}{M}$ are suspect of localized modes as has been discussed in Section 3.5.

### 6.3.2 The Vacancy

A vacancy is produced by taking out one atom, say the atom at 0. Because in our simple two-body model with the short range potential V(r) (comp. Sec. 3.4.1) the first derivative V'(1) vanishes for the 1st neighbour distance, the force exerted by one atom on any 1st neighbour vanishes. Therefore, one can take out one atom and there are no forces on the remaining atoms in their original equilibrium positions. Consequently, there are no relaxing displacements, no Kanzaki forces and no volume change. The vacancy can be represented by

$$m = 0 \ , \quad \psi = - f\hat{\psi} \ , \tag{6.4}$$

where $-\psi$ represents the spring star consisting of twelve f-springs which are taken out together with the removed atom (Fig. 6.5); $\hat{\psi}$ is the spring star with springs of strength 1.

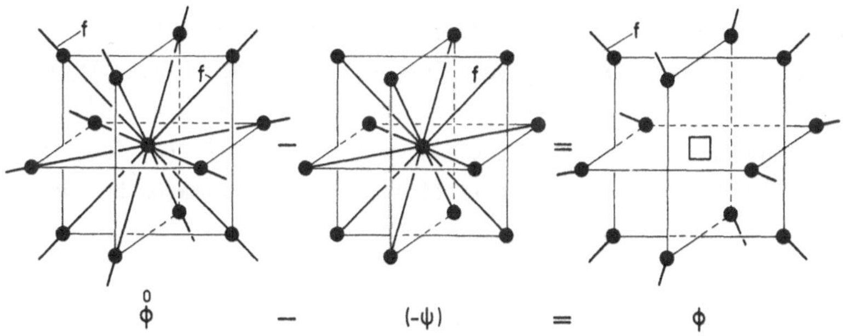


Fig. 6.5. Spring structure of a vacancy. The coupling $\Phi$ of the defect lattice (with one vacancy) is obtained from the ideal coupling $\overset{o}{\Phi}$ by removing the spring star $-\psi$

Vacancies in metals are rather difficult to assess because of the change of electronic structure in and around the vacancy. Nevertheless, attempts have been made to describe the vacancy by cutting the springs from the removed atom to its former neighbours. The difficulties of this procedure can be seen immediately for purely 1st neighbour coupling ($f_1, f_{t'}, f_t$). It has been pointed out before that cutting the transversal springs is inconsistent and leads to violation of rotational invariance, though this is sometimes overlooked. For consistency, then, one can either cut only $f_1$ or cut longitudinal and transversal springs and correct the springs between the twelve neighbours for rotational invariance[2]. If the transversal spring contributions are relatively small (comp. Table 3.2) the $f_1$ part dominates and the results are not much different from the simple model used above. If the transversal springs become larger, say 10% and more, the results, for instance the polarizability, can change by orders of magnitude and become completely ambiguous. The same will hold for bond charge models of the springs. Let us consider, for instance, Au with a relatively large $f_t/f_1$. The coupling can be separated in a direct coupling f, tetrahedral bond charge coupling with spring w and bond charges in the face-centers of the tetrahedra with spring w' (comp. Sec. 3.4.3, p. 67). The 1st neighbour springs of Au $\left(f_1 = f + w/3 + w' = a(4\overset{1}{C} + 2\overset{2}{C} + 3\overset{5}{C})/24,\ f_{t'} = -w'/9 = a(2\overset{2}{C} - \overset{5}{C})/8,\ f_t = -w/6 -2w'/9 = -a(\overset{1}{C} + 2\overset{2}{C} - 3\overset{5}{C})/12\right)$ can be obtained[3] from $f/a \cong 0.3$, $w/a \cong 1.2$, $w'/a \cong 0.3$ [$10^{12}$ dyn/$cm^2$]. Several vacancy models are possible: one can cut only the springs to the neighbouring atoms and bond charges or one can remove one or both of the bond charges when creating the vacancy. If the bond charge springs are relatively small, the models do not differ much, but if the bond charge springs are comparable with f, very different results are possible. Consequently, the basis for vacancy models is somewhat shaky, and results from these models are not reliable. But, at least, one can get a feeling for the order of magnitude.

All the above models assume that the permanent displacements about a vacancy can be neglected. This, however, is not generally true. Let us, therefore, consider again the two body model, discussed in Section 3.4.1, which gives permanent displacements if one includes at least 2nd neighbour interaction, (comp. Fig. 3.16a). The forces $\underline{F}^{m} = V'(R^{m})\hat{\underline{R}}^{m}$ on the remaining atoms at their original sites around the vacancy produce the permanent displacements in the *defect* lattice. If V(r) is short range, the forces $\underline{F}$ are small[4] and the defect lattice can be approximated by the

---

2 These corrections are $-f_t/2$ for the $f_{t'}$-springs between the neighbours and $-f_{t'}/2$ for the $f_t$-springs. The corrections, of course, depend on the model assumed and are not unique.

3 $2w/a = \overset{1}{C} + 8\overset{2}{C} - 6\overset{5}{C}$, $8w'/9a = \overset{5}{C} - 2\overset{2}{C}$, $f/a = \overset{2}{C}$.

4 $V'(R_1)$ is small because it vanishes in the 1st neighbour approximation; $V'(R > R_2)$ is small due to the short range of V.

model used in the beginning, the 1st neighbour model with twelve missing f-springs:

$$\Phi = \overset{o}{\Phi} + \psi \ , \quad \underline{u} = \frac{1}{\overset{o}{\Phi} + \psi}\underline{F} = \frac{1}{(1 + \psi\overset{o}{G})\overset{o}{\Phi}}\underline{F} = \overset{o}{G}\frac{1}{1 + \psi\overset{o}{G}}\underline{F} = \overset{o}{G}\underline{K} \ , \tag{6.5}$$

$$\underline{K} = \frac{1}{1 + \psi\overset{o}{G}}\underline{F} = \underline{F} - \frac{1}{1 + \psi\overset{o}{G}}\psi\overset{o}{G}\underline{F} = \underline{F} - t\overset{o}{G}\underline{F} \ , \tag{6.5a}$$

where the "t-matrix",

$$t = \frac{1}{1 + \psi\overset{o}{G}}\psi = \psi\frac{1}{1 + \overset{o}{G}\psi} = \psi - \psi\overset{o}{G}\psi + \psi\overset{o}{G}\psi\overset{o}{G}\psi - + \ldots \ , \tag{6.5b}$$

is a matrix confined to the subspace of $\psi$ (coordinates of the "vacancy" and its twelve neighbours). One sees that $\underline{K}$ is a somewhat renormalized version of $\underline{F}$. If $\psi$ is small, $\underline{K} = \underline{F}$ in a zero order approximation; here, however, both terms are of the same order.

For $\underline{K} = \underline{F}$ one has a very particular force pattern, for which the double force tensor,

$$P_{ik} = \sum_{\nu}\sum_{\underline{h}}{}^{\nu} F_i^{\underline{h}} X_k^{\underline{h}} = \sum_{\nu} V'(R_\nu)R_\nu \sum_{\underline{h}}{}^{\nu} \hat{X}_i^{\underline{h}} \hat{X}_k^{\underline{h}} = \frac{\delta_{ik}}{3}\sum_{\nu} Z_\nu V'(R_\nu)R_\nu = 0 \ , \quad \left(\text{comp. } (3.47)\right) \ ,$$

vanishes because of the equilibrium condition. Therefore, this force pattern produces displacements which asymptotically fall off at least proportional to $r^{-3}$; the volume change vanishes. This demonstrates that $\delta_1 V = 0$ does not necessarily mean small changes on an atomistic scale.

In the 1st and 2nd neighbour model, employed also in Section 3.4.1, $\underline{F}$ contains forces on the 1st and 2nd shell around the vacancy; due to the range of $\psi$, the correction ($\underline{F} \rightarrow \underline{K} = \underline{F} - t\overset{o}{G}\underline{F}$) contains only forces on the 1st shell. The calculation of the matrix t implies the inversion of $1 + \psi\overset{o}{G}$ in the 39-dimensional subspace of $\psi$ which can be done directly, if clumsily, with some numerical effort. We will see later how to calculate expressions such as $t\overset{o}{G}\underline{F}$ in a simple way employing symmetry arguments. For a crude estimate we use for $\overset{o}{G}$ the (local) Einstein-approximation, $\overset{o}{G}_E = 1/4f$, whereupon,

$$\underline{K} = \underline{F} - \frac{1}{1 + \psi/4f}\psi\frac{\underline{F}}{4f} = \underline{F} - \frac{\psi/4f}{1 + \psi/4f}\underline{F}^{(1)} \ ,$$

and where $\underline{F}^{(1)}$ are the 1st shell forces alone. Now, $\underline{F}^{(1)}$ is an eigenvector[5] to $\psi$

5 The spring star $\psi$ in Fig. 6.5 is subject to radial forces, the center is fixed and each spring just gives a factor f for $-\psi$ or a factor -f for $\psi$.

with eigenvalues -f and therefore $\underline{K} = \underline{F} + (1-1/4)^{-1}(1/4)\,\underline{F}^{(1)} = \underline{F} + \underline{F}^{(1)}/3$. Only $\underline{F}^{(1)}/3$ contributes to the double force tensor,

$$P_{ik} = \frac{\delta_{ik}}{3}\,\mathrm{tr}\{P\}\ ; \qquad \mathrm{tr}\{P\} = 4R_1V'(R_1) = -2R_2V'(R_2) \cong -2\mathring{R}_2V'(\mathring{R}_2)\ ,$$

and the volume change is

$$\delta_1 V = \mathrm{tr}\{P\}/\overset{1}{C} = -\frac{2\mathring{R}_2V'(\mathring{R}_2)}{4V''(\mathring{R}_1)/\mathring{a}} = -\frac{\mathring{a}^2}{2}\,\frac{V'(\mathring{R}_2)}{V''(\mathring{R}_1)}\ ,$$

where $\mathring{R}_1 = 1$, $\mathring{a} = \mathring{R}_2 = \sqrt{2}\,\mathring{R}_1$ refer to the 1st neighbour approximation, $V'(\mathring{R}_1) = 0$. For a 6 - 12 Lennard-Jones potential, $V(r) = C_{12}/r^{12} - C_6/r^6$, one obtains

$\delta_1 V/V_c = -7\cdot 2^{-7}/3 \cong -0.018$ , a very small, negative volume change of about 2%.

### 6.3.3 Simple Substitutional Defects

In the simplest case, a substitutional defect (at site 0) can be represented by a mass change, $m = P^{o}(M_s - \mathring{M})$ where $M_s$ is the mass of the substitutional atom, and by a spring change, $\psi = (f_s - f)\hat{\psi}$, where only the springs from the substitutional atom to its neighbours are supposed to change from f to $f_s$. It contains the elements of both the isotopic defect and the vacancy. One expects localized modes for $f_s/M_s \gg \Omega^2_{max}$ and resonant modes for $f_s/M_s \ll \Omega^2_{max}$.

### 6.3.4 The Octahedral Interstitial

Most simply the octahedral interstitial in a fcc lattice can be described by the mass $M_i$ of the interstitial (additive coordinates) and six additive longitudinal springs $f_i$ which connect the interstitial and its six neighbours (Fig. 6.2). One expects resonant modes if $f_i/M_i \ll \Omega^2_{max}$ and localized modes if $f_i/M_i \gg \Omega^2_{max}$. An octahedral bond charge, $M_i = 0$, only introduces additive longitudinal springs $f_i/2$ between 2nd neighbours (in $\langle 100 \rangle$ direction).

For an octahedral self-interstitial, $M_i = \mathring{M}$, the actual distance $l_i$ between interstitial and its neighbours[6] is much smaller than the equilibrium distance and one expects strong longitudinal springs $f_{i,l} = V''(l_i)$ and also sizable and negative transversal springs $f_{i,t} = V'(l_i)/l_i < 0$ (comp. Sec. 3.4.1). Transversal springs

6 If $l_i$ is much smaller than the ideal equilibrium distance $\mathring{R}_1$, $V'(\mathring{R}_1) = 0$, only the repulsive part of V(r) is important. If it is described by a power p, $V(r) \propto r^{-p}$, then $f_{i,t}/f_{i,l} = -1/(p+1)$, say - 1/13 for a 12 - 6 Lennard-Jones potential.

$f_{i,t}$ alone violate rotational invariance which can be restored[7] by adding transversal springs, $-f_{i,t}/4$, to all the twelve f-springs on the surface of the octahedron. As has been discussed in Section 3.4.1, the negative transversal springs indicate a tendency toward instability, small restoring forces or small "resonant" frequencies $\Omega_R \ll \Omega_{max}$ for certain modes. On the other hand, one anticipates well-localized modes, $\Omega_L \gg \Omega_{max}$, because $f_{i,l} \gg f$. Indeed, under these circumstances the octahedral self--interstitial exhibits resonant and localized modes, /6.4/.

The structure of resonant and localized modes can be most simply seen by studying the modes of the octahedron embedded in the otherwise fixed host crystal, a sort of augmented Einstein approximation, with a coupling denoted by $\varphi$. If this system has eigenfrequencies very large (small) compared with the maximum frequency of the host, $\Omega^2_{max} \cong 8f/\overset{o}{M}$, one can expect localized (resonant) modes. The coupling is indicated in Fig. 6.6a, coupling $f_i$ inside the octahedron, coupling f, $-f_{i,t}/4$ on the surface

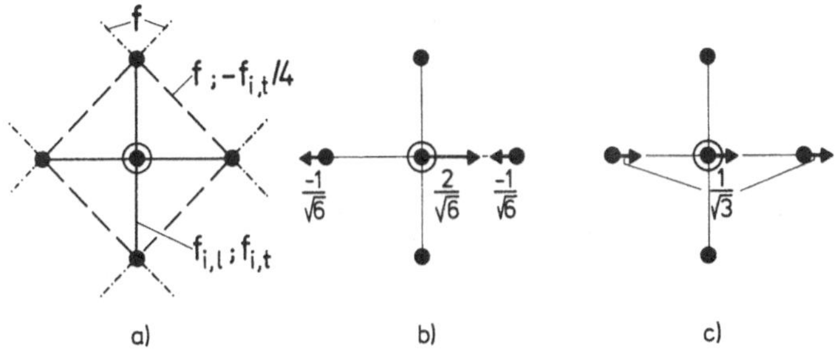


Fig. 6.6a-c. "Localized" and "resonant" modes of the octahedral self-interstitial
a) spring structure ($\overset{o}{M}\Omega^2_{max} = 8f$, $f_{i,l} = 10f$, $f_{i,t} = -f$),
b) high $\bar{\varphi} = 3f_{i,l} + (8f + 7f_{i,t})/6 \cong 30f$, $\bar{\Omega}/\Omega_{max} = \sqrt{30/8} \cong 2$,
c) low $\bar{\varphi} = (8f + 3f_{i,t})/3 \cong 5f/3$, $\bar{\Omega}/\Omega_{max} = \sqrt{5/24} \cong 0.4$

of the octahedron and f otherwise. Of course, the eigenvalues of the coupling can be evaluated exactly without great difficulties. But the trend is simpler to see if one uses approximate methods. The eigenvalues of $\varphi$ have an upper limit, $\varphi_{max}$, and a lower[8] limit, $\varphi_{min}$. For any displacement $\underline{s}$, satisfying the above condition of vanishing outside the octahedron, we must have $\varphi_{min} \leqslant (\underline{s},\varphi\underline{s}) = \bar{\varphi} \leqslant \varphi_{max}$ for normalized displacement, $(\underline{s},\underline{s}) = 1$. The idea is now to obtain large $\bar{\varphi}$ (still a lower bound to

7 Again it must be noted that the procedure is not compelling.

8 Stability requires $\varphi_{min} > 0$.

$\varphi_{max}$) by utilizing the strongest springs $f_{i,l}$, say $f_{i,l} = 10f$, $f_{i,t} = -f$, and to obtain small $\bar{\varphi}$ (still an upper bound to $\varphi_{min}$) by choosing displacements avoiding the strong springs. Fig. 6.6b shows a displacement where the springs $f_{i,l}$ enter fully; it is the mode which gives the highest frequency for a triatomic linear molecule; the corresponding $\bar{\Omega} = \sqrt{\bar{\varphi}/\overset{o}{M}}$ is rather larger than $\Omega_{max}$. Fig. 6.6c shows a displacement which does not utilize $f_{i,l}$ at all; it is the center of mass motion of a triatomic molecule; it contains the negative $f_{i,t}$ which tend to lower the value of $\bar{\varphi}$. One clearly sees the tendency to more localized and more resonant behaviour if the values of $f_{i,l}$ and $|f_{i,t}|$ become larger. For large enough negative $f_{i,t}$ the defect becomes unstable $\varphi_{min} \leqslant \bar{\varphi} < 0$.

Qualitatively the same arguments hold for the octahedral interstitial in bcc crystals.

### 6.3.5 The Tetrahedral Interstitial

In the same way one can treat the tetrahedral interstitial, but we are not going into details here. We will later only use it as a spring model of a tetrahedral defect (bond charge model) which does not change the two cubic shear moduli when present in small concentration.

### 6.3.6 The "Diatomic Defect"

A particularly simple model with spring and mass changes is the "diatomic defect", illustrated in Fig. 6.7. One (additional) 2nd neighbour longitudinal spring[9] $f_d$ in a fcc lattice is introduced and the two masses connected by $f_d$ are changed to $M_d$. This defect is highly symmetrical; it can contain resonant and localized modes si-

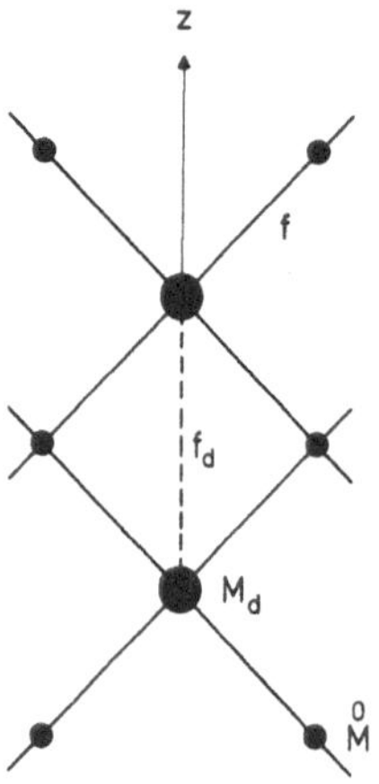


Fig. 6.7. Diatomic defect

9 Springs in <110> and <111> directions are equally simple.

multaneously, and its mathematical treatment is extremely simple. It is therefore a lucid model to demonstrate the behaviour of defects with the least mathematical effort. This defect has tetragonal symmetry; it has three equivalent orientations parallel to the cube edges and resembles in its behaviour a ⟨100⟩ self-interstitial or a diatomic molecule embedded in the lattice. As in Section 6.3.4, one can get an estimate of the frequencies by letting the molecule move in the fixed lattice. The modes are the same as in Appendix D 3. One obtains for the "oscillatory" mode $\Omega_o^2 = (2f_d + 4f)/M_d$, and for the five translational-rotational modes $\Omega_{t,r}^2 = 4f/M_d$. Therefore, one expects a resonant mode for $M_d \gg \mathring{M}$ and a localized mode for $f_d/M_d \gg 8f/\mathring{M}$; both conditions can be satisfied simultaneously.

## 6.4 Scattering by Defects

After having discussed Green's functions so extensively, the scattering of lattice waves by defects is a simple problem. The discrete lattice structure permits a treatment which is more transparent than that in a continuum, and one can readily find exactly solvable examples.[10] The language one uses in defect physics is the language of scattering theory. Therefore, knowledge of scattering theory is indispensable in discussing defects, if only to understand the more general meaning of new concepts. We will treat only stationary scattering (displacements $\propto \exp(-i\omega t)$), and we will discuss only the simplest cases for illustration. At first, we will avoid any change of degrees of freedom, i.e., we have in mind simple mass and spring changes as they occur for substitutional defects.

### 6.4.1 The Scattering Problem and Its Formal Solution

Under stationary conditions displacements and forces are periodic in time and carry the factor $\exp(-i\omega t)$. The equation of motion becomes

$$(\Phi - M\omega^2)\underline{s} = [\mathring{\Phi} - \mathring{M}\omega^2 + \chi(\omega)]\underline{s} = \underline{F}\ , \qquad \chi = \psi - m\omega^2\ , \tag{6.6}$$

with the solution $\underline{s} = G\underline{F}$.

The defect G can be expanded in powers of $\mathring{G}$, as in (6.5b),

$$G = \frac{1}{\Phi - M(\omega + i\eta)^2} = \mathring{G}\frac{1}{1 + \chi\mathring{G}} = \frac{1}{1 + \mathring{G}\chi}\mathring{G} = \mathring{G} - \mathring{G}\chi\mathring{G} + \dots\ , \tag{6.7}$$

$$G = \mathring{G} - \mathring{G}t\mathring{G} \quad \text{with the t-matrix} \quad t = \chi\frac{1}{1 + \mathring{G}\chi} = \frac{1}{1 + \chi\mathring{G}}\chi = \chi - \chi\mathring{G}\chi + \dots\ . \tag{6.8}$$

[10] Solvable in terms of Green's functions, which have to be obtained numerically.

In scattering, one considers a lattice wave,

$$\overset{i}{\underline{s}}{}^{\underline{m}} = \frac{1}{\sqrt{V_B}} \underline{e}(\underline{k}_i\sigma_i)\, e^{i[\underline{k}_i\underline{R}^{\underline{m}} - \Omega(\underline{k}_i\sigma_i)t]} \text{, the incoming or initial wave,} \tag{6.9}$$

which is a stationary solution in the perfect lattice. If a defect is present $\overset{i}{\underline{s}}$ is no longer a stationary solution. The incoming wave rather induces a stationary solution, which is a superposition of the incoming wave and a scattered outgoing wave,

$$\underline{s} = \overset{i}{\underline{s}} + \overset{s}{\underline{s}} \,, \tag{6.10}$$

which must obey (6.6). Because $(\overset{o}{\Phi} - \overset{o}{M}\omega^2)\overset{i}{\underline{s}} = 0$, one has

$$(\Phi - M\omega^2)\overset{s}{\underline{s}} = -\chi\overset{i}{\underline{s}} \text{, or} \tag{6.11}$$

$$\overset{s}{\underline{s}} = -G\chi\overset{i}{\underline{s}} = -\overset{o}{G}t\overset{i}{\underline{s}} \,, \quad \underline{s} = (1 - \overset{o}{G}t)\underline{s} = \frac{1}{1 + \overset{o}{G}\chi}\overset{i}{\underline{s}} \,. \tag{6.11a}$$

The quantities $\overset{o}{G}$ and $\overset{i}{\underline{s}}$ belong to the perfect crystal; the defect causes transitions between stationary states of the perfect crystal; these transitions are represented by the "transition"-matrix t in (6.11a). If the perturbation is small, one obtains

$$t \cong \chi \,, \quad \text{Born's approximation.} \tag{6.12}$$

The scattered wave can be viewed as produced by forces $\overset{i}{\underline{F}} = -t\overset{i}{\underline{s}}$ in the perfect lattice, i.e., by a kind of Kanzaki forces induced by $\overset{i}{\underline{s}}$ through the defect. The use of the *retarded* G which has been tacitly assumed corresponds to requiring an *outgoing* scattered wave with the energy current pointing outwards and not towards the scattering center.

The only defect property which enters the scattering is the t-matrix. The calculation is simple in principle. One can define a defect space by a projector P which encompasses all atomic coordinates influenced by the defect, e.g., one site for the isotopic defect, two sites for the diatomic defect and so forth. The perturbation operates only in this space, $P\chi P = \chi$, and it is

$$t = \chi\frac{1}{1 + \overset{o}{G}_P\chi} \,, \quad \overset{o}{G}_P = P\overset{o}{G}P \,, \tag{6.13}$$

i.e., only the components of $\overset{o}{G}$ in defect space are used. To obtain t one must calculate only the inverse of $P + P\overset{o}{G}P\chi$, which can be done directly if the number of defect coordinates is not too large. This sort of calculation does not give any physical insight. It is much more lucid to represent t by its eigenvalues and eigenvectors,

$$t = \sum_{\nu} |\nu\rangle \overset{\nu}{t} \langle\nu| , \quad \overset{s}{\underline{s}} = -\sum_{\nu} \overset{o}{G} |\nu\rangle \overset{\nu}{t} \langle\nu|\overset{i}{\underline{s}}\rangle , \tag{6.13a}$$

which exhibit the symmetries and the physics more clearly. For simple defects this is quite easy and we will demonstrate it by use of some examples.

### 6.4.2 The t-Matrix for Simple Defects

For the *isotopic defect*, P is the projection onto the defect site, $\chi = -m\omega^2 P$, and in cubic lattices $\overset{o}{G}_P = \overset{o}{G}{}^{(o)}$ is scalar (comp. the discussion in Sec. 3.5.3),

$$t = -\frac{m\omega^2}{1 - \overset{o}{A}m\omega^2} P = \overset{o}{t}P , \quad \overset{o}{A} = \overset{o}{G}{}^{(o)} ; \quad \overset{s}{\underline{s}}{}^{\underline{m}} = -\overset{o}{G}{}^{(\underline{m})} t \overset{i}{\underline{s}}{}^{(o)}$$
$$= -\overset{o}{t} \frac{e^{-i\omega t}}{V_B^{1/2}} \overset{o}{G}{}^{(\underline{m})}(\omega) \overset{i}{\underline{e}} . \tag{6.14}$$

The *diatomic defect* is also quite simple due to its high symmetry. The defect space, P, contains two atomic sites or six coordinates. The modes of the diatomic molecule discussed in Appendix D3 are are shown in Fig. 6.8: $|1\rangle$ oscillation, $|3,5\rangle$ rotation, $|2,4,6\rangle$ translation. These modes are eigenmodes of $\chi$ *and* $G_P = A$ with eigenvalues $\overset{\nu}{\chi} = \langle\nu|\chi|\nu\rangle$ and $\overset{\nu}{A} = \langle\nu|A|\nu\rangle$, indicated in Fig. 6.8; they are then also eigenmodes of t with $\overset{\nu}{t} = \overset{\nu}{\chi}/(1 + \overset{\nu\nu}{A\chi})$. For a pure spring change the defect space is only one-dimensional:

$$t = \overset{1}{t} |1\rangle\langle 1| \quad \text{with} \quad \overset{1}{t} = \frac{\overset{1}{\chi}}{1 + \overset{11}{A\chi}} , \quad \overset{1}{\chi} = 2f_d , \quad \overset{1}{A} = \overset{o}{G}{}^{(o)}(\omega) - \overset{o}{G}{}_1^{(200)}(\omega). \tag{6.15}$$

For an *octahedral bond charge* $\chi = \psi$ is a $\langle 100\rangle$-spring star containing three orthogonal, uncoupled springs $f_d = f_i/2$ between 2nd neighbours (Fig. 6.9):

$$\psi = 2f_d \left( |x\rangle\langle x| + |y\rangle\langle y| + |z\rangle\langle z| \right) = 2f_d P = f_i P. \tag{6.16}$$

The three states $|x,y,z\rangle$ define a three-dimensional defect space. These states can be adapted to cubic symmetry by considering deformations of the octahedron corresponding to the six basis strains $T(\nu = 1...6)$ of Section 4.1.3, $T(\nu)\underline{R}$, where $\underline{R}$ is restricted to the six octahedral sites. These deformations are shown in Fig. 6.10a. One sees that shears of the $\langle 100\rangle$-type do not utilize the defect springs at all and drop out. The three displacements $T(1,2,3)\underline{R}$ are radial; they possess the symmetries discussed in Section 4.4.2 and form an orthogonal basis for $\psi$ itself, which has also only radial displacements. Since

$$\langle T(\nu)\underline{R} \mid T(\nu)\underline{R}\rangle = [T^2(\nu)]_{ik} \sum_{\underline{m}}{}' X_i^{\underline{m}} X_k^{\underline{m}} = \underbrace{[T^2(\nu)]_{ii}}_{=1} \cdot \frac{Z_o}{3}\left(\frac{a}{2}\right)^2 ,$$

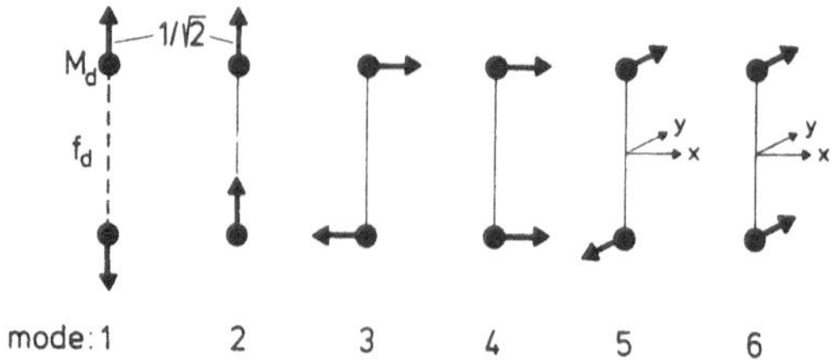


Fig. 6.8. Eigenmodes and eigenvalues of $\overset{\circ}{G}_P$, $\chi$ for the diatomic defect $\overset{\nu}{t} = \overset{\nu}{\chi}/(1+\overset{\nu\nu}{A\chi})$

| Mode no. | Eigenvalue of $\overset{\circ}{G}_P = A$ | Eigenvalue of $\chi$ |
|---|---|---|
| 1 | $\overset{1}{A} = \overset{\circ}{G}^{(0)} - \overset{\circ}{G}_l^{(200)}$ | $\overset{1}{\chi} = 2f_d - m\omega^2$ |
| 2 | $\overset{2}{A} = \overset{\circ}{G}^{(0)} + \overset{\circ}{G}_l^{(200)}$ | $\overset{2}{\chi} = -m\omega^2$ |
| 3,5 | $\overset{3}{A} = \overset{\circ}{G}^{(0)} - \overset{\circ}{G}_t^{(200)}$ | $\overset{2}{\chi}$ |
| 4,6 | $\overset{4}{A} = \overset{\circ}{G}^{(0)} + \overset{\circ}{G}_t^{(200)}$ | $\overset{2}{\chi}$ |

$\overset{\circ}{G}^{(0)}(\omega)$ is one diagonal element of the tensor $\overset{\circ}{G}^{(0)}$, $\overset{\circ}{G}^{(200)}$ refers to $\langle 100 \rangle$ neighbours; l,t refer to the longitudinal and transversal components

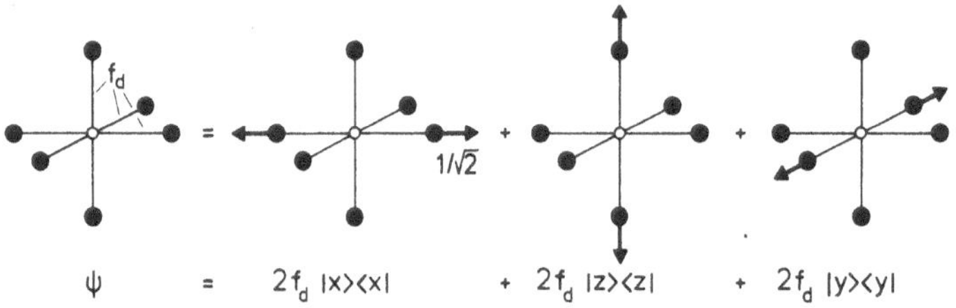


Fig. 6.9. The octahedral bond charge can be represented by the three constituent springs $2f_d$ parallel to the cube edges

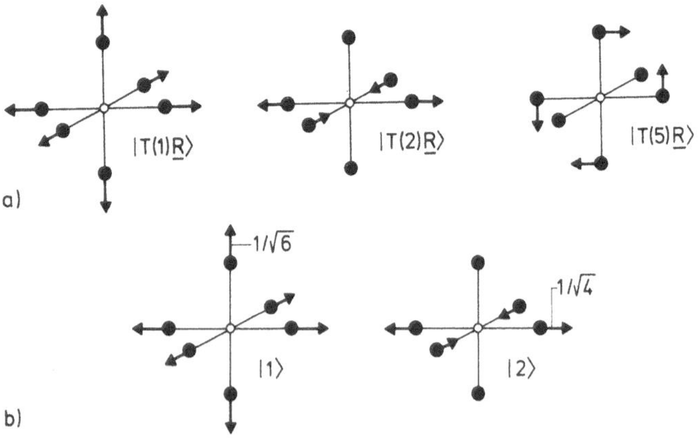


Fig. 6.10a and b. Deformations of the octahedron (a) and eigenstates of $\psi$, $\overset{\circ}{G}_P$ and t for the octahedral bond charge (b), $|\nu = 1,2,3\rangle = (\sqrt{2}/a)\,|T(\nu)\underline{R}\rangle$

where $Z_o = 6$ is the number of octahedral sites over which $\sum'$ extends, the properly normalized states adapted to cubic symmetry are

$$|\nu\rangle = \frac{\sqrt{2}}{a}|T(\nu)\underline{R}\rangle \quad \text{(Fig. 6.10b)} \; . \tag{6.16a}$$

Further

$$\psi = f_i \sum_{\nu=1}^{3} |\nu\rangle\langle\nu| = 2f_d \sum_{\nu=1}^{3} |\nu\rangle\langle\nu| \; . \tag{6.16b}$$

The states $|\nu\rangle$ have the symmetries discussed in Section 4.4.2 and $\overset{\circ}{G}_P = A$ has cubic symmetry; therefore due to symmetry $\langle\nu|A|\nu\rangle$ is diagonal with

$$\overset{1}{A} = \langle 1|A|1\rangle = \overset{\circ}{G}^{(0)} + 2\left(\overset{\circ}{G}{}_{t'}^{(101)} - \overset{\circ}{G}{}_{1}^{(101)}\right) - \overset{\circ}{G}{}_{1}^{(200)} \; , \tag{6.17a}$$

$$\overset{2}{A} = \overset{3}{A} = \langle 2|A|2\rangle = \overset{\circ}{G}^{(0)} + \overset{\circ}{G}{}_{1}^{(101)} - \overset{\circ}{G}{}_{t'}^{(101)} - \overset{\circ}{G}{}_{1}^{(200)} \; . \tag{6.17b}$$

Consequently,

$$t = \overset{1}{t}\,|1\rangle\langle 1| + \overset{2}{t}(|2\rangle\langle 2| + |3\rangle\langle 3|) \quad \text{with } \overset{1}{t} = \frac{2f_d}{1 + \overset{1}{A}2f_d}, \quad \overset{2}{t} = \frac{2f_d}{1 + \overset{2}{A}2f_d} = \overset{3}{t}. \tag{6.18}$$

A *substitutional atom with octahedral symmetry* is only slightly more difficult. The spring star has the symmetry of Fig. 6.9 and contains six springs $f_d$ from one lattice site to its six 2nd neighbours at distance a. As for the octahedral bond charge, the springs in x,y and z direction are not coupled because they are longitudinal. One pair of springs can be represented by two states $|x,x'\rangle$ with eigenvalues $f_d$ and $3f_d$ (Fig. 6.11):

$$\psi = f_d(|x\rangle\langle x| + |y\rangle\langle y| + |z\rangle\langle z|) + 3f_d(|x'\rangle\langle x'| + |y'\rangle\langle y'| + |z'\rangle\langle z'|) \; . \tag{6.19}$$

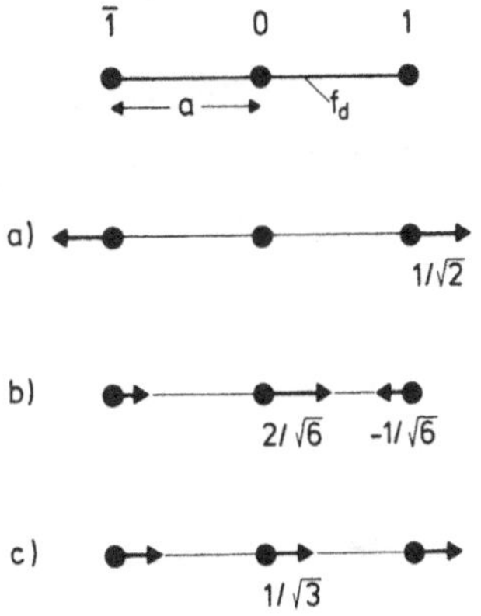


Fig. 6.11a-c. States of a spring pair $(s_x^1, s_x^o, s_x^1)$

a) $|x\rangle = \frac{1}{\sqrt{2}}(1,0,1)$ , $\psi|x\rangle = f_d|x\rangle$ ,

b) $|x'\rangle = \frac{1}{\sqrt{6}}(\bar{1},2,\bar{1})$ , $\psi|x'\rangle = 3f_d|x'\rangle$ ,

c) $|\xi\rangle = \frac{1}{\sqrt{3}}(1,1,1)$ , $\psi|\xi\rangle = 0$

Without mass change, $\chi = \psi$, the defect space is six-dimensional; P is the projector onto the six states above with $\psi \neq 0$. The even states $|x,y,z\rangle$ can be rearranged into three states $|\nu = 1,2,3\rangle$ corresponding to Fig. 6.10. With the three states $|x',y',z'\rangle$, they are eigenstates to $\psi$, $\overset{o}{G}_P = A$ and t. The eigenvalues of A correspond to those of (6.17). The three states $|x',y',z'\rangle = |4,5,6\rangle$ are degenerate; therefore

$$t = \overset{1}{t}\,|1\rangle\langle 1| + \overset{2}{t}(|2\rangle\langle 2| + |3\rangle\langle 3|) + \overset{4}{t}(|4\rangle\langle 4| + |5\rangle\langle 5| + |6\rangle\langle 6|)\ ,$$

$$\overset{1,2}{t} \text{ as in (6.18) with } f_d \text{ instead of } 2f_d; \tag{6.20}$$

$$\overset{4}{t} = \frac{3f_d}{1 + \overset{4}{A}3f_d}\ , \qquad \overset{4}{A} = \overset{o}{G}{}^{(0)} - \frac{4}{3}\overset{o}{G}{}_1^{(400)} + \frac{1}{3}\overset{o}{G}{}_1^{(200)}\ .$$

With mass change, $\chi = \psi - m\omega^2$, the defect space must be enlarged by the three states of Fig. 6.11c, $|\xi,\eta,\zeta\rangle = |7,8,9\rangle$ in order to include fully the motion of the substitutional atom, e.g., $|x_s\rangle = (0,1,0)$ and $P_s^x = |x_s\rangle\langle x_s|$. If $P_{a,b,c}$ refer to the states of Fig. 6.11a,b,c, e.g., $P_a = (|x\rangle\langle x| + |y\rangle\langle y| + |z\rangle\langle z|)$, the "fraction" of $P_s$-space ($P_s = P_s^x + P_s^y + P_s^z$) in $P_a + P_b$ is given by $P_s(P_a + P_b)P_s = P_sP_bP_s = 2P_s/3$; however, since $P_s(P_a + P_b + P_c)P_s = P_s$ the substitutional coordinate is fully contained in the three sets of states corresponding to Fig. 6.11a,b,c. The defect space $P = P_a + P_b + P_c$ is then nine-dimensional. The first three states $|\nu = 1,2,3\rangle$, where the substitutional atom is at rest, are again eigenstates of $\chi$, A, t; the remaining states $|4...9\rangle$ are not, they are eigenstates of $\psi$ but not of m or $\overset{o}{G}_P$. Still the three two-dimensional spaces built by $|x',\xi\rangle$, $|y',\eta\rangle$, $|z',\zeta\rangle$ are separated on account of symmetry. Therefore only the inverse of a two-by-two matrix is needed to calculate t.

The *substitutional atom of cubic symmetry* is represented by a spring star containing twelve longitudinal springs $f_d$ from the star center to the twelve nearest neighbours in $\langle 110\rangle$ directions:

$$\psi = f_d\hat{\psi} \text{ according to (6.4)}\ . \tag{6.21}$$

Without mass change the defect space contains six spring pairs[11] or twelve states, six even states and six odd states. The six even states can again be obtained from basis deformations of the spring star, $T(\nu = 1...6)\underline{R}$. Here only the displacement $T(1)\underline{R}$ is radial, $|1\rangle = 1/(\sqrt{2}a)\,|T(1)\underline{R}\rangle$, Fig. 6.12a. The other deformations do not lead to purely radial displacements. The states $|2...6\rangle$ can be obtained from the corresponding deformations by extracting only the radial components. This process of extraction has cubic symmetry; therefore, the states $|\nu\rangle$ possess the same sym-

[11] The non-orthogonal pairs are not coupled.

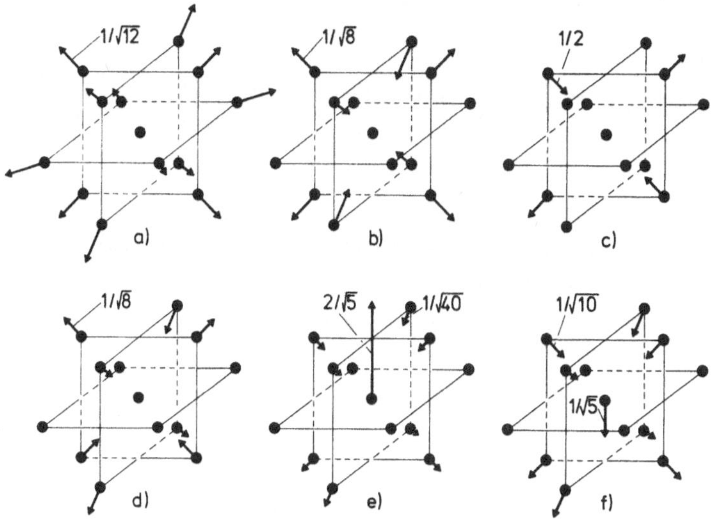


Fig. 6.12a-f. States of the cubic substitutional.
a) The normalized even state $|1\rangle = 1/(\sqrt{2}a)\ |T(1)\underline{R}\rangle$ ,
b) the normalized even state $|2\rangle$ with the symmetry of $|T(2)\underline{R}\rangle$ ,
c) the normalized even state $|5\rangle$ with the symmetry of $|T(5)\underline{R}\rangle$ ,
d) the normalized odd state $|7\rangle$; z is distinguished; substitutional at rest,
e) the normalized odd state $|10\rangle$; z is distinguished; substitutional is displaced,
f) the normalized state $|13\rangle$; z is distinguished; substitutional is displaced; springs are *not* stressed

metries as $T(\nu)\underline{R}$. Fig. 6.12b shows $|2\rangle$ and Fig. 6.12c illustrates $|5\rangle$. Then we have

$$\overset{\nu}{\psi} = f_d\ , \qquad \nu = 1,\ldots,6$$

$$\begin{aligned}
\overset{1}{A} &= G^{(0)} + 2\left(G_{zz}^{(110)} - G_{xy}^{(110)}\right) + \left(G_{yy}^{(200)} - G_{xx}^{(200)}\right) \\
&\quad - 2\left(G_{xx}^{(211)} + G_{yz}^{(211)} + 2G_{xy}^{(211)}\right) - \left(G_{xx}^{(220)} + G_{xy}^{(220)}\right) \\
\overset{2}{A} = \overset{3}{A} &= G^{(0)} - \left(G_{zz}^{(110)} - G_{xy}^{(110)}\right) + \left(G_{yy}^{(200)} - G_{xx}^{(200)}\right) \\
&\quad + \left(G_{xx}^{(211)} + G_{yz}^{(211)} + 2G_{xy}^{(211)}\right) - \left(G_{xx}^{(220)} + G_{xy}^{(220)}\right) \\
\overset{4}{A} = \overset{5}{A} = \overset{6}{A} &= G^{(0)} - \left(G_{yy}^{(200)} - G_{xx}^{(200)}\right) - \left(G_{xx}^{(220)} + G_{xy}^{(220)}\right) .
\end{aligned} \tag{6.21a}$$

Three of the odd states, $|7,8,9\rangle = |z,y,x\rangle$ have vanishing displacement of the substitutional atom. State $|\nu = 7\rangle$ is pictured in Fig. 6.12d; here the z-plane is distinguished by vanishing displacements, in $|8\rangle$ the y-plane and $|9\rangle$ the x-plane is distinguished,

$$\overset{\nu}{\psi} = f_d\ , \quad \nu = 7,8,9\ ,$$

$$\overset{7}{A} = \overset{8}{A} = \overset{9}{A} = \overset{\circ}{G}^{(0)} + \left(\overset{\circ}{G}{}^{(110)}_{xy} - \overset{\circ}{G}{}^{(110)}_{zz}\right) + \left(\overset{\circ}{G}{}^{(211)}_{xx} + \overset{\circ}{G}{}^{(211)}_{yz} + 2\overset{\circ}{G}{}^{(211)}_{xy}\right) + \left(\overset{\circ}{G}{}^{(220)}_{xx} + \overset{\circ}{G}{}^{(220)}_{xy}\right)\ . \tag{6.21b}$$

The remaining three of the odd states, $|\zeta,\eta,\xi\rangle = |10,11,12\rangle$, possess central displacements. Fig. 6.12e shows $|\zeta\rangle = |10\rangle$. One can easily establish that the eigenvalue is $5f_d$ because the projection of the center displacement on each spring is $\pm 4/\sqrt{40}$ :

$$\overset{\nu}{\psi} = 5f_d\ , \quad \nu = 10,11,12\ ,$$

$$\overset{10}{A} = \overset{11}{A} = \overset{12}{A} = \overset{\circ}{G}^{(0)} + \frac{1}{5}\left[\left(\overset{\circ}{G}{}^{(110)}_{zz} - 8\overset{\circ}{G}{}^{(110)}_{xx} - 9\overset{\circ}{G}{}^{(110)}_{xy}\right) + \left(\overset{\circ}{G}{}^{(211)}_{xx} + \overset{\circ}{G}{}^{(211)}_{yz} + 2\overset{\circ}{G}{}^{(211)}_{xy}\right) + \left(\overset{\circ}{G}{}^{(220)}_{xx} + \overset{\circ}{G}{}^{(220)}_{xy}\right)\right]\ . \tag{6.21c}$$

The simple vacancy model of Section 6.3.2 is given by $\psi = -f\hat{\psi}$ and can be represented by the above twelve states.

With mass change one needs three additional states: $|\zeta',\eta',\xi'\rangle = |13,14,15\rangle$ with $\psi = 0$ of which $|\zeta'\rangle = |13\rangle$ is shown in Fig. 6.12f. The full solution of the substitutional problem requires then the inversion of a two-by-two matrix, e.g., in the $|\zeta,\zeta'\rangle$-space. The substitutional degree of freedom is fully contained in the states $|10\rangle$ to $|15\rangle$ (comp. the substitutional of octahedral symmetry).

### 6.4.3 The Emission of Energy by the Scattered Wave

The energy emitted by the scattered wave is needed to define the scatttering cross section. It is easily calculated if one realizes that the scattered wave is produced by an effective force $\overset{i}{\underline{F}} = -t\overset{i}{\underline{s}}$, acting in the perfect lattice and restricted to the defect space. If one discusses physical quantities such as energy emission one must have real displacements $\overset{i}{\underline{s}} \to \overset{i}{\underline{s}} + \overset{i}{\underline{s}}{}^*$, $\overset{im}{\underline{s}} \to 2\overset{i}{\underline{e}}\, V_B^{-1/2} \cos(\underline{k}_i\underline{R}^m - \omega t)$ and real forces[12],

$$\overset{i}{\underline{F}} = \chi \frac{1}{1 + \overset{\circ}{G}(\omega)\chi}\overset{i}{\underline{s}} = -t(\omega)\overset{i}{\underline{s}}\ , \quad \overset{i}{\underline{F}} + \overset{i}{\underline{F}}{}^* = -t(\omega)\overset{i}{\underline{s}} - t(-\omega)\overset{i}{\underline{s}}{}^*\ .$$

We discuss first the emission of energy per unit time, J, which is caused by a force $\underline{F} + \underline{F}^*$ in the perfect lattice ($\underline{F} \propto e^{-i\omega t}$). Here

$$\underline{s} = \overset{\circ}{G}(\omega)\underline{F} + \overset{\circ}{G}(-\omega)\underline{F}^* = \overset{\circ}{G}(\omega)\underline{F} + \text{c.c.}\ , \tag{6.22a}$$

---

12 $\overset{i}{\underline{s}}{}^*$ carries the factor $e^{i\omega t}$ and corresponds to $\overset{\circ}{G}(-\omega) = \overset{\circ}{G}{}^*(\omega)$; $t(-\omega) = t^*(\omega)$ because $\chi$ is real.

$$\dot{\underline{s}} = -i\omega(\overset{\circ}{G}\underline{F} - \text{c.c.}) = 2\omega \, \text{Im}\{\overset{\circ}{G}\underline{F}\} \, . \tag{6.22b}$$

The energy emitted per unit time, the power, is given by

$$J = (\dot{\underline{s}}, \underline{F} + \underline{F}^*) = \omega\left(\frac{\overset{\circ}{G}\underline{F} - \text{c.c.}}{i}, (\underline{F} + \text{c.c.})\right) \, . \tag{6.23}$$

If one averages over one period, $(\omega/2\pi)\cdot\int_{t_o}^{t_o+\frac{2\pi}{\omega}} dt\ldots$, the average energy emitted per unit time, $\bar{J}$, becomes time independent. Terms such as $(\overset{\circ}{G}\underline{F}^*, \underline{F})$ carry the factor $e^{-2i\omega t}$ and vanish upon averaging. The final result is

$$\bar{J} = 2\omega(\underline{F}, \text{Im}\{\overset{\circ}{G}\}\underline{F}) = 2\omega \, \langle\underline{F}|\text{Im}\{\overset{\circ}{G}\}|\underline{F}\rangle \, , \tag{6.24}$$

and if one uses $\overset{\circ}{G}(\omega) = \sum_{\sigma'}\int d\underline{k}' \, |\underline{k}'\sigma'\rangle \, \tilde{\overset{\circ}{G}}(\underline{k}'\sigma', \omega) \, \langle\underline{k}'\sigma'|$ ,

$$\begin{aligned} \bar{J} &= \sum_{\sigma'}\int d\underline{k}' \, |\langle\underline{k}'\sigma'|\underline{F}\rangle|^2 \, 2\omega \, \text{Im}\{\tilde{\overset{\circ}{G}}(\underline{k}'\sigma', \omega)\} \\ &= \sum_{\sigma'}\int d\underline{k}' \, |\langle\underline{k}'\sigma'|\underline{F}\rangle|^2 \, \frac{2\pi|\omega|}{\overset{\circ}{M}} \, \delta\left(\overset{\circ}{\Omega}{}^2(\underline{k}'\sigma') - \omega^2\right) \, . \end{aligned} \tag{6.24a}$$

Consequently,

$$d\bar{J} = |\langle\underline{k}'\sigma'|\underline{F}\rangle|^2 \, \frac{2\pi|\omega|}{\overset{\circ}{M}} \, \delta\left(\overset{\circ}{\Omega}{}^2(\underline{k}'\sigma') - \omega^2\right) d\underline{k}' \tag{6.25}$$

is the energy per unit time emitted as lattice waves in the interval $d\underline{k}'$ about $\underline{k}'$ and with polarization $\sigma'$, denoted by $(\underline{k}'\sigma', d\underline{k}')$.

### 6.4.4 The Differential Cross Section

From (6.25) one obtains the differential cross section $dq(\underline{k}\sigma \to \underline{k}'\sigma')$ for scattering of an incoming wave $\underline{k}, \sigma = \underline{k}_i, \sigma_i$ into lattice waves $(\underline{k}'\sigma', d\underline{k}')$ in the following way. The incoming wave $|\underline{k}\sigma\rangle$ produces an effective force $F = -t\,|\underline{k}\sigma\rangle$ and gives an emission

$$d\bar{J} = |\langle\underline{k}'\sigma'|t|\underline{k}\sigma\rangle|^2 \, \frac{2\pi|\omega|}{\overset{\circ}{M}} \, \delta\left(\overset{\circ}{\Omega}{}^2(\underline{k}'\sigma') - \omega^2\right) \, , \quad \omega^2 = \overset{\circ}{\Omega}{}^2(\underline{k}\sigma) \, , \tag{6.26}$$

| | | |
|---|---|---|
| due to the incoming wave | $2\underline{e}(\underline{k}\sigma)\cos(\underline{k}\underline{R}^{\underline{m}} - \omega t)/V_B^{1/2}$ , | (6.27a) |
| with average kinetic energy per atom (equal potential energy) | $\frac{\overset{\circ}{M}}{2}\left(\frac{2\omega}{V_B^{1/2}}\right)^2 \overline{\sin^2\omega t} = \frac{\overset{\circ}{M}\omega^2}{V_B}$ , | (6.27b) |
| average energy density | $\frac{2\overset{\circ}{M}\omega^2}{V_B V_c}$ , and | (6.27c) |

average energy current density $\qquad \frac{2\overset{\circ}{M}\omega^2}{V_B V_c} v \;, \quad v = \left| \frac{\partial\overset{\circ}{\Omega}(\underline{k},\sigma)}{\partial \underline{k}} \right| \;.$ (6.27d)

By defintion the energy per unit time flowing through the differential cross section (a small area with normal ∥ to the current) $2\overset{\circ}{M}\omega^2 v\, dq(\underline{k}\sigma \to \underline{k}'\sigma')/V_B V_c$ equals the energy per unit time $dJ$ emitted into $(\underline{k}'\sigma', d\underline{k}')$,

$$dq(\underline{k}\sigma \to \underline{k}'\sigma') = \frac{V_B V_c}{2\overset{\circ}{M}\omega^2 v} d\bar{J} = \frac{\pi V_B V_c}{\overset{\circ}{M}|\omega| v} |\langle \underline{k}'\sigma'|t|\underline{k}\sigma\rangle|^2 \, \delta\left(\overset{\circ}{\Omega}{}^2(\underline{k}'\sigma') - \omega^2\right) d\underline{k}' \;. \tag{6.28}$$

The total cross section is

$$q = \sum_{\sigma'} \int dq = \frac{V_B V_c}{2\overset{\circ}{M}\omega^2 v} \bar{J} = \sum_{\sigma'} \int d\underline{k}' \frac{\pi V_B V_c}{\overset{\circ}{M}|\omega| v} |\langle \underline{k}'\sigma'|t|\underline{k}\sigma\rangle|^2 \, \delta\left(\overset{\circ}{\Omega}{}^2(\underline{k}'\sigma') - \omega^2\right) . \tag{6.29}$$

The matrix element,

$$\langle \underline{k}'\sigma'|t|\underline{k}\sigma\rangle = \hat{t} = e^{-i\underline{k}'\underline{R}^m} e_i(\underline{k}'\sigma') t_{il}^{mn} e_l(\underline{k}\sigma)\, e^{i\underline{k}\underline{R}^n}/V_B = \tilde{\tilde{t}}/V_B \;, \tag{6.30}$$

is usually not calculated by employing the properly normalized states $|\underline{k}\sigma\rangle$ resulting in $\hat{t}$, but by using just the plane waves $V_B^{1/2}\;|\underline{k}\sigma\rangle$ resulting in $\tilde{\tilde{t}}$:

$$dq = \frac{\pi V_B V_c}{\overset{\circ}{M}{}^2 |\omega| v} |\hat{t}|^2 \, \delta\left(\overset{\circ}{\Omega}{}^2(\underline{k}'\sigma') - \omega^2\right) d\underline{k}' = \frac{\pi V_c}{\overset{\circ}{M}{}^2 |\omega| v} |\tilde{\tilde{t}}|^2 \, \delta\left(\overset{\circ}{\Omega}{}^2(\underline{k}'\sigma') - \omega^2\right) \frac{d\underline{k}'}{V_B} . \tag{6.30a}$$

### 6.4.5 Cross Sections and Spectra for the Isotopic Defect

The evaluation of (6.30a) is simplest for the isotopic defect,

$$dq = \frac{\pi V_c}{\overset{\circ}{M}{}^2 |\omega| v} |\overset{\circ}{t}|^2 (\underline{e}',\underline{e})^2 \, \delta\left(\overset{\circ}{\Omega}{}^2(\underline{k}'\omega') - \omega^2\right) \frac{d\underline{k}'}{V_B} \;, \tag{6.31}$$

where $\overset{\circ}{t} = m\omega^2/[\overset{\circ}{G}{}^{(0)}(\omega) m\omega^2 - 1]$ does not depend on $\underline{k}',\sigma'$. For given frequency $\omega$, the scattering is determined by the polarizations $\underline{e}' = \underline{e}(\underline{k}'\sigma')$, $\underline{e} = \underline{e}(\underline{k}\sigma)$ and the $\delta$-function. Fig. 6.13 illustrates this for a particularly simple case. The total cross section, which implies $\underline{k}'$-integration over the Brillouin-Zone, is even simpler because of cubic symmetry. The quantity $\overset{\circ}{\Omega}{}^2(\underline{k}'\sigma')$ is invariant under application of all 48 cubic symmetry operations to $\underline{k}'$. Therefore, only the average of $(\underline{e}',\underline{e})^2 = e_i e_i' e_l' e_l$ over all symmetry operations enters the total cross section; the average over $e_i' e_l'$ is $\delta_{il}/3$. Consequently,

$$q(m,v,\omega > 0) = \frac{\pi V_c}{\overset{\circ}{M}{}^2 \omega v} |\overset{\circ}{t}|^2 \sum_{\sigma'} \int \frac{d\underline{k}'}{3V_B} \delta\left(\overset{\circ}{\Omega}{}^2(\underline{k}'\sigma') - \omega^2\right) = \frac{\pi V_c}{\overset{\circ}{M}{}^2 \omega v} |\overset{\circ}{t}|^2 \, Z(\omega^2) = \tag{6.31a}$$

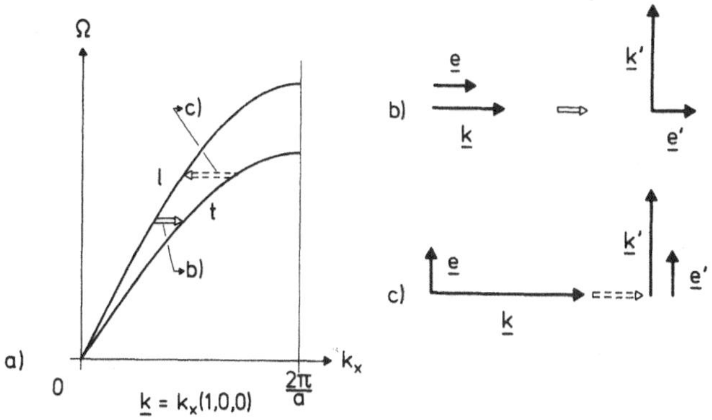


Fig. 6.13a-c. Scattering by an isotopic defect from $\underline{k}$ = [100] to $\underline{k}'$ = [001]. a) Dispersion curves in ⟨100⟩, b) scattering l → t under conservation of frequency; scattering l → t vanishes if ω becomes larger than the maximum transversal frequency; scattering l → l vanishes for all ω because $(\underline{e}_l, \underline{e}_l') = 0$, c) scattering t → l occurs for all possible transversal ω

$$= \underbrace{V_c \frac{\omega}{v}\left(\frac{m}{M}\right)^2 \omega^2 Z(\omega^2)}_{q_{BA}} \left|\frac{1}{1 - m\omega^2 \mathring{G}^{(0)}}\right|^2 \qquad (6.31a)$$

where Z is the spectrum (comp. Sec. 3.5.3) and $q_{BA}$ is Born's approximation. The dependence on the group velocity v of the incoming wave is trivial; divergencies can occur in directions of high symmetry where v vanishes at the zone boundary (Fig. 3.15). Except for that, $q_{BA}$ starts with $\omega^2$ for small ω, which is exact, and a typical value in the center of the spectrum is of the order $a^2(m/M_o)^2$. Born's approximation is valid for $|m\omega^2/\mathring{M}\Omega^2_{max}| \ll 1$. The factor $(1 - m\omega^2 G^{(0)})^{-1}$ has already been discussed in Section 3.5.3, where it was shown that $G^{(0)} = \mathring{G}^{(0)}/(1 - m\omega^2\mathring{G}^{(0)})$ for the isotopic defect. In Section 3.5.4 it was shown that for $m \gg \mathring{M}$ a resonance occurs, $1 \cong m\omega_R^2 \operatorname{Re}\{G^{(0)}(\omega_R)\}$, and that for $m/\mathring{M} = \mu = (M_d - \mathring{M})/\mathring{M} < 0$ a localized state occurs, $\omega_L > \Omega_{max}$, if $1 = m\omega_L^2\mathring{G}^{(0)}(\omega_L)$ (Section 3.5.3). Therefore, the details of the ω-dependence of q depend mainly on $\mathring{t}(m,\omega)$.

To get some feeling for this dependence we discuss the dimensionless quantity $\mathring{t}/\mathring{M}\Omega^2_{max} = \vartheta$ for the regular spectrum of Table 3.4, for which

$$\vartheta = \vartheta_1 + i\vartheta_2 = \frac{\mathring{t}}{\mathring{M}\Omega^2_{max}} = -\frac{\mu\xi}{1 - 4\mu\xi(1-2\xi) - i8\mu\xi\sqrt{\xi(1-\xi)}}\,,$$
$$\mu = \frac{m}{\mathring{M}}\,, \quad \xi = \frac{\omega^2}{\Omega^2_{max}}\,, \qquad (6.32)$$

*Explanation to Fig. 6.14a and b:*

$\mu = -1$: isotopic defect with zero mass and infinite $\omega_L$;

$\vartheta_1 = [\xi + 4\xi^2(1-2\xi)]/(1+8\xi)$,

$\vartheta_2 = -8\xi^2\sqrt{\xi(1-\xi)}/(1+8\xi)$,

$|\vartheta|^2 = \xi^2/(1+8\xi)$.

$\mu = -1/4$: the localized frequency $\omega_L$ agrees with $\Omega_{max}$, $\vartheta_2$ diverges for $\omega = \Omega_{max}$ or $\xi = 1$;

$\vartheta_1 = \xi(1+2\xi)/[4(1+3\xi)]$,

$\vartheta_2 = -\xi^{5/2}/[2\sqrt{1-\xi}(1+3\xi)]$,

$|\vartheta|^2 = \xi^2/[16(1-\xi)(1+3\xi)]$.

$\mu = 3$: corresponding to Ag impurity in Al.

$\mu = 10$: for $\mu \gg 1$ the denominator of (6.32a,b) has a sharp minimum at a resonance value $\xi_R = 1/(4\mu) \ll 1$; near the resonance

$$\vartheta_1 \cong \frac{1-\xi/\xi_R}{4[(1-\xi/\xi_R)^2 + 4\xi_R]} ,$$

$$\vartheta_2 \cong \frac{\sqrt{\xi_R}/2}{4[(1-\xi/\xi_R)^2 + 4\xi_R]} ,$$

$|\vartheta|^2 \cong (1/16)[(1-\xi/\xi_R)^2 + 4\xi_R]^{-1}$; the width of the resonance is $\delta\xi = 2\xi_R^{3/2}\ 1/(4\mu^{3/2})$ and the height is $\mu/16$

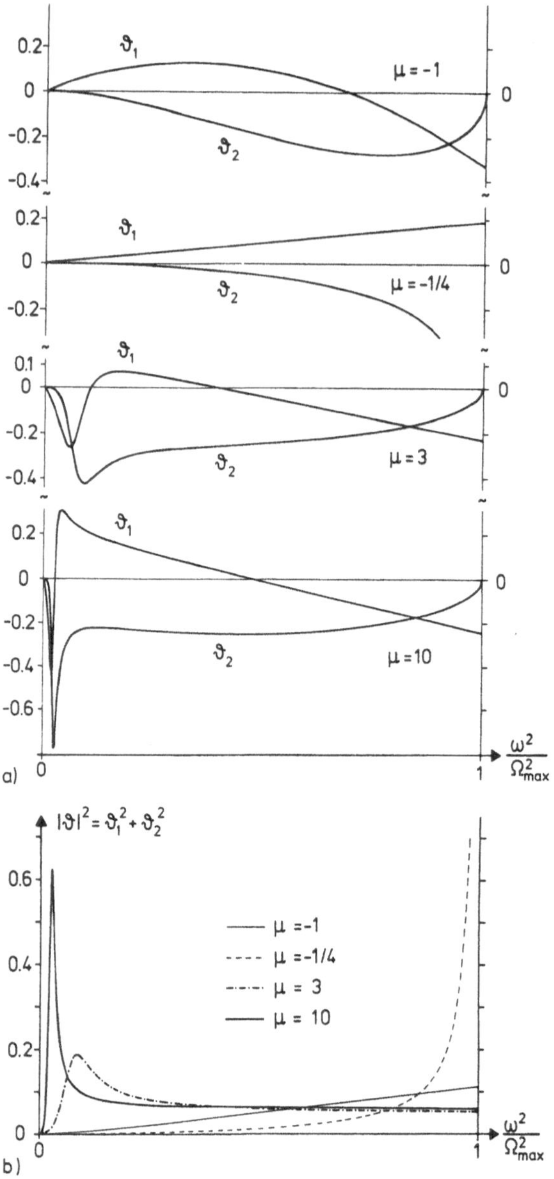


Fig. 6.14a and b. The t-matrix of an isotopic defect for several values of $\mu = m/\mathring{M}$, calculated with the regular spectrum of Table 3.4. Plotted are $\vartheta_{1/2}$, the real and imaginary part of $\mathring{t}/(\mathring{M}\omega^2)$, (a), and $|\vartheta|^2 = \vartheta_1^2 + \vartheta_2^2$, (b), as function of $\xi = \omega^2/\Omega^2_{max}$

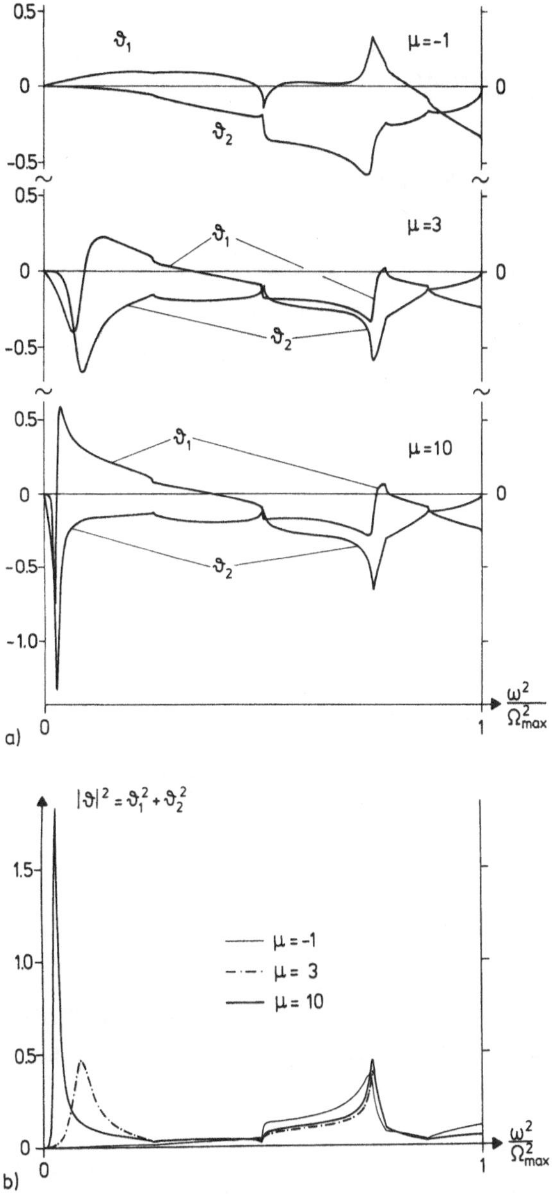


Fig. 6.15a and b. The t-matrix of an isotopic defect for the same values of $\mu = m/\overset{\circ}{M}$ as in Fig. 6.14, but calculated (numerically) with the Green's function for 1st neighbour longitudinal coupling in a fcc lattice (model of Fig. 3.15c). Plotted are $\vartheta_{1/2}$, the real and imaginary part of $\overset{\circ}{t}/(\overset{\circ}{M}\omega^2)$, (a), and $|\vartheta|^2 = \vartheta_1^2 + \vartheta_2^2$, (b), as function of $\xi = \omega^2/\Omega^2_{max}$

$$\left.\begin{matrix}\vartheta_1 \\ \\ \vartheta_2\end{matrix}\right\} = - \frac{\mu\xi}{1 - 8\mu\xi + 16\mu(\mu+1)\xi^2} \cdot \left\{\begin{matrix} 1 - 4\mu\xi(1-2\xi) \\ \\ 8\mu\xi\sqrt{\xi(1-\xi)}\end{matrix}\right. \tag{6.32a}$$

$$|\vartheta|^2 = \frac{\mu^2\xi^2}{1 - 8\mu\xi + 16\mu(\mu+1)\xi^2} ; \qquad q = \frac{\pi V_c Z\Omega^4_{max}}{\omega V} |\vartheta|^2 . \tag{6.32b}$$

For the cross section only $|\mathring{t}|^2$ is required, but later we need $\mathring{t}$ itself in order to discuss the influence of isotopic defects on the dispersion curves.

Fig. 6.14 shows (6.32a,b) and Fig. 6.15 shows $\vartheta$ for an actual $\mathring{G}^{(0)}$. One recognizes qualitative agreement: for small substitutional mass one finds small $|\vartheta|^2$ without special features; for the critical mass ($\mu < 0$) there is a divergence at $\Omega_{max}$; for large $\mu$ one recognizes the development of a resonance, $\omega_R \ll \Omega_{max}$. The spectrum of the substitutional (at site $\underline{m} = 0$) is

$$\begin{aligned} Z_d = Z^{oo}_{11} &= \frac{M_d}{\pi} \operatorname{Im}\{G^{oo}_{11}\} = \frac{M_d}{\pi} \operatorname{Im}\left\{[\mathring{G}^{(0)} - \mathring{G}^{(0)} t \mathring{G}^{(0)}]_{11}\right\} \\ &= \frac{M_d}{\pi m^2 \omega^4} \operatorname{Im}\{\mathring{t}\} = \frac{M_d}{\pi} \operatorname{Im}\left\{\frac{\mathring{A}}{1 - m\omega^2\mathring{A}}\right\} , \qquad \mathring{A} = \mathring{G}^{(0)}_{11} , \end{aligned} \tag{6.33}$$

which contains the perfect spectrum as a factor. Fig. 6.16 illustrates the spectrum employing the approximate spectrum of Table 3.4, where for $0 \leqslant \xi \leqslant 1$

$$\begin{aligned} \Omega^2_{max} Z_d &= \frac{8(\mu+1)}{\pi} \frac{\sqrt{\xi(1-\xi)}}{1 - 8\mu\xi + 16\mu(\mu+1)\xi^2} = \frac{1}{2\pi\mu} \frac{\sqrt{\xi(1-\xi)}}{(\xi-\xi_+)(\xi-\xi_-)} \\ &= \frac{Z(\xi)}{16\mu(\xi-\xi_+)(\xi-\xi_-)} \qquad \text{with } \xi_\pm = \frac{1}{4(\mu+1)} \left(1 \pm \sqrt{-\frac{1}{\mu}}\right) . \end{aligned} \tag{6.33a}$$

For very large $M_d$, $M_d/\mathring{M} \gg 1$ or $\mu \gg 1$, the complex poles, $\xi_\pm \cong (1/4\mu)(1 \pm i/\sqrt{\mu}) = \xi_R \pm i\xi_R/\sqrt{\mu}$, are at very small $\xi_R$ with even smaller imaginary parts. Only contributions near $\xi_R$ are essential and one obtains a well-defined resonance,

$$\Omega^2_{max} Z_d = \frac{1}{2\pi\mu} \frac{\sqrt{\xi}}{(\xi_R - \xi)^2 + \xi_R^2/\mu} \cong \frac{\xi_R/\pi\sqrt{\mu}}{(\xi_R - \xi)^2 + \xi_R^2/\mu} , \tag{6.33b}$$

a sharp, normalized[13] Lorentzian with halfwidth $\xi_R/\sqrt{\mu}$ and relative width $1/\sqrt{\mu}$ which is small if $\mu \gg 1$. This spectrum should be compared with that of a damped linear oscillator with mass M, small damping $2\eta$ and frequency $\Omega^2$, where MG ≅

---

13 Height ($\propto \sqrt{\mu}/\xi_R \propto \mu^{3/2}$) × width ($\propto \xi_R/\sqrt{\mu} \propto 1/\mu^{3/2}$) ≅ 1; the distribution (6.33b) is practically normalized.

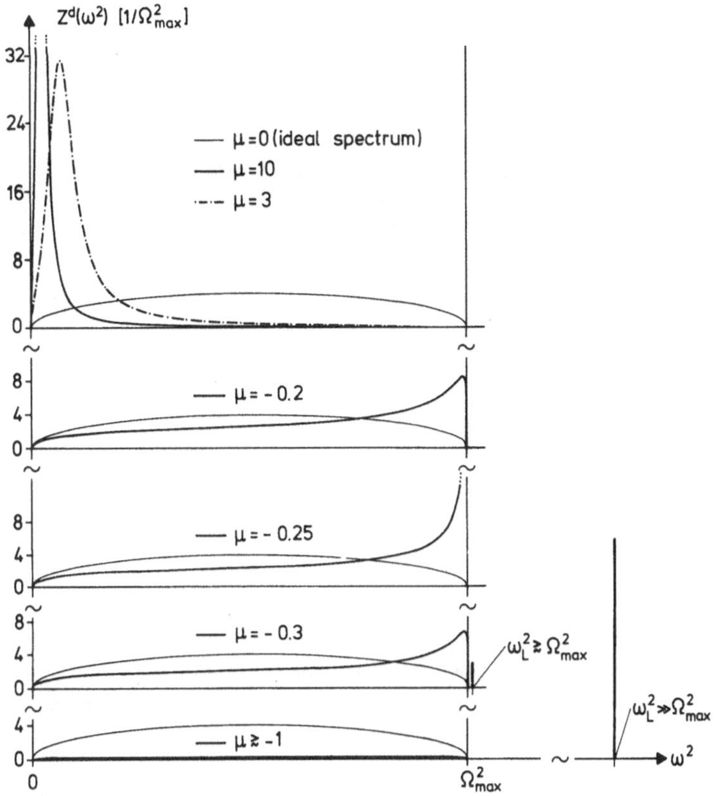


<u>Fig. 6.16.</u> Local spectra, $Z_d(\omega^2)$, for various isotopic defects, $\mu = (M_d - \mathring{M})/\mathring{M}$, calculated with the regular spectrum of Table 3.4

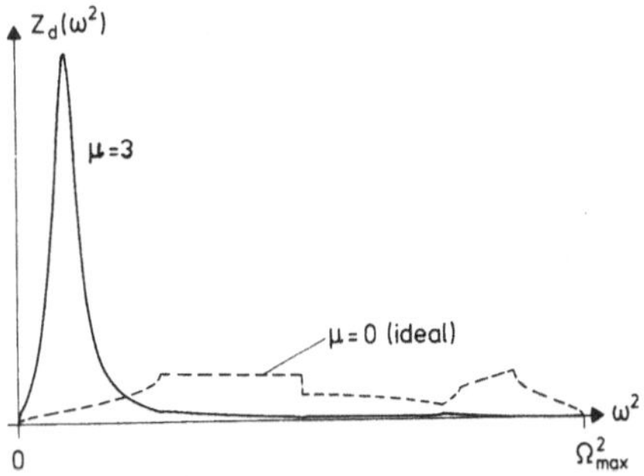


<u>Fig. 6.17.</u> Local spectrum, $Z_d(\omega^2)$, for an isotopic defect, $\mu = 3$, numerically calculated with the Green's function for 1st neighbour longitudinal coupling in a fcc lattice (model of Fig. 3.15c)

$(\Omega^2 - \omega^2 - 2i\eta\omega)^{-1}$, $\omega > 0$;

$$Z = \frac{M}{\pi}\,\mathrm{Im}\{G\} = \frac{2\eta\omega/\pi}{(\Omega^2-\omega^2)^2 + (2\eta\omega)^2} \cong \frac{2\eta\Omega/\pi}{(\Omega^2-\omega^2)^2 + (2\eta\Omega)^2} \text{ for } \eta/\Omega \ll 1 \ .$$

One sees that (6.33b) is the spectrum of a weakly damped oscillator:

$$\Omega^2 = \xi_R\Omega^2_{\mathrm{max}} = \Omega^2_{\mathrm{max}}/(4\mu)\ ,\quad 2\eta\Omega = \xi_R\Omega^2_{\mathrm{max}}/\sqrt{\mu} = \Omega^2/\sqrt{\mu}\ ,\quad 2\eta = \Omega/\sqrt{\mu}\ ;$$

the decrease of the amplitude after one period is $\exp(-\eta 2\pi/\Omega) = \exp(-\pi/\sqrt{\mu})$. For large $M_d$ the isotopic defect behaves like a very weakly damped oscillator. Physically, the damping is caused by emission of energy; it is small because at low $\omega$ the imaginary part of $\mathring{G}^{(0)}$ is small. The imaginary part of the equation of motion is for all practical purposes completely represented by (6.33b), and the corresponding real part is obtained via the Kramers-Kronig relations.

For a heavy Mößbauer atom the spectrum is practically a $\delta$-function at $\xi_R$; the atom has the thermal amplitudes of an oscillator with frequency $\omega_R$ and the classical value is reached at much lower temperatures than for the host.

For small $M_d/\mathring{M}$ one localized frequency, $\omega_L > \Omega_{\mathrm{max}}$, occurs. The spectrum contains a contribution proportional to $\delta(\omega_L^2 - \omega^2)$, indicated in Fig. 6.16. To obtain the contribution for $\omega^2 > \Omega^2_{\mathrm{max}}$, where $\mathring{A}(\omega^2)$ is real, one must use the original recipe: replace $\omega$ by $\omega + i\eta$ in (6.33) and go to the limit $\eta \to +0$.[14] The result is,

$$\begin{aligned} Z_d^{\mathrm{loc}}(\omega^2) &= -M_d\mathring{A}(\omega^2)\ \delta\left(1 - m\omega^2\mathring{A}(\omega^2)\right) = \frac{M_d\mathring{A}(\omega_L^2)}{\partial_{\omega^2} m\omega^2\mathring{A}(\omega^2)\Big|_{\omega_L^2}}\,\delta(\omega_L^2 - \omega^2) \\ &= \frac{\mu}{\mu+1}\left(1 - \frac{\sqrt{\xi_L}}{\sqrt{\xi_L - 1}}\right)^{-1}\delta(\omega_L^2 - \omega^2) \quad \text{for the regular spectrum of Table 3.4}\ , \end{aligned} \tag{6.33c}$$

where $\omega_L^2$ is obtained from $m\omega_L^2\mathring{A}(\omega_L^2) = 1$, in accordance with (3.80).

For very small $M_d/\mathring{M}$ one can use the expansion (3.81) for large $\omega^2$ and obtains $Z_d^{\mathrm{loc}} \cong \delta(\omega_L^2 - \omega^2)$. The total spectrum must be normalized ($0 \leqslant \omega^2 \leqslant \infty$); because $Z_d^{\mathrm{loc}}$ is already normalized this means that no contribution is left in the range of the ideal spectrum, $\omega^2 \leqslant \Omega^2_{\mathrm{max}}$.

For the critical mass $\mu = -1/4$ or $M_d = 3\mathring{M}/4$ we have with $\xi_\pm = 1, -1/3$

$$\Omega^2_{\mathrm{max}}\,Z_d = \frac{2}{\pi}\,\frac{\sqrt{\xi}}{\sqrt{1-\xi}\,(\xi + 1/3)} \qquad \left(\cong \frac{3}{2\pi}\,\frac{1}{\sqrt{1-\xi}} \text{ near } \xi = 1\right)\ . \tag{6.33d}$$

The spectrum becomes singular at $\Omega_{\mathrm{max}}$. This is the case where the localized fre-

---

14 For $\eta \to +0$ one has $\frac{1}{\pi}\,\mathrm{Im}\left\{\frac{N(\omega + i\eta)}{D(\omega + i\eta)}\right\} = -\,\mathrm{sgn}\{\partial_\omega D(\omega)\}\ N\delta(D)$, which can easily be checked by expansion.

quency, which exists for $\mu \leq -1/4$, is identical with $\Omega_{max}$. According to what was said above one expects a resonance behaviour for $\mu \gtrsim -1/4$. Here one obtains approximately near $\xi \lesssim 1$

$$\Omega_{max}^2 \, Z_d \cong \frac{3}{2\pi} \frac{\sqrt{1-\xi}}{(\xi_+ - \xi)} = \frac{3}{2\pi} \frac{\sqrt{1-\xi}}{\xi_+ - 1 + 1 - \xi} , \quad \xi_+ - 1 \gtrsim 0 . \tag{6.33e}$$

This spectrum is shown also in Fig. 6.16. It shows a peak at $\xi_R = 2 - \xi_+ \lesssim 1$ of height $3/(4\pi\sqrt{\xi_+ - 1})$, but is otherwise not similar to the Lorentzian at low $\omega_R$, e.g., the points of half the maximum value are at about $1 - 14(\xi_+ - 1)$ and $1 - (\xi_+ - 1)/4$. Although the peak can be high, the width is large, and the shape of $Z_d$ is unsymmetrical; the corresponding equation of motion is complicated and cannot be so easily interpreted as for low $\omega_R$, but it represents a kind of wide resonance (quasiresonance). For $\mu \lesssim -1/4$ one obtains a localized mode $\omega_L \gtrsim \Omega_{max}$ with only a small factor in front of the $\delta$-function. Most of the spectrum is still contained in $0 \leq \omega \leq \Omega_{max}$ and exhibits there a quasiresonance as for $\mu \gtrsim -1/4$ $\left(\xi_+ = 1 + 4(\mu + 1/4)^2/3 \gtrsim 1\right)$. Fig. 6.16 shows also the spectrum for $\mu = 3$, which corresponds to an Ag substitutional in Al. According to Table 6.1 the volume change by Ag impurities in Al is extremely small ($\delta_1 V/V_c = 10^{-3}$); therefore, one can hope that this impurity can be represented approximately by a pure mass change (comp. Fig. 9.3). Fig. 6.17 shows the numerical spectrum employing the 1st neighbour coupling model.

Roughly speaking the behaviour of the isotopic defect is as follows: The equation of motion is $\underline{s}^o = G^{oo}\underline{F}^o$ or $(G^{oo})^{-1}\underline{s}^o = \underline{F}^o$; its character is completely determined by $Z_d \propto \mathrm{Im}\{G^{oo}\}$ from which the real part is obtained via the Kramers-Kronig relations. For $\omega > \Omega_{max}$ the response $G^{oo}$ is real. One can then ask for a "localized" mode, a vibration which does not need an applied force to keep it going. The condition is obviously $1/G^{oo}(\omega_L) = 0$, or $G^{oo}(\omega_L) = \infty$, which one can rationalize by saying: if the response to a force is infinite one actually does not need a force to maintain the amplitude of the oscillation. For vanishing $M_d$ the localized frequency moves to infinity, the system looses one atomic degree of freedom. The amplitudes of the host are obtained in the following way: If the displacements in defect space, $P\underline{s}$ (here P projects onto the isotopic defect), are prescribed and if there are no forces otherwise, i.e.,

$$Q X \underline{s} = 0 = Q X Q \underline{s} + Q X P \underline{s} = Q \overset{o}{X} Q + Q \overset{o}{X} P s ,$$
$$\text{where } X = \Phi - M\omega^2 , \quad \overset{o}{X} = \overset{o}{\Phi} - \overset{o}{M}\omega^2 \text{ and } Q = 1 - P , \tag{6.34}$$

one obtains

$$Q\underline{s} = -\frac{1}{Q\overset{o}{X}Q} Q\overset{o}{X}P\underline{s} . \tag{6.34a}$$

This is a completely general result applicable to motions of arbitrary $\omega$ (comp.

(2.41) for the static case). For a localized mode one can simplify this expression: from $X\underline{s} = 0 = \overset{o}{X}\underline{s} + \chi\underline{s}$ one derives

$$\underline{s} + \overset{o}{G}\chi\underline{s} = 0 = \underline{s} + \overset{o}{G}P\chi P\underline{s} \; , \tag{6.34b}$$

which leads to

$$Q\underline{s} = - Q\overset{o}{G}(\omega_L)P\chi P\underline{s} \tag{6.34c}$$

and to

$$P\underline{s} + P\overset{o}{G}P\chi s = 0 \; ; \tag{6.34d}$$

eq. (6.34d) represents an eigenvalue problem for the localized modes which determines $\omega_L$. The amplitudes of the host, $Q\underline{s}$, decrease exponentially because $\omega_L > \omega_{max}$, hence the name "localized". For very small $M_d$ (very large $\omega_L$) even the neighbours of the defect are at rest in the localized mode. One is tempted to express $\underline{s}$ by $GP[PGP]^{-1}P\underline{s}$, which yields the correct prescribed $P\underline{s}$; however, for a localized mode, where G has a pole at $\omega_L$, one must consider the proper limit.

For $M_d$ larger than a critical mass no localized modes exist. A stationary motion can be maintained only by a force to cover the emitted energy. But there are *resonances* at frequencies $\omega_R$, where one needs only a relatively small force to maintain the oscillation. These must be frequencies $\omega_R$ for which both the real and the imaginary part of $1/G^{oo}$ are small. Consequently, resonant modes can only occur at both ends of the normal spectrum where the emission of energy into the lattice is small. In a resonant vibration the force applied needs only to be small; if this force is switched off, the amplitude of the oscillation decreases only slowly by emission of energy (small relative changes per period). Whereas a localized mode is stable (no emission of energy), a resonant motion is only metastable (small emission of energy, slowly decaying amplitudes). The forces to maintain resonant modes can be supplied by an incoming wave at the resonant frequency. The scattering cross section, then, reflects the large response to a resonant force and exhibits a peak. Particularly illustrative are resonant modes at low $\omega_R$ which behave like weakly damped oscillators; here $\omega_R$ is determined by $\mathrm{Re}\{1/G^{oo}(\omega_R)\} = 0$. We have seen that quasiresonant modes can occur near $\Omega_{max}$ for masses near the critical mass. For masses below the critical mass one observes a resonant together with a localized mode, here $\mathrm{Re}\{1/G^{oo}(\omega_R)\} = 0$. Above the critical mass one observes a similar resonance where $\mathrm{Re}\{1/G^{oo}(\omega_R)\}$ is only small and not zero. The quasiresonances near $\Omega_{max}$ cannot be viewed as simple, damped oscillators.

The change of the thermal energy (comp. Section 2.4.4) due to a single isotopic defect (at site 0) is

$$\delta_1 E(T) = \int d\omega^2 \, \varepsilon_{th}(\omega) \, (Z_{ii}^{nn} - \overset{o}{Z}{}_{ii}^{nn})$$

$$= \int d\omega^2 \, \varepsilon_{th}(\omega) \frac{1}{\pi} \, \mathrm{Im} \left\{ \sum_{\underline{n}} (M^{\underline{n}} G_{ii}^{nn} - \overset{oo}{M}\overset{}{G}{}_{ii}^{nn}) \right\} \tag{6.35a}$$

$$= \int d\omega^2 \, \varepsilon_{th}(\omega) \frac{1}{\pi} \, \mathrm{Im} \left\{ \overset{o}{M}(G_{ii}^{nn} - \overset{o}{G}{}_{ii}^{nn}) + mG_{ii}^{oo} \right\};$$

expressing G by $\overset{o}{G}$ [eq.(6.8)], $G = \overset{o}{G} - \overset{o}{G}t\overset{o}{G}$, and employing the t-matrix as given by (6.14) one obtains[15]

$$\overset{o}{M}\,(G_{ii}^{nn} - \overset{o}{G}{}_{ii}^{nn}) = \overset{o}{M}\,[\overset{o}{G}t\overset{o}{G}]_{ii}^{nn} = \overset{o}{M}\,[\overset{o}{G}{}^2 t]_{ii}^{nn} = t\overset{oo}{M}\,[\overset{o}{G}{}^2]_{ii}^{oo} = t\partial_{\omega^2}\overset{o}{G}{}_{ii}^{(0)} \;,$$

$$mG_{ii}^{oo} = m\,[\overset{o}{G} - \overset{o}{G}t\overset{o}{G}]_{ii}^{oo} = 3m \frac{\overset{o}{G}{}^{(0)}}{1 - m\omega^2 \overset{o}{G}{}^{(0)}} \;, \text{ where } \overset{o}{G}{}^{(0)} = \overset{o}{G}{}_{11}^{(0)} = \overset{o}{G}{}_{22}^{(0)} = \overset{o}{G}{}_{33}^{(0)} \;,$$

and therefore

$$\delta_1 E(T) = 3 \int d\omega^2 \, \varepsilon_{th}(\omega) \frac{m}{\pi} \, \mathrm{Im} \left\{ \frac{(1 + \omega^2 \partial_{\omega^2})\overset{o}{G}{}^{(0)}}{1 - m\omega^2 \overset{o}{G}{}^{(0)}} \right\} \;. \tag{6.35b}$$

For a sharp resonance, i.e., for $m = M_d \gg \overset{o}{M}$, the term $\omega^2 \partial_{\omega^2} \overset{o}{G}{}^{(0)}$ can be dropped and it is $\delta_1 E \cong 3 \int d\omega^2 \, \varepsilon_{th} Z_d$; it seems that there are three "additional" oscillatory degrees of freedom with frequencies $\omega_R$ which are essential for the low T behaviour. However, the factor of $\varepsilon_{th}$ in the integral (6.35a,b) is not a spectrum, but the difference between two spectral distributions of the same order of magnitude; it has positive and negative parts such that $\int d\omega^2 \ldots$ becomes zero. Therefore, the three "additional" degrees of freedom are fictitious; they are "missing" in the region $\omega \gg \omega_R$.

### 6.4.6 The Cross Section for the Diatomic Defect

For the diatomic defect (comp. Fig. 6.7) the evaluation of (6.28,29) is more complicated. The symmetry of the states $\nu = 1 \ldots 6$ is such that the total cross section is a sum over the partial cross sections $\overset{\nu}{q}$[16]:

$$q = \sum_{\nu=1}^{6} \overset{\nu}{q} \quad ; \quad \overset{\nu}{q} = \frac{\pi V_B V_c}{\overset{o}{M}\omega v} \sum_{\sigma'} \int d\underline{k}' \, \delta\!\left(\overset{o}{\Omega}{}^2(\underline{k}'\sigma') - \omega^2\right) |\overset{\nu}{t} \langle\underline{k}'\sigma'|\nu\rangle\langle\nu|\underline{k}\sigma\rangle|^2 \;. \tag{6.36}$$

15 From (2.27a) with $\omega > 0$, $M = \overset{o}{M}$ it follows that $\overset{o}{G}{}^2 = \partial_{\omega^2}\overset{o}{G}/\overset{o}{M}$.

16 The sum of $\langle\mu|\underline{k}'\sigma'\rangle\langle\underline{k}'\sigma'|\nu\rangle$ over all equivalent states $\underline{k}'\sigma'$ vanishes due to symmetry if $\mu \neq \nu$.

Because $\overset{\circ}{G}{}^{(0)}$ is much larger than the $\overset{\circ}{G}{}^{(200)}$ elements, which can be neglected in a crude approximation, all $\overset{\nu}{A}$ equal $\overset{\circ}{G}{}^{(0)}$ approximately. The t-matrices have the form of $\overset{\circ}{t}$ with eigenvalues $2f_d - m\omega^2$ for $\nu = 1$ and $-m\omega^2$ for all other $\nu$. In this approximation one has for $f_d = 0$

$$q = \frac{\pi V_c |\overset{\circ}{t}|^2}{\overset{\circ}{M}\omega v V_B} \sum_{\sigma'} \int d\underline{k}' \; \delta\left(\overset{\circ}{\Omega}{}^2(\underline{k}'\sigma') - \omega^2\right) 4\left(\sin^2 \frac{k_z' a}{2} \sin^2 \frac{k_z a}{2} + \cos^2 \frac{k_z' a}{2} \cos^2 \frac{k_z a}{2}\right)\left(e_z^2 e_z'^2 + e_y^2 e_y'^2 + e_x^2 e_x'^2\right), \tag{6.36a}$$

and for $m = 0$

$$q = \frac{\pi V_c |\overset{1}{t}|^2}{\overset{\circ}{M}\omega v V_B} \sum_{\sigma'} \int d\underline{k}' \; \delta\left(\overset{\circ}{\Omega}{}^2(\underline{k}'\sigma') - \omega^2\right) 4 \sin^2 \frac{k_z' a}{2} \sin^2 \frac{k_z a}{2} e_z^2 e_z'^2 , \tag{6.37}$$

$$\overset{1}{t} \cong \frac{2f_d}{1 + \overset{\circ}{G}{}^{(0)} 2f_d} .$$

We are going to make only a few remarks instead of presenting a detailed discussion:

1) The $\overset{\nu}{t}$ values determine again the details of the cross section, because the integrals are relatively smooth functions of $\omega$.

2) All partial cross sections $\overset{\nu}{q}$ start at least with $\omega^4$ (or higher powers of $\omega$, comp. Sec. 6.5.7): for $\nu = 2\ldots6$ because $\overset{\nu}{t} \propto -m\omega^2$; for $\nu = 1$ because $\sin^2 k_z'a/2 \cdot \sin^2 k_z a/2 \propto k_z'^2 k_z^2 \propto \omega^4$ if $\omega$ is small.

3) Resonances occur for $\nu = 2\ldots6$ if m is large enough and for $\nu = 1$ if m and $-f_d$ are sufficiently large.

4) If $f_d$ is large enough a localized state exists.

5) The diatomic defect can have localized and resonant states,e.g., if $f_d$ and m are sufficiently large.

6) For pure mass change and low $\omega$, Born's approximation is correct and the cross section for the diatomic defect is four times larger than the cross section (6.31a) for the corresponding isotopic defect. In this limit the scattered waves from the two defect centers superimpose, doubling the scattered amplitudes and quadrupling q.

7) For large diatomic distances the approximation $\overset{\nu}{A} = \overset{\circ}{G}{}^{(0)}$ is correct ($f_d = 0$). Then the t-matrix for the diatomic defect is a superposition of the single t-matrices: $t = \overset{\circ}{t}|\nu\rangle\langle\nu| = \overset{\circ}{t}P$, where P projects onto the two sites where the mass is changed. The $\overset{\circ}{G}{}^{(2n,0,0)}$ terms represent multiple scattering between defect sites.

## 6.5 Scattering by Defects with Additional Coordinates

If the defect changes the number of the degrees of freedom, the scattering problem requires special treatment. Interstitials are the simplest examples, e.g., the octahedral and the dumbbell interstitial with one additional atomic coordinate. We explain the procedure with these two examples in mind; the octahedral interstitial will be treated explicitly. The treatment of more complicated cases follows the same lines and the extraplation is simple.

### 6.5.1 Original, Augmented and Defect Spaces

It is useful to separate the total space of atomic displacements into original, augmented and defect space. This is sketched in Fig. 6.18, which pictures a matrix in the original space (the host lattice); each block denotes a 3×3 matrix, operating on the displacements of the corresponding atom. Indicated are a central atom, 0, and its 12 neighbours (fcc). The matrix $\overset{\circ}{M}$ contains only diagonal elements, the atomic mass, $\overset{\circ}{M}$, of the host atoms. The matrix $\overset{\circ}{\Phi}$ represents the coupling of the perfect lattice, it contains off-diagonal elements. For nearest neighbour coupling the central atom would be coupled to all its 12 neighbours, also the nearest neighbours among the 12 atoms would be linked.

Fig. 6.18b illustrates a vacancy and its defect space, P, in the 1st neighbour model of Fig. 6.5. The defect space is actually somewhat smaller than that shown in Fig. 6.18b, which is $3\times 13 = 39$-dimensional. For longitudinal springs, the actual defect space is only 12-dimensional, corresponding to the 12 removed springs or to the 12 states (6.12a,b,c); it is contained in P of Fig. 6.18b. Clearly, the use of enlarged (redundant) defect spaces cannot influence any results.

For an octahedral interstitial the space must be augmented by one atomic coordinate corresponding to three degrees of freedom or three dimensions. The quantities $\Phi$, M, G, $X = \Phi - M\omega^2$ are defined in the enlarged space, whereas $\overset{\circ}{\Phi}$, $\overset{\circ}{M}$, $\overset{\circ}{G}$, $\overset{\circ}{X} = \overset{\circ}{\Phi} - \overset{\circ}{M}\omega^2$ and $\overset{i}{\underline{s}}$ are defined only in original (host) space, $P_o$, and have vanishing components outside.[17]

If one must include additive space, it is convenient to distinguish between

original space, $P_o$ ; augmented (additive) space, $P_a$ ; defect space, P ,

and the projection $p = P_oPP_o$ of P into original space, $P_a + p = P$, $P + Q = 1$ ,

where Q is the subspace unaffected by the defect (comp. Fig. 6.19 for the octahedral interstitial). In this section we will indicate projections by subscripts, such as

---

17 $\omega$ is used instead of $\omega + i\eta$, $\eta \to +0$.

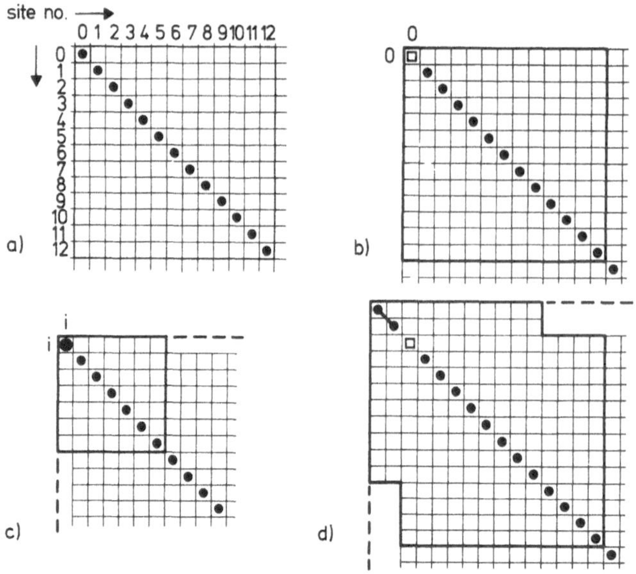


Fig. 6.18a-d. Defect spaces.
a) The space of atomic displacements (schematically): indicated are a central atom, 0, and its 12 neighbours.
b) Vacancy at 0, Fig. 6.5; if 1st neighbour interaction is assumed, the defect space contains only the vacancy and its 12 neighbour sites; their space is indicated by the thick line; it has 13 atomic coordinates = 13 × 3 coordinates; the actual defect space is smaller and contained in that space; for longitudinal 1st neighbour interaction the defect space has 12 dimensions corresponding to the 12 removed springs.
c) Octahedral interstitial, Fig. 6.2; the space must be augmented by one atomic coordinate (dashed line); for closest neighbour interaction the defect space (thick line) contains only the interstitial and its six neighbours; if the coupling is longitudinal the actual defect space has dimension 9: the atomic coordinate of the interstitial (dimension 3) and one degree of freedom from each of the six springs.
d) ⟨100⟩ dumbbell interstitial, Fig. 6.4; here it is most convenient to augment the space by the two atomic coordinates of the dumbbell atoms and introduce a vacancy at the interstitial site; in the simplest model the interstitial atoms interact only with the 8 neighbours shown in Fig. 6.4, and the vacancy "interacts" only with its 12 1st neighbours; this defect space is indicated again by thick lines; if all the springs are longitudinal, the actual dimension is 27: 6 coordinates of the dumbbell, 1 dumbbell spring, 8 springs coupling the dumbbell atoms and their neighbours, and 12 springs removed by the vacancy

$$P_a X P_a = X_{aa}\ ,\quad P_o X P_o = X_{oo}\ ,\quad P_o X P_a = X_{oa}\ ,\quad P_a X P_o = X_{ao}\ , \tag{6.38}$$

or, for Green's function in enlarged space,

$$P_a G P_a = G_{aa} \text{ and so on.} \tag{6.38a}$$

In explicit matrix notation

$$X = \begin{bmatrix} X_{aa} & X_{ao} \\ X_{oa} & X_{oo} \end{bmatrix}, \tag{6.38b}$$

or, for $XG = 1$:

$$XG = \begin{bmatrix} X_{aa}G_{aa} + X_{ao}G_{oa} & X_{aa}G_{ao} + X_{ao}G_{oo} \\ X_{oa}G_{aa} + X_{oo}G_{oa} & X_{oo}G_{oo} + X_{oa}G_{ao} \end{bmatrix} = \begin{bmatrix} P_a & 0 \\ 0 & P_o \end{bmatrix}. \tag{6.39}$$

### 6.5.2 Green's Function G

Most conveniently G is represented in terms of a "zero order" Green's function, $\overset{\circ\circ}{G}$, defined by

$$\overset{\circ\circ}{X} = X_{aa} + X_{oo} = \begin{bmatrix} X_{aa} & 0 \\ 0 & X_{oo} \end{bmatrix}, \tag{6.40}$$

$$\overset{\circ\circ}{G} = \overset{\circ\circ}{G}_{aa} + \overset{\circ\circ}{G}_{oo} = \frac{1}{X_{aa}} + \frac{1}{X_{oo}} = \begin{bmatrix} \frac{1}{X_{aa}} & 0 \\ 0 & \frac{1}{X_{oo}} \end{bmatrix} = \frac{1}{\overset{\circ\circ}{X}}. \tag{6.40a}$$

The quantity $X_{aa}$ represents the motion of the additive atoms for fixed host atoms, a kind of Einstein approximation for the interstitial region. Equivalently, $X_{oo}$ represents the motion of the host for fixed additive atoms; note that $X_{oo}$ consists of $\overset{\circ}{X}$ *and* the (defect) coupling to the fixed additive atoms[18].

The "perturbation" is given by

$$\chi = X - \overset{\circ\circ}{X} = X_{ao} + X_{oa} = \begin{bmatrix} 0 & X_{ao} \\ X_{oa} & 0 \end{bmatrix} = \psi\ ; \tag{6.41}$$

---

[18] If X represents a stable configuration, such that $X^{-1}$ exists, so do $X_{aa}$ and $X_{oo}$ in their respective subspaces. In the most general case additive changes in the host lattice would have to be included.

it has only off-diagonal elements and does not contain any masses, but only the spring coupling between interstitial space and host (therefore, in (6.41) $\chi$ can be replaced by $\psi$). The representation of G by the "zero order" Green's function and the "perturbation" is completely analogous[19] to that of G by $\overset{\circ}{G}$ and $\chi$ as given in Section 6.4.1 (for defects without change of the number of degrees of freedom); instead of (6.7) we have now

$$G = \overset{\circ\circ}{G}\,\frac{1}{1+\psi\overset{\circ\circ}{G}} = \frac{1}{1+\overset{\circ\circ}{G}\psi}\,\overset{\circ\circ}{G} = \overset{\circ\circ}{G} - \overset{\circ\circ}{G}\psi\overset{\circ\circ}{G} + \overset{\circ\circ}{G}\psi\overset{\circ\circ}{G}\psi\overset{\circ\circ}{G} - + \dots \quad . \tag{6.41a}$$

It is interesting to note that in an expansion in powers of $\psi$ "diagonal" terms alternate with "off-diagonal" terms:

$$\begin{aligned} G = {} & \overset{\circ\circ}{G}_{aa} + \overset{\circ\circ}{G}_{oo} - \overset{\circ\circ}{G}_{aa}\psi_{ao}\overset{\circ\circ}{G}_{oo} - \overset{\circ\circ}{G}_{oo}\psi_{oa}\overset{\circ\circ}{G}_{aa} + \underbrace{\overset{\circ\circ}{G}_{aa}\psi_{ao}\overset{\circ\circ}{G}_{oo}\psi_{oa}\overset{\circ\circ}{G}_{aa}}_{\text{in } P_a} + \\ & \underbrace{\overset{\circ\circ}{G}_{oo}\psi_{oa}\overset{\circ\circ}{G}_{aa}\psi_{ao}\overset{\circ\circ}{G}_{oo}}_{\text{in } P_o} - + \dots \quad . \end{aligned} \tag{6.41b}$$

### 6.5.3 The Octahedral Interstitial

A simple example is the octahedral interstitial with mass $M_i$ and longitudinal spring $f_i$ to its nearest neighbours. Fig. 6.19 sketches the various subspaces. The actual defect space is 9-dimensional (three interstitial coordinates + six longitudinal springs); again it is smaller than the space $P = p + P_a$ (21-dimensional) indicated in Fig. 6.19.

Obviously, the interstitial with *fixed* neighbours behaves like an isotropic oscillator,

$$X_{aa} = (2f_i - M_i\omega^2)P_a \;, \quad \frac{1}{X_{aa}} = G_{aa} = \frac{1}{2f_i - M_i\omega^2}\,P_a \;. \tag{6.42}$$

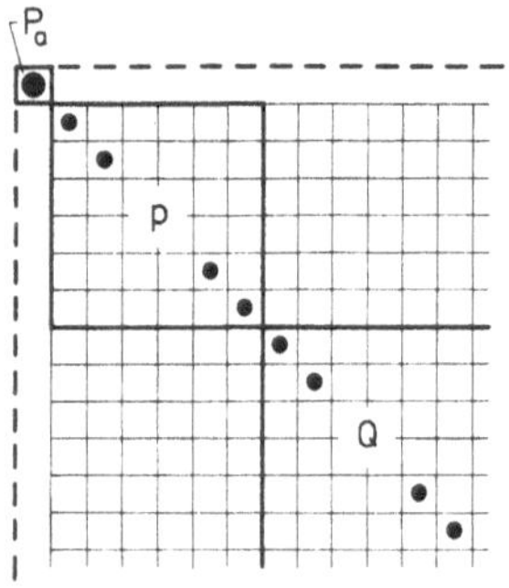


Fig. 6.19. Spaces for the octahedral interstitial of Fig. 6.18c. $P_a$: added space, containing the interstitial only, 3 dimensions. $P_a + p = P$: defect space, contains positions affected by the interstitial, 3 + 6×3 = 21 dimensions. $p = P_oPP_o$: containing all affected positions in original space, 18 dimensions, $P_o = p + Q$

19 $\overset{\circ\circ}{G}$ corresponds to $\overset{\circ}{G}$, and $\chi$ to $\chi = \psi$.

For fixed interstitial one has the familiar procedure,

$$X_{oo} = \mathring{X} + \psi_{oo}\,, \qquad \overset{oo}{G}_{oo} = \frac{1}{\mathring{X} + \psi_{oo}} = \mathring{G} - \mathring{G}t\mathring{G}\,, \qquad t = \psi_{oo}\frac{1}{1 + \mathring{G}\psi_{oo}}\,, \tag{6.42a}$$

where $\psi_{oo}$ is the coupling to the interstitial at rest. Fig. 6.20 shows the appropriate eigenmodes[20], three even and three odd modes; the value of $\psi_{oo}$ in all these modes is $f_i$ and the eigenvalues of t are:

$$\overset{\nu}{t} = \frac{f_i}{1 + \overset{\nu}{A}f_i}\,; \qquad \overset{\nu}{A} = \langle\nu|\mathring{G}|\nu\rangle\,; \qquad \overset{1}{A},\overset{2}{A} = \overset{3}{A} \text{ according to (6.17)}\,,$$

$$\overset{4}{A} = \overset{5}{A} = \overset{6}{A} = \langle 4|G|4\rangle = \mathring{G}^{(0)} + \mathring{G}_1^{(200)}\,. \tag{6.43}$$

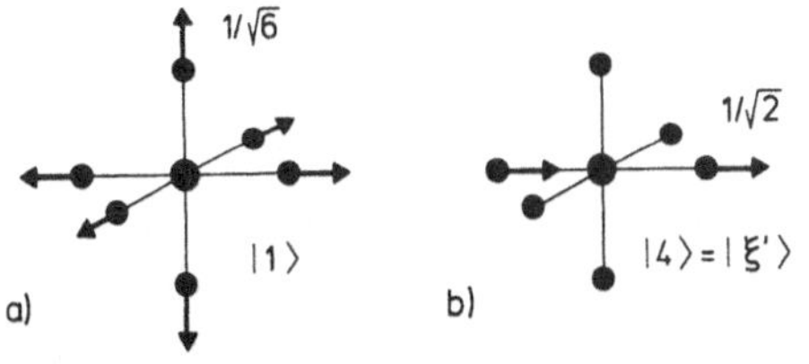


Fig. 6.20a and b. Modes for fixed octahedral interstitial.
a) The even modes $|1,2,3\rangle$ are those of Fig. 6.10; mode $|1\rangle$ is shown.
b) Of the three degenerate odd modes $|\xi',\eta',\zeta'\rangle = |4,5,6\rangle$ the mode $|\xi'\rangle$ is shown

### 6.5.4 Green's Function of the Interstitial Region, $G_{aa}$

For the local spectrum in $P_a$, which in turn determines the Mößbauer intensity, one needs $G_{aa}$, the Green's function of the interstitial region. It is obtained from (6.39),

$$X_{aa} + X_{ao}G_{oa} = P_a\,; \qquad X_{oa}G_{aa} + X_{oo}G_{oa} = 0 \quad \text{or} \quad G_{oa} = -\frac{1}{X_{oo}}X_{oa}G_{aa}\,, \tag{6.44a}$$

by elimination of $G_{oa}$ in (6.44a)

$$\left(X_{aa} - X_{ao}\frac{1}{X_{oo}}X_{oa}\right)G_{aa} = P_a \quad \text{or} \quad G_{aa} = \frac{1}{X_{aa} - X_{ao}\overset{oo}{G}_{oo}X_{oa}} = \frac{1}{X_{aa} - \psi_{ao}\overset{oo}{G}_{oo}\psi_{oa}}\,. \tag{6.44b}$$

Here, the quantity $\overset{oo}{G}_{oo}$ is only needed in defect space.

For the octahedral interstitial one has

$$P\overset{oo}{G}_{oo}P = \sum_{\nu=1}^{6} |\nu\rangle \frac{\overset{\nu}{A}}{1 + \overset{\nu}{A}f_i} \langle\nu|\,. \tag{6.45}$$

[20] In the static case only the even modes of $\overset{oo}{G}_{oo}$ will be needed (comp. Section 6.6).

Because $-X_{ao}|\nu\rangle = -\psi_{ao}|\nu\rangle$ is the force on the interstitial for the displacements of state $|\nu\rangle$, one recognizes from Fig. 6.20 that the states $\nu = 1,2,3$ drop out, and

$$X_{ao}\overset{oo}{G}_{oo}X_{oa} = \frac{f_i}{1+\overset{4}{A}f_i}\sum_{\nu=4}^{6} X_{ao}\,|\nu\rangle\langle\nu|\,X_{oa} = \frac{f_i}{1+\overset{4}{A}f_i}\,2f_i^2P_a \; ;$$

this follows from $-X_{ao}|\xi'\rangle = 2(1/\sqrt{2})f_i|x_i\rangle$, where $|x_i\rangle$ represents unit displacement of the interstitial in x-direction. Consequently, the result for $G_{aa}$ is:

$$G_{aa} = \frac{1}{2f_i - M_i\omega^2 - 2f_i^2\overset{4}{A}/(1+\overset{4}{A}f_i)}\,P_a\,, \tag{6.46a}$$

or

$$G_{aa} = \frac{1+\overset{4}{A}f_i}{2f_i - (1+\overset{4}{A}f_i)M_i\omega^2}\,P_a\,.$$

The quantity $\overset{4}{A} = \overset{o}{G}^{(0)} + \overset{o}{G}_1^{(200)}$ behaves qualitatively very much like[21] $\overset{o}{G}^{(0)} = \overset{o}{A}$. Therefore, the leading term for low frequencies is $\overset{4}{A} \cong 2\overset{o}{A} \cong 2/f_e$ (comp. (3.90a)), and for high frequencies the leading term is $\overset{4}{A} \cong \overset{o}{A} \cong -1/M\omega^2$ (comp. (3.81)). The overall order of magnitude is smaller than $1/f$, if we adopt the model with 1st neighbour longitudinal spring $f$ (comp. Section 3.5.6).

Equation (6.46a) is appropriate for weak coupling, $f_i/f \ll 1$, where the third term in the denominator is very small. It is the only term which can become imaginary and represent the emission of lattice waves. It is evident here that the interstitial behaves like an oscillator of frequency[22] $\omega_i = \sqrt{2f_i/M_i}$. If $\omega_i < \Omega_{max}$, the oscillator is weakly damped, which corresponds to a resonance; the damping is small because the coupling is small, not because $\mathrm{Im}\{\overset{4}{A}(\omega_i)\}$ is small. Therefore, this resonance can be anywhere in the ideal spectrum. If $\omega_i > \Omega_{max}$ we have a localized mode.

Eq. (6.46b) is convenient if one wants to show that the interstitial motion can represent simultaneously a resonance and a localized vibration, namely if $f_i/f \gg M_i/M \gg 1$. Resonant and localized frequencies are essentially determined by the zeros of the denominator in (6.46b). For high frequencies, where $f_i\overset{4}{A} \cong -f_i/\overset{o}{M}\omega^2$, the denominator, $2f_i - M_i\omega^2 + M_if_i/\overset{o}{M}$, becomes zero for $\omega_L^2 \cong f_i/M + 2f_i/M_i \gg f/\overset{o}{M} \cong \Omega_{max}^2$, resulting in a localized state. For low frequencies, where $f_i\overset{4}{A} \cong 2f_i/f_e \gg 1$, the denominator, $2f_i(1 - M_i\omega^2/f_e)$, vanishes at a very low frequency, $\omega_R^2 = f_e/M_i \ll \Omega_{max}^2$, which represents a resonance; the damping is small because the imaginary part of $\overset{4}{A}(\omega_R)$ at low resonance frequencies is small. Of course, for $f_i \cong f$ one obtains a resonance if $M_i \gg \overset{o}{M}$ *or* a localized state if $M_i \ll \overset{o}{M}$, and quasiresonances in between.

---

[21] A very similar $G_{aa}$, which contains only $\overset{o}{A}$, is discussed in Appendix P.

[22] Oscillation with fixed lattice.

### 6.5.5 Green's Function of the Host, $G_{oo}$

In complete analogy to $G_{aa}$ one can calculate $G_{oo}$, the Green's function of the host:

$$G_{oo} = \frac{1}{X_{oo} - X_{oa}\overset{oo}{G}_{aa}X_{ao}} = \frac{1}{\overset{o}{X} + \psi_{oo} - \psi_{oa}\overset{oo}{G}_{aa}\psi_{ao}} \; ; \quad G_{ao} = -\frac{1}{X_{aa}} X_{ao}G_{oo} \; . \tag{6.47}$$

For the octahedral interstitial one has again the states $|\nu\rangle$ as eigenstates:

$$\psi_{oo} - \psi_{oa}\frac{1}{2f_i - M_i\omega^2}\psi_{ao} = \sum_{\nu=1}^{3} |\nu\rangle f_i \langle\nu| + \sum_{\nu=4}^{6} |\nu\rangle f_i\left(1 - \frac{2f_i}{2f_i - M_i\omega^2}\right)\langle\nu| \; . \tag{6.47a}$$

For the states $\nu = 4,5,6$ the defect spring is a dynamical, $\omega$-dependent quantity which contains the complete dynamics of the interstitial. If $M_i$ is large, $G_{oo}$ passes into $\overset{oo}{G}_{oo}$ (fixed interstitial); for $M_i = 0$ one obtains the bond charge result (comp. (6.16)).

### 6.5.6 Scattering

The scattering problem can be treated as in Section 6.4.1. The incoming wave, $\overset{i}{\underline{s}}$, is only defined in host space, $P_o\overset{i}{\underline{s}} = \overset{i}{\underline{s}}$. The scattered wave is defined in enlarged space:

$$\overset{s}{\underline{s}} = -GX\overset{i}{\underline{s}} = -G(\psi_{oo} + \psi_{ao})\overset{i}{\underline{s}} \; . \tag{6.48}$$

Of interest in scattering is only the scattered wave in host space,

$$P_o\overset{s}{\underline{s}} = -P_oG(\psi_{oo} + \psi_{ao})\overset{i}{\underline{s}} = -(G_{oo}\psi_{oo} + G_{oa}\psi_{ao})\overset{i}{\underline{s}} \; . \tag{6.48a}$$

From

$$P_oGXP_a = G_{oo}\psi_{oa} + G_{oa}X_{aa} = 0 \tag{6.49}$$

we obtain

$$P_o\overset{s}{\underline{s}} = -G_{oo}(\psi_{oo} - \psi_{oa}\overset{oo}{G}_{aa}\psi_{ao})\overset{i}{\underline{s}} \; . \tag{6.50}$$

By comparison with (6.47) one sees that

$$P_o\overset{s}{\underline{s}} = -\overset{o}{G}t_s\overset{i}{\underline{s}} \; , \tag{6.51}$$

where the scattering t-matrix is given by

$$t_s = (\psi_{oo} - \psi_{oa}\overset{oo}{G}_{aa}\psi_{ao})\frac{1}{1 + \overset{o}{G}(\psi_{oo} - \psi_{oa}\overset{oo}{G}_{aa}\psi_{ao})} \; .$$

Cross sections etc. can be obtained from $t_s$ in the usual way. For the octahedral interstitial the states $|\nu = 1,\ldots,6\rangle$ are eigenstates. The needed eigenvalues are given in (6.47a) and for the A's in (6.43).

## 6.5.7 Cross Sections for Small $\omega$

For small $\omega$ the $\omega$-dependence of the total scattering cross section q can be understood from (6.29): if $\omega$ is small one can replace $\overset{o}{\Omega}(\underline{k}'\sigma')$ in the argument of the $\delta$-function by $c_{\sigma'}k'$ where $c_{\sigma'}$ is the appropriate sound velocity; roughly speaking, the $\underline{k}'$-integration results in replacing k' by $\omega/c_{\sigma'}$, and the matrix elements $|<\underline{k}'\sigma'|t|\underline{k}\sigma>|^2$ with $k' = \omega/c_{\sigma'}$, $k = \omega/c_\sigma$ determine the $\omega$-dependence of q. For the further discussion it is useful to employ the eigenrepresentation of t, $t = \overset{\nu}{t}\,|\nu><\nu|$, or $<\underline{k}'\sigma'|t|\underline{k}\sigma> = \sum_\nu \overset{\nu}{t}\, <\underline{k}'\sigma'|\nu><\nu|\underline{k}\sigma>$. If the defect has inversion symmetry the defect modes can be split into even (eigenvalue +1 for inversion, I = 1) and odd modes. All defects treated so far in this chapter have inversion symmetry whereas the tetrahedral bond charge (Fig. 5.5) has not. For example, the modes |1,3,5> of the diatomic defect are even and |2,4,6> are odd (Fig. 6.8), the modes |1,2,3> of the octahedral interstitial are even and |4,5,6> are odd. For the diatomic defect $<\nu|\underline{k}\sigma>$ is proportional to $\sin(k_z a/2)$ for even and to $\cos(k_z a/2)$ for odd modes, which are at low frequencies proportional to $\omega^1$ and to $\omega^o$ respectively. Consequently, the cross sections of even modes carry an extra factor $\omega^4$.

If one includes the $\omega$-dependence of the eigenvalues $\overset{\nu}{t}$, the following $\omega$-dependence of the total cross section is obtained:

1) *diatomic defect:* $q \propto \omega^4$ for $\nu = 1,2,4,6$; $q \propto \omega^8$ for $\nu = 3,5$; the highest power, $\omega^8$, is obtained for even states, where both $|<\underline{k}'\sigma'|\nu><\nu|\underline{k}\sigma>|^2$ and $|\overset{\nu}{t}|^2$ yield a factor $\omega^4$;

2) *fixed octahedral interstitial:* $q \propto \omega^4$ for the even states |1,2,3>; $q \propto \omega^o$ for the odd states |4,5,6>; this model is somewhat artificial because it violates translational invariance;

3) *octahedral interstitial:* $q \propto \omega^4$ for all $\nu$, (for the odd states $\overset{\nu}{t} \propto \omega^2$ at low $\omega$).

That the scattering vanishes for $\omega = 0$ or $\underline{k} = 0$ is obvious, because then the incoming wave $\overset{i}{\underline{s}}$ degenerates into a pure translation, which can certainly not be "scattered".

## 6.6 Static Scattering

The static limit of scattering theory, i.e., $\omega = 0$, can be treated in much the same way. The incoming mode is *not* a lattice wave with $\omega = 0$ and $\underline{k} = 0$ which represents a translation of odd symmetry., but a homogeneous deformation $v\underline{R}$ of even symmetry. Here we must again point out a qualitative difference between finite and infinite crystals. In infinite crystals the lattice waves form a complete set of eigenstates; however, one cannot properly construct displacements $v\underline{R}$ by superposition of lattice waves. This is also reflected in the fact that the static stability conditions are more restrictive than the positivity of the eigenvalues of $\Phi$ in an infinite crystal. Consequently, one must study elastic deformations separately. On the other hand, in finite crystals at least inversion symmetry drops out; scattering and response theory are exceedingly complicated, but homogeneous deformations can be constructed from the eigenstates of $\Phi$ and static stability follows from the positivity of the eigenvalues of $\Phi$.

## 6.6.1 Static Scattering Theory

The formalism of static scattering theory starts from the static limit of the force-free equation of motion, $\Phi \underline{s} = 0$; here the total displacement $\underline{s}$ consists of a (prescribed) homogeneous deformation $v\underline{R}$ (the "incoming wave") and the "scattered wave" $\underline{s}_d$, an additive displacement caused by the defect, which is determined by

$$(\overset{\circ}{\Phi} + \psi)(v\underline{R} + \underline{s}_d) = 0 \ . \tag{6.52}$$

Because $\overset{\circ}{\Phi} v\underline{R} = 0$, one has

$$(\overset{\circ}{\Phi} + \psi)\underline{s}_d = -\psi v\underline{R} \ , \tag{6.53}$$

and therefore

$$\underline{s}_d = -G\psi v\underline{R} = -\overset{\circ}{G} t v\underline{R} \ . \tag{6.54}$$

All quantities in this section refer to $\omega = 0$, e.g., $\overset{\circ}{G} = \overset{\circ}{G}(\omega = 0)$ is the static Green's function of the perfect lattice.

## 6.6.2 Induced Kanzaki Forces and the Induced Double Force Tensor

One recognizes from (6.54) that $-\psi v\underline{R}$ are the forces in the defect lattice which produce $\underline{s}_d$; they are illustrated in Fig. 6.21 for the octahedral interstitial. We have seen in Section 6.2 that it is advantageous to define Kanzaki forces $\underline{K}$ which produce the same displacement in the perfect crystal; obviously

$$\underline{K} = -tv\underline{R} = -t\varepsilon\underline{R} \ , \quad K_i^{\underline{m}} = -t_{is}^{\underline{ml}} X_t^{\underline{l}} v_{st} = -t_{is}^{\underline{ml}} X_t^{\underline{l}} \varepsilon_{st} \ . \tag{6.55}$$

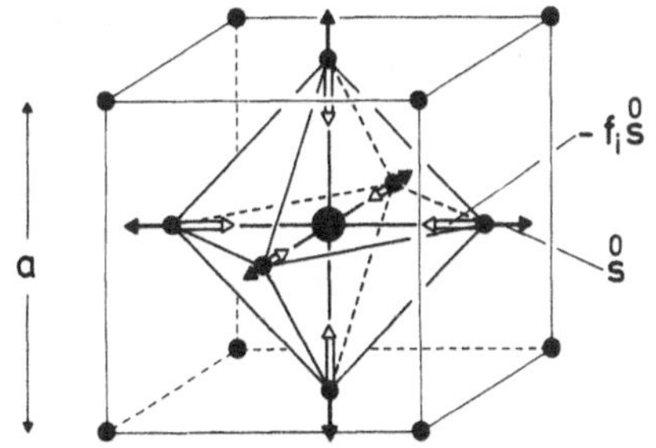


Fig. 6.21. Forces on the six octahedral atoms caused by a homogeneous dilatation: $v_{ik} = \delta_{ik}(2\overset{\circ}{s}/a)$. Without the spring star all atoms are force-free. With the spring star the six atoms are under radial forces $-f_i\overset{\circ}{s}$, which represent the term $-\psi v\underline{R}$ in (6.53)

Here v can be replaced by $\varepsilon$ because $\Phi$, $\overset{\circ}{\Phi}$, $\psi$, $\overset{\circ}{G}$, t must be rotational invariant (comp. Sec. 2.2.2).

The double force tensor of the defect is

$$P^{d}_{ki} = X^{\underline{m}}_{k} K^{\underline{m}}_{i} = \alpha_{ki,st}\varepsilon_{st} , \quad P^{d} = \alpha\varepsilon , \tag{6.56}$$

where

$$\alpha_{ki,st} = -X^{\underline{m}}_{k} t^{\underline{ml}}_{is} X^{\underline{l}}_{t} \tag{6.56a}$$

is called the (diaelastic) *polarizability* of the defect. It is a 4th rank tensor with the symmetries of the elastic moduli. The invariance against exchange of k with i, s with t follows from rotational invariance, and exchange of the pairs (ki) with (st) is possible because t is a symmetrical matrix. Translational symmetry says that $\alpha$ is invariant against a common translation, $\underline{R}^{\underline{m}} \to \underline{R}^{\underline{m}} + \underline{T}$.

In general the double force tensor of a defect does not vanish for $\varepsilon = 0$; rather one has

$$P^{d}(\varepsilon) = P^{d}(0) + \alpha\varepsilon \tag{6.57}$$

where $P^{d}(0)$ is the permanent and $\alpha\varepsilon$ the induced contribution. In analogy to the physics of dipoles in electricity and magnetism the double force tensor of a defect in an external field $\varepsilon$ contains two contributions: a temperature dependent paraelastic part, produced by orientation of the defects, vanishing e.g., if $P^{d}(0)$ has cubic symmetry but possible for dumbbells (comp. Sec. 6.7), and a diaelastic induced contribution, $\alpha\varepsilon$.

The forces $-\psi v\underline{R}$, which are illustrated in Fig. 6.21 for the octahedral interstitial, have even symmetry. Therefore, no odd modes come into play or, in other words, these forces leave the interstitial at rest. Only the even modes contribute to $\alpha$ and the octahedral interstitial is statically equivalent to the octahedral bond charge (or also to the fixed interstitial). This is generally true for defects with inversion symmetry: only the even modes couple to a homogeneous deformation. The even modes are

mode $|1\rangle$ for the diatomic defect, the other modes drop out for $\omega = 0$ ;
modes $|1,2,3\rangle$ for the octahedral interstitial ; (6.58)
modes $|1,2,3,4,5,6\rangle$ for the substitutional atom (or for a vacancy) .

A counterexample is the tetrahedral interstitial or the tetrahedral bond charge which has been discussed in detail in Section 5.6 . Here the interstitial moves under shears (to avoid any spring tensions) and it stays under dilatations.

### 6.6.3 The Change of Elastic Moduli

In Section 5.6 we have already discussed the change of elastic moduli for small concentrations of defects, $c \ll 1$; the result was

$$\delta C = -\frac{c}{V_c}\alpha \, , \qquad \delta C_{ki,st} = \frac{c}{V_c} X_k^{\underline{m}} t_{is}^{\underline{mn}} X_t^{\underline{n}} \, , \tag{6.59}$$

and in Born's approximation ($t \to \psi$)

$$\delta C_{ki,st} = \frac{c}{V_c} X_k^{\underline{m}} \psi_{is}^{\underline{mn}} X_t^{\underline{n}} = \frac{Nc}{V} X_k^{\underline{m}} \psi_{is}^{\underline{mn}} X_t^{\underline{n}} \, . \tag{6.60}$$

Because of translational invariance the defect can be located anywhere. If $\overset{\underline{\mu}}{\psi}{}^{\underline{mn}}$ refers to $\underline{\mu}$ as the position of the defect, the result (6.60) is independent of $\underline{\mu}$; therefore, we can rewrite (6.60):

$$\delta C_{ki,st} = \frac{Nc}{V} X_k^{\underline{m}} \overset{\underline{\mu}}{\psi}{}_{is}^{\underline{mn}} X_t^{\underline{n}} = \frac{c}{V} X_k^{\underline{m}} \sum_{\underline{\mu}} \overset{\underline{\mu}}{\psi}{}_{is}^{\underline{mn}} X_t^{\underline{n}} \, , \tag{6.60a}$$

where $\sum_{\underline{\mu}} \overset{\underline{\mu}}{\psi}{}^{\underline{mn}}$ depends only on $\underline{m}-\underline{n}$ and means, e.g., for the diatomic defect, that *all* springs between $\underline{R}^{\underline{m}}$ and $\underline{R}^{\underline{m}} + a(0,0,1)$ are $f_d$. Correspondingly, the coupling $c\sum_{\underline{\mu}} \overset{\underline{\mu}}{\psi}{}^{\underline{m},\underline{n}}$ represents springs $cf_d$ homogeneously distributed in the lattice. The contribution of this coupling (with translational symmetry) to the elastic data according to (5.4) would be

$$\delta C_{ki,st} = \frac{c}{V} \sum_{\underline{\mu}} X_k^{\underline{m}} \overset{\underline{\mu}}{\psi}{}_{is}^{\underline{mn}} X_t^{\underline{n}}$$

and, because the single terms do not depend on $\underline{\mu}$, we have

$$\delta C_{ki,st} = \frac{Nc}{V} X_k^{\underline{m}} \overset{o}{\psi}{}_{is}^{\underline{mn}} X_t^{\underline{n}}$$

which is identical with (6.60); hence Born's approximation results in an homogeneous crystal with an additive coupling $c\sum_{\underline{\mu}} \overset{\underline{\mu}}{\psi}{}^{\underline{mn}}$, i.e., with additive springs "$cf_d$". This is called the "virtual crystal approximation" where $\Phi$ is replaced by its average over all defect positions; it will be more thoroughly discussed in Chapter 9.

### 6.6.4 The (Diaelastic) Polarizability of Simple Defects

We illustrate the calculation of $\alpha$, as given by (6.56a), by some simple examples.

For the *diatomic defect* we have

$$\alpha_{ki,st} = -X_k^{\underline{m}} \langle {}_i^{\underline{m}} | 1\rangle \overset{1}{t} \langle 1 | {}_s^{\underline{n}} \rangle X_t^{\underline{n}} \, , \tag{6.61}$$

where, e.g., $\langle {}_i^{\underline{m}} | 1\rangle$ is the $(\underline{m},i)$-component of state $|1\rangle$ (for $| {}_i^{\underline{m}} \rangle$ comp. App. B7). The only non-vanishing element of $\alpha$ is $\alpha_{33,33}$:

$$\alpha_{33,33} = -\frac{a}{\sqrt{2}}\,\frac{2f_d}{1 + 2f_d\overset{1}{A}(\omega = 0)}\,\frac{a}{\sqrt{2}} = -\frac{a^2}{2}\overset{1}{t}\ ; \tag{6.61a}$$

therefore, only $c_{33}$ is changed:

$$\delta c_{33} = \frac{c}{V_c}\frac{a^2}{2}\overset{1}{t} \quad \text{with } \overset{1}{t} = \frac{2f_d}{1 + 2f_d\overset{1}{A}} = \frac{2f_d}{1 + f\overset{1}{A}(2f_d/f)}\ . \tag{6.61b}$$

Fig. 6.22 shows $\overset{1}{t} \propto \delta c_{33}$ vs. $f_d$. One recognizes that for very large $f_d$ nothing dramatical happens: the spring $f_d$ of Born's approximation is replaced by $1/(2\overset{1}{A}) \cong f/0.7$, which corresponds to a localized mode with very high frequency. On the other hand for negative $f_d$ and $1 + 2f_d\overset{1}{A} \gtrsim 0$, $f_d \gtrsim -1/(2\overset{1}{A}) \cong -f/0.7$, the change of $c_{33}$ becomes negative and very large. Inspection shows that this behaviour is connected with a low frequency resonance. The frequency dependent denominator is $1 + 2f_d\overset{1}{A}_1(\omega^2) + i2f_d\overset{1}{A}_1(\omega^2)$. Vanishing of the real part for low frequencies gives $1 + 2\overset{1}{A}_1(0) + 2f_d\overset{1}{A}{}_1'(0)\omega_R^2 = 0$. The denominator of (6.61b) is therefore proportional to $\omega_R^2$, and $\delta c_{33}$ becomes large if $\omega_R$ is small. That the change of C must be negative for a resonance can be seen from

$$\overset{1}{t} = \frac{2f_d}{1 + 2f_d\overset{1}{A}} = 2f_d - \frac{(2f_d)^2\overset{1}{A}}{1 + 2f_d\overset{1}{A}}\ .$$

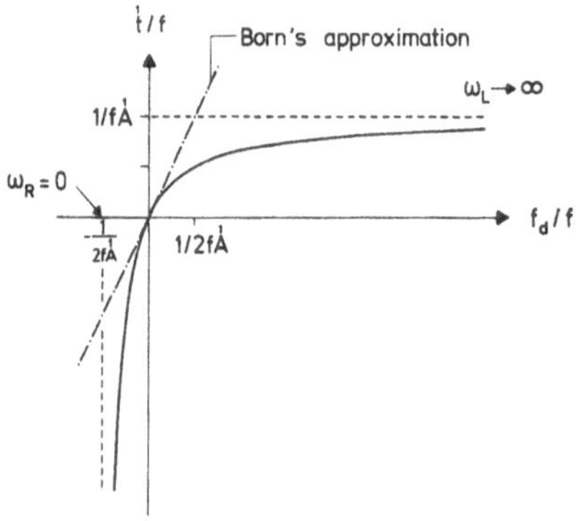


Fig. 6.22. Static t-matrix for the diatomic defect vs. defect spring, $f_d$. f is the longitudinal spring of the 1st neighbour model. For very large $f_d$ (corresponding to a high-frequency localized mode) $\overset{1}{t}$ tends to a finite value. In contrast, for negative $f_d$ ($2\overset{1}{A}f_d \gtrsim -1$) one obtains a low-frequency resonant mode, and $\overset{1}{t}$ tends to $-\infty$

In the static case $\overset{0}{G}$ is positive; therefore $\langle 1|\overset{0}{G}|1\rangle = \overset{1}{A} > 0$. G must be positive because of stability; therefore

$$\langle 1|G|1\rangle = \overset{1}{A} - \frac{2f_d\overset{1}{A}{}^2}{1 + 2f_d\overset{1}{A}} = \overset{1}{A}/(1 + 2f_d\overset{1}{A}) > 0 \quad \text{and} \quad 1 + 2f_d\overset{1}{A} > 0\ .$$

Consequently $1 + 2f_d\overset{1}{A}$ can become very small but must stay positive because of stability;the limit of stability is $1 + 2f_d^{st}\overset{1}{A} = 0$ or $\omega_R^2 = 0$. Since $\overset{1}{A} > 0$ the second term of $\overset{1}{t}$ above becomes very large and negative near the stability limit.

Obviously $\delta C$ is not cubic. However, if the three orientations of the diatomic defect are present in equal concentrations $c_1 = c_2 = c_3 = c/3$, then one has to average over the three orientations and obtains a cubic $\delta C$: $\delta c_{11} = \delta c_{22} = \delta c_{33} = ca^2 t/6V_c$.

The polarizability of a defect with cubic symmetry has also cubic symmetry. Consequently, $\alpha$ can be represented by the six basis tensors, $T(\nu = 1...6)$, of Section 4.1.3 (comp. also Sec. 4.4.2):

$$\alpha = \sum_\nu |T(\nu)\rangle \overset{\nu}{\alpha} \langle T(\nu)| \ ;$$

$$\overset{\nu}{\alpha} = \langle T(\nu)|\alpha|T(\nu)\rangle = -\langle T(\nu)\underline{R}|t|T(\nu)\underline{R}\rangle = -\langle T(\nu)\underline{R}|\nu\rangle^2 \overset{\nu}{t} \ ; \ \not\sum \ ; \tag{6.62}$$

$$\delta\overset{\nu}{C} = -\frac{c}{V_c} \langle T(\nu)\underline{R}|\nu\rangle^2 \overset{\nu}{t} \ ; \quad \overset{\nu}{t} = \overset{\nu}{\psi}/(1 + \overset{\nu}{A}\overset{\nu}{\psi}) \ .$$

For the *octahedral interstitial* (or the octahedral bond charge) one obtains ($f_i = 2f_d$)

$$\delta\overset{1}{C} = \frac{c}{V_c}\frac{a^2}{2}\overset{1}{t} \ ; \quad \overset{1}{t} = \frac{2f_d}{1 + \overset{1}{A}2f_d} \ ; \quad \overset{1}{A} = 0.18/f \ ; \tag{6.63a}$$

$$\delta\overset{2,3}{C} = \frac{c}{V_c}\frac{a^2}{2}\overset{2}{t} \ ; \quad \overset{2}{t} = \frac{2f_d}{1 + \overset{2}{A}2f_d} \ ; \quad \overset{2}{A} = 0.43/f \ ; \tag{6.63b}$$

$$\delta\overset{4}{C} = \delta\overset{5}{C} = \delta\overset{6}{C} = 0. \tag{6.63c}$$

The dependence of $\overset{1}{t}$ and $\overset{2}{t}$ on $2f_d/f$ is very similar to that of the diatomic defect. Because $\overset{2}{A} > \overset{1}{A}$, a resonance effect is possible for $\delta\overset{2}{C}$ if $2f_d \gtrsim -1/\overset{2}{A}$.

For the *substitutional atom* we have

$$\delta\overset{1}{C} = \frac{c}{V_c} 2a^2\overset{1}{t} \ ; \quad \overset{1}{t} = \frac{f_d}{1 + \overset{1}{A}f_d} \ ; \quad \overset{1}{A} = 0.24/f \ ; \tag{6.64a}$$

$$\delta\overset{2,3}{C} = \frac{c}{V_c}\frac{a^2}{2}\overset{2}{t} \ ; \quad \overset{2}{t} = \frac{f_d}{1 + \overset{2}{A}f_d} \ ; \quad \overset{2}{A} = 0.38/f \ ; \tag{6.64b}$$

$$\delta\overset{4,5,6}{C} = \frac{c}{V_c} a^2\overset{5}{t} \ ; \quad \overset{5}{t} = \frac{f_d}{1 + \overset{5}{A}f_d} \ ; \quad \overset{5}{A} = 0.33/f \ . \tag{6.64c}$$

Here a resonance effect is possible for $\delta\overset{2}{C}$.

For the *vacancy*, where $f_d$ has to be replaced by $-f$:

$$\overset{1}{\delta C} = -8c\,\frac{f}{a}\,\frac{1}{1-0.24}\ ;\quad \overset{2}{\delta C} = -2c\,\frac{f}{a}\,\frac{1}{1-0.38}\ ;\quad \overset{5}{\delta C} = -4c\,\frac{f}{a}\,\frac{1}{1-0.33} \tag{6.65}$$

and with the data of the perfect crystal,

$$\overset{1}{C} = 4f/a\ ,\quad \overset{2}{C} = f/a\ ,\quad \overset{5}{C} = 2f/a\ , \tag{6.65a}$$

one obtains the relative changes

$$\frac{1}{c}\,\frac{\overset{1}{\delta C}}{\overset{1}{C}} = -2.64\ ,\quad \frac{1}{c}\,\frac{\overset{2}{\delta C}}{\overset{2}{C}} = -3.20\ ,\quad \frac{1}{c}\,\frac{\overset{5}{\delta C}}{\overset{5}{C}} = -3.00 \tag{6.65b}$$

(read: $\frac{1}{c}\,\overset{1}{\delta C}/\overset{1}{C}$ is the relative change of $\overset{1}{C}$ in % for 1% concentration). Born's approximation can be expressed by an average spring: in the perfect crystal one has $N\cdot 6f$ normal springs from which in the defect lattice $N_d\cdot 12f$ springs must be subtracted (12 springs per vacancy). The average spring is then $f(1-2N_d/N) = f(1-2c)$, which would give in Born's approximation

$$\overset{1}{\delta C} = 4(-2cf)/a\ ;\quad \overset{2}{\delta C} = (-2cf)/a\ ;\quad \overset{5}{\delta C} = 2(-2cf)/a\ ; \tag{6.65c}$$

this agrees with (6.65) if the denominator is replaced by 1. One recognizes that Born's approximation gives a satisfactory result for the vacancy.

For the sake of completeness we give the result for the *tetrahedral interstitial*. From the discussion in Section 5.6 we know already that there are no contributions to the shear moduli. Only the compressibility changes; the interstitial does not move under a dilatation, and we have

$$\overset{1}{\delta C} = \frac{c}{V_c}\,\frac{a^2}{4}\,\overset{1}{t}\ ;\quad \overset{1}{t} = \frac{w}{1+\overset{1}{A}w}\ ;\quad \overset{1}{A} = 0.18/f\ , \tag{6.66}$$

which agrees with (5.23a) (Born's approximation) if $w/f \ll 1$.

## 6.7 Paraelastic Effects

The influence of an external strain, $\varepsilon$, on defects has been briefly discussed in Section 4.8.8. We will not discuss the Gorski effect ($\underline{r}$-dependent $\varepsilon = \varepsilon(\underline{r})$: spatial redistribution of defects by diffusion, e.g., hydrogen in bcc metals), but the Snoek effect ($\varepsilon$ independent of $\underline{r}$: redistribution of defect orientations, e.g., of dumbbell interstitials, under strain). The interaction energy W between a defect with double force tensor P and an external strain $\varepsilon$ is given by (4.99), $W = -(P,\varepsilon)$. If the defect possesses different but otherwise equivalent orientations[23], denoted by $\lambda$, the pro-

[23] Examples are self-interstitial dumbbells, where the ⟨100⟩, ⟨111⟩,⟨110⟩ dumbbells have 3,4,6 equivalent orientations (comp. /6.1/).

bability for orientation $\lambda$ is (in thermal equilibrium) proportional to $\exp(-W^\lambda/kT)$, and the average P becomes

$$\langle P\rangle = \frac{\sum_\lambda P^\lambda \exp[(P^\lambda,\varepsilon)/kT]}{\sum_{\lambda'} \exp[(P^{\lambda'},\varepsilon)/kT]} \tag{6.67}$$

If $P^\lambda$ does not depend on $\lambda$, the exponentials cancel out and $\langle P\rangle = \sum_\lambda P^\lambda/(\sum_{\lambda'} 1) = P_o$ does not depend on $\varepsilon$ (vanishing polarizability). The differences in $W^\lambda$ are usually so much smaller than kT, that the exponentials can be expanded in powers of $\varepsilon$. Most conveniently one introduces

$$P^\lambda = P_o + \delta P^\lambda \ , \tag{6.68}$$

which leads to

$$\langle P\rangle = \frac{\sum_\lambda (P_o + \delta P^\lambda) \exp[(\delta P^\lambda,\varepsilon)/kT]}{\sum_{\lambda'} \exp[(\delta P^{\lambda'},\varepsilon)/kT]} \ . \tag{6.67a}$$

In an expansion in powers of $\varepsilon$ the denominator does not contribute to the linear term, $(\sum_\lambda \delta P^\lambda = 0)$, and we obtain

$$\langle P\rangle = P_o + \sum_\lambda \delta P^\lambda(\delta P^\lambda,\varepsilon)/ZkT \ , \tag{6.67b}$$

where Z is the number of orientations, $Z = \sum_\lambda 1$. The *paraelastic polarizability* is then

$$\alpha_{ki,st} = \sum_\lambda \delta P^\lambda_{ki}\, \delta P^\lambda_{st}/ZkT \ . \tag{6.69}$$

For cubic symmetry $P_o$ must be scalar, and $\alpha$ must have the symmetries of the cubic moduli. Therefore,

$$P_o = \frac{1}{3}\,\mathrm{tr}\{P^\lambda\} \tag{6.70}$$

and, according to (6.62),

$$\overset{\nu}{\alpha} = \frac{1}{ZkT} \sum_\lambda \langle T(\nu)|\delta P^\lambda\rangle \langle \delta P^\lambda|T(\nu)\rangle \ , \ \forall_\nu \ : \tag{6.71}$$

$$\overset{1}{\alpha} = \frac{1}{ZkT} \sum_\lambda \frac{1}{3}\,(\mathrm{tr}\{\delta P^\lambda\})^2 = 0 \ , \tag{6.72}$$

because of (6.68,70) ; no reorientation under dilatation or pressure;

$$\overset{2}{\alpha} = \frac{1}{ZkT}\sum_{\lambda} \frac{1}{2}(P^{\lambda}_{11} - P^{\lambda}_{22})^2 = \frac{1}{6ZkT}\sum_{\lambda}\left[(P^{\lambda}_{11} - P^{\lambda}_{22})^2 + (P^{\lambda}_{22} - P^{\lambda}_{33})^2 + (P^{\lambda}_{33} - P^{\lambda}_{11})^2\right] ; \tag{6.73}$$

$$\overset{5}{\alpha} = \frac{1}{ZkT}\sum_{\lambda} 2(P^{\lambda}_{13})^2 = \frac{2}{3ZkT}\sum_{\lambda}\left[(P^{\lambda}_{13})^2 + (P^{\lambda}_{21})^2 + (P^{\lambda}_{32})^2\right] . \tag{6.74}$$

In $\overset{2}{\alpha}$ and $\overset{5}{\alpha}$ one can use $P^{\lambda}$ instead of $\delta P^{\lambda}$, because the contributions from $P_o$ drop out; the sums there have been supplemented such that the summands are now invariant under cubic symmetry operations. Consequently, the summands do no longer depend on $\lambda$:

$$\overset{2}{\alpha} = \frac{1}{6kT}\left[(P^{\lambda}_{11} - P^{\lambda}_{22})^2 + (P^{\lambda}_{22} - P^{\lambda}_{33})^2 + (P^{\lambda}_{33} - P^{\lambda}_{11})^2\right] , \tag{6.73a}$$

$$\overset{5}{\alpha} = \frac{2}{3kT}\left[(P^{\lambda}_{13})^2 + (P^{\lambda}_{21})^2 + (P^{\lambda}_{32})^2\right] , \quad (\text{any } \lambda) . \tag{6.74a}$$

The single P's of self-interstitial dumbbells in ⟨100⟩, ⟨111⟩, ⟨110⟩ orientations have the symmetry of the corresponding coupling matrices (comp. Sec. 3.2.4); they have longitudinal and transversal eigenvalues, $P_l$, $P_{t'}$, $P_t$.

1) For a ⟨100⟩ dumbbell[24]:

$$P_l, P_{t'} = P_t \quad ; \quad \overset{2}{\alpha} = \frac{(P_l - P_t)^2}{3kT} , \quad \overset{5}{\alpha} = 0 ; \tag{6.75}$$

in Al (comp. /6.1/): $P_l \cong 15$, $P_t \cong 16$ [eV]; $\overset{2}{\alpha} \cong 1(\mathrm{eV})^2/(3kT)$.

2) For a ⟨111⟩ dumbbell:

$$P_l, P_{t'} = P_t \quad ; \quad \overset{2}{\alpha} = 0 , \quad \overset{5}{\alpha} = \frac{2}{kT}\left(\frac{P_l - P_t}{3}\right)^2 . \tag{6.76}$$

3) For a ⟨110⟩ dumbbell:

$$P_l, P_{t'}, P_t \; ; \quad \overset{2}{\alpha} = \frac{1}{3kT}\left(\frac{P_l + P_{t'}}{2} - P_t\right)^2 , \quad \overset{5}{\alpha} = \frac{4}{3kT}\left(\frac{P_l - P_{t'}}{2}\right)^2 ; \tag{6.77}$$

in Mo (comp. /6.1/): $P_l = 32$, $P_{t'} = 9$, $P_t = 38$[eV]; $\overset{2}{\alpha} \cong 100(\mathrm{ev})^2/(kT)$, $\overset{5}{\alpha} \cong 180(\mathrm{eV})^2/(kT)$.

The diaelastic polarizabilities, treated in the foregoing section, are of the order 1 eV, in resonance cases may be 10 eV. The paraelastic values are several orders of magnitude larger. However, at low temperatures the orientations cannot change, since the activation energy for reorientation is too large; then only the diaelastic polarizability remains.

---

24 The parameter $(P_t - P_l)\varepsilon/kT$, which must be small for the expansion (6.69), becomes $10^{-3}$ for $\varepsilon = 10^{-6}$, T = 10 K.

## 6.8 Remarks on Localized and Resonant States

### 6.8.1 The Isotopic Defect

In Section 6.4.5 we have extensively treated the isotopic defect; in particular, the local spectrum $Z_d$ and the projection of the defect Green's function into the defect space

$$PGP = A\frac{1}{1+\chi A}P\ , \quad A = \overset{\circ}{G}{}^{(0)}(\omega^2)\ , \quad \chi = -(M_d - \overset{\circ}{M})\omega^2\ , \quad A = A_1 + iA_2\ , \tag{6.78}$$

have been discussed in great detail. The behaviour with respect to localized and resonant states was as follows:

$$\begin{aligned}&\textit{well localized state} \text{ for } \omega_L \gg \Omega_{max}\ ,\\ &1 + \chi(\omega_L^2)A_1(\omega_L^2) = 1 + \chi(\omega_L^2)A(\omega_L^2) = 0\ , \quad M_d \ll \overset{\circ}{M}\ ;\end{aligned} \tag{6.79a}$$

$$\begin{aligned}&\textit{localized state } (\omega_L \gtrsim \Omega_{max}) \textit{ and/or quasiresonant state } (\omega_{QR} \lesssim \Omega_{max}) \text{ for}\\ &1 + \chi(\Omega_{max}^2)A_1(\Omega_{max}^2) \cong 0\ , \quad M_d \cong \text{critical mass } \big(\text{comp. (3.80a)}\big);\end{aligned} \tag{6.79b}$$

$$\textit{resonant state } \omega_R \ll \Omega_{max}\ , \quad 1 + \chi(\omega_R^2)A_1(\omega_R^2) = 0\ , \quad M_d \gg \overset{\circ}{M}\ . \tag{6.79c}$$

For more general mass changes the behaviour of the Green's function PGP is quite similar. Consider the diatomic defect without spring change (Fig. 6.8), where PGP can be separated into contributions of the form (6.78); the real parts of the various A's show, qualitatively, the same dependence on $\omega^2$ as $\mathrm{Re}\{\overset{\circ}{G}{}^{(0)}\}$, and the above classification of resonant and localized states is applicable.

### 6.8.2 Spring Changes

Similarly one can discuss the Green's function PGP for defects which include spring changes. Consider mode $|1\rangle$ of the diatomic defect without mass change (Fig. 6.8, $M_d = \overset{\circ}{M}$), where

$$PGP = A\frac{1}{1+\chi A}\ , \quad A = \overset{\circ}{G}{}^{(0)} - \overset{\circ}{G}{}_1^{(200)}\ , \quad \chi = 2f_d\ . \tag{6.80}$$

Again the qualitative behaviour of A is not changed:

$$\begin{aligned}&\textit{well localized state} \text{ for large (positive) } 2f_d\ ,\\ &A(\omega^2 > \Omega_{max}^2) = A_1(\omega^2 > \Omega_{max}^2) < 0\ ;\end{aligned} \tag{6.80a}$$

*quasiresonant and/or localized state* for $1 + 2f_d A_1(\Omega^2_{max}) \cong 0$ ; (6.80b)

*resonant states* at $\omega_R \ll \Omega_{max}$ for negative $f_d = -\bar{f}_d < 0$ near the stability limit, $1 - 2\bar{f}_d A_1(0) \gtrsim 0$ , $1 - 2\bar{f}_d A_1(0) - 2\bar{f}_d A_1'(0)\omega_R^2 = 0$ . (6.80c)

In contrast to the mass resonance, here the resonant mode, |1>, is even and, therefore, greatly contributes to the decrease of the appropriate elastic moduli.

Also for more complicated spring and mass changes the classification (6.79) still applies.

### 6.8.3 The General Procedure

In general, to determine localized and resonant modes one starts from the Fourier transform of the equation of motion

$$\underline{s}(\omega) = G(\omega)\underline{F}(\omega) \; ; \tag{6.81}$$

if, in particular, forces are acting only in P, $P\underline{F} = \underline{F}$, one obtains the dynamics of the defect space from

$$P\underline{s} = PGP\underline{F} \quad \text{or} \quad \frac{1}{PGP} P\underline{s} = P\underline{F} \; , \tag{6.82}$$

where the influence of the host lattice is contained in $PGP = G_{dd}(\omega^2)$. Because $G_{dd}(\omega^2)$ is a symmetrical matrix in defect space, it can be expanded in terms of its $\omega$-dependent eigenfunctions and eigenvalues. Often, e.g., in all our examples given so far, the eigenvectors are already determined by symmetry and do not depend on $\omega$. The eigenmodes can then be classified into localized and resonant ones just as in Sections 6.8.1,2. In particular, localized states are determined by

$$\det\left[G_{dd}^{-1}(\omega_L^2 > \Omega_{max}^2)\right] = 0 \; , \tag{6.83}$$

i.e., the condition to have a force-free solution of (6.82). If one has calculated $\omega_L$ and the corresponding eigenmode of $G_{dd}$, $P\underline{s}_L$, in defect space, the (exponentially decreasing) displacements in the remaining lattice are obtained from

$$Q\underline{s}_L = Q\overset{o}{G}(\omega_L^2)\, P_X P\underline{s}_L \; . \tag{6.83a}$$

Resonances can be found from solutions of $\det\left[\mathrm{Re}\{G_{dd}^{-1}(\omega_R^2 \ll \Omega_{max}^2)\}\right] = 0$; the corresponding scattering states (not localized near the defect but extended over the whole lattice) can be obtained similarly to (6.83,83a).

The above discussion applies as well to defects with additional degrees of freedom (interstitial atoms). As we have demonstrated for the octahedral interstitial (Section 6.5.3), also near $\Omega_{max}$ well localized and resonant states are possible.

### 6.8.4 Analogy to Quantum Theory

The eigenstates of lattice theory are obtained from

$$\Phi\underline{s} = -\overset{\circ}{M}\underline{\ddot{s}} = \overset{\circ}{M}\omega^2\underline{s} = \overset{\vee}{\Phi}\underline{s}\ , \qquad \Phi = \overset{\circ}{\Phi} + \psi\ , \tag{6.84}$$

if, for the sake of simplicity, we consider equal masses only. We have learned, that $\overset{\vee}{\Phi} < 0$ must be excluded due to stability, that $0 \leqslant \overset{\vee}{\Phi} \leqslant M\Omega^2_{max}$ represents scattering states (extended over the whole lattice), and that $\overset{\circ}{M}\Omega^2_{max} < \overset{\vee}{\Phi}$ leads to localized states.

Often the lattice equation is compared with a "quantum theoretical (time independent Schrödinger) equation"

$$\Phi\underline{s} = E\underline{s}\ , \qquad "E" = \overset{\circ}{M}\omega^2\ , \tag{6.85}$$

where $\Phi$ represents the "Hamilton operator" with $\overset{\circ}{\Phi} > 0$ as "kinetic energy" and $\psi$ as "potential energy". In lattice theory, however, (localized) values $E = \overset{\vee}{\Phi} < 0$, which in quantum theory correspond to the familiar bound states, are not admitted, whereas the resonant states of lattice theory with $E = M\omega_R^2 \gtrsim 0$ (i.e. $\omega_R^2 \ll \Omega^2_{max}$) have a direct counterpart in quantum mechanics: the quantum theoretical resonances are "almost bound states" in the sense that small changes in a potential parameter can transform them into actual bound states, $E \lesssim 0$. The localized states of lattice theory with $\omega_L^2 \gg \Omega^2_{max}$ as well as the localized states $\omega_L^2 \gtrsim \Omega^2_{max}$ and the quasiresonant states for $\omega_R^2 \lesssim \Omega^2_{max}$ can be compared with the electronic states in a perturbed periodic potential with energies close to a band edge. This correspondence is, for example, reflected in the spatial behaviour of the lattice Green's function for localized frequencies $\omega_L^2 > \Omega^2_{max}$: as can be seen from the linear example (3.71b) $G^{(h)}(\omega^2 > \Omega^2_{max}) \propto (-1)^h \exp(-\kappa_P h)$ does not monotonically decrease with distance as $G^{(h)}(\omega^2 < 0) \propto \exp(-\kappa_P h)$ would do.

# 7. Scattering of Neutrons and X-rays by Crystals

/7.1-4/

The scattering of neutrons and X-rays by crystals is important for crystal physics. The cross sections can be expressed by the displacement-displacement correlation functions introduced in Sections 2.3.3 and 2.4.4. In turn, the correlation functions give the dispersion curves from which the force parameters and the type of forces can be deduced. For defect crystals scattering is decisive in determining the structure of point defects (mostly by X-ray scattering) and their dynamics (by neutron scattering, e.g., change of dispersion curves by resonances). For these reasons a discussion of scattering theory is mandatory if only to become acquainted with the underlying assumptions and approximations.

Within the frame of this article it is not possible to give even a short introduction into quantum mechanical scattering theory. What we can do is to briefly sketch the theory, point out the most essential steps and approximations and try to embellish the sketch with some supplementary remarks.

## 7.1 Energy and Momentum Transfers in Scattering

Because the crystal is a dynamical system it can change its energy during the scattering process. The interaction of the incoming particle with the crystal can create ($\hbar\omega < 0$) or annihilate ($\hbar\omega > 0$) "phonons". These "phonons" are lattice waves $\left(\underline{k},\sigma;\Omega(\underline{k}\sigma)\right)$ with quantized integer oscillator energies in steps of $\hbar\Omega$ and "quasimomenta[1]" $\hbar\underline{k}$. Fig. 7.1 describes a process in which one phonon $\left(\hbar\underline{\kappa},\hbar\Omega(\underline{\kappa}) = \hbar\omega\right)$ has been annihilated and has transferred its momentum and energy to the scattered particle. Of course, two or more phonons can be annihilated, but mainly we will have in mind one-phonon processes (dominant at low temperatures) where one phonon $\left(\underline{k},\Omega(\underline{k})\right)$ is annihilated ($\underline{\kappa} = \underline{k}, \omega = \Omega$) or produced[2] ($\underline{\kappa} = \underline{k}, \omega = -\Omega$). The maximum energy transfer is then $\pm\hbar\Omega_{max}$; for Cu we have from Fig. 3.15: $\hbar\Omega_{max} \cong 3 \times 10^{-2}$ eV. For the investigation of detailed structures the wavelength of the incoming particle must be

---

[1] Compare Section 4.3.4 and Chapter 5, p. 171.

[2] Production of phonons is possible at any temperature; annihilation of phonons is not possible at low temperatures where no phonons are present.

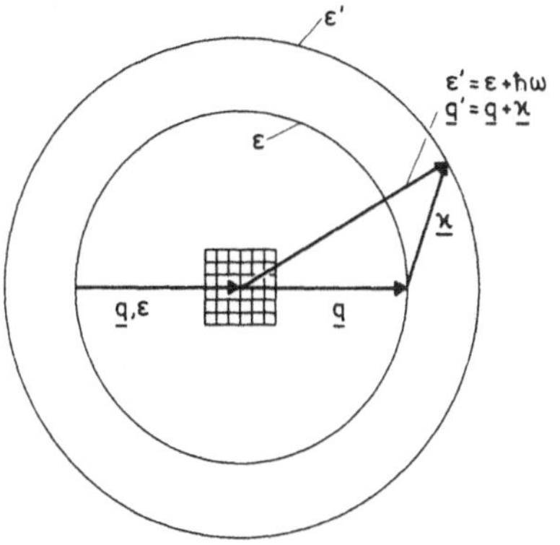


Fig. 7.1. Scattering of one particle by a crystal. Incoming particle with momentum $\hbar\underline{q}$ and energy $\varepsilon$; scattered particle with momentum $\hbar\underline{q}'$ and energy $\varepsilon'$; transferred momentum $\hbar\underline{\kappa}$, transferred energy $\hbar\omega$

at least of the order of the lattice distance a, $q = 2\pi/a$; for neutrons the energy $\varepsilon^2 = \hbar^2 q^2/2M_n$ becomes $10^{-2}$ eV corresponding to thermal neutrons of 100 K; for X-rays the energy $\varepsilon = \hbar qc$ is $3\times 10^3$ eV, where c is the velocity of light. Consequently, Fig. 7.1 refers to neutron scattering where the energy changes can be of the same order or even larger than the initial energy $\varepsilon$. For X-rays the energy changes are so small relative to the incoming energy that the surface $\varepsilon$ and $\varepsilon'$ almost coincide; in the measurements one cannot distinguish the change in energy, $q'$ approximately equals $q$, and the intensity can be observed only as a function of the direction $\hat{\underline{q}}'$. The theoretical results must be integrated over $\varepsilon'$, and one measures only the energy integrated cross section.

## 7.2 Scattering of Neutrons by a Fixed Potential

The simplest scattering problem is scattering of neutrons by a fixed potential. The "golden rule" of quantum mechanics says that the number of transitions per unit time and per unit (incoming) neutron density from a state with momentum $\hbar\underline{q}$ to final states in the interval $(\underline{q}', d\underline{q}')$ is given by[3] $(2\pi/\hbar)\,|\langle\underline{q}'|T|\underline{q}\rangle|^2\,\delta(\varepsilon'-\varepsilon)d\underline{q}'/(2\pi)^3$. The quantity $\langle\underline{q}'|T|\underline{q}\rangle$ is a transition amplitude; the scalar product is $\int d\underline{r}\,\exp(-i\underline{q}'\underline{r})\cdot$ $\cdot T\exp(i\underline{q}\underline{r})$, where $\underline{r}$ is the neutron coordinate and T the "transition operator". In "Born's approximation" T equals the potential $V(\underline{r})$ itself; T is then only a renormalized potential, just as in scattering where Born's approximation becomes exact if $\psi$ is replaced by t. The incoming wave $\exp(i\underline{q}\underline{r})$ has a (particle) current density $\hbar\underline{q}/M_n$; the (particle) density is 1 and the velocity of incoming neutrons is $\hbar q/M_n$.

---

[3] Often the name "golden rule" is associated with $(2\pi/\hbar)\,|\langle q'|T|q\rangle|^2\,\delta(\varepsilon'-\varepsilon)d\underline{q}'/(2\pi)^3$, where V is the interaction potential between incoming particle and target; this corresponds to 1st order perturbation theory.

The differential cross section, dQ, is defined by the number of neutrons $(\hbar q/M_n)$ dQ impinging on dQ which are scattered into $d\underline{q}'$ per unit time:

$$\frac{\hbar q}{M_n} dQ = \frac{1}{(2\pi)^2\hbar} |\langle \underline{q}'|T|\underline{q}\rangle|^2 \delta(\varepsilon' - \varepsilon)\, d\underline{q}' \; ; \quad \varepsilon = \frac{\hbar^2 q^2}{2M_n} \; , \quad \varepsilon' = \frac{\hbar^2 q'^2}{2M_n} \; . \tag{7.1}$$

Because of the $\delta$-function the energy of the incoming and the scattered neutrons is equal, $q' = q$; the scattering changes only the direction, and one can then ask for the differential cross section for scattering into the solid angle $d\Omega'$,

$$d\underline{q}' = q'^2 dq'\, d\Omega' = \frac{M_n q'}{\hbar^2} d\varepsilon'\, d\Omega' \; . \tag{7.2}$$

Integration over the final energies $\varepsilon'$ yields

$$\frac{\hbar q}{M_n} dQ = \frac{M_n q}{(2\pi)^2\hbar^3} |\langle \underline{q}'|T|\underline{q}\rangle|^2 d\Omega' \; , \quad \underline{q}' = q\hat{\underline{q}}' \tag{7.3}$$

$$dQ = \left(\frac{M_n}{2\pi\hbar^2}\right)^2 |\langle \underline{q}'|T|\underline{q}\rangle|^2 d\Omega' \; . \tag{7.3a}$$

For many purposes one can replace T by $(2\pi\hbar^2/M_n)A\delta(\underline{r})$, where A is the exact scattering amplitude (dimension length), which is complex in general,

$$dQ = |A|^2 d\Omega' \; . \tag{7.4}$$

This scattering amplitude is the amplitude of the outgoing scattered wave: $A(\hat{\underline{q}}')\, e^{iqr}/r$ (Fig. 7.2); well outside the range of V the scattered wave becomes the exact solution. Formally, the scattering can be described by a potential $V = 2\pi\hbar^2 A\delta(\underline{r})/M_n$ in Born's approximation. Note, however, that this procedure leads to the exact result; iteration, for instance the second Born approximation, would not improve the result. On the contrary, any "improvement" leads to a divergent result because of the singular potential proportional to $\delta(\underline{r})$.

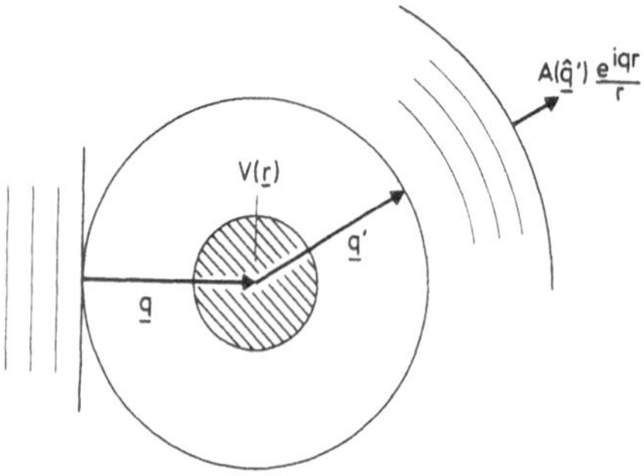


Fig. 7.2. Scattering by a fixed potential $V(\underline{r})$ produces a spherical wave with direction dependent amplitude

For thermal neutrons the scattering amplitudes are about $10^{-12}$cm (comp. /7.4/). But this description can be used as well for X-ray scattering where the amplitudes are called atomic form factors; they are of order $10^{-11}$cm (Cu).

## 7.3 Scattering of Neutrons by Crystals

The results of Section 7.2 can easily be extended to scattering by crystals if one takes into account the dynamical degrees of freedom of the lattice. The initial state $|\underline{q}\rangle$ has to be replaced by $|\underline{q},L\rangle$, where L defines the initial state of the lattice, e.g., in harmonic theory the integer occupation numbers of all oscillators. The energy must contain the lattice energy $\varepsilon \to E = E_L + \varepsilon$, $\varepsilon' \to E' = E_{L'} + \varepsilon'$, and T acts on all coordinates, the neutron coordinate $\underline{r}$ and the atomic coordinates $\underline{r}^m$, m = 1...N. The differential cross section is then

$$dQ\left(\underline{q},L \to (\underline{q}',d\underline{q}')L'\right) = \frac{M_n}{(2\pi)^2\hbar^2 q} |\langle\underline{q}',L'|T|\underline{q},L\rangle|^2 \, \delta(E' - E) \, d\underline{q}' \; . \qquad (7.6)$$

Because only the neutrons are measured, the final state, L', of the lattice does not matter; it is summed over, $\sum_{L'} \ldots$ . The initial state of the lattice, L, is not unique; the states occur with a thermal weight, $W_L \propto \exp(-E_L/kT)$, and the actual cross section for the scattered particles is the thermal average

$$dQ\left(\underline{q} \to (\underline{q}',d\underline{q}')\right) = \sum_L W_L \sum_{L'} \frac{M_n}{(2\pi)^2\hbar^2 q} |\langle\underline{q}',L'|T|\underline{q},L\rangle|^2 \, \delta(E' - E) \, d\underline{q} \; , \qquad (7.7)$$

$$\text{with } E' = E_{L'} + \varepsilon' \; , \quad E = E_L + \varepsilon \; , \quad E' - E = E_{L'} - E_L + \hbar\omega \; .$$

The *Fermi approximation* consists in replacing T by

$$T = \frac{2\pi\hbar^2}{M_n} \sum_{m=1}^{N} a_m \, \delta(\underline{r} - \underline{r}^m) = \frac{2\pi\hbar^2}{M_n} A \; , \qquad (7.8)$$

where $a_m$ is the correct individual scattering amplitude of nucleus or atom m. In this approximation multiple scattering is neglected, and the incoming wave is only scattered once by each nucleus ("kinematical theory"). This is well-justified considering the smallness of the nuclear scattering amplitudes; the mean free path of a neutron which it can travel without being scattered is of order cm.

The second assumption is that the interaction between different atoms and the neutron is well-separated. The interaction volumes in neutron scattering have nuclear size. In X-ray scattering they have the size of the electron core, and even considering thermal vibrations the interaction volumes are well-separated. In X-ray scattering, neglect of multiple scattering is also well-justified. Consequently, the Fermi approximation is quite dependable. Note again that if using (7.8) as a "potential in Born's approximation" it does not make sense to calculate higher orders to "improve" the results.

The combination of (7.8) and (7.7),

$$dQ = \sum_L W_L \sum_{L'} \frac{\hbar^2}{M_n q} |\langle \underline{q}', L'|A|\underline{q}, L\rangle|^2 \, \delta(E' - E) \, d\underline{q}' \tag{7.9}$$

contains the complete dynamics of the crystal. With (7.2) one can extract the dependence on angle $(\hat{\underline{q}}', d\Omega')$ and energy $(\varepsilon', d\varepsilon')$ of the scattered neutrons,

$$dQ = \frac{q'}{q} \sum_{L,L'} W_L \, |\langle \underline{q}', L'|A|\underline{q}, L\rangle|^2 \, \delta(E' - E) \, d\Omega' \, d\varepsilon' \; . \tag{7.10}$$

The integration over the neutron coordinate $\underline{r}$ can be carried out,

$$\langle \underline{q}', L'|A|\underline{q}, L\rangle = \langle L'|\tilde{A}(\underline{\kappa})|L\rangle \quad \text{with} \quad \tilde{A} = \sum_m a_m \, e^{i(\underline{q} - \underline{q}')\underline{r}^m} = \sum_m a_m \, e^{-i\underline{\kappa}\underline{r}^m}, \tag{7.11}$$

and with $E_{L'} - E_L + \varepsilon' - \varepsilon = E_{L'} - E_L + \hbar\omega$ we obtain

$$\frac{dQ}{d\Omega' d\varepsilon'} = \frac{q'}{q} \sum_{L,L'} W_L \, \langle L|\tilde{A}^*|L'\rangle \, \langle L'|\tilde{A}|L\rangle \, \delta(E_{L'} - E_L + \hbar\omega) \; , \tag{7.10a}$$

which leaves only matrix elements between states of the lattice. Because the states $|L'\rangle$ are complete (with respect to the lattice), $\sum_{L'} |L'\rangle\langle L'| = 1$, the sum over L' could be carried out easily, if L' were not still contained in the $\delta$-function. This can be remedied by using

$$\delta(E_{L'} - E_L + \hbar\omega) = \frac{1}{\hbar} \, \delta\!\left(\frac{E_{L'} - E_L}{\hbar} + \omega\right) = \int \frac{dt}{2\pi\hbar} \exp\left[i\left(\omega t + \frac{E_{L'} - E_L}{\hbar}\right)t\right]$$

and inserting the exponential into $\langle L'|\tilde{A}|L\rangle$; in $\langle L'|\exp(\frac{i}{\hbar} E_{L'} t) \, \tilde{A} \exp(-\frac{i}{\hbar} E_L t)|L\rangle$ one can replace the numbers $E_{L'}$ and $E_L$ by the lattice Hamiltonian $H_L$, because $|L\rangle$ and $|L'\rangle$ are eigenstates of $H_L$ with eigenvalues $E_L$, $E_{L'}$. The sum over L' can now be performed,

$$\begin{aligned} \frac{dQ}{d\Omega' d\varepsilon'} &= \frac{q'}{q} \int \frac{dt}{2\pi\hbar} \, e^{i\omega t} \sum_L W_L \, \langle L|\tilde{A}^* \exp(\tfrac{i}{\hbar} H_L t) \, \tilde{A}\cdot\exp(-\tfrac{i}{\hbar} H_L t)|L\rangle \\ &= \frac{q'}{q} \int \frac{dt}{2\pi\hbar} \, e^{i\omega t} \, \langle \tilde{A}^* \exp(\tfrac{i}{\hbar} H_L t) \, \tilde{A} \exp(-\tfrac{i}{\hbar} H_L t)\rangle_{th} \; , \end{aligned} \tag{7.12}$$

where $\langle \ldots \rangle_{th}$ denotes the thermal average. As in Section 2.3.3

$$\exp(\tfrac{i}{\hbar} H_L t) \, \tilde{A}(\underline{\kappa}, \ldots \underline{r}^m \ldots) \exp(-\tfrac{i}{\hbar} H_L t) = \tilde{A}\big(\underline{\kappa}, \ldots \underline{r}^m(t) \ldots\big) \; , \tag{7.13}$$

where $\underline{r}^m(t) = \exp(\frac{i}{\hbar} H_L t)\, \underline{r}^m(0) \exp(-\frac{i}{\hbar} H_L t)$, and

$$\frac{dQ}{d\Omega' d\varepsilon'} = \frac{q'}{q} \int \frac{dt}{2\pi\hbar} e^{i\omega t} \sum_{n,m} a_n^* a_m \langle e^{i\underline{\kappa}\underline{r}^n(0)} e^{-i\underline{\kappa}\underline{r}^m(t)} \rangle_{th} . \tag{7.14}$$

## 7.4 X-ray Scattering

We have already pointed out in Section 7.1 that for X-ray scattering the change in energy cannot be distinguished, i.e., $q' \cong q$. One can measure only the differential cross section per solid angle ($d\varepsilon' = \hbar d\omega$), integrated over $\varepsilon'$,

$$\frac{dQ}{d\Omega'} = \int dt \underbrace{\int \frac{d\omega}{2\pi} e^{i\omega t}}_{\delta(t)} \sum_{n,m} a_n^* a_m \langle e^{i\underline{\kappa}\underline{r}^n(0)} e^{-i\underline{\kappa}\underline{r}^m(t)} \rangle_{th} \tag{7.15}$$

$$= \sum_{n,m} a_n^* a_m \langle e^{i\underline{\kappa}\underline{r}^n(0)} e^{-i\underline{\kappa}\underline{r}^m(0)} \rangle_{th} .$$

Since $\underline{r}^n$ and $\underline{r}^m$ commute, we have

$$e^{i\underline{\kappa}\underline{r}^n(0)} e^{-i\underline{\kappa}\underline{r}^m(0)} = e^{i\underline{\kappa}(\underline{r}^n - \underline{r}^m)} ; \tag{7.15a}$$

the basic expression for X-ray scattering is then

$$\frac{dQ}{d\Omega'} = \sum_{n,m} a_n^* a_m \langle e^{i\underline{\kappa}(\underline{r}^n - \underline{r}^m)} \rangle_{th} , \tag{7.16}$$

where the $a_m$ are "atomic form factors" (comp. /7.1/). In a lattice, where $\underline{r}^m = \underline{R}^m + \underline{s}^m$, the average over the thermal displacements is observed.

Defects cause additive displacements and variations of form factors. The observed cross section is then the appropriate average over defect positions. The meaning of these averages is discussed in the following chapter in more detail.

In a perfect crystal, if thermal displacements can be neglected, we have

$$\frac{dQ}{d\Omega'} = \left| \sum_{\underline{m}=1}^{N} a\, e^{i\underline{\kappa}\underline{R}^{\underline{m}}} \right|^2 . \tag{7.17}$$

If $\underline{\kappa}$ is a point of the reciprocal lattice, $\underline{\kappa}\underline{R}^{\underline{m}}$ = integer multiple of $2\pi$, the cross section has a peak (Bragg peak),

$$\frac{dQ}{d\Omega'} = N^2 |a|^2 \quad \text{in Bragg peak.} \tag{7.17a}$$

Off the Bragg peaks the intensity is small (Fig. 7.3). The intensity near the Bragg

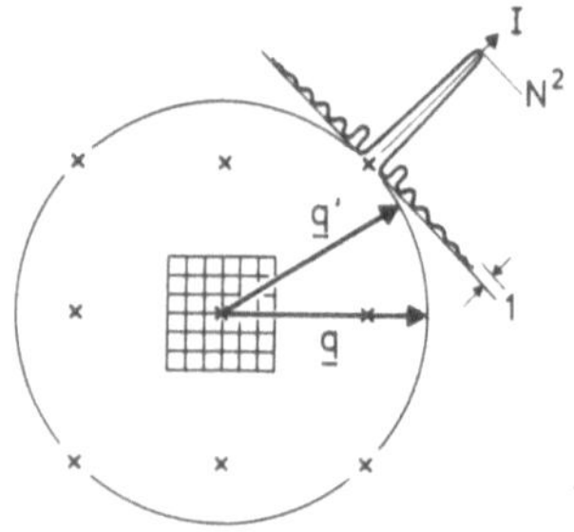


Fig. 7.3. X-ray scattering from a perfect crystal. Scattering is restricted to a sphere q' = q (Ewald sphere). The intensity is small unless q' ends in a point of the reciprocal lattice (Laue points ×) where the intensity is sharply peaked (Bragg peak)

peaks depends on the number of atoms. We will illustrate it employing a crystal with N atoms which has the shape of the original elementary cell; the atomic positions are $\underline{R}^m = m_j \underline{a}^{(j)}$ with $0 \leqslant m_j \leqslant N^{1/3} - 1$, and we obtain

$$\frac{\tilde{A}}{a} = \sum_{\underline{m}} e^{2\pi i m_j \kappa^j} = \prod_j \frac{1 - e^{2\pi i \kappa^j N^{1/3}}}{1 - e^{2\pi i \kappa^j}} = \prod_j F(\kappa^j) \quad \text{with} \quad \underline{\kappa} = 2\pi \kappa^j \underline{b}_{(j)} \ . \tag{7.17b}$$

Here $F(\xi) = \sum_{m=0}^{N^{1/3}-1} e^{2\pi i m \xi}$ has period 1, $F(\xi + 1) = F(\xi)$ ; the integral of $F^2(\xi) = \sin^2(\pi \xi N^{1/3})/\sin^2(\pi\xi)$ over one period is $N^{1/3}$, and the maximum value of $F^2(\xi)$ at $\xi = 0,1,2,\ldots$ is $F^2(0) = N^{2/3}$. Consequently, the Bragg peak has an approximate linear width of about $2\pi b N^{-1/3}$, covering a volume $(2\pi)^3/(NV_c)$ around the peak in reciprocal space.

## 7.5 The Differential Cross Section for Neutron Scattering

The evaluation of the differential cross section for scattering of neutrons by crystals requires two more assumptions in order to obtain easily interpretable results.

If we use the *harmonic approximation*, which can well be justified, then - as in classical theory - the displacement operators $\underline{s}^m(t)$, defined by $\underline{r}^m(t) = \underline{R}^m + \underline{s}^m(t)$, are linearly connected with the displacement and momentum operators at t = 0 (comp. (2.22a)). This means that the commutator of $\underline{r}^n(0)$ and $\underline{r}^m(t)$ is a number. For operators a and b, for which the commutator [a,b] = ab - ba is a number, one has

$$e^a e^b = e^{a+b+\frac{1}{2}[a,b]} = e^c \tag{7.18}$$

e.g., for $a = i\underline{\kappa}\underline{r}^n(0)$, $b = -i\underline{\kappa}\underline{r}^m(t)$. The exponent c is linear in position and momentum operators and the thermal average, now denoted by <...>, is a "Gauss average",

$$\langle \exp(c - \langle c\rangle)\rangle = \exp\left[\frac{1}{2} \langle (c - \langle c\rangle)^2\rangle\right] \ . \tag{7.19}$$

With

$$c = i\left(\underline{\kappa}\underline{r}^n(0) - \underline{\kappa}\underline{r}^m(t)\right) + \frac{1}{2}[\underline{\kappa}\underline{r}^n(0), \underline{\kappa}\underline{r}^m(t)] ,$$

$$\langle c\rangle = i\underline{\kappa}(\underline{R}^n - \underline{R}^m) + \frac{1}{2}\langle[\underline{\kappa}\underline{s}^n(0), \underline{\kappa}\underline{s}^m(t)]\rangle , \tag{7.18a}$$

$$c - \langle c\rangle = i\left(\underline{\kappa}\underline{s}^n(0) - \underline{\kappa}\underline{s}^m(t)\right)$$

we obtain from (7.18) and (7.19)

$$\begin{aligned}\langle e^{i\underline{\kappa}\underline{r}^n(0)} e^{-i\underline{\kappa}\underline{r}^m(t)}\rangle &= \langle e^c\rangle = e^{\langle c\rangle} e^{\frac{1}{2}\langle(c-\langle c\rangle)^2\rangle} \\ &= e^{i\underline{\kappa}(\underline{R}^n - \underline{R}^m)} e^{\frac{1}{2}\langle[\underline{\kappa}\underline{s}^n(0), \underline{\kappa}\underline{s}^m(t)]\rangle} e^{-\frac{1}{2}\langle(\underline{\kappa}\underline{s}^n(0) - \underline{\kappa}\underline{s}^m(t))^2\rangle} \\ &= e^{i\underline{\kappa}(\underline{R}^n - \underline{R}^m)} e^{-\frac{1}{2}\langle(\underline{\kappa}\underline{s}^n(0))^2\rangle} e^{-\frac{1}{2}\langle(\underline{\kappa}\underline{s}^m(t))^2\rangle} e^{\langle(\underline{\kappa}\underline{s}^n(0))(\underline{\kappa}\underline{s}^m(t))\rangle} ,\end{aligned} \tag{7.20}$$

and the cross section (7.14) is

$$\frac{dQ}{d\Omega' d\varepsilon'} = \frac{q'}{q}\int \frac{dt}{2\pi\hbar} e^{i\omega t} \sum_{n,m} a_n^* e^{-\frac{1}{2}\langle(\underline{\kappa}\underline{s}^n(0))^2\rangle} a_m e^{-\frac{1}{2}\langle(\underline{\kappa}\underline{s}^m(t))^2\rangle} \cdot e^{i\underline{\kappa}(\underline{R}^n - \underline{R}^m)} e^{\langle(\underline{\kappa}\underline{s}^n(0))(\underline{\kappa}\underline{s}^m(t))\rangle} . \tag{7.21}$$

Because $\langle\left(\underline{\kappa}\underline{s}^m(t)\right)^2\rangle$ does not depend on t we can introduce new temperature dependent amplitudes, $b_m = a_m \exp[-\langle(\underline{\kappa}\underline{s}^m)^2\rangle/2]$, which include the "Debye-Waller factors", $\exp[-\langle(\underline{\kappa}\underline{s}^m)^2\rangle/2]$,

$$\frac{dQ}{d\Omega' d\varepsilon'} = \frac{q'}{q}\int \frac{dt}{2\pi h} e^{i\omega t} \sum_{n,m} b_n^* b_m e^{i\underline{\kappa}(\underline{R}^n - \underline{R}^m) + \kappa_i\kappa_k l_{ik}^{nm}(t)} , \tag{7.22}$$

where

$$l_{ik}^{nm}(t) = \langle s_i^n(0)\, s_k^m(t)\rangle \tag{7.22a}$$

is the displacement-displacement correlation function (2.32c). If (7.22) is integrated over $\omega$, only the static correlation function $l_{ik}^{mn}(0)$ enters, and one comes back to the X-ray cross section (7.16). To get an order of magnitude of the average in the exponential we calculate $\langle(\underline{\kappa}\underline{s}^n)^2\rangle$ for a fcc lattice with longitudinal coupling (spring f) between 1st neighbours in the Einstein approximation,

$$\kappa_i\kappa_k \langle s_i^n s_k^n\rangle^2 = \kappa^2 \langle(s_i^n)^2\rangle = \kappa^2 \frac{\varepsilon_{th}(\Omega_E)}{M\Omega_E^2} = \kappa^2 \frac{\varepsilon_{th}(\Omega_E)}{4f} \cong \kappa^2 \frac{kT}{4f} \tag{7.23}$$

for high temperatures;

for the nearest Bragg peak, $\kappa^2 = 3(2\pi/a)^2$, we have

$$\kappa^2 \frac{kT}{4f} = 3\frac{(2\pi)^2}{a^3}\frac{kT}{\frac{1}{C}} \cong \frac{kT}{1\mathrm{eV}} \cong \frac{kT}{10^4\mathrm{K}} \quad \text{(for Cu)}. \tag{7.23a}$$

For room temperature, $\langle(\underline{\kappa}\underline{s}^n)^2\rangle$ is of the order $3\times 10^{-2}$. The change from a to b is of order 2%.

For an elemental lattice containing N particles the differential cross section (7.14) can be written as

$$\frac{dQ}{d\Omega' d\varepsilon'} = \frac{q'}{q} Na^*a \frac{1}{N}\int \frac{dt}{2\pi\hbar} e^{i\omega t} \sum_{m,n} \langle e^{i\underline{\kappa}\underline{r}^n} e^{-i\underline{\kappa}\underline{r}^m(t)}\rangle = \frac{q'}{q} Na^*aS(\underline{\kappa},\omega) \;. \tag{7.14a}$$

The quantity $S(\underline{\kappa},\omega)$ is called van Hove's scattering function; it depends only on the scattering system.

*One phonon approximation* means that one expands (7.22) in powers of $l(t)$ up to linear terms. We have seen above that this expansion is feasible and improves with decreasing temperature. In this approximation the operators $\underline{s}^n$ appear only linearly which means that only 1 phonon is created or annihilated in the process. Corrections for higher order phonon processes can be made. Then

$$\frac{dQ}{d\Omega' d\varepsilon'} = \frac{q'}{q}\frac{1}{2\pi\hbar}\sum_{n,m} b^*_n b_m e^{i\underline{\kappa}(\underline{R}^n-\underline{R}^m)} [2\pi\delta(\omega) + \kappa_i\kappa_k L^{nm}_{ik}(\omega)] \;; \tag{7.24}$$

if $\omega \neq 0$:

$$\frac{dQ}{d\Omega' d\varepsilon'} = \frac{q'}{q}\frac{1}{2\pi\hbar}\sum_{n,m} b^*_n b_m e^{i\underline{\kappa}(\underline{R}^n-\underline{R}^m)} \kappa_i\kappa_k L^{nm}_{ik}(\omega) \;, \tag{7.24a}$$

and if the results of Section 2.4.4 are used:

$$\frac{dQ}{d\Omega' d\varepsilon'} = \frac{q'}{q}\sum_{n,m} b^*_n b_m e^{i\underline{\kappa}(\underline{R}^n-\underline{R}^m)} \kappa_i\kappa_k \underbrace{\left[\frac{1}{\sqrt{M}}\delta(\Omega^2-\omega^2)\frac{1}{\sqrt{M}}\right]^{nm}_{ik}}_{\mathrm{sgn}\,\omega\ \ \mathrm{Im}\{G(\omega)\}/\pi} \cdot$$

$$\cdot\begin{cases} n_{th}(\omega) & \text{for } \omega > 0 \\ -\left(1+n_{th}(\bar{\omega})\right) & \text{for } \omega = -\bar{\omega} < 0 \end{cases} \;. \tag{7.24b}$$

Note, that for low T only production of phonons is possible: $n_{th}(\omega>0, T=0) = 0$. Because $\underline{q}' = \underline{q} + \underline{\kappa}$ and $\varepsilon' = \varepsilon + \hbar\omega$ are known, one can extract from the measurements directly the quantity

$$I = \sum_{n,m} b^*_n b_m e^{i\underline{\kappa}(\underline{R}^n-\underline{R}^m)} \mathrm{Im}\left\{\left(\underline{\kappa}, G^{nm}(\omega)\underline{\kappa}\right)\right\} \;; \tag{7.25}$$

for elemental perfect crystals, containing N atoms, where

$$I \cong b^*b \sum_{\underline{m}}\sum_{\underline{n}} e^{i\underline{\kappa}(\underline{R}^{\underline{n}}-\underline{R}^{\underline{m}})} \mathrm{Im}\left\{\left(\underline{\kappa}, \overset{\circ}{G}{}^{(\underline{n}-\underline{m})}(\omega)\underline{\kappa}\right)\right\} \;, \tag{7.25a}$$

one can assume that each point $\underline{R}^{\underline{m}}$ in the bulk, say $\underline{R}^{\underline{m}} = 0$, gives the same contribution to (7.25a),

$$I \cong b^*bN \sum_{\underline{n}} e^{i\underline{\kappa}\underline{R}^{\underline{n}}} \, \mathrm{Im}\left\{\left(\underline{\kappa}, G^{(\underline{n})}(\omega)\underline{\kappa}\right)\right\} . \tag{7.25b}$$

If $\underline{\kappa} = \underline{K} + \underline{k}$ is not too close to a reciprocal lattice point $\underline{K}$ the sum over $\underline{n}$ can be extended to infinity,

$$\begin{aligned} I &\cong Nb^*b \sum_{\underline{n}}^{\infty} e^{i\underline{\kappa}\underline{R}^{\underline{n}}} \, \mathrm{Im}\left\{\left(\underline{\kappa}, \overset{\circ}{G}{}^{(\underline{n})}\underline{\kappa}\right)\right\} \\ &= Nb^*b \, \mathrm{Im}\left\{\left(\underline{\kappa}, \tilde{\overset{\circ}{G}}(\underline{\kappa},\omega)\underline{\kappa}\right)\right\} \qquad (7.25c) \\ &= Nb^*b \sum_{\sigma} \left(\underline{\kappa}, \underline{e}(\underline{\kappa}\sigma)\right)^2 \delta\left(\Omega^2(\underline{\kappa}\sigma) - \omega^2\right)/\overset{\circ}{M} . \end{aligned}$$

The replacement of the finite sum by an infinite sum requires[4] $kL \gg 1$ where L is the linear dimension of the crystal, or, that the wavelength $\lambda = 2\pi/k$ is very small compared with L. To demonstrate this let us consider a similar sum, $\sum_{\underline{n}, R^{\underline{n}} < R} e^{i\underline{k}\underline{R}^{\underline{n}}}/R^{\underline{n}}$, for a macroscopic crystal of radius R. To show the dependence on the macroscopic dimension, R, it is sufficient to consider instead $\int_{r<R} d\underline{r}\, e^{i\underline{k}\underline{r}}/r = 4\pi(1 - \cos kR)/k^2$. In contrast to our expectation the result depends very strongly on R; changes of R by $\lambda/2$ can change the factor $(1 - \cos kR)$ from 0 to 2. However, this fast oscillation with R or k saves the argument. One will always have a width in incoming energy or a resolution of the detector which in turn gives a small width $\delta k$ in k, $\delta k/k \ll 1$. The final result is obtained by averaging over the width $\frac{1}{2\delta k}\int_{k_o-\delta k}^{k_o-\delta k} dk \ldots$, whereupon the integral becomes approximately[5]

$$\frac{4\pi}{k_o^2}\left\{1 + \frac{\sin R(k_o + \delta k) - \sin R(k_o - \delta k)}{R\delta k}\right\} .$$

The R-dependent contribution can now be dropped if $R\delta k = Rk \cdot \delta k/k \gg 1$, say $R = 10^7$ Å $= 1$ mm, $k = \pi/a \cong 1$ Å$^{-1}$, $Rk = 10^7$, $\delta k/k = 10^{-3}$, $R\delta k \cong 10^4$. It must, however, be pointed out that for very small k the replacement is not possible.

---

4 Clearly for $\underline{\kappa} = \underline{K}$, i.e., $e^{i\underline{\kappa}\underline{R}^{\underline{n}}} = 1$, the replacement is not justified, e.g., for small $\omega$, where $\mathrm{Im}\{\overset{\circ}{G}{}^{(\underline{n})}\} \propto 1/R^{\underline{n}}$, an infinite sum would diverge (comp. (3.110)).

5 $1/k^2$ can be replaced by $k_o^2$ because only the cos-term varies rapidly.

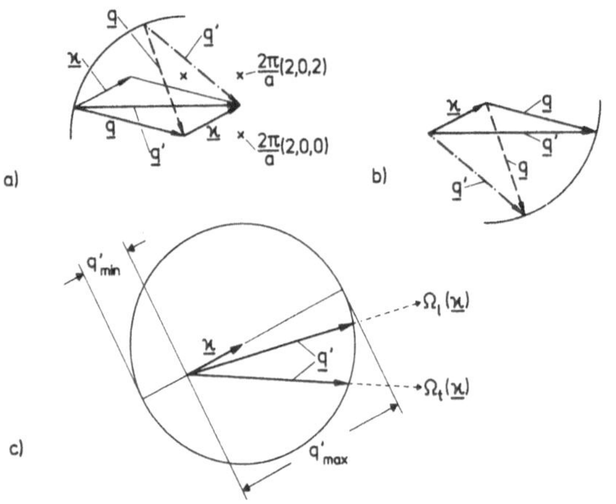


Fig. 7.4a-c. Determination of lattice frequencies by a constant -$\underline{\kappa}$-scan. a) The vectors $\underline{q}$ (initial momentum), $\underline{q}'$ (final momentum) and $\underline{\kappa}$ (momentum transfer) in reciprocal space; $\underline{\kappa}$ (thick) starts from the origin of the reciprocal (bcc) lattice indicated by ×, $\underline{\kappa} = (2\pi/a)\ (2,0,1)$; two scattering triangles $\underline{q}$, $\underline{\kappa}$, $\underline{q}'$ for fixed $\kappa$ and q are shown. b) Same situation as in (a); here equivalent scattering triangles $\kappa$, $\underline{q}$, $\underline{q}'$ are employed with $\underline{\kappa}$ as indicated in (a) by a thin line; rotation of $\underline{q}$ yields a sphere, which determines the possible final momenta, shown in (c). c) Note, that $\underline{q}'$ starts now from the origin of the reciprocal lattice; $q'_{max} = \kappa + q$, $q'_{min} = |q - \kappa|$

Fig. 7.4 illustrates, for a special case, how one can measure the dispersion curves. One can guide the measurement such that $\underline{\kappa}$ (relative to the crystal) stays constant by rotating crystal and detector; this can be represented by rotating $\underline{q}$ about the crystal center. Fig. 7.4 shows the case $\underline{\kappa} = (2\pi/a)\ (2,0,1)$ which gives the polarizations and frequencies at the Brillouin-Zone boundary in ⟨100⟩ (comp. Fig. 3.15). The energy transfer is between $\hbar\omega_{max} = \hbar^2(q'^2_{max} - q^2)/2M_n$ and $\hbar\omega_{min} = \hbar^2(q'^2_{min} - q^2)/2M_n$. Suppose that $\hbar\omega_{max} > \hbar\Omega_{l,t}$; then one obtains intensity only at those angles of rotation where $\omega = \Omega_{l,t}$, indicated in Fig. 7.3; the intensities are given by:

1) longitudinal: $\underline{\kappa} = \frac{2\pi}{a}\ (2,0,1)$ , $\underline{e}_l = (0,0,1)$ , $(\underline{\kappa},\underline{e}_l)^2 = \left(\frac{2\pi}{a}\right)^2$

2) transversal: $\underline{e}_t = (1,0,0)$ , $(\underline{e}_t,\underline{\kappa})^2 = 4\left(\frac{2\pi}{a}\right)^2$

$\underline{e}_{t'} = (0,1,0)$ , $(\underline{e}_{t'},\underline{\kappa})^2 = 0$ ;

they are in the ratio of 1 to 4. For $\underline{\kappa} = (2\pi/a)\ (1,0,0)$ only the longitudinal frequency is observed, because $(\underline{\kappa},\underline{e}_t) = 0$. On the energy gain side ($\omega > 0$) one obtains two peaks each for l and t; also on the energy loss side ($\omega < 0$) one can obtain two peaks each.

# 8. Probability, Distributions and Statistics

/8.1,2/

In the preceding chapters we have often used concepts of probability theory, such as normalized distributions and averages, tacitly assuming some basic knowledge. When discussing many defects in Chapter 9 we shall rely heavily on statistics, in the sense that we assume suitable averages to represent the physical situation adequately. One particular example, where we have assumed this all along, is the change of elastic moduli; here the stress of $N_d$ defects described by a double force tensor $P_d$, was represented by its average over defect positions. This stress is, of course, much easier to handle than the actual fluctuating stress. The question arises as to what extent this average stress reliably describes the physics.

Question like this are important in many physical problems. They are common in thermodynamics where macroscopic quantities fluctuate only slightly about their averages which are then representative. Problems of this kind turn up everywhere in defect physics, and it is certainly worthwhile discussing them. Since in most treatises they are not considered at all, we give here a short introduction into the basic problems. Again, we must emphasize that the treatment will be very crude and rudimentary. However, to understand statistics and probability theory, one need only use common sense and one can come quite far even with rudimentary concepts.

## 8.1 A Game of Dice

We will treat a very simple example in the beginning, using a die with six equivalent faces carrying the numbers 1 to 6. We throw this die N times and ask for the number n of 1's. This number is a statistical quantity; it can vary from 0 to N. The question "How many 1's are in N throws?" is not easy to answer (note: you are betting on your answer and have to think harder). Several answers might be imagined:

1) I don't know, but I can give you a probability distribution $P_N(n)$ for the event; $\sum_{n=0}^{N} P_N(n) = 1$. This is correct, but does not answer the question.

2) After having had a look at the distribution, which for large N is sharply peaked at the average $\langle n\rangle = \sum_n nP_N(n)$, you say: n equals $\langle n\rangle = N/6$. This is completely wrong; you will lose the bet!

3) A close look at the distribution reveals that $P_N$ (n near $\langle n\rangle$) is much smaller

than 1. Consequently, the bet under(2) is really a bad one, the assumption being that $P_N(n \cong \langle n\rangle) \lesssim 1$. However, if in some distance from $\langle n\rangle$ the probabilities become so small that they can be dropped, say the probability is 99.99% that n is in the interval $\langle n\rangle$ (1 ± some %), then the correct answer is: n is about $\langle n\rangle$ within some % (you can bet on that).

This last answer can be made quantitative. For this purpose we start from one throw and calculate the probabilities P (0 or 1). Let us assume that the faces are all equivalent[1]. Then the probability is given by the fraction of compatible events (0: "1 not on top"; 1: "1 on top"):

$$\frac{\text{\# of compatible events (5;1)}}{\text{\# of possible events (6)}} = \frac{5}{6} ; \frac{1}{6} .$$

This is nothing but common sense; the chances are obviously 5:1. You expect for a large number of throws about one-sixth 1's; you must be prepared to bet money on these chances. Probability distributions are always normalized:

$$\text{with } P_1(0) = q , \quad P_1(1) = p ; \quad \sum_{n=0}^{1} P_1(n) = p + q = 1 ; \tag{8.1}$$

$$\text{here } p = 1/6 , \quad q = 5/6 .$$

We proceed to the result for general N,

$$P_N(n) = \binom{N}{n} p^n q^{N-n} ; \quad \sum_{n=0}^{N} P_N(n) = (p+q)^N = 1 ; \quad \binom{N}{n} = \frac{N!}{(N-n)!n!} . \tag{8.2}$$

This result again is given by the fraction of compatible events: let us look at the problem of throwing N equivalent dice; the number of possible events is $6^N$; the number of compatible events for a certain subset of n dice showing the 1, say die no. 1,2,...,n and the N - n dice not showing the 1 is clearly $5^{N-n}$; $\binom{N}{n}$ is the number of different subsets containing n 1's.

As a rule one should first try to locate the center of the distribution or the average[2]

$$\langle n\rangle = \sum_n nP_N(n) = p\,\partial_p \sum_n P_N(n;p,q) = p\,\partial_p\,(p+q)^N = Np . \tag{8.3}$$

This average makes physical sense if the distribution has only one maximum; then often the maximum and the average are quite close. The best behaved distributions

1 Caution: the die may be rigged.

2 The quantities p,q are considered first as independent; only at the end p + q = 1 is used.

are those which have a very sharp maximum and a very small width w; in this case $\langle n\rangle$ and $n_{max}$ are close. Small width means that the probability for having n in the interval[3] $[\,\langle n\rangle - \text{some } w \leqslant n \leqslant \langle n\rangle + \text{some } w\,]$ is close to 1 or to 100%. Then one can state that

$$\text{"having n between } \langle n\rangle \left(1 \pm \text{some } \frac{w}{\langle n\rangle}\right) \text{is close to certainty" , or,} \qquad (8.4)$$
$$\text{"n is about } \langle n\rangle\text{" , or, "}\langle n\rangle \text{ is representative" ,}$$

meaning that almost all results will be close to $\langle n\rangle$. Small width means here "relatively small" or $w/\langle n\rangle \ll 1$. This discussion seems rather vague at present; however, we will derive quantitative statements below.

A good measure for the width is the mean square fluctuation or the variance

$$v = \langle(n - \langle n\rangle)^2\rangle = \langle n^2\rangle - \langle n\rangle^2 \ . \qquad (8.5)$$

The variance is always positive, $v \geqslant 0$; if $v = 0$ the distribution is a δ-distribution, $P(n) = \delta_{n,n_o}$, and the value $\langle n\rangle$ is certain, $\langle n\rangle = n_o$, $\langle n^2\rangle = n_o^2$ etc.. In our problem $\langle n^2\rangle = p\partial_p p\partial_p(p+q)^N = N^2p^2 + Npq = \langle n\rangle^2 + \langle n\rangle q$ and it is

$$v = \langle n\rangle q = Npq \ . \qquad (8.5a)$$

The width can be defined roughly by

$$w = \sqrt{v} \ , \qquad (8.6)$$

in particular

$$w = \sqrt{\langle n\rangle q} \ , \quad w/\langle n\rangle = \frac{\sqrt{q}}{\sqrt{\langle n\rangle}} = w_r = \text{relative width} \ , \qquad (8.6a)$$

and one realizes immediately that large $\langle n\rangle$ or N result in a relatively small width. Consequently, $\langle n\rangle$ is representative or n is a macroscopic quantity if $\langle n\rangle$ or N are very large.

Table 8.1 shows some data for N = 1,6,12. For N = 1 the average, 1/6, does not make such sense directly and the relative width of about 1 covers the whole interval 0,1. For N = 6 and 12 the probability is largest at $\langle n\rangle = n_{max} = 1$ and 2. For N = 6 one sees that a bet on $n \leqslant 3$ has 99.2% safety (even bet 992 ct vs. 8 ct), if N = 12 one has 87.4% safety for $n \leqslant 3$ and 97.7% for $n \leqslant 4$. An extremely improbable event would be six 1's in a row, if you start a game, the probability being $(1/6)^6 \cong 2\times10^{-5}$. There are two points of view with respect to further betting:

1) You can say, "That is really a highly improbable event; it can happen, though, and the probability for another 1 is just 1/6 again as for any decent die". This

---

3 This probability is given by $\sum_n{}' P_N(n)$ where the sum extends over the interval.

Table 8.1. Distributions $P_N(n)$ for N = 1,6,12

| | N \ n | 0 | 1 | 2 | 3 | 4 | $\langle n\rangle = \frac{N}{6}$ | $\frac{w}{\langle n\rangle} = \sqrt{\frac{5}{6\langle n\rangle}} = \sqrt{\frac{5}{N}}$ |
|---|---|---|---|---|---|---|---|---|
| $P_N(n)$ | 1 | 0.833 | 0.167 | | | | $\frac{1}{6} = 0.167$ | 2.236 |
| | 6 | 0.335 | 0.402 | 0.201 | 0.054 | | 1 | 0.913 |
| | 12 | 0.112 | 0.269 | 0.296 | 0.197 | 0.103 | 2 | 0.646 |
| $\sum_{\nu=0}^{n} P_N(\nu)$ | 1 | 0.833 | 1 | | | | | |
| | 6 | 0.335 | 0.737 | 0.938 | 0.992 | | | |
| | 12 | 0.112 | 0.381 | 0.677 | 0.874 | 0.977 | | |

attitude reveals a bad gambler (or a good mathematician).

2) A good gambler will not throw away the information given by this event. He rather will suspect that the die is rigged and bet preferentially on the 1. Even if he knows the die is "correct" from former experience, he will bet preferentially on the 1 because the low probability makes him suspicious (it might be, for instance, that the face opposite to 1 is a little sticky). As a consequence a good gambler is suspicious of extremely improbable events (and he should make sure that the jelly on the face opposite to 1 is not wiped off).

If N becomes larger the situation becomes relatively better and better. Let us discuss very large N which implies that the essential n's, near $\langle n\rangle$, are also very large. The change from $P_N(n)$ to $P_N(n+1)$ is small,

$$P_N(n + 1) = P_N(n) \frac{N - n}{n + 1} \frac{p}{q} \cong P_N(n) , \tag{8.7}$$

if n is within some widths from $\langle n\rangle$; then $P_N(n)$ can be considered a continuous function of n $\left(\sum_n \ldots \to \int dn \ldots ;\ P_N(n+1) \cong P_N(n) + P_N'(n)\right)$. The most probable value, $P_N(n_{max}+1) = P_N(n_{max})$ in (8.7), agrees with the average for large N. It is natural to try an expansion of P about $n_{max} = \langle n\rangle$. As has already been emphasized in discussing asymptotic expansions (Section 3.5.5), a direct Taylor expansion of P(n) is inferior to a logarithmic expansion of $\exp\{\ln[P(n)]\}$. To carry through this expansion we employ Stirling's asymptotic expression for the factorials of large N:

$$N! \sim N^N e^{-N}\sqrt{2\pi N} , \quad \ln(N!) \sim N[\ln(N-1)] + \frac{1}{2}\ln(2\pi N). \tag{8.8}$$

The expansion of

$$\begin{aligned}\exp\left[\ln\left(P(n)\right)\right] &= \exp\left[\ln\left(P(n_{max})\right) + \frac{1}{2}\,\partial_n^2 \ln\left(P(n)\right)\Big|_{n_{max}} (n - n_{max})^2 \right.\\ &\qquad \left. + \frac{1}{3!}\,\partial_n^3 \ln\left(P(n)\right)\Big|_{n_{max}} (n - n_{max})^3 + \dots\right] \\ &= \exp\left[\ln\left(P(n_{max})\right) + \frac{1}{2}\left(-\frac{1}{Npq}\right)(n - n_{max})^2 \right.\\ &\qquad \left. + \frac{1}{3!}\,\frac{1}{N^2}\left(\frac{1}{p^2} - \frac{1}{q^2}\right)(n - n_{max})^3 + \dots\right]\end{aligned} \tag{8.9}$$

is straightforward; if one includes only 2nd order terms, one obtains

$$P(n) \cong \frac{\exp\left(-\frac{(n - \langle n\rangle)^2}{2v}\right)}{\sqrt{2\pi v}} = \frac{\exp\left(-\frac{(n - \langle n\rangle)^2}{2w^2}\right)}{\sqrt{2\pi w^2}} . \tag{8.10}$$

This "Gaussian" distribution is normalized

$$\int_{-\infty}^{\infty} dn\; P(n) = 1 \quad , \tag{8.10a}$$

if one includes a negligible contribution from $-\infty$ to 0. In order to be valuable at all, the approximation (8.10) must be correct at least within some widths about $\langle n\rangle$, because then the probability to find n outside this interval is exceedingly small. This can be checked by the 3rd order term which is proportional to $1/\sqrt{N}$ if $(n - n_{max}) = w \propto \sqrt{N}$. From (8.9,10) one sees then that $w = \sqrt{Npq} = \sqrt{\langle n\rangle}\cdot\sqrt{q}$ is the halfwidth of the Gaussian distribution. This is sketched in Fig. 8.1 for $N = 6\cdot 10^6$; one realizes that the width, $w \cong 10^3$, is large, but relatively small as compared with $\langle n\rangle = 10^6$. Now, if one wants to make a statement about n with $99.99\% = 1 - 10^{-4}$ safety one must integrate P(n) between the limit $-\lambda w \leqslant n - \langle n\rangle \leqslant \lambda w$ and choose $\lambda$ such that the integral becomes 0.9999; this corresponds to $\lambda \cong 3$. Consequently, with

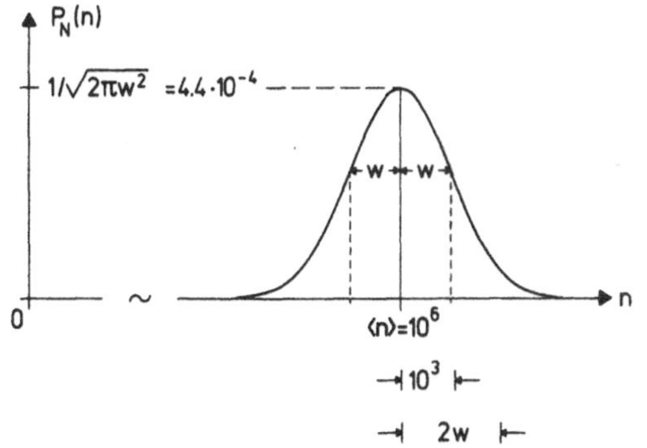


Fig. 8.1. Distribution $P_N(n)$ for $N = 6\cdot 10^6 \gg 1$, $\langle n\rangle = 10^6$, $w = 0.913\cdot 10^3$

this safety[4]:

$$n = \langle n\rangle \pm 3w = \langle n\rangle\left(1 \pm \frac{3w}{\langle n\rangle}\right) = \langle n\rangle(1 \pm 3\cdot10^{-3}) \quad \text{with 99.99\% safety} , \tag{8.11}$$

or, for only 99% safety, $\lambda \cong 2$,

$$n = \langle n\rangle(1 \pm 2\cdot10^{-2}) . \tag{8.11a}$$

These considerations also show how one makes sure that a die is correct. One must try it out, say $N = 6\cdot10^4$, $\langle n\rangle = 10^4$, $w = 10^2$; the face 1 has the required probability 1/6 with 99% safety, if after 60000 throws the number of 1's is between $\langle n\rangle(1 \pm 2\cdot10^{-2}) = 10000 \pm 200$; if the "statistical" function $n(N)$ is outside this range, the die must be rejected and should not be sold. The above is the beginning of a theory of testing products. On the other hand $n(N)/N$ equals $p$ with a large safety for large $N$ even if $p \neq 1/6$. By many throws one can then determine the weight for the various faces experimentally. If the probabilities for the different faces, $p_{1,2,\dots}$, $\sum_{\nu=1}^{6} p_\nu = 1$, can be approximated by a rational fraction $p_1 = P_1/P$ etc., the die can be represented by a more general die with $P$ faces where $P_1$, $P_2$ faces, etc., carry the numbers 1,2, etc.

## 8.2 The Simplest Tchebitchev Estimate

Whereas the estimate (8.11,11a) was derived for the special case of a Gaussian distribution, we now discuss a general estimate which is valid for arbitrary distributions. Let us consider a general probability distribution $P(Q)$ of a continuous quantity $\tilde{Q}$, where

$$P(Q)\,dQ = \text{probability for } Q \leqslant \tilde{Q} \leqslant Q + dQ \; ; \qquad \int_{-\infty}^{\infty} P(Q)\,dQ = 1 . \tag{8.12}$$

Discrete values are included, e.g., in the example of Section 8.1:

$$P(Q) = \sum_{n>0} P(n)\,\delta(n - Q) . \tag{8.12a}$$

We will be interested only in the average

$$\langle\tilde{Q}\rangle = \langle Q\rangle = \int dQ\; QP(Q) \tag{8.13}$$

and the square fluctuation

4 Of course, a Gaussian distribution is something special and the above estimate cannot be generalized. In the next section we will derive a completely general estimate.

$$v = w^2 = \int dQ\, (Q - \langle Q\rangle)^2\, P(Q) = \langle Q^2\rangle - \langle Q\rangle^2 \; . \tag{8.13a}$$

We will try to give an estimate of the kind (8.11) without knowing more about the distribution itself. If we can show that Q is a macroscopic quantity, i.e., that it fluctuates only little about the average, then the distribution is not even needed. Of course, one would like to have the distribution itself in a given statistical problem, but often the calculation of P is impossible. The calculation of low order averages such as $\langle Q, Q^2, Q^3\rangle$, however, is feasible in many cases, and they are already sufficient to decide whether Q is a well-behaved, macroscopic quantity.

First let us discuss the simplest Tchebitchev estimate. It states that

$$S(\xi) = \int_{\langle Q\rangle-\xi}^{\langle Q\rangle+\xi} dQ\, P(Q) \geqslant \underline{S}(\xi) = 1 - \frac{w^2}{\xi^2} \qquad \text{(Fig. 8.2)} \; . \tag{8.14}$$

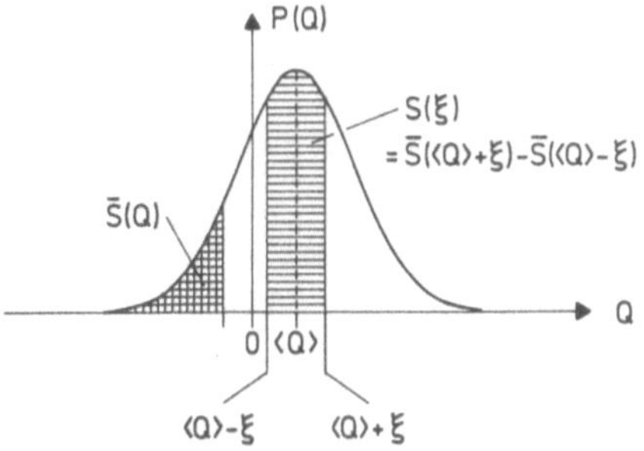


Fig. 8.2. Safety. $S(\xi)$ is given by the hatched area; it can be expressed by the auxiliary function $\bar{S}(Q)$, cross-hatched area

$S(\xi)$ is the probability or the safety that Q is in the interval $\langle Q\rangle \pm \xi$ ; $\underline{S}(\xi)$ is an optimal lower bound. If Q is a well-defined macroscopic quantity, then one must be able to find a value $\xi$ such that $\xi/\langle Q\rangle \ll 1$ (small relative interval about $\langle Q\rangle$) *and* $w^2/\xi^2 \ll 1$ (safety near 1). Obviously both conditions are met if

$$\frac{\xi}{\langle Q\rangle} = \frac{w^2}{\xi^2} \ll 1 \quad \text{or} \quad \xi^3 = w^2 \langle Q\rangle \quad \text{and} \quad \frac{\xi}{\langle Q\rangle} = \left(\frac{w}{\langle Q\rangle}\right)^{2/3} \ll 1 \; . \tag{8.15}$$

Therefore,

$$Q \text{ is macroscopic if } w/\langle Q\rangle \ll 1 \; . \tag{8.15a}$$

Take again the example of Section 8.1 with $N = 6\cdot 10^6$, $\langle n\rangle = 10^6$, $w \cong 10^3$, $w/\langle n\rangle \cong 10^{-3} \ll 1$, $\xi/\langle n\rangle = w^2/\xi^2 \cong 10^{-2} \ll 1$; here one has: $n = \langle n\rangle(1 \pm 10^{-2})$ with a safety of at least $1 - 10^{-2} = 99\%$.

The proof of (8.14) is simple. In the proof we need, in order to integrate by parts, the auxiliary function $\bar{S}(Q)$ of Fig. 8.2,

$$\bar{S}(Q) = \int_{-\infty}^{Q} dQ' \, P(Q') , \quad P(Q) = \partial_Q \bar{S}(Q) = \partial_Q [\bar{S}(Q) - 1] , \tag{8.16}$$

where $\bar{S}(Q) = 0$ for $Q = -\infty$ and $\bar{S}(Q) - 1 = 0$ for $Q = +\infty$. For this proof the value of $\langle Q \rangle$ is irrelevant and for simplicity we can assume $\langle Q \rangle = 0$ for the moment. Fig. 8.3 shows $\bar{S}(Q = \xi)$, $S(\xi) \geq 0$ which both increase monotonically. Then by definition

$$\begin{aligned} w^2 &= \int_{-\infty}^{0} dQ \, Q^2 P(Q) + \int_{0}^{\infty} dQ^2 Q \, P(Q) \\ &= \int_{-\infty}^{0} dQ \, Q^2 \partial_Q \bar{S}(Q) + \int_{0}^{\infty} dQ \, Q^2 \partial_Q [\bar{S}(Q) - 1] \quad , \end{aligned} \tag{8.17}$$

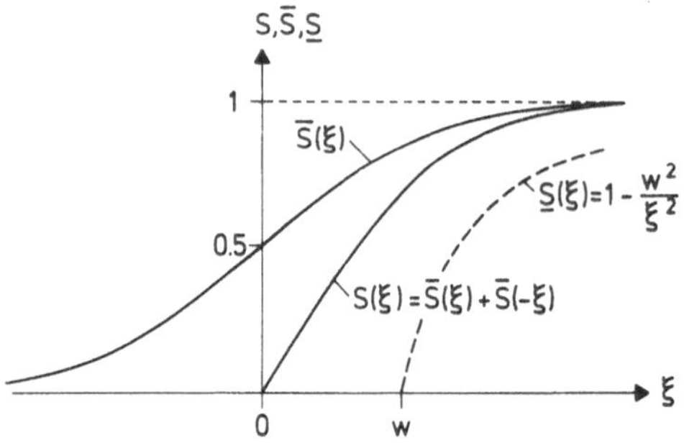


Fig. 8.3. The functions $S$, $\bar{S}$, $\underline{S}$ for $\langle Q \rangle = 0$, $P(Q)$ of Fig. 8.2

and after integration by parts we obtain

$$w^2 = - \int_{-\infty}^{0} dQ \, 2Q\bar{S}(Q) + \int_{0}^{\infty} dQ \, 2Q[1 - \bar{S}(Q)] = \int_{0}^{\infty} dQ \, 2Q \underbrace{[1 - \bar{S}(Q) + \bar{S}(-Q)]}_{1 - S(Q)} ; \tag{8.17a}$$

because the integrand is $\geq 0$, we have $\int_0^\infty dQ \ldots \geq \int_0^\xi dQ \ldots$ , i.e.,

$$w^2 \geq \int_{0}^{\xi} dQ \, 2Q[1 - S(Q)] , \tag{8.17b}$$

and finally, because $1 - S(Q)$ decreases monotonically from 1 for $Q = 0$ to 0 for $Q = \infty$,

$$w^2 \geq \int_{0}^{\xi} dQ \, 2Q[1 - S(\xi)] = \xi^2 [1 - S(\xi)] . \tag{8.17c}$$

The integration by parts can be carried out if $\langle Q^2 \rangle$ exists, i.e., if $P(Q)$ decreases more strongly than $Q^{-3}$ for large $|Q|$. That this estimate is optimal can be shown by giving an example, where the lower limit is actually reached. This is the example of Section 8.1, $P_1(n = 0,1)$ with $p = q = 1/2$, $\langle n \rangle = 1/2$, $w = 1/2$, $P_1(n) = (1/2)[\delta(n) + \delta(n-1)]$ (Fig. 8.4). The estimates with respect to safety are usually much better than $\underline{S}$; but in any case one needs (8.15a); if it is not fulfilled, the average does not contain much information and one must have the whole distribution to make reliable statements.

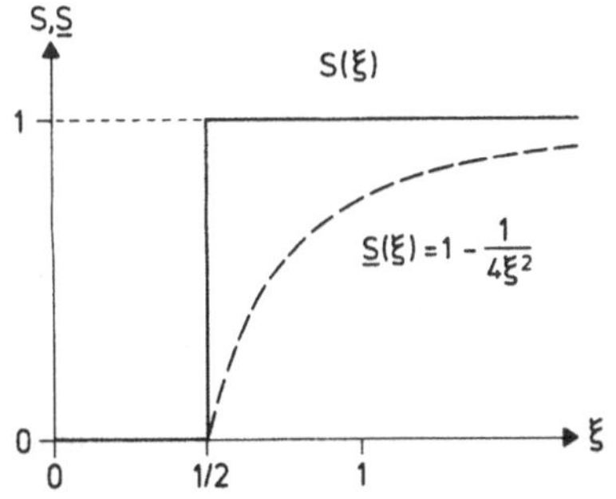


Fig. 8.4. $S(\xi)$ and $\underline{S}(\xi)$ for $P(\xi) = \frac{1}{2}[\delta(\xi) + \delta(\xi - 1)]$, $\underline{S}(\xi) = 1 - 1/(4\xi^2)$, $S(1/2) \overset{!}{=} \underline{S}(1/2)$

## 8.3 Occupation Numbers

In a binary AB-alloy a lattice site can be occupied by an atom A or B. This can be described by a variable $\tau$, the occupation number[5], attached to the lattice site. $\tau$ can assume two values, 0 and 1; $\tau = 0$ means occupation by A and $\tau = 1$ by B. For a lattice of N sites one has a set $\underline{\tau} = (\tau^1 \ldots \tau^m \ldots \tau^N)$ which represents the configuration, a set of N numbers 0 or 1 (Fig. 8.5); the number of different configurations

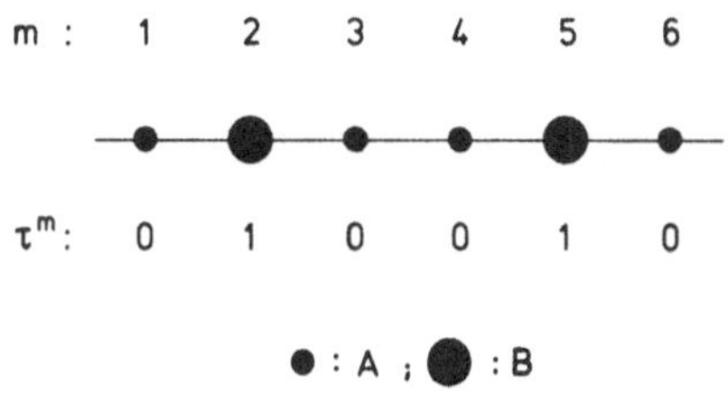


Fig. 8.5. Occupation numbers in an AB-alloy

5 In nuclear theory $\tau$ is called the isotopic spin. In the simplest case the two values of $\tau$ would refer to the state of a nucleon being a neutron ($\tau = 1$) or a proton ($\tau = 0$); this corresponds to isotopic spin 1/2 and one would use $\tau = \pm 1/2$ as "occupation number". Spin 1 with three values corresponds to a ternary alloy.

is $2^N$. Physical quantities $\tilde{Q}$ depend on the configuration, $\tilde{Q}(\underline{\tau})$, and one needs the probability distribution

$$P(\underline{\tau}) = P(\tau^1 \dots \tau^N) \quad \text{with normalization} \quad \sum_{\tau^1=0,1} \dots \sum_{\tau^N=0,1} P(\underline{\tau}) = 1 \tag{8.18}$$

for the distribution of Q. Clearly the number of B-atoms,

$$\tilde{N}_B(\underline{\tau}) = \sum_{m=1}^{N} \tau^m , \tag{8.19}$$

is a statistical quantity; it depends on the configuration and is in principle subject to fluctuations. The amount of fluctuation depends on the distribution $P(\underline{\tau})$.

We will discuss two distributions. In both, all lattice sites are equivalent. In the "microcanonical" distribution the number $N_B$ of B-atoms (and of A-atoms, $N - N_B$) is fixed. There are $\binom{N}{N_B}$ possibilities of selecting $N_B$ different lattice sites; therefore

$$P(\underline{\tau}) = \delta_{N_B,\tilde{N}_B(\underline{\tau})} \Big/ \binom{N}{N_B} \quad , \quad \text{"microcanonical" distribution.} \tag{8.20}$$

In the "grandcanonical" distribution the atomic concentration, c, of B-atoms is given and the lattice sites are not only equivalent, but also mutually independent,

$$P(\underline{\tau}) = \prod_{m=1}^{N} p(\tau^m) \quad , \quad \text{"grandcanonical" distribution ;} \tag{8.21}$$

$p(\tau)$ has only one parameter, the atomic concentration of B-atoms: $p(1) = c$, $p(0) = 1 - c$ and

$$p(\tau) = c\tau + (1-c)(1-\tau) \ . \tag{8.21a}$$

One can illustrate the situation by attaching one die to each site, the fraction c of faces carrying the 1 and the rest $1-c$ carrying 0.

Whereas in (8.20) the number $N_B$ is fixed, this number fluctuates when employing (8.21):

$$P(N_B) = \binom{N}{N_B} c^{N_B} (1-c)^{N-N_B} \quad , \quad \langle N_B \rangle = Nc \quad , \quad \langle N_B^2 \rangle - \langle N_B \rangle^2 = w^2 = Nc(1-c). \tag{8.21b}$$

Let us repeat the calculation of $\langle N_B, N_B^2 \rangle$ employing occupation numbers. According to (8.21a) one has for each site

$$\langle \tau \rangle = \langle \tau^2 \rangle = \sum_{\tau=0,1} \tau p(\tau) = c \quad , \quad \langle \tau^2 \rangle - \langle \tau \rangle^2 = c(1 - c) \ ; \tag{8.22}$$

therefore

$$\langle N_B\rangle = \sum_{m=1}^{N} \langle\tau^m\rangle = \sum_{m=1}^{N} \langle\tau\rangle = Nc \ , \tag{8.22a}$$

$$\begin{aligned}\langle N_B^2\rangle &= \sum_m \sum_n \langle\tau^m\tau^n\rangle = \sum_{m \neq n}\sum \langle\tau^m\rangle\langle\tau^n\rangle + \sum_m \langle(\tau^m)^2\rangle \\ &= N(N-1)c^2 + Nc = \langle N_B\rangle^2 + Nc(1-c) \ .\end{aligned} \tag{8.22b}$$

Imagine a crystal of 1 $cm^3$ containing about $N = 2\cdot10^{22}$ sites and a concentration $c = 5\cdot10^{-2} = 5\%$; then $\langle N_B\rangle = 10^{21}$, $(w/\langle N_B\rangle)^{2/3} \cong \langle N_B\rangle^{-1/3} = 10^{-7}$. $N_B$ is in the interval $\langle N_B\rangle(1 \pm 10^{-7})$ with a safety of at least $1 - 10^{-7} = 99.99999\%$. In such cases one can indeed say that $N_B$ equals $\langle N_B\rangle$ for all practical purposes; it is irrelevant whether one uses (8.20) with $N_B$ or (8.21) with $c = N_B/N$. However, the distribution (8.21) is not only easier to handle, it can also be more physical: if the alloy crystal is grown from the melt and only a small part of it is used, the description with fluctuating $N_B$ is more adequate. By heating the crystal to high temperatures below the melting point one hopes to overpower the atomic interaction by thermal motion such that the interaction is irrelevant and the distribution (8.21) is correct; by rapid quenching one hopes to preserve (8.21) down to low temperatures.

The above procedure can, of course, be applied as well to A as host and to B as a substitutional defect, such as a vacancy. By a corresponding heat treatment, then, one tries to make the vacancy distribution as close to the "random" distribution (8.21) as possible.

If the defects refer to interstitial sites, e.g., octahedral bond charges, the occupation numbers must also refer to the interstitial lattice.

If the alloy contains more than two kinds of atoms or if the defect on a lattice site has more than one configuration (dumbbell), the occupation number must extend over more than two values, say $\tau = 0,1,2$ for occupation by A,B,C in a ternary alloy, or $\tau = 0,1,2,3$ for occupation by a host atom, or by a dumbbell in x,y,z-orientation. Most generally, for $(l+1)$-valued $\tau$, the distribution (8.21a) is then replaced by

$$\sum_{\nu=0}^{l} c_\nu p_\nu(\tau) \ , \quad \sum_{\nu=0}^{l} c_\nu = 1 \ , \quad p_\nu(\tau) = \frac{\prod_{\mu=0}^{l}{}^{\nu} (\mu-\tau)}{\prod_{\mu=0}^{l}{}^{\nu} (\mu-\nu)} = \begin{cases} 1 & \text{for } \tau = \nu \\ 0 & \text{otherwise} \end{cases} \ ; \tag{8.23}$$

here $c_\nu$ is the concentration of type $\nu$, $\prod_\mu^\nu$ is a product over all $\mu$ except $\nu$ (Table 8.2); further

$$N_\nu = \sum_{m=1}^{N} p_\nu(\tau^m) = \text{no. of type } \nu, \quad \langle p_\nu\rangle = c_\nu \ , \quad \langle N_\nu\rangle = Nc_\nu \ . \tag{8.23a}$$

Table 8.2

| $p^n_\nu$ | $\nu = 0$ | $\nu = 1$ | $\nu = 2$ | $\nu = 3$ |
|---|---|---|---|---|
| $n = 1$ | $1 - \tau$ | $\tau$ | - | - |
| $n = 2$ | $\frac{(2-\tau)(1-\tau)}{2}$ | $(2-\tau)\tau$ | $\frac{(\tau-1)}{2}\tau$ | - |
| $n = 3$ | $\frac{(3-\tau)(2-\tau)(1-\tau)}{6}$ | $\frac{(3-\tau)(2-\tau)}{2}\tau$ | $\frac{(3-\tau)(\tau-1)}{2}\tau$ | $\frac{(\tau-2)(\tau-1)}{6}\tau$ |

## 8.4 Mass and Coupling of the Defect Crystal

For simplicity we treat only substitutional defects. The mass tensor for isotopic defects with mass $M_d$ is given by

$$M = \overset{\circ}{M} + (M_d - \overset{\circ}{M}) \sum_n P^n \tau^n \ , \quad P^n = \text{projector on site n.} \tag{8.24}$$

The coupling matrix is

$$\Phi = \overset{\circ}{\Phi} + \sum_n \overset{n}{\psi} \tau^n = \overset{\circ}{\Phi} + \Psi(\underline{\tau}) \ , \tag{8.25}$$

if the single defect coupling changes, $\overset{n}{\psi}$, can be superimposed; this is the case for small concentrations or if the changes $\overset{n}{\psi}$ from neighbouring sites do not overlap (diatomic spring defect or octahedral bond charge). Because we will focus on small concentrations, the representation (8.25) can be assumed as a rule. Then the Kanzaki forces can be superimposed as well, i.e., the static displacements are given by

$$\underline{s} = \overset{\circ}{G} \sum_m \overset{m}{\underline{K}} \tau^m \ , \tag{8.26}$$

where $\overset{m}{\underline{K}}$ is the Kanzaki force pattern of a defect at $\underline{R}^m$.

In Chapter 9 we will discuss averages of the Green's function G which is thought to represent the physics of the defect crystal. The averaging process is not simple because G contains $\underline{\tau}$ in the denominator.

## 8.5 Simple Examples of Microscopic and Macroscopic Quantities

We have already discussed one example of a statistical quantity, the number of defects, $N_d$, for given concentration c on N sites: if $\langle N_d \rangle = Nc \gg 1$, $N_d$ is macroscopic and $\langle N_d \rangle$ represents $N_d$ with small fluctuations and large safety; that is true for a macroscopic lattice, $N = 10^{22}$, and an atomic concentration as small as 1 ppm, $c = 10^{-6}$, where $Nc = 10^{16}$. On the other hand, the number of defects in a small part of the lattice for which Nc is not very large is a strongly fluctuating quantity; the average is not representative and one needs the distribution $P(N_d)$, to make definite statements. In the following discussion we will give some more illustrative examples of the same kind.

### 8.5.1 Change of "Lattice Distance" in a Linear Crystal

Fig. 8.6a illustrates the permanent displacements $\pm u$ due to a single defect (at site 0) in a linear chain. The change of distance between two neighbours 0 and 1 (Fig. 8.6b) is

$$\delta a_1 = u(\tau^0 + \tau^1) , \tag{8.27}$$

and the change of N original lattice distances is

$$\delta a_N = u(\tau^0 + \tau^N) + 2u \sum_{m=1}^{N-1} \tau^m = 2uN_d - u(\tau^0 + \tau^N) , \tag{8.28}$$

where $N_d = \sum_{m=0}^{N} \tau^m$ is the (fluctuating) total number of defects. Again if N is small,

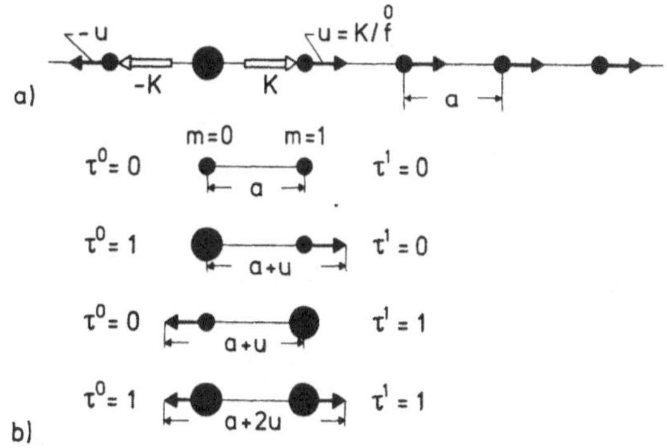


Fig. 8.6a and b. Permanent displacements in a linear crystal. a) Kanzaki force pattern of a single defect, b) the distance between the neighbours 0 and 1 depends on the configuration, $\tau^0$ and $\tau^1$

then $\delta a_1$, $\delta a_2$, etc., fluctuate greatly, whereas for large N the fluctuations about $\langle \delta a_N \rangle \cong 2u\langle N_d \rangle$ are small, which is easy to check. In X-ray scattering *large* atomic distances determine the position of the Bragg peak; therefore the observed lattice distance is $a + \delta a_N/N$, rather than $a + \delta a_1$.

## 8.5.2 Elastic Stiffness and Compliance of a Linear Chain with Defect Springs

As another example we discuss the elastic properties of a linear chain with fluctuating interatomic coupling (Fig. 8.7). To calculate the stiffness C or the compliance S we have to consider the connection between the change $\delta L$ of the original length, $L = Na$, and an external tension $\sigma$:

$$\delta L = \sigma \sum_{m=1}^{N} \frac{1}{f^m} \, , \quad \frac{\delta L}{L} = \varepsilon = \frac{\sigma}{a}\frac{1}{N}\sum_{m=1}^{N}\frac{1}{f^m} \, , \quad \varepsilon = S\sigma \, , \quad \sigma = \frac{1}{S}\varepsilon = C\varepsilon \, . \qquad (8.29)$$

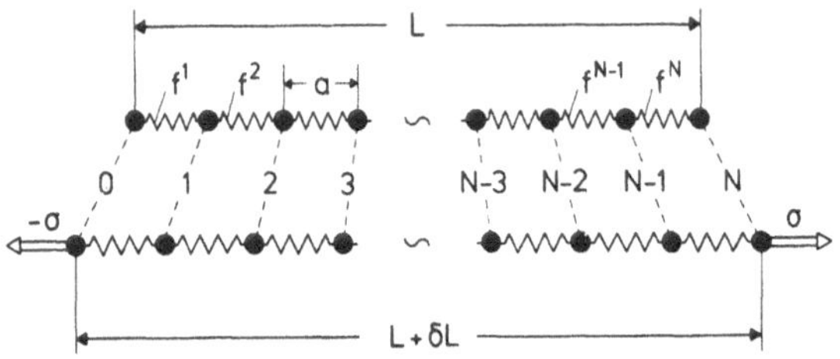


Fig. 8.7. Linear chain under external forces (tensions) $\pm\sigma$. Note, that the springs $f^1, f^2, \ldots$ need not be equal, which is not exhibited by the drawing

The springs $f^m$ between site m and site m+1 are given by

$$f^m = \overset{o}{f}(1 - \tau^m) + \overset{d}{f}\tau^m \, , \quad \frac{1}{f^m} = \frac{1 - \tau^m}{\overset{o}{f}} + \frac{\tau^m}{\overset{d}{f}} \qquad (8.30)$$

where $\overset{o}{f}$ is the spring of the host and $\overset{d}{f}$ that of the defect. Now, one con consider two kinds of procedures: either one keeps $\sigma$ constant and defines by averaging,

$$\langle\varepsilon\rangle = S^{\mathrm{eff}}\sigma \, , \quad S^{\mathrm{eff}} = \langle S\rangle \, , \qquad (8.31)$$

an effective compliance; or one keeps $\delta L$ or $\varepsilon$ constant and defines by averaging,

$$C^{\mathrm{eff}}\varepsilon = \langle\sigma\rangle \, , \quad C^{\mathrm{eff}} = \left\langle\frac{1}{S}\right\rangle \, , \qquad (8.32)$$

an effective stiffness. One expects that as in a normal homogeneous medium $C^{\mathrm{eff}} = 1/S^{\mathrm{eff}}$, which implies "$\langle\frac{1}{S}\rangle = \langle S\rangle^{-1}$" ; the latter equation holds only if S does not fluctuate at all, but if S is macroscopic it holds for all practical purposes because the distribution P(S) is concentrated in a very small interval near $\langle S\rangle$. That S is indeed a macroscopic quantity for large N, can be seen from

$$S = \frac{1}{a\overset{o}{f}} + \frac{1}{aN}\left(\frac{1}{\overset{d}{f}} - \frac{1}{\overset{o}{f}}\right)\underbrace{\sum_m \tau^m}_{N_d} \, , \qquad (8.33)$$

where $N_d$ is macroscopic (for large N), and

$$(\langle S^2\rangle - \langle S\rangle^2)/\langle S\rangle^2 = w^2/\langle S\rangle^2 \propto \frac{1}{N} \ . \qquad (8.33a)$$

From the above examples one gets the impression, that a quantity is macroscopic whenever it consists of a large number, $N \gg 1$, of small statistical contributions and that, then, no check of the fluctuation is necessary. This anticipation is faulty, as we will see in the next section.

## 8.6 The Intensity of Diffuse Scattering by a Macroscopic Alloy Crystal is a Microscopic Quantity

According to (7.11,16) the amplitude $\tilde{A}$ of X-ray or energy integrated neutron scattering is given by

$$\tilde{A}(\underline{\kappa}) = \sum_{\underline{m}=1}^{N} a_{\underline{m}}\, e^{-i\underline{\kappa}\underline{r}^{\underline{m}}} \ , \qquad \underline{\kappa} = \underline{q}' - \underline{q} \ , \qquad (8.34)$$

and the intensity, $I(\underline{q}')$, in direction $\underline{q}'$ is

$$I = \tilde{A}^*(\underline{\kappa})\, \tilde{A}(\underline{\kappa}) \ . \qquad (8.34a)$$

We will discuss the scattering by a binary AB-alloy, neglecting deviations from ideal positions ($\underline{r}^{\underline{m}} \rightarrow \underline{R}^{\underline{m}}$) and assuming a random distribution $\left(P(\underline{\tau})\text{ according to (8.21)}\right)$ of A-atoms (scattering amplitude $a_A$) and B-atoms (scattering amplitude $a_B$):

$$\tilde{A} = \sum_{\underline{m}=1}^{N} a(\tau^{\underline{m}})\, e^{-i\underline{\kappa}\underline{R}^{\underline{m}}} \ ; \qquad a(\tau) = a_A(1-\tau) + a_B\tau \qquad (8.35a)$$

$$\langle a\rangle = a_A(1-c) + a_B c \ , \qquad \langle a^2\rangle - \langle a\rangle^2 = v_1 = (a_A - a_B)^2 c(1-c) \ , \quad v_1/\langle a\rangle^2 \text{ of order } 1 \ . \qquad (8.35b)$$

To simplify the calculation we restrict ourselves to real amplitudes $a_{A,B}$. The average total amplitude is

$$\langle\tilde{A}\rangle = \langle a\rangle \sum_{\underline{m}} e^{-i\underline{\kappa}\underline{R}^{\underline{m}}} = \langle a\rangle\, S(\underline{\kappa}) \ , \qquad (8.36)$$

where $S(\underline{\kappa})$ is the "structure factor" of the crystal. As a rule, one assumes that the average intensity is representative,

$$\langle I\rangle = \sum_{\underline{m}}\sum_{\underline{n}} \langle a_{\underline{m}} a_{\underline{n}}\rangle\, e^{i\underline{\kappa}(\underline{R}^{\underline{n}}-\underline{R}^{\underline{m}})} = \sum_{\underline{m}\neq\underline{n}}\sum \langle a\rangle^2\, e^{i\underline{\kappa}(\underline{R}^{\underline{n}}-\underline{R}^{\underline{m}})} + \sum_{\underline{m}} \langle a^2\rangle \ , \qquad (8.37)$$

because $\langle a_{\underline{m}} a_{\underline{n}}\rangle = \langle a^2\rangle\, \delta_{\underline{m},\underline{n}} + \langle a\rangle^2(1-\delta_{\underline{m},\underline{n}})$, and eventually

$$\langle I\rangle = \underbrace{\langle a\rangle^2 S^*(\underline{\kappa})S(\underline{\kappa})}_{\text{"coherent scattering"}} + \underbrace{N(\langle a^2\rangle - \langle a\rangle^2)}_{\text{"diffuse scattering (incoherent)"}} \quad . \tag{8.37a}$$

The first term agrees with the (coherent) scattering from a perfect crystal with $\langle a\rangle$ as scattering amplitude; it is only important near Bragg reflections, where $\langle I\rangle \propto N^2$ with linear width $N^{-1/3}$. The second term is called diffuse (incoherent) scattering; it is proportional to N and it is the only intensity remaining off the Bragg peaks. It does not depend on direction if we additively assume that $a_{A,B}$ are isotropic.

This is now the problem: Is I macroscopic, i.e., is $(\langle I^2\rangle - \langle I\rangle^2)/\langle I\rangle^2 \ll 1$? We will treat the problem in two steps. First we look at a Bragg reflection, where I is macroscopic. Secondly, we treat the diffuse term, for simplicity assuming[6] $\langle a\rangle = 0$ whereupon diffuse scattering is the only contribution to $\langle I\rangle$; the behaviour off Bragg reflections is qualitatively identical. It turns out that I then is microscopic meaning that the diffuse intensity fluctuates wildly. Finally, we discuss under what circumstances $\langle I\rangle$, according to (8.37a), is actually observed as a macroscopic quantity. Here we must discuss the integrated intensity, integrated over a certain solid angle $\Omega$. It turns out that the intensities in two directions are uncorrelated if the angle between these directions is larger than a critical angle $\vartheta_c \propto N^{-1/3}$. If $\Omega$ is much larger than the corresponding critical $\Omega_c \cong \pi\vartheta_c^2$, then the average integrated intensity becomes macroscopic and equals $\langle I\rangle\Omega$; usually the angular resolution of the counter is much larger than $\vartheta_c$ such that $\langle I\rangle\Omega$ is indeed representative.

In the Bragg peak, i.e., for $e^{i\underline{\kappa}\underline{R}^{\underline{m}}} = 1$, the total scattering amplitude $\tilde{A} = \sum a_{\underline{m}}$ is real and from

$$\langle\tilde{A}\rangle = N\langle a\rangle \ , \quad \langle\tilde{A}^2\rangle = \langle I\rangle = \langle\tilde{A}\rangle^2 + Nv_1 \ , \quad \frac{\langle\tilde{A}^2\rangle - \langle\tilde{A}\rangle^2}{\langle\tilde{A}\rangle^2} = \frac{v_1}{N\langle a\rangle^2} \tag{8.38}$$

we see that for large N the amplitude $\tilde{A}$ and with it $I = \tilde{A}^2$ is macroscopic. The distribution of $\tilde{A}$ can also well be represented by a Gaussian, analogous to the procedure in Section 8.1,

$$P(\tilde{A}) \cong \frac{1}{\sqrt{2\pi N v_1}} \exp\left[-\frac{(\tilde{A} - \langle\tilde{A}\rangle)^2}{2Nv_1}\right] . \tag{8.38a}$$

Now we try first to simulate diffuse scattering by $\langle a\rangle = 0 = \langle\tilde{A}\rangle$ and $v_1 = \langle a^2\rangle$, whereupon

---

6 A somewhat academic assumption, but possible in principle in neutron scattering where the scattering amplitudes can have opposite sign.

$$P(\tilde{A})\, d\tilde{A} = \frac{\exp\left(-\dfrac{\tilde{A}^2}{2Nv_1}\right)}{\sqrt{2\pi N v_1}}\, d\tilde{A} \ , \quad \tilde{A}^2 = I \ , \quad \langle I \rangle = Nv_1 \ . \tag{8.39}$$

This can be rewritten into a distribution for $I > 0$,

$$P(I)\, dI = \frac{\exp\left(-\dfrac{I}{2\langle I \rangle}\right)}{\sqrt{2\pi\langle I \rangle}} \frac{dI}{\sqrt{I}} , \quad d\tilde{A} = d\sqrt{I} = \frac{1}{2\sqrt{I}} dI \quad \text{for } \tilde{A} > 0 \ , \tag{8.39a}$$

which is certainly not macroscopic; the "maximum" value (actually diverging) is at $I = 0$. That (8.39a) does not represent a sharply peaked distribution can also be seen from the variance

$$\langle I^2 \rangle - \langle I \rangle^2 = \sum_{\underline{m}_1 \dots \underline{m}_4} (\langle a_{\underline{m}_1} a_{\underline{m}_2} a_{\underline{m}_3} a_{\underline{m}_4} \rangle - \langle a_{\underline{m}_1} a_{\underline{m}_2} \rangle \langle a_{\underline{m}_3} a_{\underline{m}_4} \rangle) \ , \tag{8.40}$$

$$\langle I^2 \rangle - \langle I \rangle^2 = N(\langle a^4 \rangle - \langle a^2 \rangle^2) + 2N(N-1) \langle a^2 \rangle^2 \sim 2N^2 v_1^2 \ . \tag{8.40a}$$

The 1st term in (8.40a) corresponds to the contribution $\underline{m}_1 = \underline{m}_2 = \underline{m}_3 = \underline{m}_4$ in (8.40). The second term in (8.40a) is the contribution from two pairs: if $\underline{m}_1 = \underline{m}_3 \neq \underline{m}_2 = \underline{m}_4$ the 2nd term in (8.40) vanishes since $\langle a \rangle = 0$. The 1st terms add up to $N\langle a^2 \rangle^2$; $\underline{m}_1 = \underline{m}_4 \neq \underline{m}_2 = \underline{m}_3$ gives the same result; if $\underline{m}_1 = \underline{m}_2 \neq \underline{m}_3 = \underline{m}_4$ the two terms in (8.40) cancel. All other contributions contain at least one $\langle a \rangle$ and vanish. For large N,

$$\langle I \rangle = Nv_1 \ , \quad \langle I^2 \rangle - \langle I \rangle^2 = 2N^2 v_1^2 \ , \quad \frac{\langle I^2 \rangle - \langle I \rangle^2}{\langle I \rangle^2} = 2 \ ,$$

the relative square fluctuation is 200%; the same result is obtained from (8.39a). This result may be somewhat artificial, in view of the fact that we have neglected all the phase factors which are important in diffuse scattering off the Bragg reflections.

To discuss the behaviour off Bragg reflections we calculate first the so-called "covariance" or correlation,

$$C(\underline{\kappa},\underline{\kappa}') = \langle I(\underline{\kappa})\, I(\underline{\kappa}') \rangle - \langle I(\underline{\kappa}) \rangle \langle I(\underline{\kappa}') \rangle = \langle II' \rangle - \langle I \rangle \langle I' \rangle \ , \quad \text{with } \underline{\kappa} = \underline{q}' - \underline{q} \ , \quad \underline{\kappa}' = \underline{q}'' - \underline{q} \ , \tag{8.42}$$

which relates intensities in two directions $\underline{q}'$ and $\underline{q}''$. If this difference is small in comparison to $\langle I \rangle^2$ the two directions are uncorrelated, and the intensities are independent. From (8.34) one obtains

$$C(\underline{\kappa},\underline{\kappa}') = \sum_{\underline{m}_1 \dots \underline{m}_4} (\langle a_{\underline{m}_1} \dots a_{\underline{m}_4} \rangle - \langle a_{\underline{m}_1} a_{\underline{m}_2} \rangle \langle a_{\underline{m}_3} a_{\underline{m}_4} \rangle) \cdot \exp\left[-i\underline{\kappa}(\underline{R}^{\underline{m}_1} - \underline{R}^{\underline{m}_2}) - i\underline{\kappa}'(\underline{R}^{\underline{m}_3} - \underline{R}^{\underline{m}_4})\right] \ . \tag{8.42a}$$

For $<a> = 0$ only the terms used in (8.40) remain:

$$C = N(<a^4> - <a^2>^2) + <a^2>^2 \Bigg[ \underbrace{\sum_{\underline{m}_1 \neq \underline{m}_2} \exp\left(-i(\underline{\kappa}+\underline{\kappa}')\underline{R}^{\underline{m}_1} + i(\underline{\kappa}+\underline{\kappa}')\underline{R}^{\underline{m}_2}\right)}_{S^*(\underline{\kappa}+\underline{\kappa}')S(\underline{\kappa}+\underline{\kappa}')-N} + \underbrace{\sum_{\underline{m}_1 \neq \underline{m}_2} \exp\left(-i(\underline{\kappa}-\underline{\kappa}')\underline{R}^{\underline{m}_1} + i(\underline{\kappa}-\underline{\kappa}')\underline{R}^{\underline{m}_2}\right)}_{S^*(\underline{\kappa}-\underline{\kappa}')S(\underline{\kappa}-\underline{\kappa}')-N} \Bigg] , \tag{8.42b}$$

where $S(\kappa) = \sum_{\underline{m}} \exp(i\underline{\kappa}\underline{R}^{\underline{m}})$; for $\kappa = \underline{\kappa}'$:

$$\begin{aligned} C(\underline{\kappa},\underline{\kappa}') = <I^2> - <I>^2 &= N(<a^4> - <a^2>^2) + <a^2>^2[|S(2\underline{\kappa})|^2 + N^2 - 2N] \\ &\sim <a^2>^2[|S(2\underline{\kappa})|^2 + N^2] . \end{aligned} \tag{8.43}$$

In the Bragg peak, $\underline{\kappa}$ and with it $2\underline{\kappa}$ is a reciprocal lattice vector and we get back (8.40,41). Off a Bragg reflection, if $2\underline{\kappa}$ is not a reciprocal lattice vector, one has $C \cong <a^2>^2N^2 = <I>^2$ and obtains,

$$\frac{<I^2> - <I>^2}{<I>^2} = 1 , \tag{8.44}$$

still 100% fluctuation. Obviously, only the quantity $|S(\kappa - \underline{\kappa}')|^2$ is decisive for the correlation off Bragg reflections; in Section 7.5 it was shown that the linear width is about $2\pi b/N^{1/3}$ which corresponds to a critical angle

$$\vartheta_c \cong \frac{2\pi b}{qN^{1/3}} , \quad \text{order of } 1/N^{1/3} \text{ for } q \cong 2\pi/a . \tag{8.45}$$

If the angular distance between $\underline{q}'$ and $\underline{q}''$ is much larger than $\vartheta_c$ the two intensities are uncorrelated, i.e., $<II'> \cong <I> <I'>$. To verify (8.44) one would have to measure for one crystal the intensities in many directions, mutually separated by angles much larger than $\vartheta_c$; the relative fluctuation of these values should approximately obey (8.44).

We now consider the intensity integrated over a small solid angle $\Omega$,

$$\int_\Omega d\Omega \; I \quad \text{or} \quad \bar{I} = \frac{1}{\Omega}\int_\Omega d\Omega \; I . \tag{8.46}$$

The average of $\bar{I}$ is given by

$$<\bar{I}> = \frac{1}{\Omega}\int_\Omega d\Omega \; <I> = <I> , \tag{8.46a}$$

the old result for vanishing $\Omega$. In the fluctuation

$$\langle\bar{I}^2\rangle - \langle\bar{I}\rangle^2 = \frac{1}{\Omega^2}\iint\limits_{\Omega} d\Omega\, d\Omega' \,(\langle II'\rangle - \langle I\rangle\langle I'\rangle) \tag{8.47}$$

one can replace approximately $\langle II'\rangle - \langle I\rangle\langle I'\rangle$ by $\langle I^2\rangle - \langle I\rangle^2$ within a solid angle $\Omega_c$ corresponding to $\vartheta_c$ ($\Omega_c = \pi\vartheta_c^2$) and by zero outside, with the result

$$\langle\bar{I}^2\rangle - \langle\bar{I}\rangle^2 \cong (\langle I^2\rangle - \langle I\rangle^2)\,\frac{\Omega_c}{\Omega} \quad \text{if } \Omega \gg \Omega_c \tag{8.47a}$$

and

$$\frac{\langle\bar{I}^2\rangle - \langle\bar{I}\rangle^2}{\langle\bar{I}\rangle^2} = \frac{\langle I^2\rangle - \langle I\rangle^2}{\langle I\rangle^2}\,\frac{\Omega_c}{\Omega} \cong \frac{\Omega_c}{\Omega} = \frac{\vartheta_c^2}{\vartheta^2} = \left(\frac{2\pi b}{q\vartheta}\right)^2 \frac{1}{N^{2/3}} \;. \tag{8.47b}$$

This is small, as a rule, and therefore the integrated intensity $\bar{I} = \langle I\rangle$ is macroscopic.

In conclusion, the intensity I into a specified direction is microscopic, i.e., $\langle I\rangle$ is not representative for measurements of that kind; the fluctuation of I is of the order of 100% although it consists of many contributions of equal magnitude. By a "macroscopic" measurement of the integrated intensity, integrated over many correlation distances, one irons out the fluctuations such that the integrated intensity is macroscopic. We will not go any further into that. We only want to mention a suspicion, namely, that many statistical averages may not be representative, but are made representative only by suitable integration: an example could be the spectrum of a disordered crystal, $Z(\omega)$, which might be microscopic in contrast to the thermal energy, $E = \int d\omega\, Z(\omega)\, \varepsilon(\omega,T)$, an integrated quantity, which is macroscopic. However, this suspicion, as a rule, cannot be checked, the calculation of fluctuations being too difficult.

## 8.7 Expansions in Powers of Concentration

In applications we will treat mostly *small* concentrations of defects, $c \ll 1$. If one knows that the quantity in question allows an expansion in powers of c, one can employ the corresponding expansion of P according to (8.21):

$$P(\underline{\tau}) = \prod_{m} [(1-\tau^{\underline{m}}) + c(2\tau^{\underline{m}} - 1)]$$

$$= \prod_{\underline{m}} (1 - \tau^{\underline{m}}) \qquad \text{no defect at all}$$

(8.48)

$$+ c \sum_{\underline{n}} \prod_{\underline{m}(\neq\underline{n})} (1-\tau^{\underline{m}})(2\tau^{\underline{n}} - 1) \qquad \text{single defect at } \underline{n} \text{ possible, no defect at } \underline{m}(\neq \underline{n}),$$

$$+ \frac{c^2}{2} \sum_{\underline{n}\neq\underline{l}} \prod_{\underline{m}(\neq\underline{n},\underline{l})} (1-\tau^{\underline{m}})(2\tau^{\underline{n}} - 1)(2\tau^{\underline{l}} - 1) \qquad \text{2 defects at } \underline{n},\underline{l} \text{ possible, no defect at } \underline{m}(\neq \underline{n},\underline{l})$$

$$+ \ldots \;.$$

The average of a quantity $Q(\underline{\tau})$ becomes

$$\langle Q\rangle = Q_o + c\sum_{\underline{n}}(Q_1^{\underline{n}} - Q_o) + \frac{c^2}{2}\sum\sum_{\underline{n}\neq\underline{l}}(Q_2^{\underline{n}\underline{l}} - Q_1^{\underline{n}} - Q_1^{\underline{l}} + Q_o) + \dots \quad , \tag{8.49}$$

where $Q_o$, $Q_1^{\underline{n}}$, $Q_2^{\underline{n}\underline{l}}$ refer to no defect, 1 defect at $\underline{R}^{\underline{n}}$, 2 defects at $\underline{R}^{\underline{n}}$ and $\underline{R}^{\underline{l}}$.

As an example we will consider the elastic energy $U(\underline{\tau})$ for given surface deformations $\varepsilon$, which is $\left(\text{comp. } (5.7)\right)$

$$2U = \left(\varepsilon\underline{R},\left[\Phi - \Phi\frac{1}{q\Phi q}\Phi\right]\varepsilon\underline{R}\right) , \tag{8.50}$$

where

$$\Phi = \overset{o}{\Phi} + \Psi ; \tag{8.51}$$

$q = 1 - p$ is the projector onto the bulk, and p the projector onto the surface. Because $q\overset{o}{\Phi}\varepsilon\underline{R} = 0$

$$\begin{aligned} 2U &= \left(\varepsilon\underline{R},\left[\overset{o}{\Phi} + \Psi - \Psi\frac{1}{q\Phi q}\Psi\right]\varepsilon\underline{R}\right) \\ &= 2U_o + (\varepsilon\underline{R},T\varepsilon\underline{R}) = 2U_o + 2\delta U , \end{aligned} \tag{8.50a}$$

with the total T-matrix $T = \Psi - \Psi G\Psi$ and $G = \dfrac{1}{q\Phi q}$ . Since

$$2\delta U_1^{\underline{n}} = (\varepsilon\underline{R},\overset{\underline{n}}{t}\varepsilon\underline{R}) \quad , \quad \overset{\underline{n}}{t} = \overset{\underline{n}}{\psi} - \overset{\underline{n}}{\psi}\frac{1}{q(\overset{o}{\Phi} + \overset{\underline{n}}{\psi})q}\overset{\underline{n}}{\psi} , \tag{8.52}$$

where $\overset{\underline{n}}{\psi} = \Psi_1^{\underline{n}}$ and $\overset{\underline{n}}{t}$ refer to a single defect at $\underline{R}^{\underline{n}}$, we obtain for the average in the expansion (8.49)

$$2(\langle U\rangle - U_o) = 2\langle\delta U\rangle = c\sum_{\underline{n}}(\varepsilon\underline{R},\overset{\underline{n}}{t}\varepsilon\underline{R}) , \tag{8.53}$$

which then determines the change of elastic moduli linear in c,

$$c\sum_{\underline{n}}(\varepsilon\underline{R},\overset{\underline{n}}{t}\varepsilon\underline{R}) = V(\varepsilon,\delta C\varepsilon) . \tag{8.53a}$$

For $\underline{n}$ in the bulk, $(\varepsilon\underline{R},\overset{\underline{n}}{t}\varepsilon\underline{R})$ is independent of $\underline{n}$ and equals the value given by a single defect in an infinite crystal. Then, neglecting surface contributions, one obtains

$$(\varepsilon,\delta C\varepsilon) = \frac{Nc}{V}(\varepsilon\underline{R},t\varepsilon\underline{R}) , \tag{8.53b}$$

the result of (6.59).

That U is a macroscopic quantity can be seen most simply for small c: if one can represent the total coupling by a superposition of the coupling changes by single defects,

$$\Phi = \overset{o}{\Phi} + \sum_{\underline{m}} \overset{\underline{m}}{\psi} \, \tau^{\underline{m}} , \tag{8.54}$$

and if one employs for simplicity Born's approximation, $\overset{\underline{m}}{t} \cong \overset{\underline{m}}{\psi}$, one obtains

$$2(U - U_o) = 2\delta U = \sum_{\underline{m}} \tau^{\underline{m}} \underbrace{(\varepsilon\underline{R}, \overset{\underline{m}}{\psi}\varepsilon\underline{R})}_{\text{independent of } \underline{m} \text{ in the bulk}} \cong N_d \, (\varepsilon\underline{R}, \psi\varepsilon\underline{R}) , \tag{8.54a}$$

i.e., $\delta U$ is proportional to $N_d$, and therefore $\delta U$ is as macroscopic as $N_d$.

# 9. Properties of Crystals With Defects in Small Concentration

So far we have treated only the properties of single defects, with small excursions into the property changes of defect crystals. These changes will now be discussed more thoroughly and in more detail. In defect physics one has to start with the microscopic properties of a single defect, if one wants to understand the physics of a defect crystal, which contains necessarily many defects. The properties of a single defect must be extracted from the behaviour of crystals with such low a concentration c of defects, that they do not influence each other appreciably. The changes of crystal properties will then be somehow proportional to the number of defects, $N_d$, or to their atomic concentration, $c = N_d/N$. After having found in this way the single defect properties, one can then extrapolate to higher concentrations.

Here we will emphasize the single defect or the "small c physics", say $c < 10^{-2}$. Here the linear superposition of single defect properties can be assumed; actually the validity of this assumption determines the range of the linear behaviour in c. This assumption means, for instance, that the Kanzaki forces of the single defects can be added, $K = \sum_{\underline{\mu}} \overset{\mu}{\underline{K}} \tau^{\underline{\mu}}$, where $\overset{\mu}{\underline{K}}$ is the force pattern of a single defect at $\underline{\mu}$, or that the spring changes of single defects, $\overset{\mu}{\psi}$, can be superimposed, $\Psi = \sum_{\underline{\mu}} \overset{\mu}{\psi} \tau^{\underline{\mu}}$.

For "large" concentrations even the basic description of the situation becomes a problem. The defects are close and change their properties. The harmonic equation of motion depends in a very complicated way on the configuration, and so does the distribution $P(\underline{\tau})$ which must reflect the interaction of defects, e.g., the tendency to long range order or precipitation in alloys. Only simple models with random P have been treated, and even they are not exactly solvable, e.g., the isotopic defect

$$M = \sum_{\underline{\mu}} P^{\underline{\mu}}\left[\overset{o}{M}(1 - \tau^{\underline{\mu}}) + \overset{1}{M}\tau^{\underline{\mu}}\right] ,$$

where only the atomic mass changes, from $\overset{o}{M}$ at c = 0 to $\overset{1}{M}$ at c = 1, or a superposition of spring stars $\overset{\mu}{\psi}$,

$$\sum_{\underline{\mu}}\left[\overset{\mu}{\psi}_o(1 - \tau^{\underline{\mu}}) + \overset{\mu}{\psi}_1\tau^{\underline{\mu}}\right] ,$$

where 1st neighbour springs change from $f_o$ at c = 0 to $f_1$ at c = 1, or the combination of both.

Even though we can derive exact results for small c, linear in c, we will discuss approximations applicable to larger concentration, if only to indicate the limit of the "small c" results. The linearity in c requires some discussion; it means that, as a rule, the essential physical quantities can be expanded in powers of c; it does *not* mean that all physical results have a meaningful expansion in powers of c. We want to illustrate this very important point by some examples. The volume change and the average Kanzaki forces $\underline{K}$ are certainly proportional to c for small concentrations, the limit being given by the invalidity of the superposition, say $c \cong 10^{-2} - 10^{-3}$; analogously, the energy of a defect crystal under deformation or the elastic data will allow such an expansion.

However, the volume change, which causes an average homogeneous expansion of the defect crystal, can certainly not be treated in this way when discussing X-ray scattering; the change in volume, $N_d\delta_1 V$, corresponds to an average strain $N_d\delta_1 V/(3V) = c\delta_1 V/(3V_c)$, which in turn corresponds to an average change of "lattice distance" $\delta a/a = c\delta_1 V/(3V_c)$ (comp. the discussion in Sec. 8.5.1). The important terms in X-ray scattering are $\exp[-i\kappa_x m_x(a+\delta a)]$, where $m_x$ extends from 0 to $N^{1/3}$; typically for $\kappa_x \cong 2\pi/a$ the decisive quantity is $\exp(-i2\pi m_x \delta a/a) = \exp[-i2\pi m_x c\delta_1 V/(3V_c)]$. It is obvious that this expression cannot reasonably be expanded in powers of c, because the range of this expansion would be much too small, e.g., for $\delta_1 V/V_c \cong 1$, $N^{1/3} \cong 5\cdot 10^7$ the exponential becomes $\exp(-i\cdot 10^8 c)$, and the expansion of the exponent is valid only for $c \ll 10^{-8}$, whereas the expansion in the exponent is valid up to much larger concentrations. In this case of a non-vanishing permanent double force tensor, one should start from the average positions $\langle \underline{R}^{\underline{m}}\rangle \neq \underline{R}^{\underline{m}}$; with respect to $\langle \underline{R}^{\underline{m}}\rangle$ the average displacements vanish, and only then can one go to the limit of an infinite crystal, because its average displacements do not diverge in that limit. In this chapter we will always at first leave out the effects of permanent double force tensors; the corrections will be discussed subsequently.

In all cases we will calculate averages and use them as representative for the physical properties. This is done with all the reservations expressed in Chapter 8 with respect to possible fluctuations; e.g., the average Green's function may be microscopic, whereas the elastic data and the dispersion curves extracted from it will be macroscopic quantities.

## 9.1 X-ray Scattering

In Section 8.6 we have already discussed X-ray scattering by an alloy, without considering displacements around the "defect" atoms. To include displacements is prohibitively complicated unless the concentration is very small, which we will assume from now on. We start from (7.15), which gives the amplitude, denoted in this section by A instead of $\tilde{A}$,

$$A = \sum_{\underline{m}} a_{\underline{m}} e^{-i\underline{\kappa}\underline{r}^{\underline{m}}} , \qquad (9.1)$$

and the intensity,

$$I = AA^* = \sum_{\underline{m},\underline{n}} a_{\underline{m}} a_{\underline{n}} e^{i\underline{\kappa}(\underline{r}^{\underline{m}}-\underline{r}^{\underline{n}})} = |A|^2 , \qquad (9.1a)$$

of scattering. The atomic positions are given by

$$\underline{r}^{\underline{m}} = \underline{R}^{\underline{m}} + \underline{u}^{\underline{m}} , \quad \underline{u}^{\underline{m}} = \overset{\circ}{G}{}^{\underline{mn}}\underline{K}^{\underline{n}} = \overset{\circ}{G}{}^{\underline{mn}}\overset{\mu}{\underline{K}}{}^{\underline{n}}\tau^{\underline{\mu}} = \overset{\mu}{\underline{u}}{}^{\underline{m}}\tau^{\underline{\mu}} , \qquad (9.2)$$

where $\underline{R}^{\underline{m}}$ are the ideal positions and $\underline{u}^{\underline{m}}$ the displacements due to the Kanzaki forces, $\overset{\mu}{\underline{K}}\tau^{\underline{\mu}}$, of defects. In the simplest case of a substitutional defect, the amplitudes are either a (host amplitude) or $a_d$ (defect amplitude), and the Kanzaki force pattern is $\overset{\mu}{\underline{K}}$, if the site $\underline{\mu}$ is occupied by a defect, and zero otherwise. We assume at first that the double force tensor of $\overset{\mu}{\underline{K}}$ vanishes; then the displacements of a single defect fall off so rapidly with distance that no macroscopic changes, like $\delta_1 V$, are produced; image terms are unimportant, and the single displacements can be considered as of microscopic range.

The "coherent" scattering is defined by the average amplitude[1],

$$\langle A\rangle = \sum_{\underline{m}} \langle a_{\underline{m}} e^{-i\underline{\kappa}\underline{u}^{\underline{m}}}\rangle e^{-i\underline{\kappa}\underline{R}^{\underline{m}}} . \qquad (9.3)$$

The coherent intensity,

$$I_c = |\langle A\rangle|^2 , \qquad (9.4)$$

is for small c the same as in Section 8.6. Including $\langle\exp(-i\underline{\kappa}\underline{u}^{\underline{m}})\rangle$ would mean introducing a $\underline{\kappa}$-dependent amplitude b instead of a, analogous to the substitution in Section 7.5. The coherent intensity is restricted to the Bragg peaks.

The diffuse (or incoherent) scattering can be defined by

$$I_d = \langle AA^*\rangle - \langle A\rangle\langle A^*\rangle = \langle I\rangle - I_c = \langle(A - \langle A\rangle)(A^* - \langle A^*\rangle)\rangle ; \qquad (9.5)$$

---

1 The factor $\langle\exp(-i\underline{\kappa}\underline{u}^{\underline{m}})\rangle = \langle\exp(-i\underline{\kappa}\overset{\mu}{\underline{u}}{}^{\underline{m}}\tau^{\underline{\mu}})\rangle = \prod_{\underline{\mu}} [1 - c + c\cdot\exp(-i\underline{\kappa}\overset{\mu}{\underline{u}}{}^{\underline{m}})]$ can for small c be approximated by

$$\prod_{\underline{\mu}} \exp[-c(1 - e^{-i\underline{\kappa}\overset{\mu}{\underline{u}}{}^{\underline{m}}})] = \exp[-c \sum_{\underline{\mu}} (1 - e^{-i\underline{\kappa}\overset{\mu}{\underline{u}}{}^{\underline{m}}})] ;$$

if $\underline{m}$ is in the bulk, the displacement field has inversion symmetry, $\overset{\mu}{\underline{u}}{}^{\underline{m}} = \overset{\circ}{\underline{u}}{}^{(\underline{m}-\underline{\mu})} = -\overset{\circ}{\underline{u}}{}^{(\underline{\mu}-\underline{m})}$, and one obtains $\langle\exp(-i\underline{\kappa}\underline{u}^{\underline{m}})\rangle \cong \exp[-c \sum_{\underline{\mu}} (1 - \cos \underline{\kappa}\overset{\circ}{\underline{u}}{}^{\underline{\mu}})]$, /7.1/; in any case, this factor can be dropped if c is small.

it is important for $\underline{\kappa}$ values between the Bragg peaks. Without permanent displacements, there is no dependence on $\underline{\kappa}$, except for the (slow) dependence of the scattering amplitude on $\underline{\kappa}$, $I_d = Nc(a - a_d)^2 = N_d(a - a_d)^2$. The displacements around defects can be seen in the diffuse intensity. Therefore, the structure and symmetry of a defect can be determined by X-ray scattering. It is so far the only method to obtain the microscopic defect structure of a point defect in a metal.

We will evaluate (9.5) only for small c, using the expansion $\langle Q\rangle = Q_o + c \cdot \sum_{\underline{\mu}} (Q_1^{\underline{\mu}} - Q_o)$, where $Q_o$ corresponds to the situation without defects and $Q_1^{\underline{\mu}}$ gives the result for a single defect at $\underline{\mu}$. Then

$$\langle A\rangle \cong A_o + c \sum_{\underline{\mu}} (A_1^{\underline{\mu}} - A_o) = A_o + cA_1 \ , \qquad (9.6)$$

$$I_d = c(I_1 - A_1^* A_o - A_1 A_o^*) \ , \quad \text{neglecting terms} \propto c^2. \qquad (9.7)$$

From $I_1 \cong c \sum_{\underline{\mu}} (I_1^{\underline{\mu}} - I_o) = c \sum_{\underline{\mu}} (A_1^{\underline{\mu}} A_1^{\underline{\mu}*} - A_o A_o^*)$ one obtains

$$I_d = c \sum_{\underline{\mu}} \left[A_1^{\underline{\mu}} A_1^{\underline{\mu}*} - I_o - (A_1^{\underline{\mu}} - A_o)A_o^* - (A_1^{\underline{\mu}} - A_o)^* A_o\right] = c \sum_{\underline{\mu}} (A_1^{\underline{\mu}} - A_o)(A_1^{\underline{\mu}} - A_o)^* , \qquad (9.7a)$$

with

$$A_1^{\underline{\mu}} = a_d e^{-i\underline{\kappa}\underline{R}^{\underline{\mu}}} + \sum_{\underline{m}(\neq\underline{\mu})} a\, e^{-i\underline{\kappa}(\underline{R}^{\underline{m}} + \underline{u}^{\underline{m}}_{\underline{\mu}})} \ , \qquad (9.8a)$$

$$A_o = a\, e^{-i\underline{\kappa}\underline{R}^{\underline{\mu}}} + \sum_{\underline{m}(\neq\underline{\mu})} a\, e^{-i\underline{\kappa}\underline{R}^{\underline{m}}} \ , \qquad (9.8b)$$

$$A_1^{\underline{\mu}} - A_o = A_d^{\underline{\mu}} = (a_d - a)e^{-i\underline{\kappa}\underline{R}^{\underline{\mu}}} + \sum_{\underline{m}(\neq\underline{\mu})} a\, e^{-i\underline{\kappa}\underline{R}^{\underline{m}}}\left(e^{-i\underline{\kappa}\underline{u}^{\underline{m}}_{\underline{\mu}}} - 1\right) , \qquad (9.8c)$$

where $\underline{u}^{\underline{m}}_{\underline{\mu}}$ is the displacement of atom $\underline{m}$ due to a single defect at $\underline{\mu}$. Again one can argue that in the bulk $|A_d^{\underline{\mu}}|^2$ does not depend on $\underline{\mu}$,

$$I_d \cong Nc \left|(a_d - a)e^{-i\underline{\kappa}\underline{R}^{o}} + \sum_{\underline{m}(\neq o)} a\, e^{-i\underline{\kappa}\underline{R}^{\underline{m}}}(e^{-i\underline{\kappa}\underline{u}^{\underline{m}}_{o}} - 1)\right|^2 , \qquad (9.9)$$

and that $\sum_{\underline{m}} \ldots$ can be extended to $\infty$. The phase factor $e^{-i\underline{\kappa}\underline{R}^{o}}$ is irrevalent for $I_d$, and finally we have with $\underline{R}^{o} = 0$,

$$I_d = Nc \left|a_d - a + \sum_{\underline{m}\neq o}^{\infty} a\, e^{-i\underline{\kappa}\underline{R}^{\underline{m}}}(e^{-i\underline{\kappa}\underline{u}^{\underline{m}}} - 1)\right|^2 = Nc|\tilde{A}_d|^2 , \qquad (9.10)$$

where $\underline{u}^{\underline{m}}$ are the displacements by one defect in the origin. This result is rather illustrative. The diffuse amplitude is just the difference between the defect and the perfect scattering amplitude. If the displacements are neglected, the result ($I_d = Nc|a_d - a|^2$) is that of (8.37a). For other substitutional defects one obtains instead of $a_d - a$:

| | |
|---|---|
| $-a$ | for the vacancy, $a_d = 0$ ; |
| $-a + a(e^{-i\underline{\kappa}\underline{R}^H} + e^{i\underline{\kappa}\underline{R}^H})$ | for the dumbbell interstitial at site 0, where $\pm\underline{R}^H$ are the positions of the dumbbell atoms. |

For a normal interstitial one has

$$I_d = Nc\left|a_d + \sum_{\text{all } \underline{m}} a\, e^{-i\underline{\kappa}\underline{R}^{\underline{m}}}(e^{-i\underline{\kappa}\underline{u}^{\underline{m}}} - 1)\right|^2 ,$$

where one must sum over *all* host lattice sites.

It is easy to include permanent doubleforces, which produce a volume change $\delta_1 V$. The reference lattice must now include the average homogeneous dilatation of the defect lattice $a \to a + \delta a$, $\delta a/a = N_d\delta_1 V/(3NV_c) = c\delta_1 V/(3V_c)$ . The displacements in $I_d$ are most important near the defect, where the image displacements are small in comparison to $\overset{\infty}{\underline{s}}$ (comp. Sec. 4.8.5). Therefore, $\underline{u}^{\underline{m}}$ in $I_d$ can be replaced by $\overset{\infty}{\underline{s}}{}^{\underline{m}}$.

## 9.2 The Average Green's Function /9.1/

The most important quantity of a defect crystal is the defect Green's function $G^{\underline{m}\underline{n}}(\omega)$, which is probably subject to large fluctuations, at least for microscopic distances $\underline{R}^{\underline{m}} - \underline{R}^{\underline{n}}$. However, the change of dispersion curves derived from the imaginary part of the average Green's function $\langle G^{\underline{m}\underline{n}}(\omega)\rangle$ or the elastic data derived from an effective coupling, $\Phi^{eff} = \langle G(0)\rangle^{-1}$, will be macroscopic. That G itself is the decisive quantity, can be seen in the formulation of the various problems; as a rule, it is the response G which enters into physics. We emphasize again that the result of this averaging process is an effective lattice, which is homogeneous (translational symmetry). If in a cubic lattice the defects have cubic symmetry, the effective lattice also possesses cubic symmetry; the same is true if one averages over all equivalent orientations of a non-cubic defect, e.g., the three orientations of the ⟨100⟩ dumbbell self-interstitial.

Green's function is defined by (no additional degrees of freedom)

$$G = \frac{1}{\Phi - M\omega^2} = \frac{1}{\overset{\circ}{\Phi} - \overset{\circ}{M}\omega^2 + X(\omega)} , \qquad (9.11)$$

where we have written $\omega$ instead of $\omega + i\eta$ for short. The deviation from the perfect situation is given by X, which is a statistical quantity, $X(\underline{\tau})$. If we express G by $\overset{\circ}{G}$ and the transition matrix T, including all defects (comp. (6.8)),

$$G = \overset{\circ}{G} - \overset{\circ}{G}T\overset{\circ}{G}\ , \quad T = X\frac{1}{1+\overset{\circ}{G}X} = \frac{1}{1+X\overset{\circ}{G}}X = X - X\overset{\circ}{G}X + \dots\ ,$$

$$X = T\frac{1}{1-\overset{\circ}{G}T} = \frac{1}{1-T\overset{\circ}{G}}T\ , \tag{9.11a}$$

only T is a statistical quantity. Therefore,

$$\langle G(\omega)\rangle = \overset{\circ}{G} - \overset{\circ}{G}\langle T\rangle\overset{\circ}{G} = \frac{1}{\overset{\circ}{\Phi} - \overset{\circ}{M}\omega^2 + \Sigma(\omega)}\ , \quad \Sigma = \langle T\rangle\frac{1}{1-\overset{\circ}{G}\langle T\rangle} = \frac{1}{1-\langle T\rangle\overset{\circ}{G}}\langle T\rangle. \tag{9.12}$$

We will discuss cases only where lattice symmetry is conserved. If we start from an infinite lattice with cubic symmetry, $\Sigma(\omega)$ possesses all the symmetries of the infinite cubic lattice. In particular, the spatial Fourier transform,

$$\langle\tilde{G}(\underline{k},\omega)\rangle = \left[\tilde{\overset{\circ}{\Phi}}(\underline{k}) - \overset{\circ}{M}\omega^2 + \tilde{\Sigma}(\underline{k},\omega)\right]^{-1}\ , \tag{9.13}$$

exists, of which the eigenvectors $\underline{e}(\underline{k}\sigma)$ for $\underline{k}$ in ⟨100⟩ ,⟨111⟩ and ⟨110⟩ direction are known from symmetry,

$$\langle\tilde{G}(\underline{k},\omega)\rangle = \sum_{\sigma} |\underline{e}(\underline{k}\sigma)\rangle\, \tilde{G}(\underline{k}\sigma,\omega)\, \langle\underline{e}(\underline{k}\sigma)|\ ,$$

$$\text{with } \tilde{G}(\underline{k}\sigma,\omega) = \left[\tilde{\overset{\circ}{\Phi}}(\underline{k}\sigma) - \overset{\circ}{M}\omega^2 + \Big(\underline{e}(\underline{k}\sigma),\tilde{\Sigma}(\underline{k}\sigma)\underline{e}(\underline{k}\sigma)\Big)\right]^{-1}. \tag{9.13a}$$

If the change X is a superposition of changes $\chi$ due to a single defect, which is correct for small enough c, we have

$$X = \sum_{\underline{\mu}} \chi^{\underline{\mu}}\tau^{\underline{\mu}}\ , \quad T = \sum_{\underline{\mu}} \chi^{\underline{\mu}}\tau^{\underline{\mu}}\Big(1 + \overset{\circ}{G}\sum_{\underline{\mu}'}\chi^{\underline{\mu}'}\tau^{\underline{\mu}'}\Big)^{-1}\ . \tag{9.14}$$

The total transition matrix, T contains $\underline{\tau}$ in the denominator, which makes the calculation of the exact average impossible. But for small c the expansion (8.49) can be employed:

$$\langle T\rangle \cong c\sum_{\underline{\mu}} T_1^{\underline{\mu}} = c\sum_{\underline{\mu}} \psi^{\underline{\mu}}\frac{1}{1+\overset{\circ}{G}\chi^{\underline{\mu}}} = c\sum_{\underline{\mu}} t^{\underline{\mu}} = T^1\ , \tag{9.15}$$

$$\Sigma \cong T^1\frac{1}{1+\overset{\circ}{G}T^1} \cong T^1 = \Sigma_{STA}\ , \tag{9.15a}$$

if only terms linear in c are considered. This approximation is obviously correct for small enough c; it contains only the t-matrices of the single defects and is appropriately called the single-t-matrix approximation, STA. It contains as a special case Born's approximation, $t \cong \chi$:

$$\Sigma_{VCA} \cong c \sum_{\underline{\mu}} \overset{\underline{\mu}}{\chi} = \langle X \rangle , \tag{9.15b}$$

which is called the virtual crystal approximation, VCA. The VCA describes the defect crystal by average masses, $\overset{\circ}{M} + c(M_d - \overset{\circ}{M})$, and average coupling, $\overset{\circ}{\Phi} + c \sum_{\underline{\mu}} \overset{\underline{\mu}}{\psi}$; therefore $\Sigma_{VCA} = \langle X \rangle$ is real, in contrast to $\Sigma_{STA}(\omega)$, which is complex for $\omega \neq 0$.

For an isotopic defect we have according to (6.14):

$$\Sigma_{STA} = ct = c(t_1 + it_2) , \quad t = - \frac{m\omega^2}{1 - \overset{\circ}{G}^{(0)}(\omega^2)m\omega^2} , \quad m = M_d - \overset{\circ}{M} , \tag{9.16}$$

$$\Sigma_{VCA} = -cm\omega^2 . \tag{9.16a}$$

Let us first discuss (9.16a),

$$\langle \tilde{G}(\underline{k}) \rangle_{VCA} = [\tilde{\overset{\circ}{\Phi}}(\underline{k}) - (\overset{\circ}{M} + mc)\omega^2]^{-1} , \quad \Omega^2(\underline{k}) = \frac{\overset{\circ}{M}}{\overset{\circ}{M} + mc} \overset{\circ}{\Omega}{}^2(\underline{k}) , \tag{9.17}$$

which represents a crystal with changed masses and unchanged coupling. In the elastic theory the sound velocities[2] change from $v_o$ to $v = v_o/\sqrt{1 + mc/\overset{\circ}{M}}$. For simplicity let us consider the simple completely isotropic case (3.110) where $v_o$ does not depend on $\underline{k}$ and polarization, $\tilde{\overset{\circ}{\Phi}}_{il}(\underline{k}) = \delta_{il}\overset{\circ}{M}v_o^2k^2$ for small k, and where

$$\overset{\circ}{G}_{il}(\underline{R},\omega) = \frac{\delta_{il}}{\overset{\circ}{M}V_B} \int d\underline{k} \, \frac{e^{ikR}}{v_o^2k^2 - \omega^2} = \frac{\delta_{il}}{\overset{\circ}{M}V_B} \, \frac{2\pi^2 \exp(i\omega R/v_o)}{v_o^2 R} \tag{9.18}$$

in the elastic limit.

The value in the exponent is determined by the pole, $k_p = \omega/v_o$, of the integrand. With (9.17) one obtains

$$\langle G_{il}(\underline{R},\omega) \rangle_{VCA} = \frac{\delta_{il}}{\overset{\circ}{M}V_B} \, \frac{2\pi^2 \exp[i\omega R/v(c)]}{v_o^2 R} \tag{9.18a}$$

a quite "normal" Green's function where only the sound velocity is changed, $v(c) = v_o\sqrt{\overset{\circ}{M}/(\overset{\circ}{M} + mc)}$, the pole being at $\omega/v(c)$. However, in the STA the pole becomes complex $k_p = k_1 + ik_2$, and $\exp(ik_pR) = \exp[(ik_1 - k_2)R]$ decreases exponentially with distance, which does not occur in a normal homogeneous crystal. This "damping" has nothing to do with absorbing energy or damping of displacements. It is rather a decrease in coherence: a periodic force in the origin causes a displacement at $\underline{R}$ which is completely in phase with the force; the phase shift (in this example $\omega R/v$) is fixed; this coherence is destroyed by scattering at defects, which introduce statistical phase shifts. Then $\langle G \rangle$ can be considered as that part of G which is

---

[2] called v here in order to avoid confusion with concentration, c.

still in phase with the exciting force; $k_2$ vanishes, if the phase in a single scattering event is conserved, which means real t (e.g., small $\omega$ or validity of Born's approximation). These considerations apply as well to $g(\underline{r},t)$ and to beams or pulses of sound waves. The average Green's function represents the part of the original pulse which is still coherent; the rest is the part which has been removed from the pulse by (Rayleigh) scattering. The frequency dependence is obtained from

$$k_p = \sqrt{\frac{\overset{\circ}{M}\omega^2 - ct(\omega)}{\overset{\circ}{M}v^2}} \quad . \tag{9.18b}$$

For the isotopic defect and small $\omega$ it is $\mathrm{Im}\{t(\omega)\} \propto \omega^4\, \mathrm{Im}\{\overset{\circ}{G}{}^{(0)}(\omega^2)\} \propto \omega^5$, hence $k_2 \propto \omega^4$; for the diatomic spring defect $\big($comp. (6.15)$\big)$ we have $\mathrm{Im}\{t\} \propto \mathrm{Im}\{\overset{\circ}{G}{}^{(0)} - \overset{\circ}{G}{}_1^{(200)}\} \propto \omega^3$ and $k_2 \propto \omega^2$ (for the expansion of $\overset{\circ}{G}{}^{(h)}$ in powers of $\omega$ compare Section 3.5.4, p. 83 and p. 86). The proportionality to $\omega^3$ is valid for all[3] spring defects; for very small $\omega$ this damping is small and can be neglected.

The result (9.18a) allows one to discuss the possibilities of expansions in powers of c. One clearly sees that an expansion of $\exp(i\sqrt{1+mc/\overset{\circ}{M}}\,\omega R/v_o)$ in powers of c such as

$$\exp(i\sqrt{1+mc/\overset{\circ}{M}}\,\omega R/v_o) = \exp(i\omega R/v_o)\cdot\left(1 + \frac{i\omega R}{2v_o}\frac{mc}{\overset{\circ}{M}} + \dots\right) ,$$

does not make sense physically; it would be valid only for small $mc\omega R/(2\overset{\circ}{M}v_o)$. This expansion would correspond to a direct expansion of $\langle G\rangle$ in powers of c:

$$\langle G\rangle = \overset{\circ}{G} - \overset{\circ}{G}\overset{1}{T}\overset{\circ}{G} \dots = \overset{\circ}{G} + \overset{1}{G}{}' + \dots , \tag{9.19}$$

whereas

$$\langle G\rangle_{STA} = \frac{1}{\overset{\circ}{G}{}^{-1} + \overset{1}{T}} = \overset{\circ}{G}\frac{1}{1+\overset{1}{T}\overset{\circ}{G}} = \overset{\circ}{G} - \overset{\circ}{G}\overset{1}{T}\frac{1}{1+\overset{\circ}{G}\overset{1}{T}}\overset{\circ}{G} = \overset{\circ}{G} + \overset{1}{G} . \tag{9.19a}$$

The poles of $\langle G\rangle$ are its most important quantities. They can be expanded in powers of c, and the correct result for small c is given by the STA, where the expansion is done in the denominator. Though the direct expansion (9.19) is not consistent, it can nevertheless be used as a recipe, if one is to understand that $\overset{1}{G}{}'$ of (9.19) has to be replaced by the correct $\overset{1}{G}$ of (9.19a) in the end. We will employ this recipe as a useful tool to shortcut derivations.

We would like to mention two other approximations which are thought to hold for larger concentrations. Both pass into the STA for low c, and they can be used to check the limits of the STA. In the average-T-matrix approximation (ATA) one starts

[3] The proof is trivial if one realizes that all modes $|\nu\rangle$ with $\overset{\nu}{t} \neq 0$ are orthogonal to a common translation, whereupon terms proportional to $\omega$ in $\mathrm{Im}\{\overset{\nu}{t}\}$ drop out.

from the VCA. Seen from the virtual crystal, occupation of a lattice site by a host or a defect atom *both* causes a change, i.e., both cases correspond to a "defect" in the virtual crystal. Let us illustrate this employing an isotopic defect, where the VCA mass is

$$\overset{o}{M}(1 - c) + M_d c = \overset{o}{M} + mc = M_{VCA} ; \tag{9.20}$$

the mass change for $\tau = 0$ (occupation by a host) is

$$m_o = \overset{o}{M} - M_{VCA} = -mc , \tag{9.20a}$$

and for $\tau = 1$:

$$m_1 = M_d - M_{VCA} = m(1 - c) . \tag{9.20b}$$

The mass on one site is

$$M(\tau) = M_{VCA} + m_o(1 - \tau) + m_1\tau , \quad \langle M(\tau)\rangle = M_{VCA} + m_o(1 - c) + m_1 c = M_{VCA} . \tag{9.21}$$

The scattering by one site depends on its occupation, the t-matrix being

$$t(\tau) = t_o(1 - \tau) + t_1\tau , \tag{9.22}$$

$$t_o = - \frac{m_o\omega^2}{1 - G^{(o)}_{VCA} m_o \omega^2} , \quad t_1 = - \frac{m_1\omega^2}{1 - G^{(o)}_{VCA} m_1 \omega^2} , \tag{9.22a}$$

$$\langle t\rangle = t_o(1 - c) + t_1 c . \tag{9.22b}$$

In the ATA the scattering by a single defect in the virtual crystal is treated exactly,

$$\Sigma_{ATA} = \sum_{\underline{\mu}} \overset{\underline{\mu}}{\sigma} , \quad \text{where} \quad \overset{\underline{\mu}}{\sigma} = \langle \overset{\underline{\mu}}{t}\rangle \frac{1}{1 - G_{VCA}\langle \overset{\underline{\mu}}{t}\rangle} . \tag{9.23}$$

It is easy to establish that $\Sigma_{ATA}$ passes into $\Sigma_{STA}$ for small c. Analogously, one can proceed for more general defects. For an isotopic defect, $G_{VCA}(\omega) = \overset{o}{G}\left(\sqrt{\overset{o}{M}_{VCA}/M}\ \omega\right)$ is obtained simply by scaling the original $\overset{o}{G}$; in general $G_{VCA}$ must be calculated anew for every concentration, which implies the necessity for much numerical work.

Another (numerically complicated) approximation is the so-called coherent potential approximation, CPA. Here, the starting point is an "effective crystal" with complex, frequency dependent coupling and masses. The effective data are determined by the requirement that $\langle t(\omega)\rangle$ vanishes. For $\omega = 0$ the CPA coupling is real; the CPA springs connect all coordinates in defect space.

If the defects have permanent Kanzaki forces, corresponding to a change in volume, the reference lattice must include the homogeneous dilatation produced by the defects.

## 9.3 The Change of Elastic Data

From the static limit of (9.12),

$$\langle G(0)\rangle = \frac{1}{\overset{o}{\Phi} + \Sigma(0)} , \tag{9.24}$$

one recognizes that $\overset{o}{\Phi} + \Sigma(0)$ can be viewed as an effective coupling, from which the elastic moduli can be calculated:

$$\Phi^{\mathrm{eff}} = \overset{o}{\Phi} + \Sigma(0) . \tag{9.25}$$

Because $\Phi^{\mathrm{eff}}$ is translational and rotational invariant (comp. (2.5b,c)), the results of Section 5.2 for *finite* homogeneous crystals (N sites) can be used,

$$VC_{ik,mn} = X_k^{\underline{m}} \left[\Phi^{\mathrm{eff}}\right]_{in}^{\underline{mn}} X_m^{\underline{n}} = V\overset{o}{C}_{ik,mn} + X_k^{\underline{m}} \left[\Sigma(0)\right]_{in}^{\underline{mn}} X_m^{\underline{n}} , \tag{9.26}$$

and the change of the elastic data is given by

$$\delta C_{ik,mn} = X_k^{\underline{m}} \left[\Sigma(0)\right]_{in}^{\underline{mn}} X_m^{\underline{n}}/V . \tag{9.26a}$$

In the STA, where $\Sigma(0) = c\sum_{\underline{\mu}} \overset{\underline{\mu}}{t}$, the contribution from a defect in the bulk is independent of $\underline{\mu}$ and equals the result in an infinite perfect lattice. This leads to

$$\delta C_{ik,mn} = \frac{Nc}{V} X_k^{\underline{m}} t_{in}^{\underline{mn}} X_m^{\underline{n}} , \quad \delta_1 C_{ik,mn} = \frac{1}{V} X_k^{\underline{m}} t_{in}^{\underline{mn}} X_m^{\underline{n}} \quad \text{for 1 defect} , \tag{9.26b}$$

i.e., the STA expression for $\delta C$ is identical with the result (6.59), which was obtained by employing the concept of polarizability of defects.

The influence of a permanent volume change, $\delta_1 V$, can be obtained as follows. We have seen in Section 4.8.5 that the volume change $\delta_1 V$ by a single defect can be split into two contributions, $\delta_1 V = \delta_1 V^{\infty} + \delta_1 V^{I}$. The change $\delta_1 V^{\infty}$ is produced by the solution $\overset{\infty}{\underline{s}}$ in the infinite medium; this corresponds, in the average, to a pressure on the surface which has to be relaxed for a finite crystal with free surface, and this relaxation produces an additive change $\delta_1 V^{I}$. We have also seen that $\overset{\infty}{\underline{s}}$ does not cause a change in density, because the volume change on the external surface is compensated by an equal change on an internal surface near the defect (compare Fig. 4.22 and (4.87): $\langle \mathrm{div}\{\overset{\infty}{\underline{s}}\}\rangle = 0$); but $\delta_1 V^{\infty}$ causes a change of the lattice distance (in the average), because it changes the external volume. In contrast, the change $\delta_1 V^{I}$ from the relaxation of a genuine pressure causes both changes in lattice distance *and* a corresponding change of springs. The elastic moduli of a homogeneous crystal are proportional to the ratio "linear combination of springs (depending on the lattice distance a), f(a), over lattice distance", symbolically

$$C = \alpha \frac{f(a)}{a} . \tag{9.27}$$

The volume change $\delta_1 V^\infty = N\delta V_c^\infty \left( = N\delta(a^3/4)^\infty \text{ in fcc lattices}\right)$ does not contribute to the change of f in (9.27),

$$\delta_1 C^\infty \doteq \alpha\, f(a)\, \delta a^\infty\, \partial_a\left(\frac{1}{a}\right) = -C\, \frac{\delta a^\infty}{a} = -C\, \frac{\delta_1 V^\infty}{3NV_c} , \tag{9.27a}$$

whereas by $\delta_1 V^I$ both a's in (9.27) are affected (comp. Sec. 4.9),

$$\delta_1 C^I = \alpha\, \delta a^I\, \partial_a\left(\frac{f(a)}{a}\right) = \frac{\delta_1 V^I}{N} \frac{\partial C}{\partial V_c} = -\frac{\delta_1 V^I}{NV_c} K \frac{\partial C}{\partial p} . \tag{9.27b}$$

The total change in $\overset{\circ}{C}$ due to the permanent displacements is

$$\delta_1 \overset{\circ}{C} = -\overset{\circ}{C} \frac{\delta_1 V^\infty}{3NV_c} - \frac{\delta_1 V^I}{NV_c} K \frac{\partial \overset{\circ}{C}}{\partial p} \qquad \text{for a single defect ,} \tag{9.28}$$

$$\delta \overset{\circ}{C} = -\frac{c}{3} \frac{\delta_1 V^\infty}{V_c} \overset{\circ}{C} - c\, \frac{\delta_1 V^I}{V_c} K \frac{\partial \overset{\circ}{C}}{\partial p} \qquad \text{for many defects in small concentration } c \ll 1 . \tag{9.28a}$$

These changes must be added to (9.26b). To get a feeling for the order of magnitude, let us consider a Cu-vacancy. If we use the model with only one longitudinal (nearest neighbour) spring, (6.65), we have from (9.26b) $(1/c)\delta C/C \cong -3$ for all moduli. The volume change is $\delta_1 V = -0.2\ V_c$, and from Table 4.8 we obtain $\delta_1 V^\infty = 0.69\ \delta_1 V = -0.14\ V_c$, $\delta_1 V^I = \delta_1 V - \delta_1 V^\infty = -0.06\ V_c$; therefore, according to (9.27a), $(1/c)\cdot(\delta C^\infty/C) \cong 0.05$, and with the data of Table 4.9 (9.28a) yields

$$\frac{1}{c}\frac{\delta K}{K} = 0.35 , \quad \frac{1}{c}\frac{\delta\mu'}{\mu'} = 0.64 , \quad \frac{1}{c}\frac{\delta\mu}{\mu} = 0.49 .$$

## 9.4 Neutron Scattering Theory

In Section 7.5 we have already derived the expression for the scattered intensity (comp. (7.25)),

$$I = \sum_{\underline{n},\underline{m}} b_{\underline{n}} b_{\underline{m}} \left(\underline{\kappa}, \mathrm{Im}\{G^{\underline{n}\underline{m}}(\omega)\}\, \underline{\kappa}\right) e^{i\underline{\kappa}(\underline{R}^{\underline{n}} - \underline{R}^{\underline{m}})} ; \tag{9.29}$$

for simplicity we assume real amplitudes, the normal case in neutron scattering. The scattering amplitudes $b_{\underline{m}}$, the atomic positions $\underline{R}^{\underline{m}}$ and Green's function G can all be statistical quantities, over which the appropriate average has to be taken. In the following discussion we will consider several simple cases for illustration.

### 9.4.1 Isotopic Defects with Small Mass Change

The simplest case is that of an isotopic defect with small mass change where one can treat arbitrary concentrations. The VCA is already an excellent approximation, which can be used throughout and can be considered as the starting $\overset{o}{G}$, Though the masses are approximately the same, the scattering amplitudes can be very different, e.g., natural Cu[4] is an isotopic mixture of 69% $Cu^{63}$ with $b^{63} \cong 0.7\cdot10^{-12}$cm and 31% $Cu^{65}$ with $b^{65} \cong 1.1\cdot10^{-12}$cm. In (9.29) the Green's function G can be replaced by $\overset{o}{G}$ with the average mass $\overset{o}{M} = 63.6$; the remaining average implies only the factor $b_{\underline{n}} b_{\underline{m}}$, and with

$$\langle b_{\underline{n}} b_{\underline{m}} \rangle = \langle b\rangle \langle b\rangle (1 - \delta_{\underline{nm}}) + \delta_{\underline{nm}} \langle b^2\rangle$$

one obtains

$$\langle I\rangle = \langle b\rangle^2 \sum_{\underline{nm}} \left(\underline{\kappa}, \mathrm{Im}\{\overset{o}{G}{}^{\underline{mn}}(\omega)\}\underline{\kappa}\right) e^{i\underline{\kappa}(\underline{R}^{\underline{n}}-\underline{R}^{\underline{m}})} + (\langle b^2\rangle - \langle b\rangle^2) \sum_{\underline{m}} \left(\underline{\kappa}, \mathrm{Im}\{G^{\underline{mm}}(\omega)\}\underline{\kappa}\right) . \tag{9.30}$$

The first term can be treated as in (7.25a,b,c); in the 2nd term, $\overset{o}{G}{}^{\underline{mm}}$ can be replaced by $\overset{o}{G}{}^{(0)}$ of the infinite crystal:

$$\langle I\rangle = N\left[\langle b\rangle^2 (\underline{\kappa}, \mathrm{Im}\{\tilde{\overset{o}{G}}(\underline{\kappa},\omega)\}\underline{\kappa}) + (\langle b^2\rangle - \langle b\rangle^2)(\underline{\kappa}, \mathrm{Im}\{\overset{o}{G}{}^{(0)}(\omega)\}\underline{\kappa})\right]$$

$$= \frac{N}{\overset{o}{M}}\left[\langle b\rangle^2 \sum_{\sigma} \left(\underline{\kappa}, \underline{e}(\underline{\kappa}\sigma)\right)^2 \delta\left(\overset{o}{\Omega}{}^2(\underline{\kappa}\sigma) - \omega^2\right) + (\langle b^2\rangle - \langle b\rangle^2)\kappa^2\pi^2\overset{o}{Z}(\omega^2)\right] , \tag{9.30a}$$

for Cu: $\langle b\rangle^2 = 0.68\cdot10^{-24}\mathrm{cm}^2$ , $\langle b^2\rangle - \langle b\rangle^2 \cong 0.03\cdot10^{-24}\mathrm{cm}^2$ .

If at constant $\underline{\kappa}$ one scans through $\omega$, one expects to observe three sharp phonon lines of $\delta$-shape over a background which has the shape of the spectrum $\overset{o}{Z}$. For small concentration c of one isotope, the 2nd term in (9.30a) is proportional to c and can be neglected in zeroth order. In that case one would observe the phonon lines of the abundant isotope slightly shifted due to the mass change.

### 9.4.2 Small Concentrations[5]

For the discussion of more general defects than an isotopic defect with small mass change we restrict ourselves again to small concentrations. The treatment becomes particularly easy, if we use the recipe given in (9.19,19a): we "expand" in powers

---

[4] Actually Cu is not a very good example, experimentally, because of its relatively large cross section for neutron absorption ($2\cdot10^{-24}\mathrm{cm}^2$).

[5] The summation convention is not used in this section; we have omitted $\not{\Sigma}$.

of c, but replace $\overset{1}{G}'$ of (9.19) by $\overset{1}{G}$ of (9.19a) in order to deal only with physically reasonable Green's functions. The scattering amplitudes, however, can be directly expanded in powers of c. The expansion of $\langle b_{\underline{n}} b_{\underline{m}} G^{\underline{nm}}\rangle$ up to linear terms in c becomes ($b_{host} = b$, $b_{defect} = b_d$)

$$\langle b_{\underline{n}} b_{\underline{m}} G^{\underline{mn}}\rangle = b^2 \overset{\circ}{G}{}^{\underline{mn}} + c\sum_{\underline{\mu}}\left(\overset{\mu}{b}_{n}\overset{\mu}{b}_{m}\overset{\mu}{G}{}^{\underline{mn}} - b^2\overset{\circ}{G}{}^{\underline{mn}}\right), \tag{9.31}$$

where $\overset{\mu}{b}_{\underline{m}} = b + (b_d - b)\delta^{\underline{\mu m}}$ is the scattering amplitude if one defect is at $\underline{\mu}$, and where $\overset{\mu}{G} = \overset{\circ}{G} - \overset{\circ}{G}\overset{\mu}{t}\overset{\circ}{G}$ is the corresponding G; $\overset{\circ}{G} + c\sum_{\underline{\mu}}(\overset{\mu}{G} - \overset{\circ}{G}) = \overset{\circ}{G} + \overset{1}{G}'$ is replaced, when needed, by $\overset{\circ}{G} + \overset{1}{G} = G_{STA}$. We then have

$$\begin{aligned}\langle b_{\underline{n}} b_{\underline{m}} G^{\underline{mn}}\rangle = b^2 \overset{\circ}{G}{}^{\underline{mn}} + c\sum_{\underline{\mu}}\Big[ & b^2(\overset{\mu}{G}{}^{\underline{mn}} - \overset{\circ}{G}{}^{\underline{mn}}) + b(b_d - b)(\delta^{\underline{m\mu}} + \delta^{\underline{n\mu}})\overset{\mu}{G}{}^{\underline{mn}} \\ & + (b_d - b)^2\,\delta^{\underline{m\mu}}\,\delta^{\underline{n\mu}}\,\overset{\mu}{G}{}^{\underline{mn}}\Big] .\end{aligned} \tag{9.32}$$

The 1st two terms contain the expansion of $b^2\langle G^{\underline{mn}}\rangle = b^2(\overset{\circ}{G} + \overset{1}{G}' + \dots)$, and if one replaces $b^2(\overset{\circ}{G} + \overset{1}{G}')$ by $b^2 G^{\underline{mn}}_{STA} = b^2(\overset{\circ}{G} + \overset{1}{G})$, they contain the correct poles of $\langle G\rangle$ up to terms linear in c. Usually the remaining terms can be neglected. For instance, the contribution of the last term to $\langle I\rangle$ is $c(b_d - b)^2\sum_{\mu}(\underline{\kappa}, \mathrm{Im}\{\overset{\mu}{G}{}^{\underline{\mu\mu}}(\omega)\}\underline{\kappa})$. For an isotopic defect $(M_d/\pi)\,\mathrm{Im}\{\overset{\mu}{G}{}^{\underline{\mu\mu}}\}$ is the defect spectrum, $Z_d(\overset{\mu}{\omega}{}^2)$; its contribution $Nc(b_d - b)^2\,\pi\, Z_d(\omega^2)/M_d \propto N_d = Nc$ can be neglected against the contribution from $b^2 G_{STA}$, which is proportional to $N \gg N_d$. Only in the case of a very sharp resonance ($M_d \gg M$) at not too small concentrations one must be careful near the resonance frequency, because then the resonance can interfere with the phonon term. For an isotopic defect (comp. (6.14)), where $\overset{\mu}{t}{}^{\underline{mn}} = t\delta^{\underline{m\mu}}\delta^{\underline{n\mu}}$ with the scalar quantity $t = -m\omega^2/(1 - m\omega^2\overset{\circ}{G}{}^{(0)})$, one has

$$\delta^{\underline{m\mu}}\,\overset{\mu}{G}{}^{\underline{mn}} = \delta^{\underline{m\mu}}\left[\overset{\circ}{G} - \overset{\circ}{G}\overset{\mu}{t}\overset{\circ}{G}\right]^{\underline{mn}} = \delta^{\underline{m\mu}}\,(1 - \overset{\circ}{G}{}^{(0)}t)\overset{\circ}{G}{}^{\underline{mn}} ,$$

because $\delta^{\underline{m\mu}}\left[\overset{\circ}{G}\overset{\mu}{t}\overset{\circ}{G}\right]^{\underline{mn}} = \delta^{\underline{m\mu}}\,\overset{\circ}{G}{}^{\underline{m\mu}}t\overset{\circ}{G}{}^{\underline{\mu n}} = \delta^{\underline{m\mu}}\,\overset{\circ}{G}{}^{(0)}t\overset{\circ}{G}{}^{\underline{mn}}$; therefore,

$$\langle b_{\underline{n}} b_{\underline{m}} G^{\underline{mn}}\rangle = \left[b^2(\overset{\circ}{G} + \overset{1}{G}') + 2cb(b_d - b)(1 - \overset{\circ}{G}{}^{(0)}t)\overset{\circ}{G}\right]^{\underline{mn}} + Nc(b_d - b)^2\,\delta^{\underline{mn}}\,\overset{m}{G}{}^{\underline{mm}} .$$

In the term proportional to $(b_d - b)$ one can replace $\overset{\circ}{G}$ by $\overset{\circ}{G} + \overset{1}{G}'$, if one neglects terms $\propto c^2$, and upon replacing $\overset{\circ}{G} + \overset{1}{G}'$ by $G_{STA}$ one obtains[6]

$$\langle b_{\underline{n}} b_{\underline{m}} G^{\underline{mn}}\rangle = \left[b + c(b_d - b)(1 - \overset{\circ}{G}{}^{(0)}t)\right]^2 G^{\underline{mn}}_{STA} + Nc(b_d - b)^2\,\delta^{\underline{mn}}\,\overset{m}{G}{}^{\underline{mm}} , \tag{9.32a}$$

---

[6] ELLIOTT and TAYLOR,/9.2/, have derived a similar result, thought to be valid for higher concentrations; the terms linear in c are those of (9.32a).

again neglecting terms $\propto c^2$.

For small concentrations, $\langle b_{\underline{n}} b_{\underline{m}} G^{\underline{mn}} \rangle$ can be replaced by $b^2 G_{STA}$, where according to (9.15)

$$G_{STA} = \frac{1}{\overset{o}{\Phi} - \overset{o}{M}\omega^2 + c\sum_{\underline{\mu}} \overset{\underline{\mu}}{t}} . \tag{9.33}$$

The coupling $\overset{o}{\Phi}^{\underline{mn}} = \overset{o}{\Phi}^{(\underline{m}-\underline{n})} = \overset{o}{\Phi}^{(\underline{h})}$ has translational symmetry. This is also true for $\sum_{\underline{\mu}} \overset{\underline{\mu}}{t}^{\underline{mn}}$ (in the infinite crystal),

$$\sum_{\underline{\mu}} \overset{\underline{\mu}}{t}^{\underline{mn}} = \sum_{\underline{\mu}} \overset{o}{t}^{\underline{m}-\underline{\mu},\underline{n}-\underline{\mu}} = \sum_{\mu} \overset{o}{t}^{\underline{m}-\underline{n}+\underline{\mu},\underline{\mu}} , \tag{9.34}$$

where the superscript 0 indicates the defect position at the origin. The matrix $\tilde{\overset{o}{\Phi}}(\underline{k})$ is obtained from

$$\tilde{\overset{o}{\Phi}}(\underline{k}) = \sum_{\underline{h}} \overset{o}{\Phi}^{(\underline{h})} e^{-i\underline{k}\underline{R}^{\underline{h}}} = \overset{o}{M}\overset{o}{\Omega}^2(\underline{k}). \tag{9.35}$$

Correspondingly one has

$$\tilde{t}(\underline{k},\omega) = \sum_{\underline{h}} \sum_{\underline{\mu}} e^{-i\underline{k}\underline{R}^{\underline{h}}} \overset{o}{t}^{\underline{h}+\underline{\mu},\underline{\mu}}(\omega) , \tag{9.35a}$$

which results in

$$\tilde{G}_{STA}(\underline{k},\omega) = \frac{1}{\tilde{\overset{o}{\Phi}}(\underline{k}) - \overset{o}{M}\omega^2 + c\tilde{t}(\underline{k},\omega)} = \frac{1}{\overset{o}{M}\overset{o}{\Omega}^2(\underline{k}) - \overset{o}{M}\omega^2 + c\tilde{t}(\underline{k},\omega)} . \tag{9.36}$$

As we have seen in Section 7.6, the intensity in neutron scattering becomes now

$$\langle I \rangle_{STA} = N\, b^2 \left(\underline{\kappa}, \mathrm{Im}\{\tilde{G}_{STA}(\underline{\kappa},\omega)\}\underline{\kappa}\right) . \tag{9.37}$$

The difference between (9.37) and (7.25c) is that $c\tilde{t}$ is $\omega$-dependent and complex, in general. Therefore, $\langle I \rangle_{STA}$ cannot be resolved into three $\delta$-functions. If the concentration is very small, the scanning at constant $\underline{\kappa} = \underline{k}$ should only be slightly[7] altered by the defects, which means that $\tilde{t}(\underline{k},\omega)$ can be replaced by $\tilde{t}\left(\underline{k},\overset{o}{\Omega}(\underline{k})\right) = t_1 + it_2$. In the main symmetry directions the eigenvectors of $G_{STA}$ are determined by symmetry alone, if t has cubic symmetry, which is true for all defects treated so far (also for dumbbells which must be averaged over all orientations). The contribution of one polarization to $\langle I \rangle$ is then

---

7 Barring particular cases where t diverges, as discussed in /3.10/.

$$\langle I\rangle = N\,b^2(\underline{\kappa},\underline{e})^2 \frac{-ct_2}{(\overset{o}{M}\overset{oo}{\Omega}^2 + ct_1 - \overset{o}{M}\omega^2)^2 + c^2t_2^2} . \tag{9.38}$$

Instead of a δ-function as in the perfect crystal, $\langle I\rangle \propto \delta(\overset{o}{M}\overset{oo}{\Omega}^2 - \overset{o}{M}\omega^2)$, one obtains a Lorentz curve with width $ct_2$ and a shift in $\omega^2$ of $ct_1/\overset{o}{M}$.

### 9.4.3 Dispersion Curves of Crystals with Isotopic Defects

To illustrate the result (9.37) we employ again the isotopic defect; according to (6.14) we have

$$\overset{o}{t} = t\,\delta^{\underline{m}o}\,\delta^{\underline{n}o} , \quad \tilde{t}(\underline{k}) = t = -\frac{m\omega^2}{1 - \overset{o}{G}^{(0)}(\omega^2)m\omega^2} = t_1 + it_2 , \quad m = M_d - \overset{o}{M}, \tag{9.39}$$

where $\tilde{t}$ and $\tilde{\Sigma}_{STA} = c\tilde{t}$ do not depend on $\underline{k}$. For small $M_d$ one obtains a contribution to $\langle I\rangle$ at the localized frequency $\omega_L > \Omega_{max}$, independent of $\underline{k}$ if $\omega_L \gg \Omega_{max}$. Inside the normal spectrum, $\omega < \Omega_{max}$, one observes a shift and broadening of phonon lines for small c.

We will discuss only the mass resonance, $M_d \gg \overset{o}{M}$, which is particularly illustrative (comp. Section 3.5.4, p. 84). The resonance frequency, $\omega_R$, is low, and the interesting features occur near $\omega_R$. Therefore, the expansions $\overset{o}{\Omega}^2(\underline{k}) = v_o^2k^2$, $G^{(0)}(\omega) \cong \alpha + i\pi a\omega/\overset{o}{M}$ (comp. (3.90)) and $m \cong M_d \gg \overset{o}{M}$ can be employed, leading to

$$\tilde{G}_{STA} \cong \frac{1}{\overset{o}{M}v_o^2k^2 - \overset{o}{M}\omega^2 + c\tilde{t}} , \text{ where} \tag{9.40}$$

$$\tilde{t} \cong -\frac{M_d\omega^2}{1 - \alpha M_d\omega^2 - i\pi a\omega^3 M_d/\overset{o}{M}} = -\frac{M_d\omega^2\omega_R^2}{\omega_R^2 - \omega^2 - i2\eta_R(\omega)\omega_R} , \tag{9.40a}$$

with

$$\omega_R^2 = \frac{1}{\alpha M_d} = \frac{f_e}{M_d} , \quad \omega_R = \text{resonance frequency} ,$$

$$\eta_R(\omega) = \frac{M_d\pi a\omega^3\omega_R}{2\overset{o}{M}} , \quad \eta_R(\omega_R) = \frac{\pi M_d a\omega_R^4}{2\overset{o}{M}} = \text{resonance damping} .$$

To get a feeling for the order of magnitude, we use the values for a fcc lattice with 1st neighbour longitudinal coupling (spring f):

$$\overset{o}{M}\Omega_{max}^2 = 8f , \quad \alpha = \frac{1}{f_e} = \frac{3.36}{\overset{o}{M}\Omega_{max}^2} = \frac{0.42}{f} , \quad a\Omega_{max}^3 = 1.32.$$

Then,

$$\omega_R^2 = \frac{\overset{\circ}{M}}{3.36\ M_d}\,\Omega_{max}^2\ , \qquad \frac{\eta_R(\omega_R)}{\omega_R} = \frac{1.32\pi}{2}\,\frac{M_d}{\overset{\circ}{M}}\left(\frac{\omega_R}{\Omega_{max}}\right)^3 \cong 0.34\left(\frac{\overset{\circ}{M}}{M_d}\right)^{1/2} \ . \tag{9.41}$$

For small c we have

$$\frac{\tilde{c}t(\overset{\circ}{\Omega}{}^2)}{M} = -\frac{cM_d\omega_R^2\overset{\circ}{\Omega}{}^2}{\overset{\circ}{M}}\cdot\frac{\omega_R^2 - \overset{\circ}{\Omega}{}^2 + 2i\eta_R(\overset{\circ}{\Omega})\omega_R}{(\omega_R^2 - \overset{\circ}{\Omega}{}^2)^2 + 4\eta_R^2(\overset{\circ}{\Omega})\omega_R^2} \ . \tag{9.42}$$

One realizes that the maximum of the Lorentz curve, $\Omega^2(\underline{k})$, is shifted from $\overset{\circ}{\Omega}{}^2(\underline{k})$ to $\overset{\circ}{\Omega}{}^2 + ct_1$ by $ct_1 \propto \overset{\circ}{\Omega}{}^2 - \omega_R^2$, corresponding to a shift in $\Omega$ by $ct_1(\overset{\circ}{\Omega}{}^2)/(2\overset{\circ}{\Omega})$. This is sketched in Fig. 9.1 for $\underline{k}$ in a main symmetry direction, where $v_o$ can be a lon-

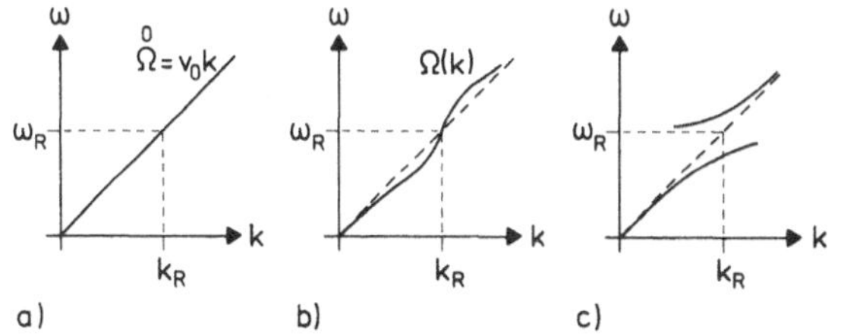


Fig. 9.1a-c. Influence of heavy masses on the dispersion curves in the elastic limit. a) Dispersion curve $\overset{\circ}{\Omega} = v_o k$, $v_o$ = sound velocity of the selected polarization, $\omega_R$ = resonance frequency, $k_R = \omega_R/v_o$ (depends on polarization), b) small c: maxima of the Lorentzians versus k, c) "large" c: maxima of double peak structure versus k

gitudinal or transversal sound velocity; only one dispersion curve is shown. For $\overset{\circ}{\Omega} < \omega_R$ the "phonon" frequency, $\Omega$, is lowered, whereas $\Omega > \overset{\circ}{\Omega}$ for $\overset{\circ}{\Omega} > \omega_R$, and $\Omega = \overset{\circ}{\Omega}$ at $\overset{\circ}{\Omega} = \omega_R$. The width (proportional to $-ct_2$) vanishes if $\overset{\circ}{\Omega}$ is either large or small as compared with $\omega_R$; it becomes largest near $\omega_R$. If, however, c becomes larger, this simple picture becomes invalid. The shape of the scanning curve is no longer a Lorentzian, and it must be discussed more carefully. It turns out that for large enough c the Lorentzian changes into a double peak structure (shown in Fig. 9.1c) which looks like the (in optics) well-known hybridization of a resonance level crossed by a dispersion curve (of light).

For the sake of simplicity we discuss only the scanning for $k = k_R$, defined by $\overset{\circ}{\Omega}(k_R) = \omega_R$, where, near $\omega = \omega_R$, we can replace the factor $\omega^2$ of $\tilde{t}$ by $\omega_R^2$ and also $\eta_R(\omega)$ by $\eta_R(\omega_R)$ ,

$$\overset{\circ}{M}G_{STA} \cong \frac{1}{\omega_R^2 - \omega^2 - \frac{cM_d\omega_R^4}{\overset{\circ}{M}}(\omega_R^2 - \omega^2 - i2\eta_R\omega_R)^{-1}} \; . \tag{9.43}$$

The imaginary part is easy to obtain:

$$\overset{\circ}{M}\, \mathrm{Im}\{G_{STA}\} = \frac{c\,\frac{2M_d\eta_R\omega_R^5}{M_o}}{(\omega_R^2 - \omega^2)^4 + (\omega_R^2 - \omega^2)^2\left(4\eta_R^2\omega_R^2 - 2c\,\frac{M_d\omega_R^4}{\overset{\circ}{M}}\right) + \left(\frac{cM_d\omega_R^4}{\overset{\circ}{M}}\right)^2} \; . \tag{9.44}$$

The coefficient $4\eta_R^2\omega_R^2 - 2cM_d\omega_R^4/\overset{\circ}{M}$ of $(\omega_R^2 - \omega^2)^2$ in (9.44) decides the behaviour of $\mathrm{Im}\{G_{STA}\}$ for small $(\omega_R^2 - \omega^2)^2$. If this coefficient is positive, $c < c_H$, $\mathrm{Im}\{G_{STA}\}$ has a maximum for $\omega = \omega_R$; otherwise it has a minimum ($c > c_H$) and shows a double peak structure (hybridization). The critical concentration, $c_H$, for hybridization is

$$c_H = \frac{2\overset{\circ}{M}\eta_R^2\omega_R^2}{M_d\omega_R^4} \cong 0.23\left(\frac{\overset{\circ}{M}}{M_d}\right)^2 \quad \text{(for Ag in Al in this approximation:} \quad c_H \cong 1.4\%)$$

The behaviour of the scanning curve is sketched in Fig. 9.2 (note that for $\omega$ near $\omega_R$ one has $(\omega_R^2 - \omega^2) \cong 2\omega_R(\omega_R - \omega)$).

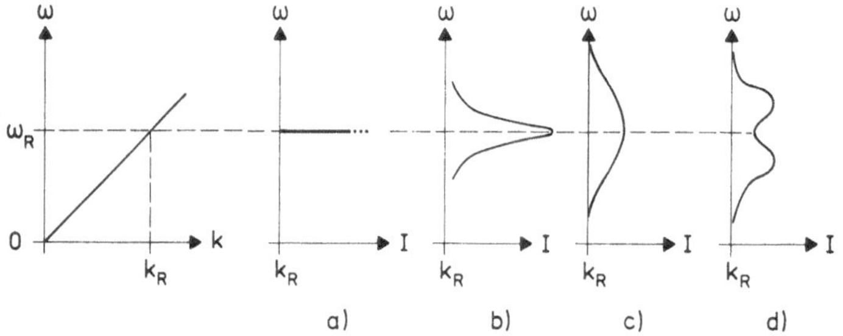


Fig. 9.2a-d. Scanning curves for $k = k_R$, depending on concentrationc, $\overset{\circ}{M}\,\mathrm{Im}\{G_{STA}\}$ (schematically).
a) For $c \to 0$: $I \propto \delta(\omega_R^2 - \omega^2) = \delta(\omega_R - \omega)/2\omega_R$ ,
b) for $c < c_H$: Lorentzian; near $\omega_R$: $I \propto [\text{const} - (\omega_R - \omega)^2]$,
c) for $c = c_H$: broad maximum; near $\omega_R$: $I \propto [\text{const} - (\omega_R - \omega)^4]$ ,
d) for $c > c_H$: double peak structure; near $\omega_R$: $I \propto [\text{const} + (\omega_R - \omega)^2]$

The theory has been worked out in more detail for substitutional Ag in Al, /9.3/, which can well be represented by a mass change alone because its volume change is so small (comp. the remarks in Section 6.3.1). The agreement of theory and experi-

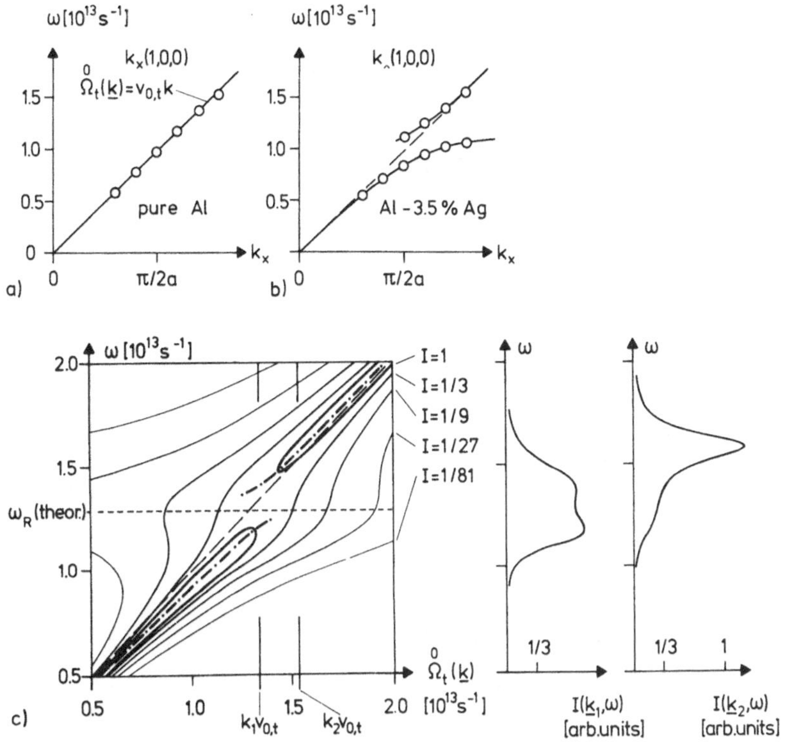


Fig. 9.3a-c. Scans through transverse dispersion curve in [100] near $k_R$ for 3.5% substitutional Ag in Al. a) Dispersion curve of pure Al and experimental points,O. b) Experimental maxima, O. c) Equal intensity contours of $\mathrm{Im}\{G_{STA}\}$ for 3.5% Ag in Al, treated as a pure isotopic defect; ----- $\omega = \overset{0}{\Omega}_t = v_{0,t}k$ ; -·-·-·- maximum intensity

ment is good.[8] Theoretical results are shown in Fig. 9.3. In the case of hybridization there are no longer phonons with some width; the phonons are no more well-defined near $k_R$. Instead one has a mixing of phonons and resonance vibrations of the heavy mass.

8 Deviations between experiment and theory are mainly due to the fact that the substitutional Ag cannot be described by a pure mass change; there are small spring changes near the defect.

# Appendix

## A. Translation and Rotation in Three Dimensions

If an atom at $\underline{r}$ (Fig. A.1) is *translated* by the vector $\underline{T}$, its new position is

$$\tilde{\underline{r}} = \underline{r} + \underline{T} \quad \text{or in components} \quad \tilde{x}_i = x_i + T_i \; .$$

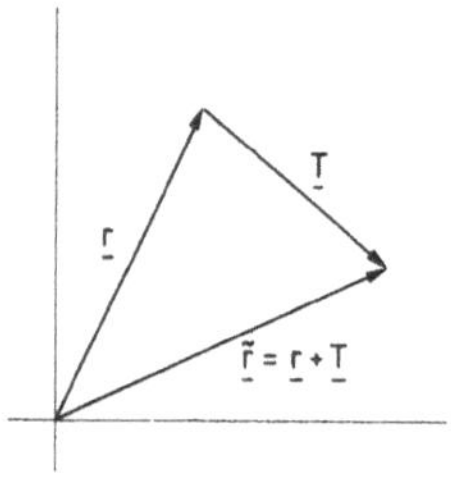


Fig. A.1

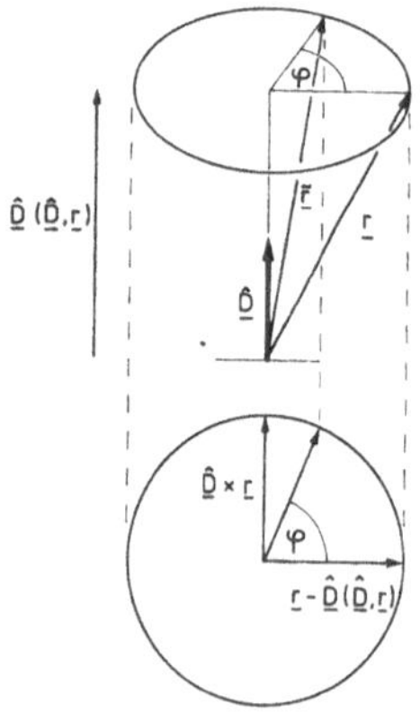


Fig. A.2

*Rotation* $D(\hat{\underline{D}},\varphi)$ about an axis $\hat{\underline{D}}$ through the origin by the angle $\varphi$ leads to the new position (Fig. A.2):

$$\tilde{\underline{r}} = \hat{\underline{D}}(\hat{\underline{D}},\underline{r}) + \cos\varphi\,[\,\underline{r} - \hat{\underline{D}}(\hat{\underline{D}},\underline{r})] + \sin\varphi\,(\hat{\underline{D}}\times\underline{r}) = D(\hat{\underline{D}},\varphi)\underline{r}$$

(where $\hat{\underline{D}}\times\underline{r}$ denotes the vectorial product and D is a $3\times3$ matrix) or in components

$$\tilde{x}_i = \sum_{k=1}^{3} D_{ik}(\varphi)x_k = D_{ik}(\varphi)x_k$$ (summation convention, comp. footnote to Section 2.1, p. 5).

With the help of the 3rd rank alternating tensor $\varepsilon_{ilk}$ (exchange of two subscripts changes the sign),

$$\varepsilon_{ilk} = -\varepsilon_{lik} = -\varepsilon_{kli} \quad \text{and} \quad \varepsilon_{123} = 1 \quad (\varepsilon_{321} = -1 \ , \quad \varepsilon_{112} = 0 \ \text{ etc.}),$$

D can be written $D_{ik}(\varphi) = \hat{D}_i\hat{D}_k + \cos\varphi\,(\delta_{ik} - \hat{D}_i\hat{D}_k) + \sin\varphi\;\varepsilon_{ilk}\hat{D}_l$. For small ("infinitesimal") angles $\varphi$ one obtains (up to terms linear in $\varphi$)

$$D_{ik}(\varphi) \cong \delta_{ik} + \varphi\,\omega_{ik}$$

$$\omega_{ik} = \varepsilon_{ilk}\hat{D}_l = -\omega_{ki} \quad \text{(antisymmetric)}.$$

The translation or rotation of a function $f(\underline{r})$ is achieved by applying the inverse transformation $\left(\underline{T}^{-1} = -\underline{T},\ D^{-1}(\varphi) = D(-\varphi)\right)$ to the argument $\underline{r}$. This is most simply illustrated from the "particle density" $\rho(\underline{r}) = \delta(\underline{r} - \underline{R})$ of one atom (considered as a point) at $\underline{R}$. If, for instance, the atom is translated by $\underline{T}$, i.e., $\underline{R} \to \underline{R} + \underline{T}$, the new density is

$$\tilde{\rho}(\underline{r}) = \delta\left(\underline{r} - (\underline{R} + \underline{T})\right) = \delta\left((\underline{r} - \underline{T}) - \underline{R}\right) = \rho(\underline{r} - \underline{T}) \quad .$$

## B. Vectors and Linear Operators

### B.1 Vectors in Three-Dimensional Space

A vector $\underline{a}$ in three-dimensional space is determined by length a and direction $\hat{\underline{a}}$ (unit vector, length 1), $\underline{a} = a\hat{\underline{a}}$. We denote them by underlining $\underline{a}$, $\underline{r}$, $\underline{u}^1, \ldots$, or according to Dirac: $|\underline{a}\rangle$ or $|a\rangle$ , $|\underline{r}\rangle$ , $|\underline{u}^1\rangle = |1\rangle$ , ... . The scalar product of two vectors $\underline{a}$ and $\underline{b}$ is written as $(\underline{a},\underline{b})$ , $\langle\underline{a}|\underline{b}\rangle$ , $\langle a|b\rangle$ or simply $\underline{a}\ \underline{b}$ , whatever is convenient; it is $(\underline{a},\underline{b}) = a \cdot b \cos\vartheta$ where $\vartheta$ is the angle between $\underline{a}$ and $\underline{b}$: $(\underline{a},\underline{a}) = a^2 > 0$, and for "orthogonal" vectors $(\underline{a},\underline{b}) = 0$.

Any vector $\underline{r}$ can be represented by a linear combination of three (non-planar) basis vectors $\underline{u}^j = |j\rangle$ , j = 1,2,3 (Fig. B.1):

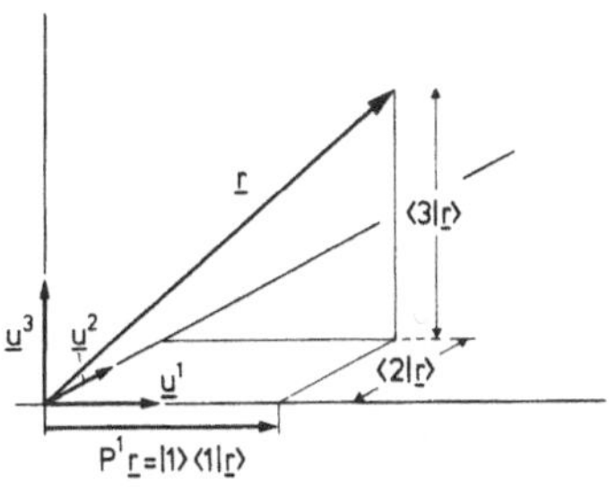


Fig. B.1

$$\underline{r} = x_1\underline{u}^1 + x_2\underline{u}^2 + x_3\underline{u}^3 = x_j\underline{u}^j = x_j|j\rangle \ .$$

Most convenient is an orthonormal basis, where the basis vectors are mutually orthogonal unit vectors:

$$(\underline{u}^j,\underline{u}^{j'}) = \langle j|j'\rangle = \delta_{jj'} = \begin{cases} 1 & \text{if } j = j' \\ 0 & \text{otherwise} \end{cases} \ .$$

Consequently $x_j = (\underline{u}^j,\underline{r}) = \langle j|\underline{r}\rangle$ and

$$\underline{r} = \underbrace{|1\rangle\langle 1|}_{P^1}\underline{r}\rangle + \underbrace{|2\rangle\langle 2|}_{P^2}\underline{r}\rangle + \underbrace{|3\rangle\langle 3|}_{P^3}\underline{r}\rangle = |j\rangle \underbrace{\langle j|\underline{r}\rangle}_{x_j} = |j\rangle\, x_j$$

where e.g., $P^1$ is the projection "operator" and $P^1|\underline{r}\rangle$ the vectorial projection onto the direction $|1\rangle$. The meaning of the operation $P^1|\underline{r}\rangle = P^1\underline{r} = \underline{u}^1(\underline{u}^1,\underline{r}) = |1\rangle\langle 1|\underline{r}\rangle$ is most obvious in Dirac's notation. Similarly, the projection $P^a$ on an arbitrary direction $\hat{\underline{a}} = |\hat{\underline{a}}\rangle$ is $P^a = |\hat{\underline{a}}\rangle\langle\hat{\underline{a}}|$.

Analogously $P^{1,2} = |1\rangle\langle 1| + |2\rangle\langle 2|$ projects onto the plane spanned by $|1\rangle$ and $|2\rangle$ and, more general, $P^{a,b} = |\hat{\underline{a}}\rangle\langle\hat{\underline{a}}| + |\hat{\underline{b}}\rangle\langle\hat{\underline{b}}|$ with $\langle\hat{\underline{a}}|\hat{\underline{b}}\rangle = 0$ is the projector onto a two-dimensional subspace, the plane spanned by $|\hat{\underline{a}}\rangle$ and $|\hat{\underline{b}}\rangle$. Of course, the sum of projectors onto a complete set of orthogonal basis vectors $\hat{\underline{a}},\hat{\underline{b}},\hat{\underline{c}}$ or $\underline{u}^1,\underline{u}^2,\underline{u}^3$ yields unity e.g.,

$$\sum_j P^j = |j\rangle\langle j| = 1 \ .$$

## B.2 Linear Operators in Three-Dimensional Space

A linear operator M in three-dimensional space transforms a vector $\underline{r}$ into $\tilde{\underline{r}} = M\underline{r}$; linearity means that

$$M(\alpha\underline{r} + \beta\underline{r}') = \alpha(M\underline{r}) + \beta(M\underline{r}') \ .$$

Therefore it is sufficient to know the transformation $M|j\rangle$ of the basis vectors $|j\rangle$, and M can be represented by a 3×3 matrix $M_{ij} = \langle i|M|j\rangle = (\underline{u}^j,M\underline{u}^j)$:

$$M = \underbrace{|i\rangle\langle i|}_{=1}\, M\, \underbrace{|j\rangle\langle j|}_{=1} = |i\rangle\, M_{ij}\, \langle j| \ , \quad \tilde{x}_i = M_{ij}x_j \ .$$

In general the matrix $\langle i|M|j\rangle$ is not symmetrical in i and j; rather one has

$$\langle j|M|i\rangle = \langle i|M'|j\rangle$$

which defines the adjoint operator M'; the matrix $[M']_{ij}$ is the "transposed" of the matrix $M_{ij}$: $[M']_{ij} = M_{ji}$.

## B.3 Symmetrical Operators

If M' = M, the operator is called symmetrical. A simple example is the projection $P^{\underline{a}}$ on $\hat{\underline{a}}$:

$$P^{\underline{a}} = |\hat{\underline{a}}\rangle\langle\hat{\underline{a}}| \ , \quad P^{\underline{a}}_{ij} = \langle i|\hat{\underline{a}}\rangle\langle\hat{\underline{a}}|j\rangle = \hat{a}_i\hat{a}_j$$

$$P^{\underline{a}} = \begin{bmatrix} \hat{a}_1^2 & \hat{a}_1\hat{a}_2 & \hat{a}_1\hat{a}_3 \\ \hat{a}_2\hat{a}_1 & \hat{a}_2^2 & \hat{a}_2\hat{a}_3 \\ \hat{a}_3\hat{a}_1 & \hat{a}_3\hat{a}_2 & \hat{a}_3^2 \end{bmatrix} .$$

If in particular $\hat{\underline{a}} = \underline{u}^1$, i.e., $P^{\underline{a}} = P^1$:

$$P^1_{ik} = \begin{bmatrix} 1 & 0 & 0 \\ 0 & 0 & 0 \\ 0 & 0 & 0 \end{bmatrix} .$$

Similarly $P^{\underline{a},\underline{b}}_{ij} = \hat{a}_i\hat{a}_j + \hat{b}_i\hat{b}_j$, in particular

$$P^{1,2}_{ij} = \begin{bmatrix} 1 & 0 & 0 \\ 0 & 1 & 0 \\ 0 & 0 & 0 \end{bmatrix} .$$

The most general example is the quadratic form

$$F(\underline{r}) = x_i M_{ik} x_k/2, \ M_{ik} = M_{ki} .$$

Here $F(\underline{r})$ = const. defines the surface of an ellipsoid[1] (Fig. B.2), and $M\underline{r}$ is the gradient, perpendicular to the surface. For the three principal axes, $M\underline{r}$ and $\underline{r}$ have the same direction. Conversely, one can determine the principle axes by requiring $M\underline{r}$ to be parallel to $\underline{r}$, or

$$M\underline{r} = C\underline{r} \ ,$$

[1] For simplicity we do not consider the most general case, which would include, e.g., hyperboloids.

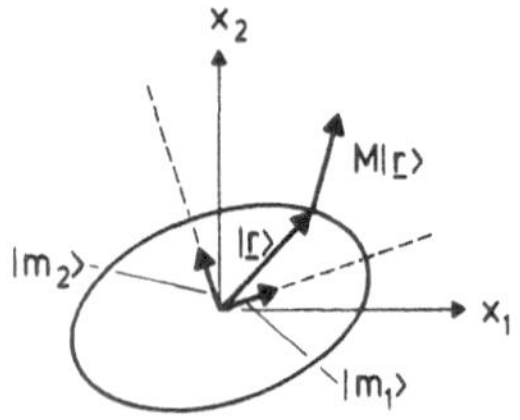


Fig. B.2

where C is a number; this eigenvalue problem defines three eigenvectors $|m_\nu\rangle$, $\nu = 1,2,3$, and three eigenvalues $\overset{\nu}{M}$

$$M|m_1\rangle = \overset{1}{M}|m_1\rangle \ , \quad \text{etc.}$$

The three eigenvalues are the three roots of the secular equation

$$\det\,(M_{ik} - C\,\delta_{ik}) = 0 \ .$$

The eigenvectors can always be chosen as orthogonal , $\langle m_\nu|m_\mu\rangle = \delta_{\nu\mu}$; if they are chosen as basis vectors one sees

$$M = \sum_\nu |m_\nu\rangle\,\overset{\nu}{M}\,\langle m_\nu| \ , \qquad M_{ij} = \sum_\nu \langle i|m_\nu\rangle\,\overset{\nu}{M}\,\langle m_\nu|j\rangle$$

because the matrix $\langle m_\nu|M|m_\mu\rangle$ is diagonal. In this system

$$F(\underline{r}) = \frac{1}{2}\,(\overset{1}{M}x_1^2 + \overset{2}{M}x_2^2 + \overset{3}{M}x_3^2) \ ,$$

and for $F(\underline{r}) = 1/2$ one sees that $\overset{\nu}{M}{}^{-1/2}$ are the three principal lengths. If two eigenvalues are equal this is called two-fold degeneracy (rotational ellipsoid), and for complete degeneracy, $\overset{1}{M} = \overset{2}{M} = \overset{3}{M}$, the ellipsoid becomes a sphere. Except for the geometrical interpretation by an ellipsoid (which requires the same sign for all $\overset{\nu}{M}$) the above results hold for any real symmetrical matrix M: M has three eigenvectors and three eigenvalues; the eigenvalues are real and the eigenvectors can be chosen as an orthonormal basis.

Let us again discuss projectors as examples for symmetrical operators. If one considers a projection $P\underline{r}$, repeated operation with P does not change the result, $[P]^2\underline{r} = P\underline{r}$ or $[P]^2 = P$. This also must hold for the eigenvalues $(\overset{\nu}{P})^2 = \overset{\nu}{P}$, which means that $\overset{\nu}{P}$ is either 1 or 0. Examples: $P^{\underline{a}}$ has one eigenvector $\hat{\underline{a}}$ with eigenvalue 1 and two eigenvectors perpendicular to $\hat{\underline{a}}$ with eigenvalue 0 ; $P^1_{ij}$ is diagonal and exhibits these three eigenvalues in its diagonal elements; $P^{\underline{a},\underline{b}}$ has the eigenvectors $\hat{\underline{a}}$ and $\hat{\underline{b}}$ (or any linear combination) with eigenvalue 1 and the vector perpendicular to $\hat{\underline{a}}$ , $\hat{\underline{b}}$ with vanishing eigenvalues; again $P^{1,2}$ exhibits these three eigenvalues in its diagonal elements. In general $[P]^2 = P$ or eigenvalues 1 and 0 define a projector.

An important quantity, which can be derived from a symmetrical operator M, is the trace of M, tr{M}:

$$\mathrm{tr}\{M\} = M_{jj} = \sum_{\nu} \langle j|m_\nu\rangle \overset{\nu}{M} \langle m_\nu|j\rangle = \sum_{\nu} \underbrace{\langle m_\nu|j\rangle \, \langle j|m_\nu\rangle}_{=1} \overset{\nu}{M} = \sum_{\nu} \overset{\nu}{M} \; .$$

It is the sum of the eigenvalues. For projectors, tr{P} is always integer and equals the dimension of the subspace belonging to P:

$$\mathrm{tr}\{P^{\underline{a}}\} = \mathrm{tr}\{P^1\} = 1 \; , \quad \mathrm{tr}\{P^{\underline{a},\underline{b}}\} = \mathrm{tr}\{P^{1,3}\} = 2 \; , \quad \mathrm{tr}\{1\} = 3 \; .$$

## B.4 Rotations, Orthogonal and Unitary Operators, Complex Vector Space

Rotations $D(\hat{\underline{D}},\varphi)$ obviously do not change a scalar product,

$$(\underline{a},\underline{b}) = (\tilde{\underline{a}},\tilde{\underline{b}}) = (D\underline{a},D\underline{b}) = (D'D\underline{a},\underline{b}) = (\underline{a},D'D\underline{b}) \; ,$$

which means that

$$D'D = DD' = 1 \; , \quad D' = D^{-1} = D(-\varphi) \neq D \; , \quad D'_{ij} = D_{ij}(-\varphi) \neq D_{ij} \; , \quad D_{ij}D_{kj} = \delta_{ik} .$$

Such operators are called *orthogonal*. Just as symmetrical operators, rotations have eigenvectors and eigenvalues (one eigenvector with eigenvalue 1 is the rotational axis $\hat{\underline{D}}$). They are, however, complex and we have to leave the familiar *real* three--dimensional space for the *complex* three-dimensional space, the elements of which are linear superpositions of three basis vectors with complex coefficients, i.e., $\underline{r} = |j\rangle x_j$ with complex $x_j$. That this is necessary can be seen from a rotation about the z-axis, $|3\rangle$,

$$D = \begin{bmatrix} \cos\varphi & \sin\varphi & 0 \\ -\sin\varphi & \cos\varphi & 0 \\ 0 & 0 & 1 \end{bmatrix} , \quad D_{ij} = \langle i|D|j\rangle \; ,$$

$$D(|1\rangle \pm i|2\rangle) = (\cos\varphi \pm i \sin\varphi)(|1\rangle \pm i|2\rangle) = e^{\pm i\varphi}(|1\rangle \pm i|2\rangle) \; , \quad D|3\rangle = |3\rangle \; ,$$

where the eigenvalues $e^{\pm i\varphi}$, 1 and the eigenvectors are complex. To deal with this situation one introduces a new (hermitean) scalar product:

$$(\underline{a},\underline{b}) = (\underline{b},\underline{a})^* \; ; \quad (\alpha\underline{a},\beta\underline{b}) = \alpha^*\beta(\underline{a},\underline{b}) \; ; \quad (\underline{a},\underline{a}) = (\underline{a},\underline{a})^* > 0 \; , \text{ real} \; ,$$

e.g., the vector $1/\sqrt{2}\,(|1\rangle + i|2\rangle)$ is now a unit vector. Therefore $1/\sqrt{2}\,(|1\rangle \pm i|2\rangle)$, $|3\rangle$ are the properly normalized eigenvectors of D and they form an orthonormal basis.

The complex space is a straightforward extension of real space: all relations of Section B.1 still hold if one uses the hermitean scalar product. The analogue of the adjoint operator M', $[M']_{ij} = M_{ji}$, is the hermitean adjoint $M^+$ defined by $(\underline{a},M\underline{b}) = (M^+\underline{a},\underline{b})$; its matrix elements are $[M^+]_{ij} = M^*_{ji}$. The generalization of *symmetrical* operators, $M = M'$, are *hermitean* operators obeying $M = M^+$; they have real eigenvalues and their eigenvectors form an orthonormal basis.

The analogue of an *orthogonal* operator D, $DD' = D'D = 1$, is a *unitary* operator U defined by $UU^+ = U^+U = 1$, which does not change a hermitean scalar product:

$$(U\underline{a},U\underline{b}) = (U^+U\underline{a},\underline{b}) = (\underline{a},U^+U\underline{b}) = (\underline{a},\underline{b}) \ .$$

The eigenvalues obey $\check{U}^*\check{U} = 1$, or $|\check{U}| = 1$. An example for a unitary operator is the rotation about the z-axis: if one chooses its complex eigenvectors as basis vectors, one has

$$D(\varphi) = \begin{bmatrix} e^{i\varphi} & 0 & 0 \\ 0 & e^{-i\varphi} & 0 \\ 0 & 0 & 1 \end{bmatrix} , \quad D^+(\varphi) = \begin{bmatrix} e^{-i\varphi} & 0 & 0 \\ 0 & e^{i\varphi} & 0 \\ 0 & 0 & 1 \end{bmatrix} = D(-\varphi) = D^{-1}(\varphi) \ .$$

## B.5 Functions of Hermitean or Unitary Operators

A function F(M) of a hermitean or unitary operator M can readily be defined with aid of the eigenvectors and eigenvalues of M:

$$F(M) = \sum_\nu |m_\nu\rangle \, F(\check{M}) \, \langle m_\nu| \ ,$$

$$\text{e.g.,}\ F(M) = M = \sum_\nu |m_\nu\rangle \, \check{M} \, \langle m_\nu| \ , \qquad F(M) = M^2 = \sum_\nu |m_\nu\rangle \, \check{M}^2 \, \langle \underline{m}_\nu| \ ,$$

$$\text{or}\quad F(M) = \sqrt{M} = \sum_\nu |m_\nu\rangle \sqrt{\check{M}} \, \langle m_\nu| \ ;$$

$\sqrt{M}$ is also hermitean if M is "positive", i.e., all $\check{M} > 0$.

In the same way one can define functions of two operators A,B if they "commute": $AB = BA$; in this case A and B have common eigenvectors, $A,B|\nu\rangle = \check{A},\check{B}|\nu\rangle$:

$$F(A,B) = \sum_\nu |\nu\rangle \, F(\check{A},\check{B}) \, \langle\nu| \ .$$

If $AB \neq BA$ one, of course, has to regard the sequence of the operators. As a special example needed in Chapter 7 we consider $\exp(A + B) \neq \exp(A)\cdot\exp(B)$, as can be seen from an expansion in powers of A,B:

$$\exp(A+B) = 1 + A + B + \frac{1}{2}(A+B)^2 + \ldots = 1 + A + B + \frac{1}{2}(A^2 + AB + BA + B^2) + \ldots$$

$$\exp(A)\cdot\exp(B) = (1+A+\frac{1}{2}A^2+\ldots)(1+B+\frac{1}{2}B^2+\ldots) = 1+A+B+\frac{1}{2}(A^2+2AB+B^2)+\ldots ,$$

$$\text{hence}\quad \exp(A+B) - \exp(A)\cdot\exp(B) = \frac{1}{2}(AB - BA) + \ldots \neq 0 .$$

Actually, if AB - BA is a pure number, one can sum up this expansion with the result $\exp(A+B) = \exp(A)\cdot\exp(B)\cdot\exp[-(AB - BA)/2]$ (for a proof see /2.8/). In Chapter 7 we need this relation for the "coordinate" $A = \alpha q$ (q = displacement) and the "momentum" $B = \beta p$, where $pq - qp = (\hbar/i)(\partial_q q - q\partial_q) = \hbar/i$ and therefore $AB - BA = -\alpha\beta(\hbar/i)$:

$$\exp(\alpha q)\cdot\exp(\beta p) = \exp(\alpha q + \beta p)\cdot\exp(i\hbar\alpha\beta/2) .$$

## B.6 Change of Basis

If one passes from one set of orthonormal basis vectors, $(\underline{u}^j)$, to another, $(\underline{\tilde{u}}^j)$, this transition is achieved by an orthogonal (real space) or unitary matrix (complex space): $\underline{\tilde{u}}^{j'} = U_{jj'}\underline{u}^j$, i.e., a change of the basis corresponds to a rotation of the coordinate system. The coordinates of the same vector $\underline{r}$ are in both bases connected by

$$\underline{r} = \tilde{x}_{j'}\underline{\tilde{u}}^{j'} = x_j\underline{u}^j , \text{ or } \tilde{x}_{j'} = (\underline{\tilde{u}}^{j'},\underline{r}) = \underbrace{(\underline{\tilde{u}}^{j'},\underline{u}^j)}_{U^+_{j'j}}x_j \quad \text{(Fig. B.3)} .$$

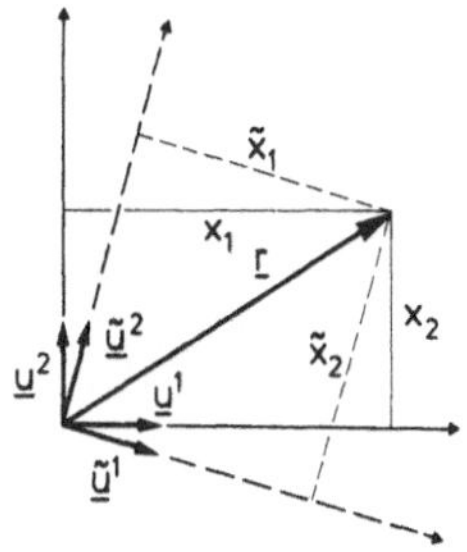


Fig. B.3

The scalar product is invariant because U is unitary:

$$(\underline{a},\underline{b}) = a^*_j b_j = \tilde{a}^*_{j'}\tilde{b}_{j'} = \underbrace{U_{jj'}U^+_{j'k}}_{=\delta_{jk}} a_j b_k .$$

The matrix elements of an operator

$$M = |i\rangle\, M_{ij}\, \langle j| = |\tilde{i}'\rangle\, \tilde{M}_{i'j'}\, \langle\tilde{j}'|$$

transform like a product of two vector components:

$$\tilde{M}_{i'j'} = U^{*}_{ii'} U_{jj'} M_{ij} = U^{+}_{i'i} M_{ij} U_{jj'} \; .$$

Therefore the trace is invariant because it transforms like a scalar product.

Often it is convenient to choose the eigenvectors of M, ($|m_j\rangle$), as basis vectors; here the transformation is given by $U_{jj'} = \langle m_{j'}|j\rangle$. Of course, M here is diagonal ("eigen-representation" of M): $\tilde{M}_{ij} = \overset{i}{M}\delta_{ij}$,

$$M = \sum_i |m_i\rangle\, \tilde{M}_{ii}\, \langle m_i| = \sum_i |m_i\rangle\, \overset{i}{M}\, \langle m_i| \; ,$$

and the diagonal elements are the eigenvalues of M.

## B.7 Higher-Dimensional Vector Spaces

So far we have dealt with three-dimensional space only; the extension to higher-dimensional spaces is straightforward: all the preceding results can immediately be taken over. As a simple example consider the 3N-dimensional space of displacements $|\underline{s}\rangle$ of an arbitrary set of N atoms; a possible choice of basis vectors is, for instance, to take the vectors $|u^n_i\rangle = |^n_i\rangle$, n = 1,...,3N , i = 1,2,3 , which means that the n-th atom is displaced in i-direction (cartesian coordinate system) by unit length while all other atoms remain in their equilibrium positions: $s^n_i = \langle u^n_i|\underline{s}\rangle = \langle ^n_i|\underline{s}\rangle$. As simple examples for 3N-dimensional operators consider the projectors

$$P_x(1) = |u^1_x\rangle\langle u^1_x| \; , \quad P_x(1)|\underline{s}\rangle = s^1_x|u^1_x\rangle \; , \quad \mathrm{tr}\{P_x(1)\} = 1 \; ,$$

which projects out $s^1_x$ (one-dimenisonal subspace), or

$$P(1) = |u^1_i\rangle\langle u^1_i| \; , \quad P(1)|\underline{s}\rangle = s^1_i|u^1_i\rangle \; , \quad \mathrm{tr}\{P(1)\} = 3 \; ,$$

which projects out $\underline{s}^1$ (three-dimensional subspace).

# C. Fourier Transforms

## C.1 Definition

The Fourier transform $F(\omega)$ of a function $f(t)$ is defined as

$$F(\omega) = \int_{-\infty}^{\infty} dt'\, f(t')\, e^{i\omega t'} \; ;$$

the inverse transformation

$$f(t) = \int_{-\infty}^{\infty} \frac{d\omega}{2\pi} F(\omega)\, e^{-i\omega t}$$

is plausible from the representation of the δ-function:

$$\delta(t - t') = \int_{-\infty}^{\infty} \frac{d\omega}{2\pi} e^{i\omega(t'-t)} \quad , \text{ e.g.,}$$

$$f(t) = \int_{-\infty}^{\infty} \frac{d\omega}{2\pi} \int_{-\infty}^{\infty} dt'\, f(t')\, e^{i\omega(t'-t)} = \int_{-\infty}^{\infty} dt'\, f(t') \underbrace{\int_{-\infty}^{\infty} \frac{d\omega}{2\pi} e^{i\omega(t'-t)}}_{\delta(t-t')} = f(t) \; .$$

## C.2 Examples

| $f(t)$ | $F(\bar{\omega})$ |
|---|---|
| $f(-t)$ | $F^*(\omega)$ |
| $\partial f/\partial t$ | $-i\omega F(\omega)$ |
| $\exp(-i\Omega t)$ | $2\pi\, \delta(\omega - \Omega)$ |
| $\sin(\Omega t),\ \Omega > 0$ | $i\pi\, [\, \delta(\omega - \Omega) - \delta(\omega + \Omega)] = 2\pi i\omega\, \delta(\omega^2 - \Omega^2)$ |
| $\cos(\Omega t)$ | $\pi\, [\, \delta(\omega - \Omega) + \delta(\omega + \Omega)] = 2\pi\, \lvert\omega\rvert\, \delta(\omega^2 - \Omega^2)$ |

In the latter two formulas we have used the fact that $\delta(\alpha\omega) = \delta(\omega)/|\alpha|$ and therefore $(\Omega > 0)$

$$\delta(\omega^2 - \Omega^2) = \delta\Big((\omega - \Omega)(\omega + \Omega)\Big) = \frac{1}{2\Omega}\Big[\delta(\omega - \Omega) + \delta(\omega + \Omega)\Big]$$

$$= \frac{1}{2|\omega|}\Big[\delta(\omega - \Omega) + \delta(\omega + \Omega)\Big] = \frac{1}{2\omega}\Big[\delta(\omega - \Omega) - \delta(\omega + \Omega)\Big] \; .$$

## C.3 Applications

Fourier transformation is a useful tool in solving linear differential equations: because the Fourier transforms $F^{(n)}(\omega)$ of derivatives $(\partial/\partial t)^n f(t)$ are simply $F^{(n)}(\omega) = (-i\omega)^n F(\omega)$, the differential equation becomes upon Fourier transformation algebraic. As an example consider the oscillator-equation

$$\ddot{s}(t) + 2\eta\dot{s}(t) + (\Omega^2 + \eta^2)s(t) = f(t) \ , \qquad \eta > 0$$

which by Fourier transformation, $s(t) = \int_{-\infty}^{\infty} \frac{d\omega}{2\pi} S(\omega) \exp(-i\omega t)$, becomes

$$\left[-\omega^2 - 2i\eta\omega + (\Omega^2 + \eta^2)\right] S(\omega) = F(\omega) \ ;$$

hence one solution is

$$s(t) = \int_{-\infty}^{\infty} \frac{d\omega}{2\pi} \frac{F(\omega)}{\Omega^2 - (\omega + i\eta)^2} e^{-i\omega t} \ .$$

## C.4 Green's Function of the One-Dimensional Oscillator

If one chooses $f(t) = \delta(t)/M$, $F(\omega) = 1/M$, one obtains the Green's function of the one-dimensional oscillator:

$$g(\eta,t) = \int_{-\infty}^{\infty} \frac{d\omega}{2\pi} \underbrace{\frac{1}{M} \frac{1}{\Omega^2 - (\omega + i\eta)^2}}_{G(\eta,\omega)} e^{-i\omega t} \ .$$

In the complex $\omega$-plane, $G(\eta,\omega)$ has two poles at $\omega = \pm\Omega - i\eta$. The path of integration along the real axis is denoted by C (Fig. C.1). The quantity $e^{-i\omega t} = e^{-i\omega' t} e^{\omega'' t}$

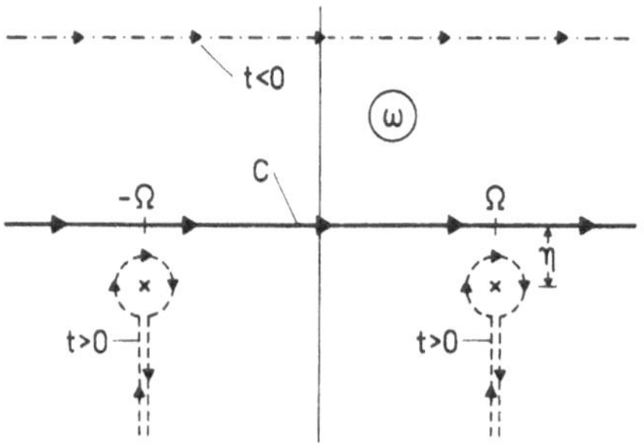


Fig. C.1

vanishes exponentially in the upper half plane ($\omega'' > 0$) for $t < 0$ and in the lower half for $t > 0$. Consequently, for $t < 0$ one can shift C to $i\infty$ and obtains $g(\eta,t<0) = 0$. For $t > 0$ one can shift the path to $-i\infty$, but one is left with the contribution of the two poles:

$$g(\eta,t) = -2\pi i \left[-\frac{1}{2M\Omega}\left(e^{-i(\Omega-i\eta)t} - e^{-i(-\Omega-i\eta)t}\right)\right] \Theta(t)$$

$$= \frac{\sin \Omega t}{M\Omega}\, e^{-\eta t}\Theta(t) \quad \text{where} \quad \Theta(t) = \begin{cases} 1 & \text{for } t \geq 0 \\ 0 & \text{for } t < 0 \end{cases} .$$

In the limit $\eta \to +0$ :

$$g(\eta \to +0,t) = g^+(t) = \frac{\sin \Omega t}{M\Omega}\,\Theta(t) .$$

The same result is obtained with $\eta = 0$ if one chooses a path $C_+$ in the upper half plane (above the poles $\pm\Omega$) (Fig. C.2). The advanced solution $g^-(t)$ is obtained by

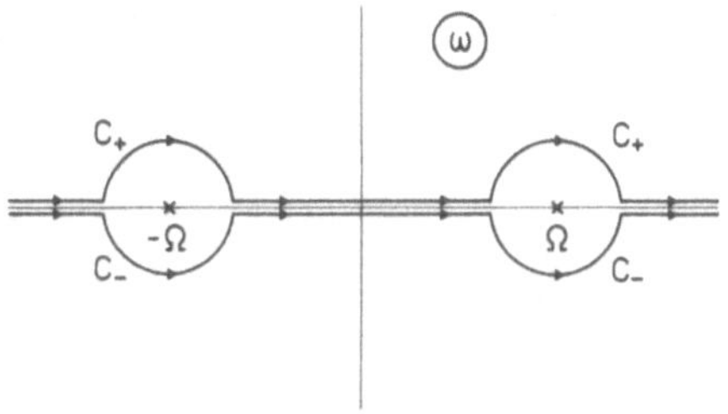


Fig. C.2

the limit $\eta \to -0$ or, equivalently, using $C_-$ as the path of integration

$$g^+(t) = \lim_{\eta\to+0} \int_{-\infty}^{\infty} \frac{d\omega}{2\pi}\, G(\eta,\omega)\, e^{-i\omega t} = \int_{C_+} \frac{d\omega}{2\pi}\, G(0,\omega)\, e^{-i\omega t}$$

$$g^-(t) = \lim_{\eta\to-0} \int_{-\infty}^{\infty} \frac{d\omega}{2\pi}\, G(\eta,\omega)\, e^{-i\omega t} = \int_{C_-} \frac{d\omega}{2\pi}\, G(0,\omega)\, e^{-i\omega t} = \int_{-\infty}^{\infty} \frac{d\omega}{2\pi}\, G^*(\eta \to +0,\omega)\, e^{-i\omega t} .$$

Consequently, one has

$$g^+(t) - g^-(t) = 2i \int_{-\infty}^{\infty} \frac{d\omega}{2\pi}\, \mathrm{Im}\{G(\eta\to+0,\omega)\}\, e^{-i\omega t} = \int_{C_+-C_-} \frac{d\omega}{2\pi}\, G(0,\omega)\, e^{-i\omega t}$$

$$= (-2\pi i)\cdot\frac{1}{2\pi}\cdot\frac{-1}{2M\Omega}\left(e^{-i\Omega t} - e^{-i(-\Omega)t}\right) = \frac{\sin \Omega t}{M\Omega} ,$$

because the integrations along the real axis cancel and only the two poles contribute; this is a special example for the more general relation (2.17a),

$$\mathrm{Im}\{G(\eta\to 0,\omega)\} = \mathrm{Im}\{G(\omega)\} = \frac{\pi\, \mathrm{sgn}\,\omega}{M}\, \delta(\omega^2 - \Omega^2) ,$$

which is obtained in the same way, if one considers

$$2i \int_{-\infty}^{\infty} d\omega \; \mathrm{Im}\{G(\eta \to +0,\omega)\} \; A(\omega) = \int_{C_+ - C_-} d\omega \; G(0,\omega) \; A(\omega)$$

with an arbitrary function $A(\omega)$:

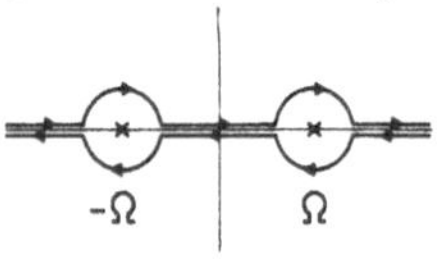


=

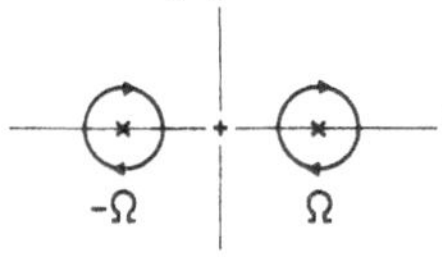


$$= -2\pi i \left[ \frac{\delta(\omega - \Omega)}{-2M\Omega} + \frac{\delta(\omega + \Omega)}{2M\Omega} \right] = 2i \frac{\pi}{M} \delta(\omega^2 - \Omega^2) \; .$$

Similarly one confirms from

$$2 \int_{-\infty}^{\infty} d\omega \; \mathrm{Re}\{G(\eta \to +0,\omega)\} A(\omega) = \int_{C_+ + C_-} d\omega \; G(0,\omega) \; A(\omega)$$

that $\mathrm{Re}\{G(\omega)\} = \frac{1}{M} \frac{P}{\Omega^2 - \omega^2}$ , because

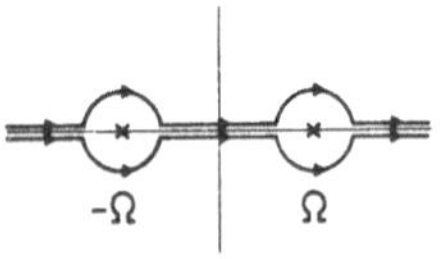


= 2(

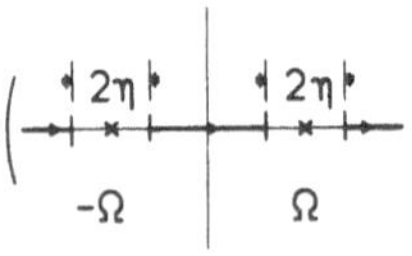


$$\Big) = \frac{2}{M} \left( \frac{1}{2\Omega} \frac{P}{\Omega - \omega} + \frac{1}{2\Omega} \frac{P}{\Omega + \omega} \right) = \frac{2}{M} \frac{P}{\Omega^2 - \omega^2} \; ;$$

the integration about the poles cancel and the remaining integral along the real axis

$$\lim_{\eta \to +0} \left( \int_{-\infty}^{-\Omega-\eta} \cdots + \int_{-\Omega+\eta}^{\Omega-\eta} \cdots + \int_{\Omega+\eta}^{\infty} \cdots \right)$$

is by definition Cauchy's principal value.

## D Diatomic Molecule

To illustrate the formalism of Section 2.4 we treat the diatomic molecule: two atoms, $-1 = \bar{1}$ and 1, with masses $M^{\bar{1}} = m$, $M^1 = M$ coupled by a spring f.

### D.1 One-Dimensional Motion

In one dimension both atoms are restricted to move in spring direction (Fig. D.1).

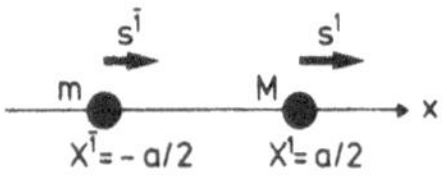


Fig. D.1

The equation of motion is

$$M\ddot{\underline{s}} = -\Phi\underline{s} + \underline{F} \; .$$

Here $\underline{s} = \begin{pmatrix} s^1 \\ s^{\bar{1}} \end{pmatrix}$ is two-dimensional;

$$\Phi = f\begin{bmatrix} 1 & \bar{1} \\ \bar{1} & 1 \end{bmatrix}, \quad M = \begin{bmatrix} M & 0 \\ 0 & m \end{bmatrix} \quad \text{and} \quad \sqrt{M} = \begin{bmatrix} \sqrt{M} & 0 \\ 0 & \sqrt{m} \end{bmatrix}$$

are $2 \times 2$ matrices. The eigenvalues and eigenvectors of $\Phi$ are:

$$\overset{1}{\Phi} = 0 \;, \quad |1\rangle = \frac{1}{\sqrt{2}}\begin{pmatrix} 1 \\ 1 \end{pmatrix} \quad \text{(translation of the molecule)} \;,$$

$$\overset{2}{\Phi} = 2f \;, \quad |2\rangle = \frac{1}{\sqrt{2}}\begin{pmatrix} 1 \\ -1 \end{pmatrix} \quad \text{(stretching of the spring)} \;,$$

hence $\Phi = 2f\,|2\rangle\langle 2|$ .

With $\sqrt{M}\underline{s} = \underline{\tilde{s}}$ , $\frac{1}{\sqrt{M}}\underline{F} = \underline{\tilde{F}}$ , $\frac{1}{\sqrt{M}}\Phi\frac{1}{\sqrt{M}} = D = \Omega^2$ one has $\ddot{\underline{\tilde{s}}} = -D\underline{\tilde{s}} + \underline{\tilde{F}}$. The dynamical matrix

$$D = \Omega^2 = \frac{f}{\sqrt{mM}}\begin{bmatrix} \sqrt{m/M} & \bar{1} \\ \bar{1} & \sqrt{M/m} \end{bmatrix}$$

has the eigenvalues and eigenvectors

$$\overset{1}{D} = 0 \;, \quad |\tilde{1}\rangle = \frac{1}{\sqrt{M+m}}\begin{pmatrix} \sqrt{M} \\ \sqrt{m} \end{pmatrix} = \frac{\sqrt{2}}{\sqrt{M+m}}\sqrt{M}\,|1\rangle \;,$$

a center of mass motion,

$$\underline{s} = \frac{1}{\sqrt{M}}|\tilde{1}\rangle = \frac{1}{\sqrt{M+m}}\begin{pmatrix} 1 \\ 1 \end{pmatrix} \quad \text{, and}$$

$$\overset{2}{D} = \left(\frac{1}{M} + \frac{1}{m}\right)f = \overset{2}{\Omega}{}^2 \;, \quad |\tilde{2}\rangle = \frac{1}{\sqrt{M+m}}\begin{pmatrix} \sqrt{m} \\ -\sqrt{M} \end{pmatrix},$$

an oscillation,

$$\underline{s} = \frac{1}{\sqrt{M}}|\tilde{2}\rangle = \frac{1}{\sqrt{M+m}}\begin{pmatrix} \sqrt{m/M} \\ -\sqrt{M/m} \end{pmatrix}, \quad Ms^1 + ms^{\bar{1}} = 0 \;,$$

where the center of mass is fixed at 0. If one applies an invariant force pattern $\underline{F}_{inv} = \sigma\begin{pmatrix} 1 \\ -1 \end{pmatrix} = \sqrt{2}\sigma\,|2\rangle$, the static solution is

$$\underline{s} = |2\rangle\frac{1}{\overset{2}{\Phi}}\langle 2|\underline{F}_{inv}\rangle = \sqrt{2}\sigma\,\frac{1}{\overset{2}{\Phi}}|2\rangle = \frac{\sigma}{2f}\begin{pmatrix} 1 \\ -1 \end{pmatrix}, \quad f(s^1 - s^{\bar{1}}) = \sigma = \text{spring tension}$$

(Fig. D.2).

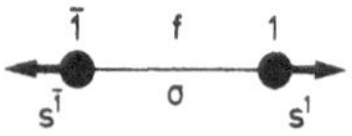


Fig. D.2

The displacement is such that the geometrical center of 1 and $\bar{1}$ is conserved; superimposing a translation, $\underline{s} \to \underline{s} + \sqrt{2}\, T|1>$, still yields a correct solution of $\Phi\underline{s} = \underline{F}_{inv}$. On the other hand, the static response is also represented by $G(\omega \to 0)$. For $\underline{s} = G(\omega)\underline{F}_{inv}$ one has

$$\underline{s} = \frac{1}{\sqrt{M}} \frac{|\tilde{2}><\tilde{2}|}{\overset{2}{\Omega}{}^2 - (\omega + i\eta)^2} \frac{1}{\sqrt{M}} \underline{F}_{inv} \quad \text{because} \quad <\tilde{1}| \frac{1}{\sqrt{M}} |\underline{F}_{inv}> = <\frac{1}{\sqrt{M}}\tilde{1}|\underline{F}_{inv}> = 0 \; .$$

This means that invariant forces do not produce center of mass motion. Consequently, the static displacement given by

$$G(\omega = 0)\underline{F}_{inv} = \frac{1}{\overset{2}{\Omega}{}^2} \frac{1}{\sqrt{M}} |\tilde{2}><\tilde{2}| \frac{1}{\sqrt{M}} \underline{F}_{inv} = \frac{\sqrt{2}}{\overset{2}{\Omega}{}^2} \frac{1}{\sqrt{M}} |\tilde{2}><\tilde{2}| \frac{1}{\sqrt{M}} |2>$$

conserves the center of gravity as shown in Fig. D.3 for M = 3m.

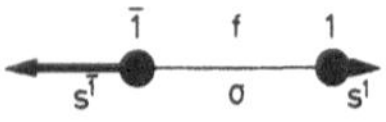


Fig. D.3

Therefore the responses

$$\frac{1}{\Phi} = \frac{|2><2|}{\overset{2}{\Phi}} \quad \text{and} \quad G(\omega = 0) = \frac{1}{\sqrt{M}} \frac{|\tilde{2}><\tilde{2}|}{\overset{2}{D}} \frac{1}{\sqrt{M}}$$

are not identical; they differ by a translation which is of no consequence physically. If the masses are equal:

$$\frac{1}{\Phi} = \frac{|2><2|}{\overset{2}{\Phi}} = G(\omega = 0) = \frac{|2><2|}{M\overset{2}{D}} = \frac{1}{4f}\begin{bmatrix} 1 & \bar{1} \\ \bar{1} & 1 \end{bmatrix}, \quad \overset{2}{\Phi} = M\overset{2}{D} = 2f \; .$$

## D.2 The Time-Dependent Green's Function

For equal masses, M = m, the time-dependent Green's function g(t) is, according to (2.27),

$$g(t) = \Theta(t) \frac{\sin \Omega t}{M\Omega} = \Theta(t) \sum_{\nu} |\nu> \frac{\sin \overset{\nu}{\Omega} t}{M\overset{\nu}{\Omega}} <\nu| = \Theta(t)\left( \frac{t}{M} \underbrace{|1><1|}_{P} + \frac{\sin \overset{2}{\Omega} t}{M\overset{2}{\Omega}} \underbrace{|2><2|}_{Q} \right) .$$

Here P is the projector onto translation corresponding to (2.37), $P\underline{F}_{inv} = 0$. In general the motion consists of translation and oscillation, e.g., for $\underline{F} = \begin{pmatrix} 1 \\ 0 \end{pmatrix} \delta(t)$:

$$\frac{\Theta(t)}{\sqrt{2}}\left(\frac{t}{M}\,|1\rangle + \frac{\sin \overset{2}{\Omega} t}{M\overset{2}{\Omega}}\,|2\rangle\right) = \frac{\Theta(t)}{2M}\begin{pmatrix} t + \sin(\overset{2}{\Omega} t)/\overset{2}{\Omega} \\ t - \sin(\overset{2}{\Omega} t)/\overset{2}{\Omega} \end{pmatrix} \text{(comp. Fig. D.4).}$$

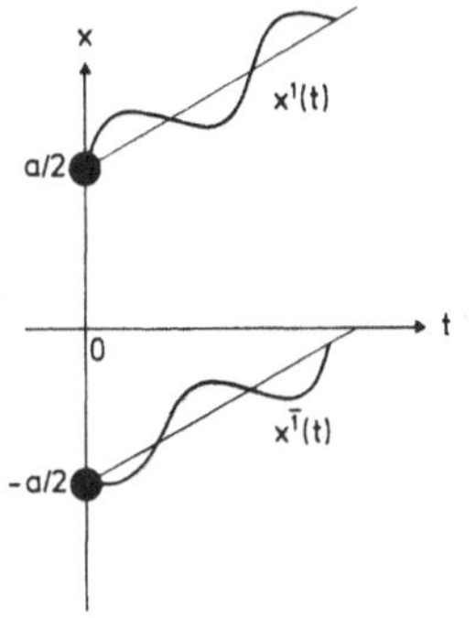


Fig. D.4

## D.3 Three-Dimensional Motion

In three dimensions $\Phi$ is a $6 \times 6$ matrix and must have six eigenvectors and -values; the oscillation $|2\rangle$ is the only eigenvector with non-vanishing eigenvalue (2f). The translation $|1\rangle$ in x-direction, the two translations in y,z-direction and the two rotations about the y- and z-axis are five mutually perpendicular eigenvectors with vanishing eigenvalue (Fig. D.5):

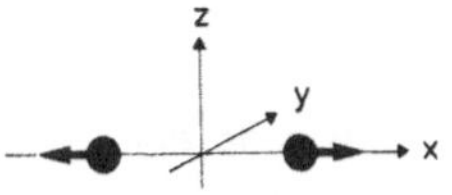


oscillation,

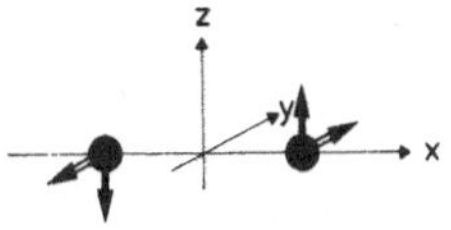


rotations about z-axis ( ⇒ ) and y-axis ( → ),

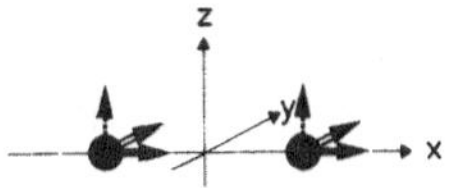


translation along x, y and z

Fig. D.5

## E. Fourier Series

In the Fourier expansion $f(x) = (1/2\pi)\int dk\, e^{ikx}F(k)$ (analogous to Appendix C $(t,\omega \rightarrow x,k)$) of a *periodic* function $p(x) = p(x-a)$ only periodic contributions can enter; i.e., $\exp(-ikx) = \exp[-ik(x-a)]$, $ka$ = integer multiple of $2\pi$ or $ka = K^h a = h2\pi$ with integer h. The inversion of $p(x) = \sum_h \tilde{p}^h e^{-iK^h x}$ is given by

$$\tilde{p}^{h'} = \int_{-a/2}^{a/2} \frac{dx}{a} e^{iK^{h'}x} p(x) ,$$

the integration extends over any interval a. The validity of the inversion can be seen from the identity

$$p^{h'} = \sum_h \int_{(a)} dx\, e^{i(K^{h'}-K^h)x} \tilde{p}^h \quad \text{because} \quad \int_{(a)} \frac{dx}{a} e^{i(K^{h'}-K^h)x} = \delta^{h'h} .$$

The Fourier integral of Appendix C can be viewed as the limit of an infinite period a:

$$k = \frac{2\pi}{a} h \ \text{ as a "continuous" variable,} \quad \sum_h \ldots \rightarrow \int dh = \frac{a}{2\pi} \int dk , \quad \frac{a}{2} \tilde{p}^h = \tilde{p}(k).$$

In three dimensions, with three non-planar periods $\underline{a}^{(j)}$, the $\underline{K}$-selection in $(2\pi)^{-3}\int d\underline{k}\, e^{-ikr} F(\underline{k})$ is: $(\underline{K},\underline{a}^{(j)})$ = integer multiple of $2\pi$ for $j = 1,2,3$; this condition is evaluated in Appendix F.

In complete analogy one can express a displacement pattern $(s^m)$ by

$$s^m = \int_{-\pi b}^{+\pi b} \frac{dk}{2\pi b} e^{ikx^m} , \quad \tilde{s}(k) = \sum_m s^m e^{-ikx^m} , \quad x^m = ma .$$

The periodic $\tilde{s}(k)$ is the Fourier transform of the displacement. If one considers

$$S^m = \Phi^{mn} s^n$$

and its Fourier transform

$$\tilde{S}(K) = \Phi^{mn} e^{-ikx^m} \int \frac{dk'}{2\pi b} e^{ik'x^n} \tilde{s}(k') = \int \frac{dk'}{2\pi b} \underbrace{e^{-ikx^m} \Phi^{mn} e^{ik'x^n}}_{\tilde{\tilde{\Phi}}(k,k')} \tilde{s}(k')$$

one recognizes how to Fourier transform a matrix $\Phi^{mn}$. If in particular $\Phi^{mn}$ has translational symmetry, $\Phi^{mn} = \Phi^{(m-n)}$, we have

$$\tilde{\tilde{\Phi}}(k,k') = \underbrace{\sum_h \Phi^{(h)} e^{-ikx^h}}_{\tilde{\Phi}(k)} \cdot \underbrace{\sum_h e^{i(k'-k)x^n}}_{2\pi b\ \delta_p(k'-k)} \quad \text{and} \quad \tilde{S}(k) = \tilde{\Phi}(k)\tilde{s}(k) .$$

Only the quantity $\tilde{\Phi}(k)$ enters the "Fourier transform" of a matrix with translational symmetry. The extension to three dimensions is straightforward.

Obviously $\tilde{\Phi}(k)$ is an eigenvalue of $\Phi$ and the eigenvectors are proportional to $\exp(ikX^m)$. The standard normalization of $|k>$ is such that

$$|s> = \int dk\ |k> <k|s> \ , \quad \text{i.e.,} \quad \int dk\ |k> <k| = \text{"1"} \ ,$$

which is obtained by choosing

$$<k|s> = <k|m> <m|s> = \frac{1}{\sqrt{2\pi b}} \tilde{s}(k) \quad \text{with} \quad <k|m> = \frac{e^{-ikx^m}}{\sqrt{2\pi b}} \ , \quad <m|s> = s^m \ \text{and}$$

$$<k'|k> = <k'|m> <m|k> = \sum_m \frac{e^{i(k-k')x^m}}{2\pi b} = \delta_p(k' - k) \ .$$

In three dimensions $2\pi b$ has to be replaced by $V_B$.

## F. Non-Orthogonal Basis

If one has three non-planar vectors $\underline{a}^{(j)}$ as a basis, one can represent[2] any vector $\underline{r}$ by $\underline{r} = |\underline{a}^{(j)}>x_j$. For an orthonormal system one would obtain the component $x_1$ from the scalar product $<\underline{a}^{(1)}|\underline{r}>$ which here,

$$<\underline{a}^{(1)}|\underline{r}> = <\underline{a}^{(1)}|\underline{a}^{(j)}>x_j = g^{1j}x_j \ ,$$

does not yield the component $x_1$ directly, rather $x_j = [g^{-1}]^{j1} <\underline{a}^{(1)}|\underline{r}>$. Things become much more transparent if one introduces a second,"reciprocal" basis $\underline{b}_{(j)}$ defined by

$$<\underline{b}_{(j)}|\underline{a}^{(1)}> = \delta^1_j = \begin{cases} 1 & \text{if} \ \ j = 1 \\ 0 & \text{if} \ \ j \neq 1 \end{cases} .$$

Then

$$|\underline{r}> = |\underline{a}^{(j)}>x_j = |\underline{a}^{(j)}> <\underline{b}_{(j)}|\underline{r}> = |\underline{b}_{(j)}>x^j = |\underline{b}_{(j)}> <\underline{a}^{(j)}|\underline{r}> \ ,$$

$$x_j = <\underline{b}_{(j)}|\underline{b}_{(1)}>x^1 = g_{j1}x^1 \ , \quad x^1 = g^{1j}x_j \ .$$

A scalar product can be expressed in various ways:

---

2 The use of lower and upper indices is standard. As a rule the summation convention refers to one upper and one lower index. We use this notation only in this appendix.

$$\langle \underline{r}|\underline{R}\rangle = x_j X^j = x_j g^{jl} X_l = x^j X_j = x^j g_{jl} X^l \ .$$

The condition $\langle \underline{K}|\underline{a}^{(j)}\rangle = 2\pi h^j$ of Appendix E is identical with $\underline{K} = \underline{b}_{(j)} 2\pi h_j$. A two-dimensional example is given in Fig. F.1.

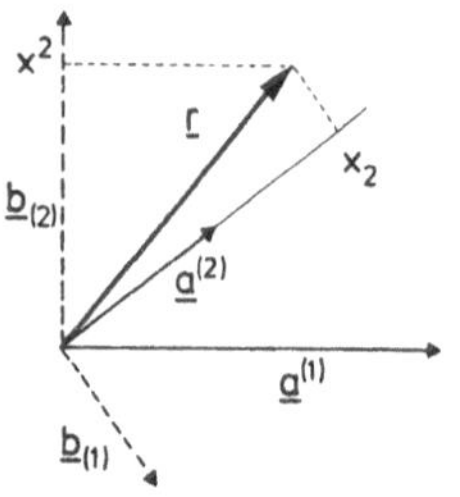


$$\underline{a}^{(1)} = 2\ (1,0)$$

$$\underline{a}^{(2)} = \frac{1}{5}\ (4,3)$$

$$\underline{b}_{(1)} = \frac{1}{6}\ (3,4)$$

$$\underline{b}_{(2)} = \frac{5}{3}\ (0,1)$$

Fig. F.1

## G. Singularities in the Spectrum

The contribution to the spectrum by one branch is

$$Y(\lambda) = \int \frac{d\underline{k}}{V_B}\, \delta\Big(\Lambda(\underline{k}) - \lambda\Big)\ , \qquad \Lambda(\underline{k}) = \Omega^2(\underline{k})\ , \qquad \lambda = \omega^2\ .$$

The actual spectrum would be $Z(\lambda) = \sum_\sigma Y_\sigma(\lambda)/3$. We treat only one branch and omit the branch index $\sigma$. Here $V_B Y(\lambda) d\lambda$ is the volume between the surface $\Lambda(\underline{k}) = \lambda$ and $\Lambda(\underline{k}) = \lambda + d\lambda$ in one Brillouin-Zone. Therefore Y can be expressed by a surface integral over the surface $\Lambda(\underline{k}) = \lambda$:

$$Y(\lambda) = \int\limits_{\Lambda(\underline{k}) = \lambda} \frac{dS_{\underline{k}}}{|\mathrm{grad}\ \Lambda(\underline{k})|} \frac{1}{V_B}\ .$$

The derivation is sketched in Figures G.1 and G.2.

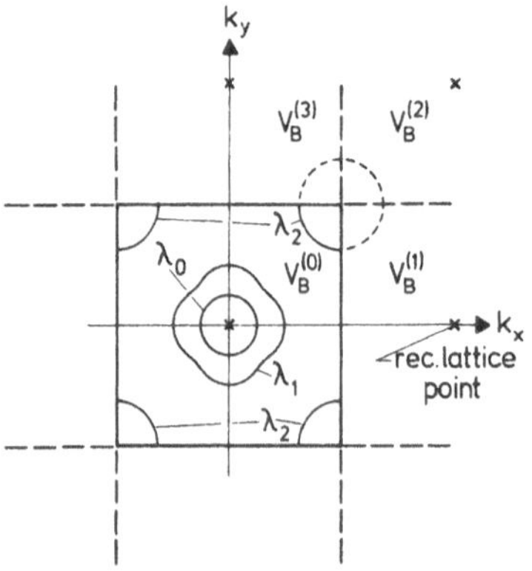


Fig. G.1. Two-dimensional example. $\Lambda = \lambda_o$ and $\Lambda = \lambda_1$ are closed surfaces in $V_B^{(O)}$, $\Lambda = \lambda_2$ is closed if one employs adjacent $V_B$'s, $V_B^{(1)}$, $V_B^{(2)}$, $V_B^{(3)}$, about reciprocal lattice points points ( × )

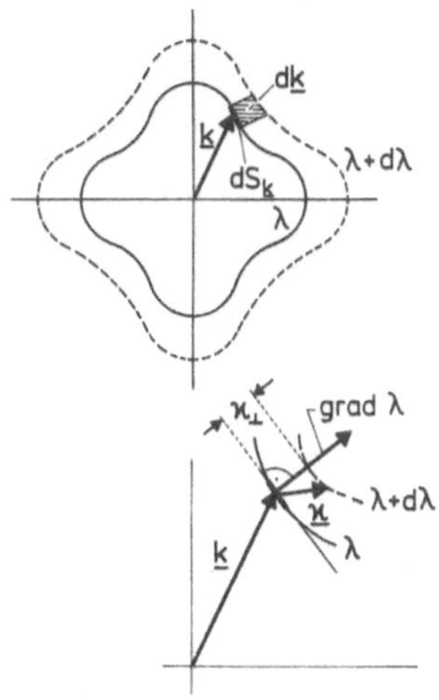


Fig. G.2. Constant frequency contours.

$\Lambda(\underline{k}) = \lambda$ ———

$\Lambda(\underline{k}) = \lambda + d\lambda$ -----

$\Lambda(\underline{k} + \underline{\kappa}) = \lambda + d\lambda$

$\kappa_\perp |\mathrm{grad}\ \Lambda| = d\lambda$

$d\underline{k} = \kappa_\perp\ dS_{\underline{k}} = \dfrac{dS_{\underline{k}}}{|\mathrm{grad}\ \Lambda|}\ d\lambda$

One recognizes that singularities (comp. /2.3/) can occur if grad $\Lambda = 0$. Points $\underline{k}_c$ for which grad $\Lambda(\underline{k}_c) = 0$ are called critical points, where most simply $\Lambda(\underline{k})$ has a maximum or a minimum or a saddle point.

We will illustrate this by a simple example, the sc lattice with 1st neighbour springs $f_l = f_t = f$ (compare the coupling of Fig. 3.15d):

$$\tilde{\Phi}_{ij}(\underline{k}) = \delta_{ij}\ 2f(3 - \cos k_x a - \cos k_y a - \cos k_z a) = \delta_{ij}\ M\Omega^2(\underline{k})\ ,$$

$$\Lambda_{min} = 0 \leqslant \Omega^2(\underline{k}) = \Lambda(\underline{k}) \leqslant \Lambda_{max} = 12f/M\ .$$

The problem is completely degenerate; the contributions of each branch to the spectrum are the same, $Z(\lambda) = Y(\lambda)$. Further $Y(\lambda)$ is symmetrical about $\Lambda_{max}/2$. The critical values are simply (Fig. G.3):

| $\underline{k}_c$ | type of extremum | number of points in $V_B$ | $\Lambda_c = \Lambda(\underline{k}_c)$ |
|---|---|---|---|
| $\underline{k}_o = 0$ | minimum | 1 | $\Lambda_{min} = 0$ |
| $\underline{k}_1 = \frac{\pi}{a}(1,0,0)$ | saddlepoint[3] | 3 | $\Lambda_{max}/3 = \Lambda_s$ |
| $\underline{k}_2 = \frac{\pi}{a}(1,0,1)$ | saddlepoint | 3 | $2\Lambda_{max}/3 = \Lambda_{s'}$ |
| $\underline{k}_3 = \frac{\pi}{a}(1,1,1)$ | maximum | 1 | $\Lambda_{max}$ |

[3] There are six equivalent $k_1$-points on the faces of the cube. Because each point belongs to two cubes, it has weight 1/2; for the other points similar arguments apply.

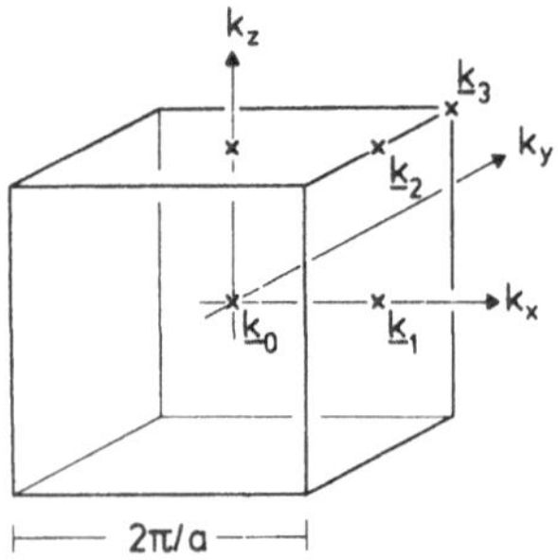


Fig. G.3

First let us consider the contribution from points near $\underline{k}_o = 0$, $\lambda \ll \Lambda_{max}$, for which:

$$\Lambda(\underline{k}) \cong \frac{fa^2}{M}(k_x^2 + k_y^2 + k_z^2) = \Lambda_{11}k^2 ,$$

$$Y(\lambda \gtrsim 0) = \int \frac{4\pi k^2\, dk}{V_B} \underbrace{\delta(\Lambda_{11}k^2 - \lambda)}_{\delta(k^2 - \lambda/\Lambda_{11})/\Lambda_{11}} = \int \frac{2\pi\sqrt{k^2}\, dk^2}{V_B} \frac{\delta(k^2 - \frac{\lambda}{\Lambda_{11}})}{\Lambda_{11}} = \frac{2\pi}{V_B\Lambda_{11}^{3/2}}\sqrt{\lambda} .$$

The surfaces of constant small $\lambda$ are indicated by the sphere about $\underline{k} = 0$ in Figs. G.1 and G.4.

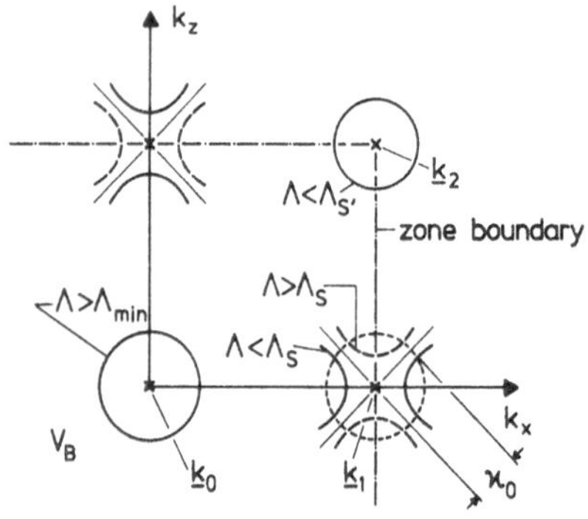


Fig. G.4

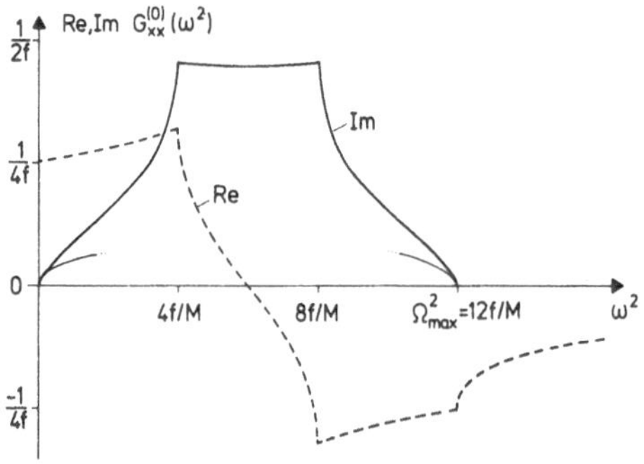


Fig. G.5

Near the maximum, $\lambda \lesssim \Lambda_{max}$, $\underline{k} = \underline{k}_3 + \underline{\kappa}$ with small $\kappa$, one has

$$\Lambda(\underline{k}_3 + \underline{\kappa}) \cong \Lambda_{max} - \Lambda_{11}\kappa^2$$

$$Y(\lambda \lesssim \Lambda_{max}) = \int \frac{4\pi\kappa^2\, d\kappa}{V_B}\delta(\Lambda_{max} - \Lambda_{11}\kappa^2 - \lambda) = \frac{2\pi}{V_B\Lambda_{11}^{3/2}}\sqrt{\Lambda_{max} - \lambda} = \frac{6\sqrt{3}}{\pi^2}\frac{\sqrt{\Lambda_{max} - \lambda}}{\Lambda_{max}^{3/2}} .$$

The spheres of constant $\lambda \lesssim \Lambda_{max}$ correspond to the circles about the corner of the Brillouin-Zone in Fig. G.1. The computed spectrum (Fig. G.5) clearly exhibits the square root behaviour at $\lambda = 0, \Lambda_{max}$.

Near $\underline{k}_1$ we have a saddlepoint,

$$\Lambda(\underline{k}_1 + \underline{\kappa}) = \Lambda_s + \Lambda_{11}(-\kappa_x^2 + \kappa_y^2 + \kappa_z^2) \ .$$

Fig. G.4 sketches the constant $\lambda$-surfaces near $\underline{k}_1$. The contributions to Y near $\Lambda_s$ are *not* restricted to $\underline{k}$-values near $\underline{k}_1$. But one can try to extract the singular part, $Y_s(\lambda \cong \Lambda_s)$, from the vicinity of $\underline{k}_c$, $\kappa^2 \leqslant \kappa_o^2$:

$$Y_s = \int\limits_{\substack{0<\kappa_o^2 - \kappa^2 \\ =\kappa_o^2-\rho^2-\kappa_x^2}} d\underline{\kappa}\ \delta\Big(\Lambda_s - \lambda + \Lambda_{11}\underbrace{(\kappa_y^2 + \kappa_z^2}_{\rho^2} - \kappa_x^2)\Big) = \frac{\sqrt{2}\pi\kappa_o}{\Lambda_{11}} \begin{cases} + \dfrac{\pi}{\sqrt{2}\Lambda_{11}^2\kappa_o}(\Lambda_s - \lambda) & , \ \lambda \gtrsim \Lambda_s \\ - \dfrac{2\pi}{\Lambda_{11}^{3/2}}\sqrt{\Lambda_s - \lambda} & , \ \lambda \lesssim \Lambda_s \end{cases}$$

Because the contributions to $Y(\lambda \cong \Lambda_s)$ are not confined to the vicinity of $\underline{k}_1$, the value $\sqrt{2}\pi\kappa_o/\Lambda_{11}$ does not determine the actual value $Y(\Lambda_s)$. However, the (singular) square root behaviour, which does not depend on $\kappa_o$, is correct. Consequently, for $\lambda \leqslant \Lambda_s$, one obtains[4]

$$Y(\lambda \lesssim \Lambda_s) = Y(\Lambda_s) - \frac{3}{V_B}\frac{2\pi}{\Lambda_{11}^{3/2}}\sqrt{\Lambda_s - \lambda} \ .$$

Near $\Lambda_{s'}$, one obtains via the same procedure

$$Y(\lambda \gtrsim \Lambda_{s'}) = Y(\Lambda_{s'}) - \frac{3}{V_B}\frac{2\pi}{\Lambda_{11}^{3/2}}\sqrt{\lambda - \Lambda_{s'}} \ .$$

In the numerical spectrum one sees also quite clearly the $\sqrt{\ }$-behaviour near $\Lambda_s$, $\Lambda_{s'}$ (Fig. G.5).

One recognizes that one could construct an approximate spectrum by using only the singular parts: one uses for $0 \leqslant \lambda \leqslant \Lambda_s$ the two functions $Y(\lambda) = a\sqrt{\lambda}$, $Y = A - 3a\sqrt{\Lambda_s - \lambda}$ which are joined smoothly; A is the approximate value for $Y(\Lambda_s)$. The approximate $\Lambda_s$, $\Lambda_{s'}$ are connected by a parabola such that the approximate spectrum is normalized.

One can construct other approximate spectra. A very old procedure is that of Debye which extrapolates the behaviour for small $\lambda$,

$$Y_D = a\sqrt{\lambda}\ \Theta(\Lambda_D - \lambda) \ , \quad a\frac{2}{3}\Lambda_D^{3/2} = 1 \ , \quad \Lambda_D \cong 1.3\ \Lambda_{max} \ ,$$

---

[4] The total contribution of all equivalent points in $V_B$, including the correct weight, is $3Y_s$.

where the cutoff $\Lambda_D$ is determined by normalization ($\sqrt{\Lambda_D} = \Omega_D$ = Debye frequency).

Another procedure would be to approximate Y by

$$Y_o(\lambda) = \frac{8}{\pi\Lambda_o^2}\sqrt{\lambda(\Lambda_o - \lambda)}\ , \qquad a = \frac{8}{\pi\Lambda_o^{3/2}}\ , \qquad \Lambda_o \cong 1.6\ \Lambda_{max}\ .$$

What one should keep in mind is that for small $\lambda$ the spectrum, $Z(\lambda)$, is proportional to $\sqrt{\lambda}$ (corresponding to $z(\omega) \propto \omega^2$) and that the spectrum at $\Lambda_{max}$ is proportional to $\sqrt{\Lambda_{max} - \lambda}$ (corresponding to $z(\omega) \propto \sqrt{\Omega_{max} - \omega}$) as a rule. There are deviations, however, from that rule. An example is the spectrum corresponding to the lattice model of Fig. 3.15d (three equal 1st neighbour springs in a fcc lattice, $f_l = f_t = f_{t'} = f$) which exhibits a logarithmic divergency at $\Omega^2_{max} = 16f/M$. This model yields:

$$\Lambda(\underline{k}) = \frac{4f}{M}\left[3 - \cos\frac{k_xa}{2}\cos\frac{k_ya}{2} - \cos\frac{k_ya}{2}\cos\frac{k_za}{2} - \cos\frac{k_xa}{2}\cos\frac{k_ya}{2}\right]\ .$$

The maximum value is reached, for instance, at $\underline{k} = (2\pi/a)\ (1,0,0)$, $\Lambda_{max} = 16f/M$. One recognizes that on the line $k_x = 2\pi/a$, $k_z = 0$ one has $\Lambda(k_y) = \Lambda_{max}$ and grad $\Lambda = 0$. Instead of a critical point one has a critical line. Here it is a critical line of maxima, which gives rise to a genuine divergency, exhibited in Fig. 3.29c. An analogous line of critical points occurs in the model of Fig. 3.15c, where along the same line the frequency for polarization [010] is constant, namely $\sqrt{4f/M}$. For symmetry reasons the full gradient vanishes along this line. In this case, however, the critical points are saddlepoints, which results in a step (comp. Fig. 3.29b). The corresponding "step" in Fig. 3.29a is not a real step; the frequencies depend on $k_y$ but only very weakly because of the smallness of the transversal springs.

The expansion about a critical point for other models will be

$$\Lambda(\underline{k}_c + \underline{\kappa}) = \Lambda_c + \Lambda_{ij}\kappa_i\kappa_j\ ,$$

which can be transformed into

$$\Lambda(\underline{k}_c + \underline{\kappa}) = \Lambda_c + \Lambda_{11}\kappa_1^2 + \Lambda_{22}\kappa_2^2 + \Lambda_{33}\kappa_3^2\ .$$

The treatment is now strictly analogous to the simple case above. One merely has to replace $\Lambda_{11}^{3/2}$ by $(\Lambda_{11}\Lambda_{22}\Lambda_{33})^{1/2}$. Further complications can occur if at a critical point two branches are degenerate (exception $\underline{k}_c = 0$). Here a Taylor expansion in terms of $\underline{\kappa}$ might not exist ("non-analytical" critical point).

In one dimension the common singularities are of $1/\sqrt{\ldots}$-type (comp. Fig. 3.28). In two dimensions one has step discontinuities for maxima and minima, and logarithmic divergencies at saddlepoints.

## H. The Low-$\omega$ Expansion of $\mathrm{Re}\{G^{(0)}(\omega)\}$

The expansion of

$$MG_1^{(0)}(\omega^2) = \int d\lambda' \, Z(\lambda') \frac{P}{\lambda' - \lambda} = \alpha + \beta\lambda + \ldots , \qquad \lambda = \omega^2 ,$$

is cumbersome. In an "expansion" of the integrand such as

$$P\left(\frac{1}{\lambda' - \lambda}\right) = P\left(\frac{1}{\lambda'} + \frac{\lambda}{(\lambda')^2} + \ldots\right) = \frac{\lambda' - \lambda}{(\lambda' - \lambda)^2 + \eta^2}$$

$$= \frac{\lambda'}{(\lambda')^2 + \eta^2} - \lambda \cdot \partial_{\lambda'} \frac{\lambda'}{(\lambda')^2 + \eta^2} + \ldots , \qquad \eta \to +0 ,$$

the various terms are tedious to calculate, except for the 1st term where $\eta = 0$ does not cause difficulties: the remaining integral, $\alpha = \int d\lambda' \, Z(\lambda')/\lambda'$, exists because $Z(\lambda')/\lambda' = a/\sqrt{\lambda'} + b\sqrt{\lambda'} + \ldots$ in three dimensions. The 2nd expansion coefficient, however, is not simply given by $\int d\lambda' \, Z(\lambda')/(\lambda')^2$ (i.e., $\eta = 0$), because this integral diverges at the lower limit: $Z(\lambda')/(\lambda')^2 = a/(\lambda')^{3/2} + b/\sqrt{\lambda'} + \ldots$ . What one can calculate directly is the difference $\beta - \beta_0$ between the coefficients of two spectra $Z(\lambda')$ and $Z_0(\lambda')$, $\beta - \beta_0 = \int d\lambda' \, [Z(\lambda') - Z_0(\lambda')]/(\lambda')^2$, provided that $a = a_0$, i.e.,

$$Z(\lambda') - Z_0(\lambda') = \sqrt{\lambda'}\,[\underbrace{a - a_0}_{=0} + (b - b_0)\lambda' + \ldots] ,$$

because then $Z - Z_0$ starts with $(\lambda')^{3/2}$ and the integral $\int d\lambda' \, [Z(\lambda') - Z_0(\lambda')]/(\lambda')^2$ exists. If the coefficient $\beta_0$ is known one can express $\beta$ by $\beta_0$ and a simple integral.

From this representation one can learn how to obtain $\beta$ directly from $Z$ by using the following trick. The function $(Z - Z_0)/\sqrt{\lambda'} = \lambda'(b - b_0) + \ldots$ starts with $\lambda'$ and one can integrate by parts:

$$\beta - \beta_0 = -2\int_0^\infty d\lambda' \, \frac{Z - Z_0}{\sqrt{\lambda'}} \, \partial_{\lambda'} \frac{1}{\sqrt{\lambda'}} = \int_0^\infty \frac{2d\lambda'}{\sqrt{\lambda'}} \, \partial_{\lambda'} \frac{Z - Z_0}{\sqrt{\lambda'}}$$

$$= \underbrace{\int_0^\infty \frac{2d\lambda'}{\sqrt{\lambda'}} \, \partial_{\lambda'} \frac{Z}{\sqrt{\lambda'}}}_{\text{exists: } \partial_{\lambda'}(Z/\sqrt{\lambda'}) = b + \ldots} - \underbrace{\int_0^\infty \frac{2d\lambda'}{\sqrt{\lambda'}} \, \partial_{\lambda'} \frac{Z_0}{\sqrt{\lambda'}}}_{\beta_0} .$$

The remaining integral can be split and the integral over $Z_0$ just cancels $\beta_0$, as can be seen directly, e.g., for the Debye-spectrum of Table 3.4. Consequently,

$$\beta = 2 \int_0^\infty \frac{d\lambda'}{\sqrt{\lambda'}}\, \partial_{\lambda'} \frac{Z(\lambda')}{\sqrt{\lambda'}}$$

for all Z which do not have (e.g., logarithmic) divergencies. Divergencies must be treated separately.

## I. The Operator $D = \Omega^2$ in Continuum Theory

In lattice theory the equation of motion according to Section 2.4.1 is

$$\sqrt{M}\,\ddot{\underline{\tilde{s}}} = \underbrace{-\frac{1}{\sqrt{M}}\,\Phi\,\frac{1}{\sqrt{M}}}_{D}\,\underbrace{\sqrt{M}\,\underline{s}}_{\tilde{\underline{s}}}$$

The eigenvectors and eigenvalues of D are determined by[5]

$$D\overset{\nu}{\tilde{\underline{s}}} = \overset{\nu}{D}\overset{\nu}{\tilde{\underline{s}}}\ , \qquad \overset{\nu}{D} = (\overset{\nu}{\tilde{\underline{s}}}, D\overset{\nu}{\tilde{\underline{s}}}) = (\overset{\nu}{\underline{s}}, \Phi\overset{\nu}{\underline{s}}) \quad \text{if} \quad (\overset{\nu}{\tilde{\underline{s}}}, \overset{\nu}{\tilde{\underline{s}}}) = 1\ ,$$

and $\overset{\nu}{D} > 0$ if $\Phi$ is a positive matrix.

In continuum theory (4.33) without forces becomes

$$\sqrt{\rho_o(\underline{r})}\,\ddot{s}_i(\underline{r},t) = \underbrace{\frac{1}{\sqrt{\rho_o}}\,\partial_k\, C_{ik,mn}\,\partial_m \frac{1}{\sqrt{\rho_o}}}_{-D_{in}}\,\underbrace{\sqrt{\rho_o}\, s_n(\underline{r},t)}_{\tilde{s}_n(\underline{r},t)}\ .$$

The eigenfunctions $\overset{\nu}{\tilde{\underline{s}}}(\underline{r},t)$ and eigenvalues $\overset{\nu}{D}$ of D are given by the differential equation $D_{ik}\overset{\nu}{\tilde{s}}_k(\underline{r}) = \overset{\nu}{D}\overset{\nu}{\tilde{s}}_i(\underline{r})$ and by boundary conditions on the surface S. If the real eigenfunctions are normalized

$$\int_V d\underline{r}\ \overset{\nu}{\tilde{s}}_i(\underline{r})\ \overset{\nu}{\tilde{s}}_i(\underline{r}) = 1\ ,$$

the eigenvalue can be written as

$$\overset{\nu}{D} = \int d\underline{r}\ \overset{\nu}{\tilde{s}}_i(\underline{r})\ D_{ik}\ \overset{\nu}{\tilde{s}}_k(\underline{r}) = -\int d\underline{r}\ \overset{\nu}{s}_i\ \partial_k\ C_{ik,mn}\ \underbrace{\partial_m\ \overset{\nu}{s}_n}_{\overset{\nu}{\varepsilon}_{mn}}$$

$$= \int d\underline{r}\ \underbrace{\overset{\nu}{\varepsilon}_{ki}\ C_{ik,mn}\ \overset{\nu}{\varepsilon}_{mn}}_{> 0} - \int_S dS_k\ \overset{\nu}{s}_i\ \overset{\nu}{\sigma}_{ik}\ .$$

---

[5] In this appendix the summation convention extends only over cartesian indices i,k,m,n, but not over $\nu$, $\sum_\nu$.

The surface integral vanishes for proper boundary conditions, e.g., $\underline{s} = 0$ or $\sigma d\underline{S} = 0$ on the surface. Consequently, $\check{D} > 0$ on account of elastic stability.

## J. Transformation of Tensors Under Rotations

### J.1 General Transformation of a n-th Rank Tensor

A tensor of n-th rank, $t^{(n)} = (t_{i_1 \dots i_n})$, transforms in analogy to Appendix B like the product of n "vector" components or n coordinates,

$$\tilde{t}_{i_1 \dots i_n} = D_{i_1 i_1'} \dots D_{i_n i_n'} \, t_{i_1' \dots i_n'}$$

where $\tilde{t}$ is the transformed tensor and D the rotation. In this notation a vector is a tensor of rank 1, a scalar has rank 0.

One can find linear combinations of the components of $t^{(n)}$ which only transform among themselves. Simple examples are:

the trace of $t^{(2)}$, $t_{ii}$ , transforms like a scalar ,

a partial trace of $t^{(3)}$, $t_{ikk}$ , transforms like a vector.

Quite generally, one can split a tensor $t^{(n)}$ into sets of components which only transform among themselves and cannot be separated further (irreducible sets). One advantage of such a separation is the following: if one has a linear relation between two tensors in a rotationally invariant theory, e.g., $\sigma = C\varepsilon$ for isotropy, only subsets with equal transformation behaviour can be connected by a scalar because the transformation behaviour for both tensors must be the same. It turns out that there are two such sets for symmetrical tensors of 2nd rank; therefore C contains two independent moduli in an isotropic theory.

The type of sets is well known. The basic sets[6] correspond to integer spins ($l = 0,1,2,3 \dots$) with $2l+1$ z-components, $-l \leqslant m \leqslant l$. The components are usually chosen such that they are eigenstates to a rotation $D(z,\varphi)$ about the z-axis by an angle $\varphi$ with eigenvalue $\exp(im\varphi)$ (comp. App. B.4).

In the following we will briefly discuss what "spins" are hidden in the tensors we have used so far and how one can extract them from the original tensor.

---

6 These sets are called irreducible; the $(2l+1) \times (2l+1)$ matrices are called irreducible representations of the rotation D. These problems are treated rigorously in the group theory of the rotation group which we do not want to discuss in detail.

## J.2 Scalars and Vectors

Scalars and vectors transform according to $l = 0$ and $l = 1$. The proper vector components are $t^{(1)}_{x \pm iy} = t^{(1)}_x \pm it^{(1)}_y$ for $m = \pm 1$ and $t^{(1)}_z$ for $m = 0$. Tensors of even rank contain at least one scalar, e.g., $t^{(2)}_{ii}$ , $t^{(4)}_{iikk}$ and tensors of odd rank contain at least one vector, e.g., $t^{(3)}_{ikk}$ , $t^{(5)}_{ikkmm}$.

## J.3 Tensors of 2nd Rank

A tensor of 2nd rank, e.g., the tensor of deformation $v = v_{ik}$ (Sec. 4.1.1), has nine components. The separation

$$v_{ik} = \frac{v_{ik} + v_{ki}}{2} + \frac{v_{ik} - v_{ki}}{2} = \varepsilon_{ik} + \omega_{ik}$$

into a symmetrical ($\varepsilon$, six components) and an antisymmetrical part ($\omega$, three components) is invariant, which means that the symmetry does not change by a rotation, i.e., from $\varepsilon_{ki} = \varepsilon_{ik}$ follows $\tilde{\varepsilon}_{ki} = \tilde{\varepsilon}_{ik}$. Therefore the components of $\varepsilon$ and $\omega$ transform into themselves. Further one can separate $\varepsilon$ into one dilatation ($l = 0$) and five shears ($l = 2$)

$$v_{ik} = \frac{\varepsilon_{ll}}{3}\delta_{ik} + \left(\varepsilon_{ik} - \frac{\varepsilon_{ll}}{3}\delta_{ik}\right) + \omega_{ik} = v^{(0)}_{ik} + v^{(2)}_{ik} + v^{(1)}_{ik} \; ,$$

$l$: 0 2 "1"

which is again an invariant separation into three irreducible sets. The m-components of the spin 2 set would be: $v^{(2)}_{x+iy,x+iy} = v^{(2)}_{xx} - v^{(2)}_{yy} + i(v^{(2)}_{xy} + v^{(2)}_{yx}) = \varepsilon_{xx} - \varepsilon_{yy} + 2i\varepsilon_{xy}$ for $m = 2$, $\varepsilon_{xz} + i\varepsilon_{yz}$ for $m = 1$, $v^{(2)}_{zz} = \varepsilon_{zz} - \mathrm{tr}\{\varepsilon\}/3$ for $m = 0$ etc. The tensor $\omega$ has only three components. If we use the alternating 3rd rank "tensor" of Appendix A it is

$$\omega_{ik} = \varepsilon_{iks}\omega_s \; , \qquad \omega = \begin{bmatrix} 0 & \omega_3 & -\omega_2 \\ -\omega_3 & 0 & \omega_1 \\ \omega_2 & -\omega_1 & 0 \end{bmatrix} ,$$

where the "vector" $\underline{\omega} = (\omega_1, \omega_2, \omega_3)$ is given by $\omega_s = \varepsilon_{smn}v_{mn}/2 = \varepsilon_{smn}\omega_{mn}/2$. One can easily show that $\underline{\omega}$ transforms like a vector under proper rotations, but it does not change sign under inversion ($\tilde{\omega}_{ik} = \omega_{ik}$ for inversion). Such a vector which behaves normally for proper rotations and does not change under inversion is called a pseudovector. It must be noted that it is assumed tacitly that $\varepsilon_{ski}$ is invariant under in-inversion, i.e., it does not transform as a tensor. If $\varepsilon_{sik}$ would transform like a tensor, it would stay invariant under proper rotations and change sign under inversion, its transformation behaviour would be that of a pseudoscalar and $(\omega_s)$ as defined above would transform like a vector. Usually $\varepsilon_{sik}$ is meant as scalar.

## J.4 Symmetrical Tensors of 3rd Rank

A symmetrical tensor of 3rd rank, $t_{ikm}^{(3)}$, has ten independent components. It contains one spin 1 part, $t_{ikk}^{(3)}$, and one spin 3 contribution, e.g.,

$$t^{(3)}_{x+iy,x+iy,x+iy} = t^{(3)}_{xxx} + 3it^{(3)}_{xxy} - 3t^{(3)}_{xyy} - it^{(3)}_{yyy} \text{ for } m = 3 \text{ etc. },$$

which takes care of the ten components, three from $l = 1$ and seven from $l = 3$.

## J.5 The 4th Rank Tensor of Elastic Moduli

The 4th rank tensor $C_{ik,mn}$ has 21 independent components. The separation into a tensor of total symmetry, T, and the tensor $\Gamma$, which represents the deviations from Cauchy's relations, is also invariant under rotations. The tensor T (15 components) contains one symmetrical 2nd rank tensor $T_{ik}^{(2)} = T_{ikmm}$ which can be separated again into a scalar $T_{ii}^{(2)} = T_{iimm} = (C_{ii,mm} + 2C_{im,im})/3$ corresponding to $l = 0$ and a trace-free 2nd rank tensor ($l = 2$). The remaining nine components of T correspond to spin 4 ($l = 4$), e.g., $T_{zzzz}$ to $m = 0$. The Cauchy tensor $\Gamma^{(2)}$ or $\Gamma$ given in (4.35c) can be split into spin $l = 0$ and $l = 2$. In cubic crystals one has (comp. Section 4.4)

$$C = c_{12}A_1 + c_{44}A_2 + c_aA_3 = \underbrace{\frac{c_{12} + 2c_{44}}{3}(A_1 + A_2) + c_aA_3}_{T} + \underbrace{\frac{c_{12} - c_{44}}{3}(2A_1 - A_2)}_{\Gamma} .$$

## J.6 Cubic Rotations

The transformation properties of a tensor under rotations by *discrete* angles, e.g., under rotations of the cubic symmetry group (cubic rotations), can be discussed similarly. The separation of $\varepsilon$ in Section 4.1 into one dilatation and two types of shears is a good example. This separation is invariant under cubic rotations. The irreducible sets are well-known and can be found in textbooks (e.g., /2.5/).

# K. Rotational Symmetries and Isotropic Behaviour of a Tensor

To discuss the problem of Section 4.5.1 we work in the eigensystem of the rotations where the z-axis corresponds to the axis of rotation and the other axes are chosen as $(1/\sqrt{2})[\underline{u}^{(x)} \pm i\underline{u}^{(y)}]$ according to Appendix B.4. A tensor of rank l transforms like the product of coordinates $z, x \pm iy$; under a rotation by the angle $\varphi$ the components $t_m$ transform as $\tilde{t}_m = t_m e^{im\varphi}$, where $-l \leq m \leq l$ (e.g., $t_{zzzz}$, $t_{x+iy,x+iy,x-iy,x-iy}$ belong to $m = 0$, $t_{x+iy,x+iy,x+iy,x+iy}$ to $m = 4$). If the tensor is invariant under a rotation by the angle $2\pi/n$, $\tilde{t}_m = t_m \exp(i2\pi m/n) = t_m$, it follows that all coefficients $t_m$ must vanish for which $m/n$ is not an integer. If $l/n$ is not integer, i.e., $n \geqslant l+1$, then $m/n$ is not an integer either, except for $m = 0$. This means that all coefficients $t_m$ must vanish, except for $t_o$ which is invariant under rotations. Consequently, if $n > l$, the tensor is invariant against rotation by an arbitrary angle $\varphi$. A sixfold axis makes the 4th rank tensor of elastic moduli isotropic (hexagonal) and a fourfold axis (cubic) does not.

## L. Dislocation Loops

Dislocation loops are important two-dimensional defects which can be described by two-dimensional double force patterns. They are defined by a cut along a surface $\underline{S}$ which is surrounded by a line (the loop) and the displacement $\underline{b}$ of one side of the cut against the other as shown in Fig. L.1. The displacement field depends only on

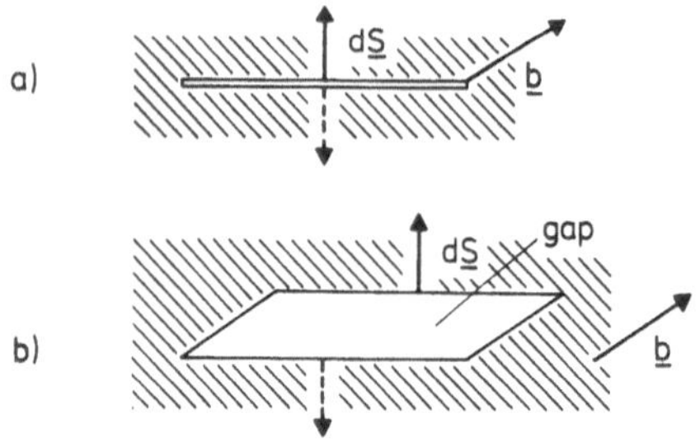


Fig. L.1a and b. Creation of a dislocation loop. a) 1st, make a cut in the crystal along a surface $\underline{S}$, here for simplicity assumed plane. b) 2nd, shift the "upper" surface by $\underline{b}$. 3rd, introduce material to fill the gap for $\underline{b}d\underline{S} > 0$; remove material correspondingly if $\underline{b}d\underline{S} < 0$. 4th, weld together along the new surfaces and you are left with a dislocation loop around the edges of the surface

the loop itself and not on the special cut bounded by the loop. Singularities are along the loop line. A dislocation loop is the most general case of internal stresses in a continuum. In a lattice the loop will be a quasi-stable configuration of low energy if the displacement $\underline{b}$ is a lattice vector (the Burgers vector) because then there is no disorder along the new surfaces and no surface contribution to energy. As a rule only the shortest lattice vectors have been found experimentally.

We will show that the appropriate displacements are ($d\underline{S}'$ refers to integration over $\underline{r}'$)

$$s_i(\underline{r}) = - \underbrace{\int_S dS'_l\, b_m\, C_{lm,sk}}_{P_{sk}}\, \partial_s\, G_{ik}(\underline{r} - \underline{r}')$$

corresponding to translational-rotational invariant force densities

$$f_k(\underline{r}) = - \int_S dS'_l\, b_m\, C_{lm,sk}\, \partial_s\, \delta(\underline{r} - \underline{r}') \, .$$

The solution so far has singularities along the surface S. Now we compare the solutions with different surfaces $S^1$ and $S^2$ (Fig. L.2); employing Gauss' theorem and

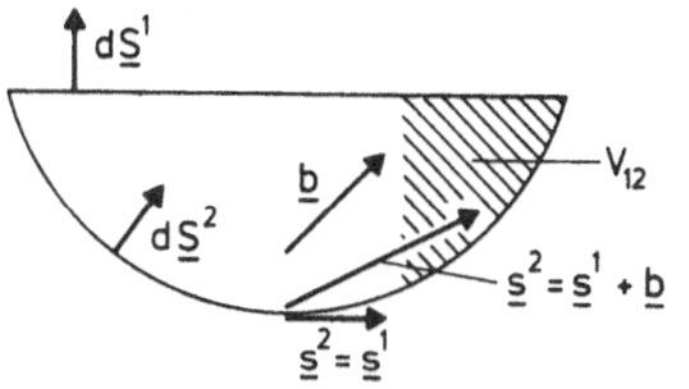


Fig. L.2. Jump of displacement. The integral for $\underline{s}^1 - \underline{s}^2$ extends over the closed surface $S^1 + S^2$. From $\underline{s}^2(\underline{r}) = \underline{s}^1(\underline{r}) + \underline{b}\, \Theta_{V_{12}}(\underline{r})$ one concludes that $\underline{s}^2$ jumps by $\underline{b}$ when passing through the surface $S^2$ because $\underline{s}^1$ is continuous in $S^2$. The same holds for the solution $\underline{s}^1$ when passing through $S^1$

using $\partial'_l\, G_{ik}(\underline{r} - \underline{r}') = - \partial_l\, G_{ik}(\underline{r} - \underline{r}')$ we obtain

$$s_i^1 - s_i^2 = \int_{V_{12}} d\underline{r}'\, b_m \underbrace{C_{lm,sk}\, \partial_l\, \partial_s\, G_{ik}(\underline{r} - \underline{r}')}_{-\delta_{im}\delta(r - r') \text{ from definition of } G} = - b_i\, \Theta_{V_{12}}(\underline{r}) \, , \quad \text{or}$$

$$\underline{s}^1(\underline{r}) - \underline{s}^2(\underline{r}) = \begin{cases} -\underline{b} & \text{if } \underline{r} \text{ inside } V_{12} \\ 0 & \text{if } \underline{r} \text{ outside } V_{12} \end{cases} ;$$

therefore the displacement field $\underline{s}$ jumps by $\underline{b}$ when passing through the defining surface S. It also shows that the displacement is essentially determined by the loop itself and not by the special choice of S.

For the sake of simplicity we will discuss only a plane loop, Fig. L.3, where $P_{sk} = C_{sk,lm}\, S_l\, b_m$. The volume change (4.78) becomes now

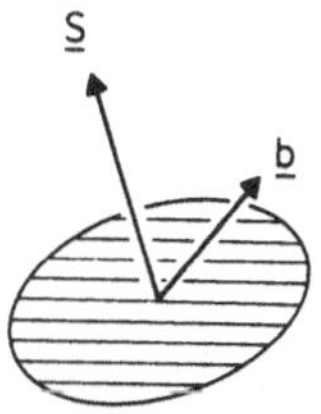


Fig. L.3. Plane loop. Indicated are the surface S with normal $\hat{\underline{S}} = \underline{S}/S$ and the Burgers vector $\underline{b}$

$$\Delta V = S_{ii,sk}P_{sk} = S_{ii,sk}C_{sk,lm}S_l b_m = S_l b_l = (\underline{S},\underline{b})$$

which is the "volume of the gap" when creating the loop, independent of crystal symmetry.

Loops like that in Fig. L.3 can be produced by the collapse of a planar arrangement of $N_v$ vacancies, where $(\underline{S},\underline{b}) = -N_v V_c$. The $N_v$ atoms, brought to the surface to form $N_v$ vacancies, produce a volume change $N_v V_c$. If the vacancies stay single and if they do not relax, then $N_v V_c$ is the volume change; if they do relax by $\delta_1 V$ (comp. Sections 4.8.4 and 4.8.5), the volume change due to creating $N_v$ vacancies is $N_v(V_c + \delta_1 V)$. If, however, the vacancies flock together and relax into a dislocation ring, the volume change is zero; as a rule $\delta_1 V/V_c$ is small and can be neglected in a zero order approximation. A similar consideration applies to interstitials; if $N_i$ atoms are removed from the surface and stay in the crystal as single interstitials, the volume change is $N_i(-V_c + \delta_1 V)$, where $\delta_1 V$ typically is $2V_c$, resulting in a volume change $N_i V_c$; the volume change again vanishes if the interstitial rearrange in a float and form a dislocation ring. Consequently, the volume does not appreciably change if vacancies and/or interstitials are formed and arranged into loops; if they are produced singly and if $\delta_1 V = 0$ for a vacancy and $\delta_1 V = 2V_c$ for an interstitial, then each vacancy or interstitial increases the *external* volume (the total volume of the finite crystal) by $V_c$.

Another point must be mentioned with respect to the interaction energy of two dislocations considered as defects. In contrast to the point defects, which are defined by (Kanzaki) *forces* $\underline{K}$, the dislocation loops are defined by *displacements* (across the loop plane). Whereas for the energy of point defects one must use $(\underline{s},\Phi s)/2 - (\underline{K},\underline{s})$ as starting point resulting in $-(\underline{K},\underline{GK})/2$ as total energy, for dislocations one must start with $(\underline{s},\Phi\underline{s})/2$ alone, which means a change in sign.

## M. Core Displacements for Simple Lattice Models of a Dilatation Center

In Section 4.8.5 we have seen that the radial displacements of atoms at distance l from an isotropic defect (at the origin, comp. Fig. 4.21) are approximately given by $s_r(l) \cong \Delta V^\infty/(4\pi l^2)$. For comparison we now calculate these displacements from *lattice* models for an isotropic defect which yield the same double force tensor (and, therefore, the same $\Delta V^\infty$). For the perfect crystal we employ the lattice model of Fig. 3.15c, i.e., longitudinal 1st neighbour coupling (spring f) in a fcc lattice, where $c_{11} = 2f/a$, $c_{12} = c_{44} = f/a$ (comp. Table 5.2) and $\bar{c}_{11} = (12/5)f/a$. The variational value $\overline{\Delta V^\infty}$ for the volume change due to an isotropic double force tensor $P_{ik} = P_o\delta_{ik}$ (comp. Sections 4.8.5 and 4.8.7) is

$$\overline{\Delta V^{\infty}} = \frac{P_o}{\bar{c}_{11}} = P_o \frac{5a}{12f} \ ;$$

for the following estimates we will use $\overline{\Delta V^{\infty}}$ instead $\Delta V^{\infty}$ because the variational value is already a good approximation (comp. Table 4.8).

1) Substitutional defect (Fig. M.1)

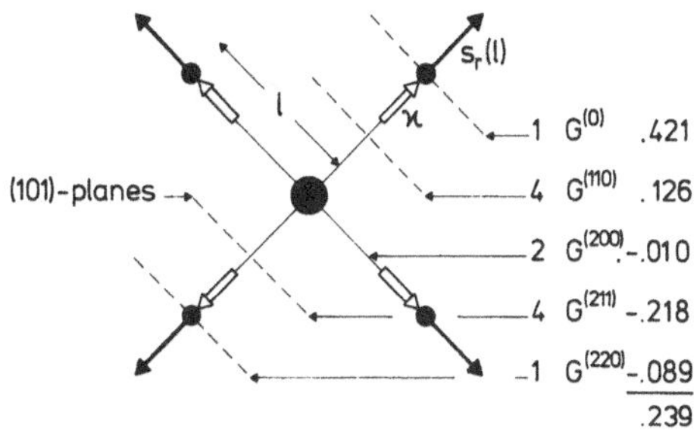


Fig. M.1. Dilatation center (●) at a lattice site of a fcc lattice (•). The contribution of the nearest neighbours of the defect to $s_r(1)$ are given; the atoms are located in (101) planes. Indicated are: the number of atoms in these planes (1st column), a characteristic G for each plane (2nd column) and the contribution of this plane to $fs_r(1)/\kappa$ (3rd column)

We assume that the force pattern is given by radial forces on the nearest neighbours,

$$\underline{K}^{\underline{m}} = \kappa \hat{\underline{R}}^{\underline{m}} \quad \text{for the twelve 1st neighbours,} \quad |\underline{R}^{m}| = 1 = a/\sqrt{2} \ .$$

The corresponding double force tensor is

$$P_{ik} = \kappa \sum_{\underline{m}}{}' \hat{X}_i^{\underline{m}} \hat{X}_k^{\underline{m}} = 4\kappa 1 \delta_{ik} = P_o \ \delta_{ik} \ ,$$

where $\sum_{\underline{m}}'\ldots$ extends over the nearest neighbours. Consequently, if one considers $\kappa$ as a parameter to be adapted to a given $P_o$, $\kappa$ is determined by $\kappa = P_o/41$. The lattice theoretical result for the displacements of the nearest neighbours is

$$\underline{s}^{\underline{m}} = G^{(\underline{m}-\underline{n})} \underline{K}^{\underline{n}} = s_r(1) \hat{\underline{R}}^{\underline{m}} \ , \qquad s_r(1) = \kappa \sum_{\underline{n}}{}' \langle \hat{\underline{R}}^{\underline{m}} | G^{(\underline{m}-\underline{n})} | \hat{\underline{R}}^{\underline{n}} \rangle \ ,$$

i.e., the 1st neighbours are radially displaced by the same amount $s_r(1)$. With the numerical values of Table 3.6 for $G^{(\underline{h})}$ we obtain $s_r(1) = 0.239\kappa/f$ which has to be compared with

$$s_r(1) \cong \frac{\overline{\Delta V^{\infty}}}{4\pi 1^2} = \frac{5\sqrt{2}}{12\pi} \frac{\kappa}{f} = 0.19 \frac{\kappa}{f} \ .$$

Obviously, the agreement is satisfactory.

2) Octahedral interstitial (Fig. M.2)

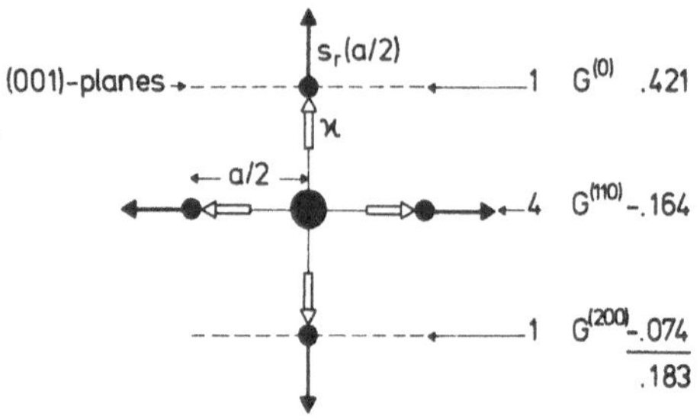


Fig. M.2. Dilatation center (●) at an octahedral site of a fcc lattice (•)

With a force pattern as above (radial forces on the nearest neighbours of the interstitial, i.e., on the six atoms forming an octahedron about the site $\underline{R} = \frac{a}{2}(0,0,1)$) the radial displacement is calculated in complete analogy to above and one obtains $s_r(a/2) = 0.183\kappa/f$. The double force tensor is $P_{ik} = \kappa a \delta_{ik}$ and the approximate displacement is

$$s_r(a/2) \cong \frac{\overline{\Delta V^\infty}}{4\pi(a/2)^2} = \frac{5}{12\pi}\frac{\kappa}{f} = 0.14\,\frac{\kappa}{f}\,.$$

Again the agreement between exact and approximate result is not bad.

## N. Interaction Between Two Dilatation Centers in a Cubic Lattice

From (4.100) we have for the interaction energy of two isotropic double force tensors $P^a_{ij} = P^a_o\,\delta_{ij}$, $P^b_{mn} = P^b_o\,\delta_{mn}$:

$$W(\underline{R}) = -P^a_o P^b_o \int \frac{d\underline{k}}{(2\pi)^3}\, e^{i\underline{k}\underline{R}}\left(\hat{\underline{k}}, \tilde{G}(\hat{\underline{k}})\hat{\underline{k}}\right) \quad \text{with} \quad \tilde{G}(\hat{\underline{k}}) = \frac{1}{\rho_o \tilde{D}(\hat{\underline{k}})}\,;$$

for cubic symmetry $\tilde{D}(\hat{\underline{k}})$ is given by

$$\rho_o \tilde{D}_{ik}(\hat{\underline{k}}) = C_{im,kn}\hat{k}_m\hat{k}_n = \overline{C}_{im,kn}\hat{k}_m\hat{k}_n + c_a\,\delta_{ikmn}\,\hat{k}_m\hat{k}_n\,, \qquad \tilde{D} = \overline{\tilde{D}} + \tilde{D}_a\,,$$

where we have separated $\tilde{D}$ into an isotropic part, $\overline{\tilde{D}}$, determined by Voigt's averages, and an anisotropic part, $\tilde{D}_a$. In an expansion of $\tilde{G}$ in powers of the anisotropy,

$$\tilde{G}(\hat{\underline{k}}) = \frac{1}{\rho_o \overline{\tilde{D}}(\underline{k})\,(1 + \overline{\tilde{G}}\tilde{D}_a)} = \overline{\tilde{G}} - \overline{\tilde{G}}\tilde{D}_a\overline{\tilde{G}} + \ldots\,,$$

$\overline{\tilde{G}}$ is the isotropic Green's function corresponding to $\overline{C}$; it consists of a longitudinal and a transversal part (with respect to $\hat{\underline{k}}$, comp. Section 4.8.1, p. 142):

$\bar{\bar{G}}_l = 1/\bar{c}_{11}$, $\bar{\bar{G}}_t = 1/\bar{c}_{44}$. Including only linear terms in $c_a$,

$$(\hat{\underline{k}},\tilde{G}(\hat{\underline{k}})\hat{\underline{k}}) \cong (\hat{\underline{k}},[\bar{\bar{G}} - \bar{\bar{G}}\tilde{D}_a\bar{\bar{G}}],\hat{\underline{k}}) = \bar{\bar{G}}_l - \bar{\bar{G}}_l^2(\hat{\underline{k}},\tilde{D}_a\hat{\underline{k}}) \ ,$$

we obtain

$$W(\underline{R}) \cong \frac{p_o^a p_o^b}{\bar{c}_{11}^{\,2}} c_a \int \frac{d\underline{k}}{(2\pi)^3} e^{i\underline{k}\underline{R}}(\hat{k}_x^4 + \hat{k}_y^4 + \hat{k}_z^4) \ .$$

To evaluate the remaining integrals, e.g., $\int d\underline{k}\ e^{i\underline{k}\underline{R}}\hat{k}_x^4/(2\pi)^3$, we introduce a "small" quantity $\eta > 0$, which guarantees convergence of the integral,

$$\int \frac{d\underline{k}}{(2\pi)^3}\ e^{i\underline{k}\underline{R}}\ \frac{k_x^4}{k^4} = \lim_{\eta \to +0} \int \frac{d\underline{k}}{(2\pi)^3}\ e^{i\underline{k}\underline{R}}\ \frac{k_x^4}{(k^2+\eta^2)k^2} =$$

$$= \lim_{\eta \to +0} \partial_X^4 \int \frac{d\underline{k}}{(2\pi)^3}\,\frac{e^{i\underline{k}\underline{R}}}{(k^2+\eta^2)k^2} \quad ;$$

the $\underline{k}$-integration, which can easily be performed in the complex k-plane, yields $[1 - \exp(-\eta R)]/(4\pi\eta^2 R)$, and therefore

$$\int \frac{d\underline{k}}{(2\pi)^3}\ e^{i\underline{k}\underline{R}}\ \hat{k}_x^4 = \lim_{\eta \to +0} \partial_X^4 \frac{1}{4\pi\eta^2 R}(1 - e^{-\eta R}) = -\frac{1}{8\pi}\,\partial_X^4\, R = \frac{3}{8\pi R^3}\,(1 - 6\hat{X}^2 + 5\hat{X}^4) \ .$$

Eventually, including the contributions of the $\hat{k}_y^4$- and $\hat{k}_z^4$-terms, we obtain

$$W(\underline{R}) \cong \frac{p_o^a p_o^b}{\bar{c}_{11}^{\,2}} c_a \frac{3}{8\pi R^3}\left[-3 + 5(\hat{X}^4 + \hat{Y}^4 + \hat{Z}^4)\right] \ ,$$

which is identical with the result (4.100c).

## O. Two Possible Displacements with Identical Surface Forces for the Unstable Isotropic Model

In the elastically unstable model (4.102) a double force P in the origin produces the displacement $s_i = -P_{ik}\partial_k(1/4\pi c_{44} r)$:

$$P_{ik} = 4\pi c_{44}\delta_{ik} \quad \text{produces} \quad s_i = -\partial_i \frac{1}{r} = \frac{x_i}{r^3} \ , \qquad \varepsilon_{ik} = \frac{1}{r^3}(\delta_{ik} - 3\hat{x}_i\hat{x}_k) \ , \qquad \varepsilon_{mm} = 0 \ ;$$

$$P_{ik} = 8\pi c_{44}\begin{bmatrix} 0 & 0 & 0 \\ 0 & 0 & 0 \\ 0 & 0 & 1 \end{bmatrix} \quad \text{produces} \quad s_z = -2\partial_z \frac{1}{r} = \frac{2z}{r^3} \ , \qquad s_y = s_x = 0 \ ;$$

$$\varepsilon_{zi} = \frac{1}{r^3}(2\delta_{iz} - 6\hat{x}_z\hat{x}_i) \ , \qquad \varepsilon_{mm} = \varepsilon_{zz} \ .$$

Both result in the same stress tensor,

$$\sigma_{iz}(\underline{r}) = -2c_{44}\partial_i\partial_z \frac{1}{r} = \frac{2c_{44}}{r^3}(\delta_{iz} - 3\hat{x}_i\hat{x}_z) \ .$$

If one cuts along a plane perpendicular to the z-axis and considers the space not containing the force centers, one can obviously obtain *identical* surface forces (along the cut) from two *different* solutions $\underline{s}(\underline{r})$ of the force-free equation of motion. Consequently, if one considers the above double forces as "image forces" introduced to compensate for given surface stresses (i.e., to keep the surface force--free), one realizes that in this model the solution for a free surface is not unique. To prove the uniqueness of the solution one would have to assume a *positive* energy density (comp. footnote to Section 4.8.7, p. 157), whereas in this unstable model the energy density is not necessarily positive.

## P. Simple "Shell" Examples for Resonant and Localized Modes of Defects with Additional Degrees of Freedom

For simplicity we treat only one-dimensional examples; they can be directly extended to three dimensions. The notation is that of Sections 3.4 and 6.4,5.

1) Single shell at 0 (Fig. P.1)

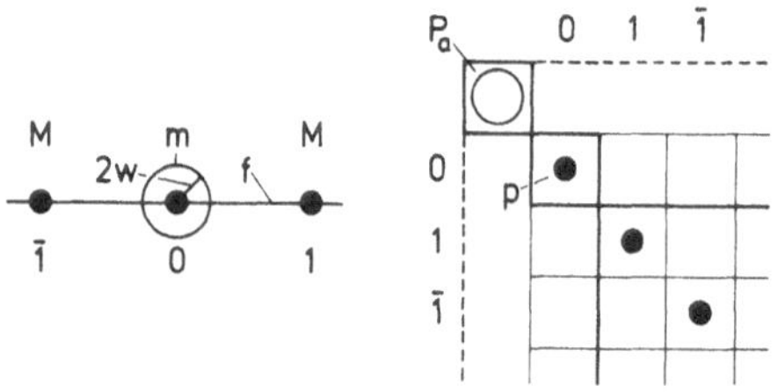


Fig. P.1. Single shell at 0. The shell is bound by a spring 2w to its nucleus; the nuclei are coupled by a (1st neighbour) spring f. Indicated are the augmented space, $P_a$, and the projection of the defect space into original space, p

If a single shell (at site 0) is introduced into a linear chain (spring f), the equations of motion are (comp. (3.65)):

$$M\omega^2 s^n = f(2s^n - s^{n+1} - s^{n-1}) + \delta^{no}\, 2w(s^o - u^o) \ ,$$

$$m\omega^2 u^o = 2w(u^o - s^o) \ .$$

Via the 2nd equation $u^o$ can be expressed by $s^o$ and the equation for the nuclei becomes

$$M\omega^2 s^n = f(2s^n - s^{n+1} - s^{n-1}) + \chi^{nm} s^m$$

where the (frequency-dependent) "coupling change" $\chi^{mn}$ due to the "shell defect" is

$$\chi^{nm} = \delta^{no}\,\delta^{mo}\,\overset{\circ}{\chi}\ , \quad \overset{\circ}{\chi} = -\frac{2wm\omega^2}{2w - m\omega^2}\ ;$$

for very large 2w, $\overset{\circ}{\chi} \cong -m\omega^2$ represents an isotopic defect. The corresponding t-matrix is

$$t^{nm} = \delta^{no}\,\delta^{mo}\,\overset{\circ}{t}\ , \quad \overset{\circ}{t} = \frac{\overset{\circ}{\chi}}{1 + \overset{\circ\circ}{A\chi}} = \frac{1}{1/\overset{\circ}{\chi} + \overset{\circ}{A}} = \left(-\frac{1}{m\omega^2} + \frac{1}{2w} + \overset{\circ}{A}\right)^{-1}, \quad \overset{\circ}{A}(\omega) = \overset{\circ}{G}{}^{(0)}(\omega).$$

For small w ($w \ll f$) the value of $\overset{\circ}{t}$ is appreciable only if $m\omega^2 \cong 2w$ where $|\overset{\circ}{t}|^2 = |\overset{\circ}{A}{}^{-1}|^2$ has its maximum. The three-dimensional model gives identical results.

The shell-shell Green's function, $G^{ss}$ (corresponding to $G_{aa}$ in Section 6.5.4), can be obtained by elimination of $s^o$ in the shell equation: from the force-free equation of the lattice, where $2ws^o$ is a "defect term" and where $2wu^o$ can be considered as an effective force, one can express $s^o$ by $u^o$ and the lattice Green's function. The calculation is simple and results in $2w(u^o - s^o) = 2w/(1 + 2w\overset{\circ}{A})$, from which one obtains

$$G^{ss} = \frac{1}{-m\omega^2 + 2w/(1 + 2w\overset{\circ}{A})} = \frac{1}{2w - m\omega^2 - (2w)^2\overset{\circ}{A}/(1 + 2w\overset{\circ}{A})}\ ;$$

$G^{ss}$ is practically identical to the $G_{aa}$ of the octahedral interstitial, (6.44), and it can be discussed in the same way. For $w \ll f$ one has a resonance at $\omega_s^2 = 2w/m$ if $\omega_s^2 < \Omega_{max}^2$, and a localized state at $\omega_s^2$ is obtained if $\omega_s^2 > \Omega_{max}^2$.

2) Two shells at 0 and 1 (Fig. P.2)

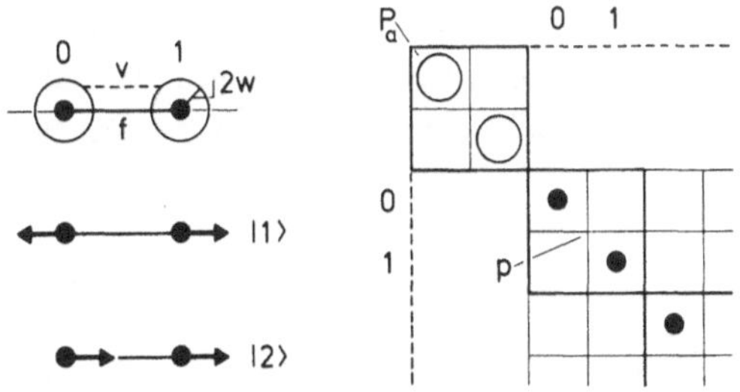


Fig. P.2. Two shells at 0 and 1 coupled by a spring v to each other and by springs 2w to their nuclei. The eigenmodes of $\chi$ are indicated. The augmented space, $P_a$, and the projection of the defect space into original space, p, are two-dimensional

If one adds a second shell at site 1, which is coupled to the shell at 0 by a spring v, the equations of motion become

$$M\omega^2 s^n = f(2s^n - s^{n+1} - s^{n-1}) + \delta^{no}\,2w(s^o - u^o) + \delta^{n1}\,2w(s^1 - u^1)\ ,$$

$$m\omega^2 u^o = 2w(u^o - s^o) + v(u^o - u^1)\ ,$$

$$m\omega^2 u^1 = 2w(u^1 - s^1) + v(u^1 - u^o)\ ,$$

and upon elimination of $u^{\circ}$, $u^1$ we obtain

$$M\omega^2 s^n = f(2s^n - s^{n+1} - s^{n-1}) + \chi^{nm} s^m ,$$

where

$$\chi = |1\rangle \overset{1}{\chi} \langle 1| + |2\rangle \overset{2}{\chi} \langle 2| \quad \text{with} \quad \frac{1}{\overset{1}{\chi}} = \frac{1}{2v - m\omega^2} + \frac{1}{2w} , \quad \frac{1}{\overset{2}{\chi}} = -\frac{1}{m\omega^2} + \frac{1}{2w} ;$$

the modes $|1\rangle$ and $|2\rangle$ are shown in Fig. P.2. The corresponding t-matrix is

$$t = |1\rangle \overset{1}{t} \langle 1| + |2\rangle \overset{2}{t} \langle 2| , \quad \overset{1,2}{t} = \frac{1}{1/\overset{1,2}{\chi} + \overset{1,2}{A}} \quad \text{with} \quad \overset{1}{A} = \overset{\circ}{G}^{\circ\circ} - \overset{\circ}{G}^{\circ 1} , \quad \overset{2}{A} = \overset{\circ}{G}^{\circ\circ} + \overset{\circ}{G}^{\circ 1} .$$

The case $\overset{2}{\chi}$, $\overset{2}{t}$ corresponds essentially to $\overset{1}{\chi}$, $\overset{1}{t}$ discussed above. In the case $\overset{1}{\chi}$, $\overset{1}{t}$ the inter-shell oscillation, $m\omega^2 = 2v$ (for $w = 0$), becomes weakly coupled to the lattice if $w \ll f$, which leads to a sharp resonance at $\omega_s^2 = 2(v+w)/m$ (if $\omega_s < \Omega_{max}$), where again $|\overset{1}{t}|^2 = |\overset{1}{A}^{-1}|^2$ has its maximum. One recognizes that $\overset{1}{t}$ vanishes for $m\omega^2 = 2v$, i.e., for $\omega^2$ only slightly smaller than $\omega_s^2$; in a constant-$\kappa$-scan one would obtain a zero for $\omega^2 = 2v/m$ and this zero would be partly responsible for a double peak structure of the scan for $\overset{\circ}{\Omega}^2$ near $\omega_s^2$ (comp. Section 9.3). The frequency $\omega_s$ always appears as a sharp peak, the reason again being the smallness of the coupling, w, to the lattice. The inter-shell oscillation behaves like a very weakly damped oscillator if $\omega_s < \Omega_{max}$, and for $\omega_s > \Omega_{max}$ one obtains an undamped (localized) oscillation.

3) Single shell at 0 with spring 2v to the origin

If a single shell at 0 is not bound to the *nucleus* but rather to the *site* 0 by a spring 2v, the model is no longer translationally invariant; the resulting t-matrices are analogous to $\overset{1}{t}$ above. The defect term in the lattice equation of motion is unchanged, and the shell equation of motion is $2w(u^{\circ} - s^{\circ}) + (2v - m\omega^2)u^{\circ} = 0$. The elimination is again easy:

$$\frac{1}{\overset{\circ}{\chi}} = \frac{1}{2v - m\omega^2} + \frac{1}{2w} \left( = \frac{1}{\overset{1}{\chi}} \right) , \quad \overset{\circ}{t} = \frac{1}{\dfrac{1}{2v - m\omega^2} + \dfrac{1}{2w} + \overset{\circ}{A}} ;$$

the discussion is analogous to that of $\overset{1}{t}$.

# References

## Chapter 2


2.1 G. Leibfried: Gittertheorie der mechanischen and thermischen Eigenschaften der Kristalle. In *Encyclopedia of Physics*, Vol. 7, Part 1 (Springer, Berlin, Göttingen, Heidelberg 1955)
2.2 M. Born, K. Huang: *Dynamical Theory of Crystal Lattices* (University Press, Oxford 1954)
2.3 A.A. Maradudin, E.W. Montroll, G.H. Weiss, I.P. Ipatova: Theory of Lattice Dynamics in the Harmonic Approximation. In *Solid State Physics*, Suppl. 3, 2nd ed. (Academic Press, New York 1971)
2.4 W. Cochran: The Dynamics of Atoms in Crystals. In *The Structures and Properties of Solids*, Vol. 3 (Arnold, London 1973)
2.5 W. Ludwig: *Festkörperphysik I/II*. (Akademische Verlagsgesellschaft, Frankfurt 1970)
2.6 G.K. Horton, A.A. Maradudin: *Dynamical Properties of Solids*, Vol. I (North Holland, Amsterdam 1974)
2.7 W. Jones, N.H. March: *Theoretical Solid State Physics*, Vol. I/II (Wiley, London, New York, Sydney, Toronto 1973)
2.8 A. Messiah: *Quantenmechanik I* (de Gruyter, Berlin, New York 1976) Chap. 12


## Chapter 3


Compare also Refs. 2.1-7 for general discussion
3.1 R.A. Johnson: J. Phys. F 3, 295 (1973)
3.2 *Interatomic Potentials and Simulation of Lattice Defects*, ed. by P.C. Gehlen, J.R. Beeler Jr., R.I. Jaffee (Plenum Press, New York, London 1972)
3.3 B. Dorner, M. Steiner: J. Phys. C 9, 15 (1976)
3.4 R.M. Nicklow, G. Gilat, H.G. Smith, L.J. Raubenheimer, M.K. Wilkinson: Phys. Rev. 164, 922 (1967)
3.5 G. Gilat, R.M. Nicklow: Phys. Rev. 143, 487 (1965); R. Stedman, G. Nillsson: Phys. Rev. 145, 492 (1966)
3.6 W.A. Kamitakahara, B.N. Brockhouse: Phys. Lett. A 29, 639 (1969); W. Drexel, W. Gläser, F. Gompf: Phys. Lett. A 28, 531 (1969)
3.7 J.W. Lynn, H.G. Smith, R.M. Nicklow: Phys. Rev. B 8, 3493 (1973)
3.8 B.N. Brockhouse, T. Arase, G. Caglioti, K.R. Rao, A.D.B. Woods: Phys. Rev. 128, 1099 (1962); R. Stedman, L. Almqvist, G. Nillsson: Phys. Rev. 162, 549 (1967)
3.9 J.C. Phillips: Rev. Mod. Phys. 42, 317 (1970)
3.10 B. Splettstößer: Z. Phys. B 26, 151 (1977)
3.11 K. Schroeder: *Diffusion Reactions of Point Defects*. Berichte der Kernforschungsanlage Jülich, Jül - 1083 - FF (1974)
3.12 H.R. Schober, M. Mostoller, P.H. Dederichs: Phys. Stat. Sol. (b) 64, 173 (1974)


## Chapter 4


Comp. also Refs. 2.1-7 for general discussion
4.1 L.D. Landau, E.M. Lifshits: *Lehrbuch der Theoretischen Physik*, Bd. VII. (Akademie-Verlag, Berlin 1966)

4.2 R.F.S. Hearmon: *Introduction to Applied Anisotropic Elasticity.* (University Press, Oxford 1961)
4.3 A.E.H. Love: *A Treatise on the Mathematical Theory of Elasticity.* (University Press, Cambridge 1959)
4.4 H.B. Huntington: The Elastic Constants of Crystals. In *Solid State Physics,* Vol. 7. (Academic Press, New York 1958)
4.5 R. Siems: *Wechselwirkungen zwischen Defekten in Kristallen.* Berichte der Kernforschungsanlage Jülich, Jül - 545 - FN (1968)
4.6 P.H. Dederichs, J. Pollmann: *Elastisches Verschiebungsfeld und Wechselwirkungsenergie von Punktdefekten in anisotropen, kubischen Kristallen.* Berichte der Kernforschungsanlage Jülich, Jül - 836 - FF (1972)
4.7 H.J. Kanzaki: J. Phys. Chem. Sol. 2, 24 (1957)
4.8 J.D. Eshelby: The Continuum Theory of Lattice Defects. In *Solid State Physics,* Vol. 3 (Academic Press, New York 1956)
4.9 A.S. Nowick, B.S. Berry: *Anelastic Relaxation in Crystalline Solids.* (Academic Press, New York 1972)
4.10 J. Völkl: *The Gorski Effect.* Berichte der Bunsen-Gesellschaft 76, 797 (1972).
4.11 R.D. Mindlin: Phys. Rev. 7, 195 (1936)
4.12 Y. Hiki, A.V. Granato: Phys. Rev. 144, 411 (1966)
4.13 J.D. Eshelby: Energy Relations and the Energy-momentum Tensor in Continuum Mechanics. In *Inelastic Behaviour of Solids,* ed. by M.F. Kanninen, W.F. Adler, A.R. Rosenfield, R.I. Jaffee (Mc Graw-Hill, New York 1970)

Chapter 5

Comp. Refs. 2.1-3 for general discussion
5.1 Y. Fujii, N.A. Lurie, R. Pynn, G. Shirane: Phys. Rev. B 10, 3647 (1974)
5.2 Y. Endoh, G. Shirane, J. Skalyo Jr.: Phys. Rev. B 11, 1681 (1975)

Chapter 6

The reviews 6.1,2 contain comprehensive lists of references
6.1 W. Schilling: *Self-Interstitial-Atoms in Metals.* Proc. of the Int. Conf. on the Properties of Atomic Defects in Metals, Argonne, Ill., USA, Oct. 18-22, 1976, J. Nucl. Mat. (in print)
6.2 P.H. Dederichs, C. Lehmann, H. R. Schober, A. Scholz, R. Zeller: *Lattice Theory of Point Defects.* Proc. of the Int. Conf. on the Properties of Atomic Defects in Metals, Argonne, Ill., USA, Oct. 18-22, 1976, J. Nucl. Mat. (in print)
6.3 U. Gonser (ed.): *Mößbauer Spectroscopy.* Topics in Applied Physics, Vol. 5, (Springer, Berlin, Heidelberg, New York 1975)
6.4 R. Zeller: *Schwingungsverhalten von Zwischengitteratomen.* Berichte der Kernforschungsanlage Jülich, Jül - 1259 - FF (1975)

Chapter 7

Compare, e.g., Ref. 2.3, Chap. 7 and the references given there
7.1 M.A. Krivoglaz: *Theory of X-ray and Thermal Neutron Scattering.* (Plenum Press, New York 1969)
7.2 S.W. Lovesey, T. Springer (eds.): *Dynamics of Solids and Liquids by Neutron Scattering.* Topics in Current Physics, Vol. 3 (Springer, Berlin, Heidelberg, New York 1977)
7.3 W. Marshall, S.W. Lovesey: *Theory of Thermal Neutron Scattering.* (Clarendon Press, Oxford 1971)
7.4 L. Koester: Neutron Scattering Lengths and Fundamental Neutron Interactions In *Neutron Physics,* Springer Tracts in Modern Physics, Vol. 80 (Springer, Berlin, Heidelberg, New York 1977)

Chapter 8

8.1 W. Feller: *An Introduction to Probability Theory and its Applications,* Vol. I,

3rd ed. (Wiley, New York 1957)
8.2 D. Morgenstern: *Einführung in die Wahrscheinlichkeitsrechnung und mathematische Statistik*, 2nd ed. (Springer, Berlin, Heidelberg, New York 1968)

## Chapter 9

9.1 R.J. Elliott, I.A. Krumhansl, P.L. Leath: Rev. Mod. Phys. 46, 465 (1974)
9.2 R.J. Elliott, D.W. Taylor: Proc. Roy. Soc. (London) A 296, 161 (1967)
9.3 A. Zinken, U. Buchenau, H.J. Fenzl. H.R. Schober: Solid State Commun. 22, 693 (1977)

# Subject Index

# Springer Series in Solid-State Sciences

Editors: M. Cardona, P. Fulde, H.-J. Queisser

PRINCIPLES OF MAGNETIC RESONANCE
Springer Series in Solid-State Sciences, Vol.1
by *C.P. Slichter*
Approx. 420 pp (1978)
ISBN 3-540-08476-2

PRINCIPLES OF MAGNETIC RESONANCE is a textbook intended for graduate students or others beginning research in magnetic resonance or electron spin resonance. It is intended for physicists, chemists, applied scientists, or others who have had a one year graduate course in quantum mechanics from one of the standard textbooks. The text aims at developing a physical understanding of magnetic resonance and familiarity with the principal theoretical techniques needed to read resonance articles in scientific journals. Homework problems are provided to give practice in utilizing the techniques. The author seeks to develop depth of understanding of the most important topics rather than giving a comprehensive account of all of resonance. The new edition differs from the original one principally by the addition of chapters on spin temperature in magnetic resonance, double resonance, techniques for line-narrowing in solids (the Waugh-Mansfield approach). There are two new appendices.

INTRODUCTION TO SOLID-STATE THEORY
Springer Series in Solid-State Sciences, Vol.2
by *O. Madelung*
Approx. 550 pp (1978)
ISBN 3-540-08516-5

This book is intended to serve as a textbook in solid-state theory for graduate students of physics and material science. In addition, it should provide the theoretical background needed by physicists doing research in both pure solid-state physics and solid-state physics as applied to electrical engineering. The fundamentals of solid-state theory are developed starting from one unifying point of view: from the description by *delocalized* and *localized* states and--within the concept of delocalized states --by *elementary excitations*. The development of solid-state theory within the last ten years has shown that by a systematic introduction of these concepts, large parts of the theory can be described in a unified way. At the same time, this form of description gives a "pictorial" formulation of many elementary processes in solids which facilitates its understanding.

The book is a revised and partly rewritten version of a German textbook published a few years ago.

DYNAMICAL SCATTERING OF X-RAYS IN CRYSTALS
Springer Series in Solid-State Sciences, Vol.3
by *Z.G. Pinsker*
Approx. 530 pp (1978)
ISBN 3-540-08564-5

This book presents the first complete treatment of the dynamical scattering of X-rays in perfect and elastically distorted crystals. The theory is systematically developed, experimental methods are discussed, and significant results are illustrated. In comparison to the Russian edition (Nauka 1974), the presentation is substantially enlarged and supplemented by the theory of scattering in elastically distorted crystals and the solution of the multibeam problem. Reference is made to papers published as recently as 1977.

The book is aimed at scientists and engineers concerned with the technology of single-crystal materials.

INELASTIC ELECTRON TUNNELING SPECTROSCOPY
Springer Series in Solid-State Sciences, Vol.4
ed. by *T. Wolfram*
Approx. 250 pp (1978)

This book represents the proceedings of the International Conference on Inelastic Electron Tunneling Spectroscopy and the Symposium on Electron Tunneling, held at the University of Missouri, Columbia, MO, on May 25-27, 1977.